Irrigation Systems: Theory and Practices

Irrigation Systems: Theory and Practices

Edited by **Davis Twomey**

R CALLISTO
REFERENCE

New York

Published by Callisto Reference,
106 Park Avenue, Suite 200,
New York, NY 10016, USA
www.callistoreference.com

Irrigation Systems: Theory and Practices
Edited by Davis Twomey

International Standard Book Number: 978-1-63239-437-8 (Hardback)

Printed in the United States of America.

Contents

Preface

The purpose of the book is to provide a glimpse into the dynamics and to present opinions and studies of some of the scientists engaged in the development of new ideas in the field from very different standpoints. This book will prove useful to students and researchers owing to its high content quality.

This book addresses the importance of irrigation and provides a theoretical and practical analysis of the same. It consists of a broad spectrum of topics and information based on agricultural water productivity in stressed environments and systems of irrigation practices across the globe which are extensively discussed and illustrated. The various topics elucidated in this book include impacts of irrigation on the physiology of plants, process of creation of drought tolerant variety, deficit irrigation practices and genetic manipulation, plant-water-soil-atmosphere relationships, agronomic practices in challenging environments, detailed account on the onslaught of global warming, agro-meteorological effects and climate change, casing practices for sustainable irrigation practices, etc. Therefore, with the aforementioned topics, the book efficiently examines the field of irrigation which is important for the sustenance of agriculture.

At the end, I would like to appreciate all the efforts made by the authors in completing their chapters professionally. I express my deepest gratitude to all of them for contributing to this book by sharing their valuable works. A special thanks to my family and friends for their constant support in this journey.

Editor

Part 1

Agricultural Water Productivity in Stressed Environments

Deficit (Limited) Irrigation – A Method for Higher Water Profitability

Saeideh Maleki Farahani[1] and Mohammad Reza Chaichi[2]
[1]Department of Crop Production and Plant Breeding
Faculty of Agricultural Sciences Shahed University
[2]Department of Crop Production and Plant Breeding
Faculty of Agricultural Sciences University of Tehran
Iran

1. Introduction

Increasing world population and limitation of water and soil resources make the control of resource usage essential. Policymaking for the future must be based on a more profitable use of water and soil and it is necessary to consider economical, political and social aspects in order to reach a better condition in water and soil resources. Agricultural management, macro and micro policy should be based on sustainable use of limited water and soil resources. In some cases expanding farmlands needs vast investment while some times it is not possible. Plant production per given amount of water should be basis for organizing possibilities and invests to increase water profitability (Fereres and Soriano, 2007; Blum, 2009). The necessity of planning to increase the water use efficiency is inevitable from world population growth and water amount.

Development pressure irrigation system, crop production based on crop rotation, plant nutrition and pest control are all for better use of water and soil resources.

Undoubtedly future water management should be based on more production per given amount of water. Deficit or limited irrigation is one of the irrigation methods which has been designed for more efficient use of water in some crops (English, 1990). Environmental conditions, type of crop and available possibilities have particular importance in water management regarding deficit irrigation (English, 1990).

In this method a plant won't encounter moisture deficiency during growth and development under normal condition, in other words, plant absorbs water requirements for metabolic functions easily. However, when a drought stress happens to a plant either in all its or at least in one of its growth stages, it won't be able to do metabolic functions due to water limitation or unbalanced water situation.

Drought stress is described by its intensity and duration which have interaction with plant growth stage (Samarah and Al-Issa, 2006; Farooq *et al.*, 2009). For example even a medium drought stress at anthesis time of wheat or barley causes more reductive effect on yield than a drought stress during grain filling (English and Nakamura, 1989; Martyniak, 2008; Katerij *et* al 2009; Maleki farahani, 2009). Effect of severe short stress is more than a medium long stress, because under medium stress the plant is able to reduce bad effects of stress by stimulating some metabolic and morphologic mechanisms. Therefore it can be said that environmental stress including drought stress at any plant growth stage which has more contribution to the yield has determinant effects on yield reduction.

2. Deficit irrigation

Deficit irrigation is a water management method in which water will be saved with accepting little yield reduction without any severe damage to the plant (English 1990). Medium stress may be a delay in irrigation for a few days or reduced water consumption in each irrigation, but plant shouldn't encounter severe drought stress at any mentioned situation.

The principal attitude in deficit irrigation methods are using saved water for expanding farmlands, saving water for using in critical growth stage or using for cultivating of cash crops like summer plants.

3. Crop production response to given water

Generally yield increases sharply per given water unit in production curve. After a sharp incline in yield, there is a fairly increase until it reaches maximum yield and after that yield will be constant with more given water. The zone for applying deficit irrigation is when yield increases slowly with each given water unit. Selection of exact point for water amount in deficit irrigation depends on following factors:

1. Type of crop
2. Possibilities for farmland expansion
3. Energy usage per area unit for farmland preparation
4. Costs of sowing, cultivation operations and harvesting

4. Methods for application of deficit irrigation

Selecting the methods depends on available possibilities and soil texture. Considering soil conditions, deficit irrigation is possible in two ways:

In soils with light texture (sandy soil), soil doesn't have high water holding capacity, thus in such a situation irrigation periods may be constant or its frequency increases, however, in deficit irrigation the water amount reduces compared to normal irrigation in each irrigation (English, 1990).

Accordingly an experiment conducted by Jorat et al (2011) on two forage sorghum cultivars. The irrigation treatments consisted of IR_{70}: irrigation after 70mm accumulative evaporation from evaporation pan class A (control), IR_{100}: irrigation after 100mm accumulative evaporation from evaporation pan class A and IR_{130}: irrigation after 130mm accumulative evaporation from evaporation pan class A which were assigned to the main plots. The sowing density of 15, 20 and 25 plants per square meter and two sorghum varieties (Speedfeed and Pegah) were allocated as factorial arrangement to the subplots. The results indicated that the highest forage yield was produced by Speedfeed variety at the control (IR_{70}), medium water stress (IR_{70}) and severe water stress (IR_{70}) treatments with 25 plants per square meter density. The plant height followed an increasing trend as sowing density increased and decreased as water stress got more severe. The stem and leaf dry matter followed the same trend as forage yield in response to water stress and sowing density. The leaf/stem ratio increased as sowing density increased.

Also in another study on chickpea the deficit irrigation was induced by reduction of volume of water in each consecutive irrigation. In this study which was conducted by Chaichi et al (2004), five chickpea accessions were treated by different irrigation gradient systems during generative growth stage. The irrigation gradient treatments were 5, 10, 15 and 20 percent of reduced water supplies compared to control (moisture kept at field capacity throughout the experimental period) at two-week intervals. Irrigation treatments started from flowering commencement and finished when plants reached physiological maturity. The volume of irrigation water in every other day intervals was determined by soil texture and soil moisture curve based on a preliminary experiment, which was 300 ml. Irrigation treatments were: 1: Control: soil moisture kept at field capacity level (±5%) throughout the experimental period by irrigating of 300 ml of water every other day, 2: Irrigation with 5% reduction of water supply compared to control in a two-week interval from flowering commencement to physiological maturity, 3: Irrigation with 10% reduction of water supply compared to control in a two-week interval from flowering commencement to physiological maturity, 4: Irrigation with 15% reduction of water supply compared to control in a two-week interval from flowering commencement to physiological maturity, 5: Irrigation with 20% reduction of water supply compared to control in a two-week interval from flowering commencement to physiological maturity.

Irrigation treatments were applied to simulate the pattern of available moisture reduction in dry land farming areas.

	First period	Second period	Third period	Fourth period	Fifth period
Irrigation Gradient	May, 10 May, 23	May, 24 June, 6	June, 7 June, 21	June, 22 July, 4	July, 5 July, 18
Control	cc 300	cc 300	cc 300	cc 300	cc 300
5%	cc 285	270 cc	cc 255	240 cc	cc 225
10%	270 cc	240 cc	210cc	180 cc	150 cc
15%	255 cc	210 cc	165 cc	120 cc	75 cc
20%	240 cc	180 cc	120 cc	60 cc	0 cc

Table 1. Irrigation schedule and volume of irrigation water for chickpea accessions in 2001

Chickpea accessions were sown on March 6, 2001 outside the greenhouse and were normally irrigated to commencement of flowering. On May 10, 2001 pots were transferred to a controlled greenhouse and irrigation treatments were applied. Temperature and humidity was kept constant (temperature 23 ± 2 °C and humidity 65% ±5%).

Seed production per plant was significantly (P<0.05) affected by both chickpea genotypes and interaction of irrigation systems x chickpea genotypes. Based on the mean seed production per plant, chickpea genotypes could be classified in three categories of high yielding accessions (4488 and 4283), medium yielding accession (5132 and 4348) and low yielding accessions (5436). The medium and low yielding accession produced 18 and 45 percent less seed yield per plant compared to high yielding ones, respectively.

At irrigation gradients of 5 and 10% there was 39% less seed production and at irrigation gradients of 15 and 20% there was a 54% reduction compared to control. Nonsignificant difference in seed production at 15 and 20% irrigation systems indicates that chickpea accessions have a relative tolerance to drought stress and can produce an acceptable minimum production under unfavorable moisture conditions. Accession No. 4283 was the best seed producer at control, however, it showed a severe sensitivity to water stress especially at irrigation system of 20% when it produced the least amount of seed among chickpea genotypes. Accession No. 4488 not only had the highest mean (over all irrigation system) seed production among all chickpea cultivars, it also had fairly stable seed production ability under all irrigation systems. By producing of bigger seeds with less number per pod, and producing more pods per plant, accession No. 4488 was the best seed producer among other genotypes. The lower number of branches provided with less leaf area ultimately reduced its evapotranspiration under stressed conditions. Accession No.4488 was followed by No. 5132, which despite lower mean seed production had a better stability under all irrigation systems. This genotype followed the same vegetative and generative growth pattern of accession No.4488.

Accessions No. 4283 and 4488 produced the most biomass and seed yield (respectively) averaged over all irrigation treatments. Accession No. 4283 showed a severe reaction to irrigation gradient compared to other accessions, while accession No. 4488 was more stable in biomass and seed production across all irrigation gradients.

In heavy texture soils (clay soil) with high water holding capacity, irrigation intervals should be scheduled so that irrigation intervals will be increased while the plant will not encounter severe drought stress. In heavy soils, deficit irrigation is also possible by reducing water amount in each irrigation if the irrigation intervals are kept constant.

In both methods, water consumption has to be less than normal condition per farm area unit. There are some factors which influence the efficiency of deficit irrigation including land leveling when irrigation is applied in surface and the existence of possibilities for conducting water in short time so that it can distribute uniformly in the farm.

In a study performed by Heidari Zooleh et al (2011) on Foxtail Millet they used alternate irrigation systems with different intervals in a pot experiment. Their treatments consisted of different irrigation methods and intervals. There were three irrigation intervals: I1: Control, irrigated every 2 days, I2: Mild water stress, Irrigated every 3 days, I3: Sever water stress, irrigated every 4 days, There were three methods of water application, viz: Conventional irrigation (M1): the whole root system was relatively evenly dried, Fixed irrigation (M2): fixed irrigation group by which water was always applied to one part of root system during the whole experimental period, Alternate irrigation (M3): watering was alternated between two halves of root system of the same pot. The watered and dried halves of root system were alternately replaced each irrigation interval. Irrigation intervals were determined according to factors such as greenhouse temperature and humidity. At each irrigation event, enough water was allowed to be absorbed by the soil in each pot, and any excess water was allowed to drain. The pots were weighed before and after each irrigation event to determine the water consumption by the plant in each pot. They found The I1 had the highest dry forage yield, while I2 did not have significant difference compared with I1, but I3 had a significant reduction of dry forage yield compared with I1 . For example under conventional irrigation, I2 and I3 had a dry biomass reduction of 5% and 34% compared with I1, respectively. Less water was used by M2I3 and M3I3 compared with M1I3 but dry forage yields were not affected. Under conventional irrigation, irrigation interval of 3 and 4 days had a dry biomass reduction of 5% and 34% compared with irrigation interval of 2 days, respectively. In addition, less water was used by M2I2 and M3I2 compared with M1I2 but dry forage yields were not affected. The most important point is that M2I2 significantly reduced dry forage yield compared with M3I1, while M3I2 did not have a significant reduction compared with M1I1, M2I1 and M3I1. These suggest that alternate irrigation of root is the best irrigation method among other irrigation methods. Also There was significant difference between M2I3 and M1I1 in terms of WUE and the difference among the other treatments were not significant. M2I3 had a WUE increase of 40% compared with M1I1. There was positive and significant correlation between WUE and leaf to stem ratio. By increasing irrigation interval, water consumption was reduced evident in the I2 in fixed and alternate irrigation. Reductions in water consumption, but not in biomass, with fixed and alternate irrigation compared with conventional irrigation method suggests that these two irrigation methods can be used for saving soil water. This is especially so with alternate irrigation under mild water stress (M3I2) that did not reduce forage dry weight when compared with M3I1. Under irrigation interval of 3 days, fixed and alternate irrigation used 29% and 20% less water compared with conventional irrigation, respectively. There was positive and significant correlation between water consumption and fresh forage yield, dry forage yield, plant height, leaf area, leaf dry weight, leaf relative water content (sampling stage 1, 2), root dry weight, root volume, root surface area and root length, while there was negative and significant correlation between water consumption and leaf to stem ratio and specific leaf weight (SLW). Overall their results showed that fresh and forage yield were reduced by increasing irrigation interval. Under conventional irrigation, irrigation interval of 3 and 4 days had a dry biomass reduction of 5% and 34% compared with irrigation interval of 2 days, respectively. Under irrigation interval of 3 and 4 days, less water was used by the alternate and fixed irrigation compared with conventional irrigation, but plant growth in terms of dry biomass, plant height, leaf to stem ratio, specific leaf weight, leaf area, root dry weight, root volume, root surface area and root length, was not affected. Under irrigation interval of 3 days, fixed and alternate irrigation used 29% and 20% less water compared with conventional irrigation, respectively. However, water stress increased specific leaf weight, but reduced leaf area, leaf dry weight and leaf relative water content. Root growth was less sensitive than shoot to water stress. Under mild water stress, alternate irrigation performed better than fixed irrigation compared with all irrigation methods under non-water stress, so they suggested to use alternate irrigation under mild water stress to achieve acceptable yield along with efficient use of water. In the other study water deficit irrigation systems applied on pearl millet (*Pennisetum americanum* L.) by reducing water amount in each time and irrigation times (Rostamza et al., 2011). The irrigation treatments were 40%, 60%, 80% and 100% depletion of available soil water (I_{40}, I_{60}, I_{80} and I_{100}, respectively). The results indicated that water stress affected total dry matter (TDM), leaf aria index (LAI), water (WUE) and nitrogen utilization efficiency (NUE). The highest TDM of 21.45 t/ha was observed at I_{40}. Furthermore, NUE and LAI were higher at I_{40}. WUE increased as the water depletion increased and reached to a maximum of 3.44 kg DM m^{-3} at severe stress. In forage quality, TDN% reached to the highest value of 54.7% in non stress water treatment. However, CP% increased by soil water depletion and more N fertilizer application. The highest profit was observed when more water and N fertilizer was applied. They concluded pearl millet in semi-arid area can be cultivated with acceptable forage yield by saving irrigation water compared to traditional forms and reducing nitrogen supply.

5. Suitable crops for water management under deficit irrigation

Crop selecting has special importance in this method. As a general rule plants which their fresh yields are are consumed are not eligible to apply deficit irrigation systems on them. Summer crops including sugarbeet, potato and some forage crops and vegetables are not suitable. While small grains including wheat, barley, triticale and drought stress tolerant oil seeds specially safflower and canola are important crops that applying deficit irrigation is possible for them and among industrial crops, the cotton can be indicated (English, 1990). However, it is necessary to notice that drought stress doesn't induce specially at pod setting stage by applying deficit irrigation.

6. Environmental conditions and deficit irrigation

Identification of environmental conditions is of great importance for applying deficit irrigation; some of these conditions are listed as following:

Soil: soil texture and structure along with topography have determinant role to apply deficit irrigation. In relatively light soils applying deficit irrigation is not as easy as heavy soils. As well as in soils without enough organic matter, this method is not applicable due to low water holding capacity.

Pressure irrigation equipments are most important factors when the farmland is unleveled. In salty soil due to intensity of osmotic potential as a result of water deficiency the selection of irrigation method and type of crop have special importance.

7. Weather conditions

Drought stress is intensified by warm weather, as Maleki Farahani *et al* (2010b) found in their research that under deficit irrigation the barley 1000 seed weight decreased by 12% although in year with fairly higher temperature during grain filling 1000 seed weight decreased by 35%, thus applying deficit irrigation is more successful in autumn-winter crops than summer crops. Sanjani et al (2008) found yield of cow pea and sorghum decreased by about 50% in additive intercropping system of grain sorghum and cowpea under limited irrigation. The limited irrigation (moisture stress) treatments consisted of IR1: normal weekly irrigation (control), IR2: moderate moisture stress during vegetative and generative growth, IR3: moderate moisture stress during vegetative and severe during generative growth, IR4: severe moisture stress during vegetative and moderate during generative growth. Also Soltani et al (2007) evaluated 11 new corn hybrids under water deficit irrigation by applying different amount of water including irrigation after 70, 100 and 130 mm evaporation from A evaporation pan. Their findings revealed that all hybrids produced significantly less yield after medium or sever water stress as average yield over 11 hybrids was 7.5, 5.4 and 4.9 t/ha in 70, 100 and 130 mm treatments respectively. However, corn seed inoculation by phosphate soluibilizing microorganisms (*Arbuscular Mycorrhiza* and *Pseudomonas fluorescence*) showed satisfying results when applied along with above three irrigation levels (70, 100 and 130 mm) (Ehteshami et al.,2007). They stated that phosphate soluibilizing microorganisms can interact positively in promoting plant growth as well as P uptake in corn plants, leading to plant tolerance improving under water deficit irrigation systems. Summer farming will be successful if the temperature doesn't rise over the required optimum plant temperature. In tropical weather condition because of salt transformation due to soil water evaporation, it may intensify the salinity and drought stress after applying deficit irrigation. As a general recommendation, this method is more successful in autumn- winter crops than summer crops because of salts being washed downward, lower evapotranspiration and higher precipitation.

8. Crop growth stage

Success in applying deficit irrigation is highly dependent on asynchronism of sensitive growth stages and drought stress (Kirda, 2000). Plant growth and development stages in which important yield components are determined shouldn't encounter drought stress. For example, spikelet differentiation and anthesis have important role in wheat yield, therefore for wheat cultivation, deficit irrigation should set in a manner to avoid drought stress in both mentioned stages (English and Nakamura, 1989; Ghodsi et al, 2005; Ghodsi et al., 2007). Irrigation frequency and irrigation time should be regulated based on crop growth stage and their sensitivity of them to drought stress. For example, it is suggested to perform two light irrigations at grain filling of wheat without producing optimal moisture condition.

There is a need to find detrimental effect of water stress in crops while limited irrigation is applied in different growth stages of crops. There are evidences that some experiments regarding deficit irrigation have been done in some crops like wheat, turnip, sorghum and etc. Ghodsi et al (2007) performed a field experiment on different bread wheat varieties to find the most critical growth stages to water stress. They conducted a field experiment in Torogh Agricultural Research Station (Mashhad, Iran) in 2000/01 and 2001/02 cropping seasons, using a split plot design based on a randomized complete block design with 3 replications. Main plots were assigned to 7 levels of water stress treatments D1, full irrigation; D2, cessation of watering from one leaf stage to floral initiation, and in other treatments, cessation of watering under rain shelter D3, one leaf stage to floral initiation; D4, floral initiation stage to early stem elongation; D5, early stem elongation stage to emergence of flag leaf; D6, emegence of flag leaf stage to anthesis; D7, anthesis stage to late grain filling (soft dough). Sub-plots were assigned to four bread wheat cultivars: Roshan, Ghods, Marvdasht and Chamran. Results of combined analysis of variance showed, biological yield, grain yield, yield components, harvest index and other traits were significantly affected by water stress treatments. Under D5, D6 and D7 treatments, grain yield decreased compared to D1 by 36.7, 22.8 and 45.6%, respectively. There were also significant differences between genotypes for yield and yield components. Significant correlation coefficients were found between grain yield and number of spike per m^2, number of grains per spike, harvest index, spike weight at anthesis and seed set percentage. Under water stress conditions, grain yield was more affected by number of grain per unit area. Results showed, susceptibility of developmental stages of bread wheat to water stress were different. Exposing to water stress in each developmental stages, lead to decrease in yield. Grain filling (D7) and stem elongation (D5) stages were the most critical stages under water stress conditions. The effect of water stress in early pre-anthesis (D6) and tillering (D3) stages was also considerable. The results of this study illustrated that imposing moisture stress in critical growth stages (Commencement of stem elongation, anthesis and grain filling) would significantly decrease grain yield; however, imposing moisture stress in initial growth stages would not have such a significant effect on grain yield. Furthermore, wheat cultivars reacted differently to different moisture stress treatments. Chamran cultivar had a higher grain yield and was more tolerant to moisture stress during critical growth stages. On the other hand, it was demonstrated that application of lower moisture stress treatments (D3 and D4) relatively increased water use efficiency (WUE), however, severe moisture stress treatments (D5, D6 and D7) decreased WUE. Genetic differences also played a significant role in variation in WUE among different cultivars. Roshan and Chamran cultivars exhibited the lowest and the highest WUE, respectively. It was also illustrated that there were some differences in moisture stress treatments for radiation use efficiency (RUE). D1, D2 and D3 treatments showed the highest RUE, while the lowest RUE belonged to D5 and D6 treatments.

In other study that has been conducted by Keshavarzafshar et al (2011) the reponse of forage turnip were evaluated to water deficit. In this study a field trial was conducted in Research Farm of College of Agriculture, University of Tehran, in Karaj/Iran (N 35°56″, E 50°58″), during 2009. The climate type of this site was arid to semiarid with the annual average climate parameters as follows: air temperature 13.5°C, soil temperature 14.5°C, and with a rainfall of 262 mm per year. The soil texture of the experimental field was Clay loam (33% sand, 36% silt and 31% clay) with pH= 8.2 and Ec = 3.41ds/m. The organic carbon content of the surface layer soil (0–15 cm) was 1.02 %. The soil had no salinity and drainage problem, and water table was more than 7 m deep. Turnip seeds were plantd on March 3rd, 2009. Plant to plant spacing was 10 cm and plant rows were 70 cm apart. The depth of sowing was 2 cm. The crop was harvested on June 15th, 2009. After elimination of border effects, one square meter area was hand harvested in each plot. After harvest, fresh yields of roots and leaves were measured and samples were dried in oven at 70° C to a constant weight for dry matter content. Three replicated samples of each treatment were taken for forage quality analysis.

Their results showed that highest tuber yield of 930.8 Kg/ha was produced at no water stress treatment (IR_N) while the lowest yield of 307 kg/ha was produced at control (IR_0). The most efficient irrigation regime in regard to tuber production was IR_1 causing 59% more tuber dry matter compared to control. As the severity of the water stress reduced, at IR_2 and IR_3, the efficiency of extra water application followed a decreasing trend.

In the most severe water stress condition (IR_0), 100%F_{Ch} treatment demonstrated the best performance in tuber biomass production (almost five fold more than control). under favorable moisture condition (IR_N), application of integrated fertilizer (50% F_{Ch}+F_{Bi}) produced the highest tuber yield which was 18% more than control. In other irrigation levels, no significant difference between these two treatments, 100% F_{Ch} and 50% F_{Ch}+F_{Bi}

was observed. As the severity of water stress increased, the total biomass followed a decreasing trend. The highest biomass production of 3640 kg/ha was achieved by IR_N irrigation regime which was nearly five fold more than control (IR_0). The highest efficiency of biomass production per unit water utilization was achieved in IR_1 in which with only one irrigation at sowing time, the biomass production reached 2091 Kg/ha (100% increment compared to IR_0). In IR_2 by an extra irrigation at tuber formation stage, the added biomass was only 472 kg/ha more than IR_1, showing a much less efficiency in biomass production per unit water application.

Interaction effect of irrigation regimes and P fertilizers on total biomass yield of turnip was significant ($p < 0.01$). In IR_0 treatment, application of 100% F_{Ch} and 50% $F_{Ch}+F_{Bi}$ increased biomass yield compared to control. Except for IR_0, in other irrigation regimes application of F_{Bi} treatment had no significant effect on biomass production of turnip.

The effects of irrigation regimes and P fertilizers on tuber protein yield of turnip were significant ($p < 0.01$). Water stress caused a significant decrement in crude protein yield. The highest yield of crude protein (129.4 kg/ha) was obtained by IR_N while the lowest yield (48.6 kg/ha) was obtained from IR_0 (nearly threefold increment). By one irrigation at sowing time (IR_1), the yield of crude protein highly increased (52 % increase compared to the control). However, the extra irrigation at tuber formation stage (IR_2) and third irrigation at stem elongation stage (IR_3) performed a lower efficiency in increasing the protein yield of turnip tuber.

As the water stress severity decreased, the digestibility of tuber dry matter followed an increasing trend. The lowest percent of DMD (62.9%) was obtained by IR_0 and the highest percent (66. 9 and 68.5) was achieved by IR_3 and IR_N, respectively.

Application of phosphorous chemical fertilizer (100% F_{Ch}) had positive effect on dry matter digestibility of turnip tuber and increased it by more than 10 percent compared to control. However, other fertilizers had no significant effect on this trait.

By decreasing the severity of water stress, the ADF percent of turnip tuber followed a decreasing trend. The highest tuber ADF was observed in IR_0 (30%) and the lowest percent was achieved in IR_N (23.4 %).

The interaction effect of irrigation regimes and phosphorous fertilizers on ADF percent of turnip tuber was significant ($P < 0.01$). In the most severe water stress condition (IR_0), application of sole bio fertilizer (F_{Bi}) and integrated fertilizer (50% $F_{Ch}+F_{Bi}$) increased tuber ADF compared to control. However, in other irrigation regimes, application of 100% F_{Ch} and 50% $F_{Ch}+F_{Bi}$ resulted in lower ADF percent compared to control. Overall, in all irrigation regimes, chemical P fertilizer had the most positive effect on decreasing ADF of turnip tuber.

Also as the water stress severity decreased, the tuber ME followed an increasing trend. The ME in IR_0 was 8.7 while in IR_N it was 9.6 MJ/kg dry matter.

Finally they concluded that turnip tuber yield was adversely affected by water stress and it is very sensitive to water stress at germination, establishment and early growth stages.

Considering to find most sensitive growth stages to water deficit, the following study was performed by Khalili et al (2006) on grain sorghum variety Kimia. The Experiment was initiated in Research Farm of College of Agriculture, University of Tehran located in Karaj/Iran during summer 2004. The main plots were allocated to five different irrigation regimes which applied drought stress on sorghum (soil moisture approached wilting point before the next irrigation) at different vegetative and generative growth stages. The irrigation regimes comprised of: 1) Full irrigation (IR1) (control): The plots in this treatment were irrigated at weekly intervals up to the end of the growing period. 2) Moderate drought stress in both vegetative and generative stages (IR2): The plots allocated to this treatment were irrigated on weekly basis until the plants reached well establishment at 6 to 8-leaf growth stage and then the irrigation was ceased until 10 to 12-leaf stage where the plots received irrigation. Again irrigation was ceased until the early flowering stage (5 to 10% flowering) which the plant received another irrigation. The next irrigation was applied when the plants were in early milky grain stage and since then no irrigation was applied until the plants reached the physiological maturity. 3) Moderate drought stress in vegetative stage (after 6-8 leaf stage) and severe drought stress in generative stage (IR3): Irrigation treatment was identical to IR2 up to early flowering stage and then no irrigation was applied until plants reached the physiological maturity. 4) Severe drought stress at vegetative stage and moderate stress at generative stage (IR4): At vegetative growth stage the irrigation treatment was similar to IR2 except that no irrigation was applied at 10 to 12-leaf growth stage. However, the irrigation treatment followed exactly the same as IR2 in generative part of the plant growth. 5) Severe drought stress in both vegetative and generative growth stages (IR5): The Irrigation treatment followed the same trend as IR4 at vegetative and IR3 at generative stages of plant growth. 1 The statistical analysis of the data showed that there was a significant difference ($p<0.01$) in grain yield production due to different irrigation regimes. The highest grain yield of 5871 kg/ha was obtained from control plots while the lowest grain yield of 500 kg/ha (less than ten times) was produced in severe drought stress both in vegetative and generative growth stages. As the drought stress in generative stage of the plant increased, grain yield followed a decreasing trend. In the severe drought stress regime in generative stage (IR3), the reduction of the kernel weight and one thousand kernel weight could be accounted for grain yield decrement. This shows the importance of water availability in generative stage of the plant growth (especially grain filling stage). The severe reduction of grain yield in irrigation regimes of IR2, IR3 and IR5 indicated the plant sensitivity to drought stress at different phenological stages. Grain production decreased over 50% in these treatments compared to control, however, in IR4 treatment, this reduction was only about 30%.

The results of this experiment indicate the importance of irrigation at early flowering and milky grain stages of the plant growth which could produce not only a proper grain yield, but also contribute in significant water conservation compared to control (full irrigation). The number of irrigations in IR4 treatment was reduced by 50% (from 18 to 9) compared to control, which from ecological and economical point of the views is very important in dry areas. The statistical evaluations showed that there is a statistically significant positive correlation between kernel weight, kernel length, one thousand kernel weight; biological yield and harvest index with grain yield production. Drought stress especially in generative growth stages caused a severe decrement in grain yield which could be because of decreasing of one thousand kernel weight, kernel length decrement and consequentlydecreasing the number of grains per kernel. Also the lower number of grains in each kernel may be due to disordered pollination and finally decrement the number of fertilized flowers. By applying a regular irrigation on sorghum from germination to plant establishment stage (7-8 leaf) and then limited irrigations just at 10-12 leaf, early flowering and milky grain stages, the number of irrigations will be decreased from 18 to 9 times. Despite of 30%grain yield reduction in this system; still it is beneficial from ecological and economical point of the views for arid environments. So, Khalili et al (2008) suggested that by severe moisture stress at vegetative along with providing the minimum water requirements in generative growth stages of grain sorghum, the water consumption efficiency of the plant will be improved and a reasonable grain yield is achievable.

9. Economical aspect of deficit irrigation

Benefits: Beneficial effects of deficit irrigation are evaluated as different economic and social aspects. Researches have indicated that regardless to equal energy use either in normal or deficit irrigation, the amount of production per given water unit usually is more under deficit irrigation than normal irrigation. With water saving and providing possibilities for farmland expansion, the equipment use efficiency increases, therefore labor and machinery will be used in more efficient way (English and Raja, 1996).

Also the farmer income will increase by cultivating of high demanded vegetable and summer crops through saved water in deficit irrigation. Furthermore, results of applying deficit irrigation by reducing irrigation times have shown quality enhancement of subsequent produced seeds. Seeds which produced under deficit irrigation condition germinated earlier and had greater germination percentage in drought and salinity stress which induced either by polyethylene glycol or NaCl compared to seed produced under normal irrigation (Maleki Farahani *et al.*, 2010b). Moreover, the grain nutritional quality enhanced after implementing deficit irrigation (Maleki Farahani *et al.*, 2011). Deficit irrigation increased barley N content by 12% as well as Zn and Mn 27% and 7% compared to control. Also 4% increment was observed in P concentration an important element for seed germination.

In macro view, increment of agricultural production and efficiency of labor and machinery resulting from application of deficit irrigation can be assumed as benefits.

Disadvantages: Lack of knowledge about sensitive plant growth stages, insufficient planning for water use and distribution not only can affect the benefits of deficit irrigation but also can cause damage for the farmers. Drought stress in every critical growth stage will make irrecoverable damages for crop (English, 1990).

Deficit irrigation is not the same as complementary irrigation. In complimentary irrigation which is usually performed in dry land farming systems, one or two irrigations are applied at critical growth stages in which raining don't take place. However, in deficit irrigation the farmer's attitude should be based on relative reduction of water in an irrigated farming system. If the time and amount of water in this method are not determined properly, an irrecoverable damage will suffer the crop. More emphasis is on proper planning in this method to prevent probable damages.

10. The role of policymakers in development of deficit irrigation

Development and recommendation of new methods won't have favorable results if they aren't based on evaluation and planning. In first point of view, deficit irrigation won't be welcome by farmers because of relative reduction of yield. In agricultural farming systems, which are managed by deficit irrigation, the net income is less than normal irrigation because expenses for land preparation and weed control are equal in both systems.

Generally, subsidizing and farmers supporting are not inevitable in case of policy for deficit irrigation. Subsidization may be indirect as providing of inputs like chemicals to the farmers who manage their farms with deficit irrigation method. Moreover water can be available with lower price for the mentioned farmers to compensate yield reduction.

In years which water source deficiency may take place because of lower precipitation, the development of deficit irrigation is a preference. Repetition of deficit irrigation in a long period of time may be set as farmers culture. Media plays key role in explaining deficit irrigation to be accepted by farmers. Planning for better use of water resources is inevitable.

11. Conclusion

Deficit irrigation methods are those irrigation methods that yield increases per given water unit (water productivity). Beside the water productivity, quality of the crop could be improved by more tolerance to drought and salt stress as well as more nutritional quality. The performance of these method is better in large lands and in years with lower precipitation which water is limited. In general it can be apply by either fixed irrigation frequency and reduced water in each irrigation or reduced irrigation frequency and fixed amount of water in each irrigation time. In both ways the basic principle is water usage reduction compared to normal irrigation, so that none of the critical plant growth stage encounters drought stress.

Soil texture, weather conditions, type and growth stage of plant and available possibilities have important role for applying and selecting of deficit irrigation method. Governmental support through subsidizing can play an important role in deficit irrigation development.

12. References

Blum, A., 2009. Effective use of water (EUW) and not water-use efficiency (WUE) is the target of crop yield improvement under drought stress. Field Crops Research, 112:119–123.

Chaichi, M. R., Rostamza, M. and Esmaeilan, K. S. 2004. Tolerance evaluation of chickpea accessions to drought under different irrigation systems during generative growth stage. Journal of Agricultural Science and Natural Resources, 10: 55-64.

Ehteshami, S. M. R., Aghaalikhani, M. Khavazi, K. and Chaichi, M. R. 2007. Effect of phosphate soluibilizing microorganisms on quantitative and qualitative characteristics of maize (Zea mays L.) under water deficit stress. Pakistan Journal of Biological Sciences, 10: 3585- 3591.

English, M. J. 1990. Deficit irrigation. I. Analytical framework. Journal of Irrigation and Drainage Engineering, 116: 399–412.

English, M. J and Nakamura, B. 1989. Effect of deficit irrigation and irrigation frequency on wheat yield. . Journal of Irrigation and Drainage Engineering, 115: 172-184.

English, M. J and Raja, S. N. 1996. Perspectives on deficit irrigation. Agricultural Water Management, 32: 1-14.

Farooq, M., Wahid, A., Kobayashi, N., Fujita, D and Basra, S., 2008. Plant drought stress: effects, mechanisms and management. Agronomy for Sustainable Development, 1-28.

Fereres, E. and Soriano, M. A. 2007. Deficit irrigation for reducing agricultural water use. Journal of Experimental Botany, 58: 147–159.

Ghodsi, M., Chaichi, M. R., Mazaheri, D. and Jalal Kamali, R., 2005. Determination of susceptibility of developmental stages in bread wheat to water stress and its effect on yield and yield components. Seed and Plant, 20: 489-509.

Ghodsi, M., Jalal Kamali, R., Mazaheri, D. and Chaichi, M. R. 2008. Water and radiation use efficiency in different developmental stages in four bread wheat cultivars under moisture stress conditions. Desert, 12: 129-137.

Heidari Zooleh, H., Jahansooz. M. R.,Yunusa, I., Hosseini, S. M. B., Chaichi, M. R. and Jafari, A. A. 2011. Effect of Alternate Irrigation on Root-Divided Foxtail Millet (*Setaria italica*). Australian Journal of Crop Science, 5: 205-213.

Jorat, M., Raei, Y., Moghaddam, H. and Chaichi, M. R. 2011. The effect of Limited Irrigation of sowing density on two forage Sorghum Varieties. M.Sc. Thesis Tabriz University, Tabriz, Iran.

Katerij, N., Mastrorilli, M.,Van Hoorn, J., Lahmer, F., Hamdy, A and Oweis, T., 2009. Durum wheat and barley productivity in saline–drought environments. European Journal Agronomy, 31: 1-9.

Keshavarzafshar, R., Chaichi, M. R., Moghadam, H. and Ehteshami, S. M. R. Effect of phosphorous fertilizers and irrigation regimes on yield and forage quality of Turnip. M.Sc. Thesis University of Tehran, Tehran, Iran.

Khalili, A., Akbari, N. and Chaichi, M. R. 2006. Limited irrigation and phosphorus fertilizer effects on yield and yield components of grain sorghum (Sorghum bicolor L.var.Kimia). M.Sc. Thesis Lorestan University, Lorestan, Iran.

Kirda, C. 2000. Deficit irrigation scheduling based on plant growth stages showing water stress tolerance. Publishing and Multimedia Service, Information Division, FAO, Viale delle Terme di Caracalla, 00100 Rome, Italy.

Martyniak, L., 2008. Response of spring cereals to a deficit of atmospheric precipitation in the particular stages of plant growth and development. Agriculture and Water Management, 95: 171–178.

Maleki Farahani,S. 2009. Evaluation of deficit irrigation and fertilizer on yield and grain properties of barley (*Hordeum vulgare* cv. Turkman). PhD thesis, University of Tehran, Tehran. Iran.

Maleki Farahani, S., Chaichi, M. R., Mazaheri, D., Tavakkol Afshari, R. and Savaghebi, G. 2010a. Barley grain chemical analysis under different irrigation and fertilizing systems. Asian Journal of Chemistry, 22, 3: 2397-2406.

Maleki Farahani, S., Mazaheri, D., Chaichi, R., Tavakkol Afshari, R. and Savaghebi, G. 2010b. Effect of seed vigour on stress tolerance of barley (*Hordeum vulgare*) seed at germination stage. Seed Science and Technology, 38: 494-507.

Maleki Farahani, S., Chaichi, M. R., Mazaheri, D., Tavakkol Afshari, R. and Savaghebi, G. 2011. Barley grain mineral content as affected by different fertilizing systems and drought stress. Journal of Agricultural Science and Technology, 13: 315-326.

Rostamza, M., Chaichi, M. R., Jahansooz, M. R. and Alimadadi, A. 2011. Forage quality, water use and nitrogen utilization efficiencies of pearl millet (*Pennisetum americanum* L.) grown under different soil moisture and nitrogen levels. Agricultural Water Management, 98: 1607- 1614.

Samarah, N. and Al-Issa, T.A. 2006. Effect of planting date on seed yield and quality of barley grown under semi-arid Meditarranean conditions. Journal of Food, Agriculture and Environment, 4: 222-225.

Sanjani, S. and Chaichi, M. R. 2oo8. Investigation of yield and yield components in additive intercropping of grain sorghum (*Sorghum bicolor*) and cowpea (*Vigna unguiculata*) in limited irrigation condition. M.Sc. Thesis University of Tehran, Tehran, Iran.

Soltani, M., Azizi, F., Chaichi, M. R. and Heidari Sharifabad. H. 2007. Evaluation the effect of limited irrigation on physiological, morphological and yield of 11 new corn hybrids. M.Sc. Thesis Azad Islamic University, Science and Research branch, Tehran, Iran.

2

Effects of Irrigation on the Flowering and Maturity of Chickpea Genotypes

Kamel Ben Mbarek, Boutheina Douh and Abdelhamid Boujelben

High Agronomic Institute Chott-Mariem

Tunisia

1. Introduction

In Tunisia, chickpea (*Cicer arietinum L.*), particularly Kabuli genotypes, is the second pulse crop after faba bean. It is grown, in spring rainfed conditions (Wery, 1990), in humid and sub humid regions, mainly at Bizerte, Mateur, Béja, Jendouba and Nabeul areas (DGPA, 2006). Feeble production, about 13.518 tons with a reduced grain yield, nearly 0,67 t.ha^{-1} (DGPA, 2008), characterized by inter annual fluctuations, does not satisfy national needs. Tunisian government makes recourse to massive annual imports, about 19.000 tons (AAC, 2006), which account 141% of the national production. To satisfy national needs of this foodstuff, it would be useful to undertake researches to increase chickpea production through drought and thermal tolerant stress genotypes and extension of this species culture area to the semi-arid zones. Spring chickpea culture is subjected to drought stress, generally, combined with a thermal stress. These two abiotic stresses explain, partly, the production irregularity and the chickpea grain yield instability in our regions. Kumar and Abbo, (2001) reported that throughout the world, 90% of the chickpea cultures are rainfed and final dryness is the principal abiotic stress which blocks the production increase. Golezani *et al,* (2008) indicated that, in many areas of leguminous culture, such as chickpea, the climate is characterized by extremely variable precipitations and rather often deficit. Under such environmental conditions, scientists and farmers try to identify crops and soil management techniques for an adequate water use efficiency. Both temperature and moisture supply during the growing period had a strong influence on chickpea plant phenology (Silim and Saxena, 1993). Nayyar et al., (2006) reported that the flowering and pod setting stages appear to be the most sensitive stages to water stress. McVicar et al., (2007) noticed that the moisture stress is required to encourage seed set and to hasten maturity. If weather turns warm and dry, plants will be delayed in maturity and produce lower yields. However, Summerfield and Roberts (1988) announced that the chickpea flowering time is variable depending on season, sowing date, latitude, and altitude. According to Roberts et al., (1985), time to flowering was a function of temperature and photoperiod in chickpea. Ellis et al., (1994) further noticed that in some chickpea genotypes, time to flowering was influenced by photoperiod and temperature, whereas in others, flowering time was determined solely by photoperiod. Gumber and Sarvjeet (1996) studied the chickpea genetics of time to flowering and found that it was controlled by two genes. Kumar and van Rheenen (2000) announced the presence of one major gene (Efl-1/ efl-1) plus polygenes for this trait. Or et al., (1999) also supported this result, but they associated the major gene with sensitivity to photoperiod (Ppd/ppd).

This present study aims to evaluate the effects of amounts of irrigation on flowering and maturity of eight kabuli chickpea genotypes conducted in spring culture under Tunisian semi-arid edapho-climatic conditions.

2. Material and methods

2.1 Edapho-climatic conditions of the experimental site

The experiment was conducted at the Higher Institute of Agronomy of Chott Mariem, Tunisia (Longitude 10°38E, Latitude 35°55N, altitude 15 m) from May to July 2008 (three months). The climate is typically Mediterranean with 370 mm annual rainfall and an average of 6 mm day^{-1} evaporation from a free water surface. The minimum and maximum temperatures have respective mean values 14 and 23 °C. Relative hygroscopy and wind speed are respectively 70 % and 2,3 m/s. This zone is characterized by seven months annually dryness period (mid-March – beginning of October) (Fig. 1). It is defined by reduced and rare precipitations, high evaporation and maximum temperatures. During trial, temperature and relative hygroscopy variations are followed using a thermohygrographe beforehand calibrated (Fig. 2).

Soil is characterized by 52, 5% of total porosity, 20,5% of field capacity and 8,2% of permanent fading point. It is a silt-clay-sandy type (USDA, 1951), alkaline, relatively poor in organic matter (3,5%) and low salinity. The soil electric conductivity, measured at 25 °C temperature, is 0,27 ms.cm^{-2}.

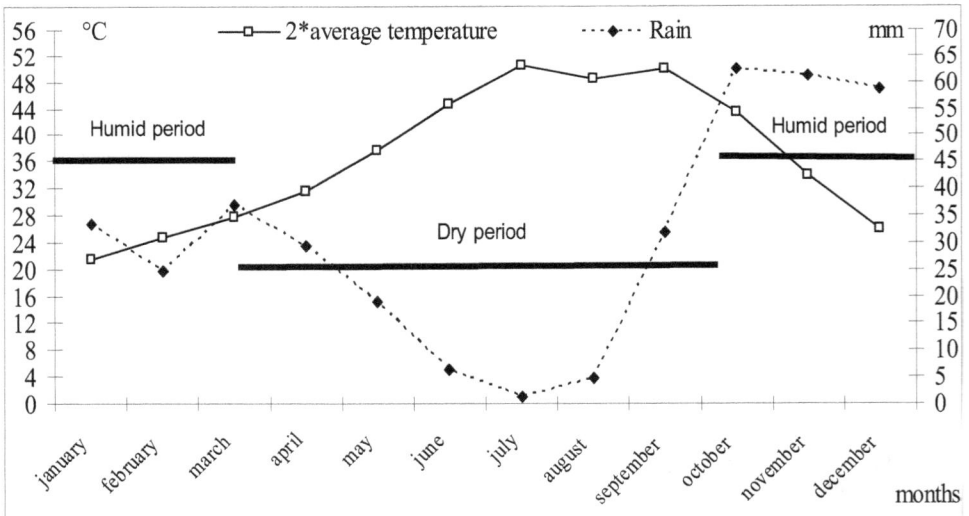

Fig. 1. Ombrothermic diagram of the Chott Mariem zone

2.2 Vegetable material, sowing and harvest dates

The vegetable material is composed of eight kabuli chickpea genotypes. Six of them, namely: Béja1, Amdoun1, Nayer, Kasseb, Bochra and Chétoui (ILC3279), are commercial varieties registered by the National Tunisian Agronomic Research Institute (INRAT) in the obtaining vegetable Tunisian catalogue. The two others, improved lines, FLIP96-114C and FLIP88-42C, were pleasantly provided by the ICARDA within the framework of the "International Vegetable Testing Program (LITP) " Alep; Syria (Table 1).

N°	Name	Pedigree	Origin
1	Béja1	INRAT 93-1	Tunisian
2	Amdoun1	Be-sel-81-48	Tunisian
3	Nayer	FLIP 84 - 92 C	Tunisian
4	Kasseb	FLIP 84 - 460 C	Tunisian
5	Bochra	FLIP 84 - 79 C	Tunisian
6	FLIP96-114C	X93 TH 74/FLIP87-51CXFLIP91-125C	ICARDA/ICRISAT
7	FLIP88-42C	X85 TH 230/ILC 3395 x FLIP 83-13C	ICARDA/ICRISAT
8	Chetoui	ILC3279	Tunisian

Table 1. Kabuli chickpea (*Cicer arietinum* L.) genotypes

Culture is conducted, in *situ*, under controlled conditions, in pots 24 cm diameter and 24 cm height. Pots, filled with arable land, are arranged under hemispherical greenhouse covered with polyethylene (180 μ thickness) and aired on the two sides. Sowing is realized on April 16, which is four weeks delayed date compared to the normal spring sowing (Malhotra and Johansen, 1996) at a rate of three chickpea seeds per pot. After plant establishment, the plants were culled with only one seedling left in the pot. Harvest took place at the end of July of the same year.

2.3 Irrigation

Water irrigation, coming from the Nebhana dam, is characterized by 1,09 ms.cm^{-2} electric conductivity (measured at 25 °C temperature). It contains 0,70 g.l^{-1} of dry residue of which 0,25 g.l^{-1} are sodium chlorides. The easily usable reserve (EUR), evaluated with 464 ml, is calculated according to the formula stated by Soltner, (1981)

$$EUR = 1/2\left[(Fc - pF)/100\right] * D_{ap} * V$$

With Fc: Field capacity; pF: Permanent fading point; D_{Ap}: Apparent density; V: Pot soil volume.

Studied factor is water regime mode with four treatments or amounts irrigation (DI) namely: 100%, 75%, 50% and 25% of the EUR in a randomized block experimental design with three replications. Irrigations are achieved on the basis of the crop evapotranspiration (ETc) (Ben Mechlia, 1998). Reference crop evapotranspiration (ET$_0$) was calculated from Blanney-Criddel formula (Doorenbos and Pruitt, 1977). Crop coefficient (Kc) and adopted chickpea physiological phases durations are those used by FAO (Allen et al., 1998).

2.4 Studied parameters

Parameters studied are:

- Early flowering date (EFlDt, in days after sowing (DAS)): blooming date of the first flowers,

- 50% flowering date (FlDt, in DAS): blooming date of 50% of flowers,
- Flowering phase duration (FlDr, in days): the time passed between the blooming of the first and the last flowers,
- Early pods maturity date (EMtDt, in DAS): yellowing date of the first pods,
- 50% pods maturity date (MtDt, in DAS): yellowing date of 50% pods,
- Pods maturity duration (MtDr, in days): the time passed between yellowing of the first and the last pods.

XLSTAT and SPSS (version 10) Software were adopted to achieve statistical analyses. From obtained data, variance analysis (ANOVA,) and means comparison (Student-Newman-Keuls test at 5% level) were performed.

3. Results and discussion

3.1 Evaluation of the farming site climatic conditions

The chickpea (*Cicer arietinum* L.) biological cycle lasted 104 days. During the biological cycle, the relative hygroscopy varied from 47,5 to 73%. It fell with less than 50% at the beginning and the end of the pods maturity phase duration. Mean temperatures recorded during growth, initial, development, filling and maturity phases are respectively of 24 °C, 26 °C, 30 °C and 33 °C (Fig. 2).

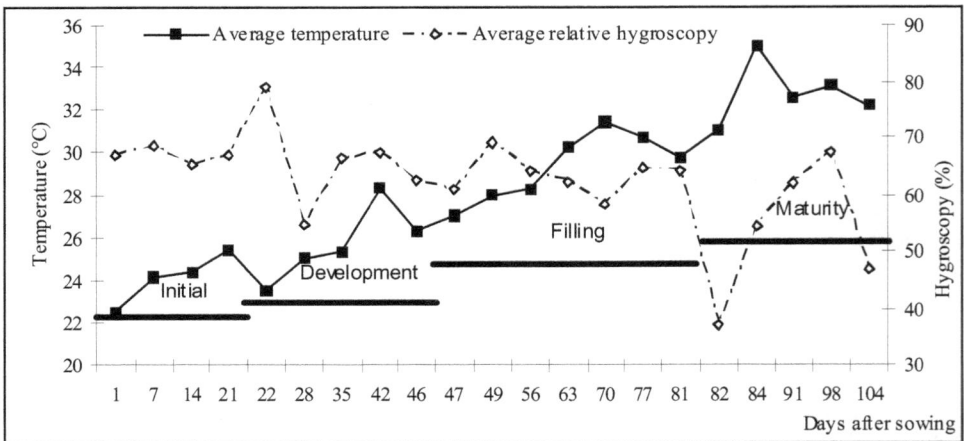

Fig. 2. Averages temperatures and relative hygroscopy of the chickpea (*Cicer arietinum* L) farming cycle conducted *in situ*

Summerfield et al., (1981) observed that chickpea reproductive phase suffers considerably from high temperatures (35/18 °C, day/night). Under such thermal conditions, grain yield is reduced to 33% per comparison to that under lenient conditions such as 30/10 °C day/night. According to Wery et al., (1994), critical temperature during the reproductive phase which includes flowering, filling and enlargement seeds chickpea is evaluated with 30 °C. Recorded temperatures showed that critical temperature was exceeded only during the pods maturity phase duration (Fig. 2). This reveals that this chickpea culture did not suffer from thermal stress.

Crop coefficient (Kc) varies according to the chickpea culture growth phases. The crop evapotranspiration (ETc) is relatively low during the initial and pods maturity phases duration, whereas it is greatest during the filling and enlargement seeds phases (Fig. 3). Slama, (1998) indicated that the chickpea fears drought stress and crop water requirements are high during the reproductive growth phase, in particular, the flowering and filling seeds stages.   A chickpea culture water requirement is evaluated to 392 mm (Fig. 4). They are divided into 8,4% during the initial phase, 24,5% during the development, 61,7% during the filling and the enlargement seeds and 5,4% during pods maturity. With the amount irrigation 100% of EUR, cumulated water irrigation provided appear equivalent to the culture water needs (Fig. 4). It appears that the chickpea culture did not undergo drought stress. These results are in conformity with those indicated by Slama, (1998) which stated that the chickpea culture water requirements vary, according to genotypes, from 300 to 400 mm. With 75 % of EUR amounts irrigation, that equivalent to 300 mm, chickpea culture submit to drought stress during the semi-filling and seed maturity phases.   With 50 % of EUR amounts irrigation, that equivalent to 200 mm, chickpea culture submit to drought stress during the semi-development, filling and seed maturity phases (Fig. 4).   According to Nayyar et al., (2006), flowering and filling seeds seem the most sensitive chickpea growth phases to drought stress. With 25 % of EUR amount irrigation, equivalent to 100 mm, drought stress affected chickpea seedlings during all vegetative and reproductive culture phases (Fig. 4). Saxena (1987) noticed that, *in situ*, chickpea water consumption depends on the ground moisture and the discounted grain yield.

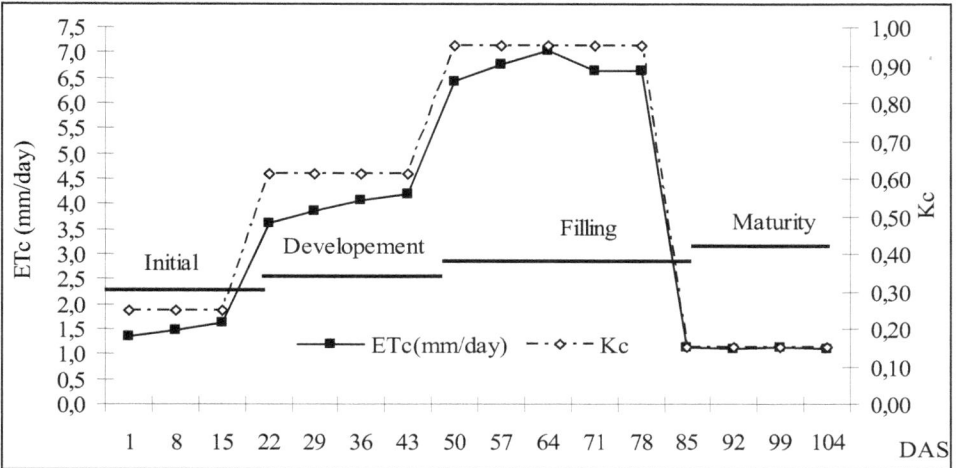

Fig. 3. Crop evapotranspiration (Etc) and crop coefficient (Kc) according to the chickpea (*Cicer arietinum* L.) phonologic growth phases

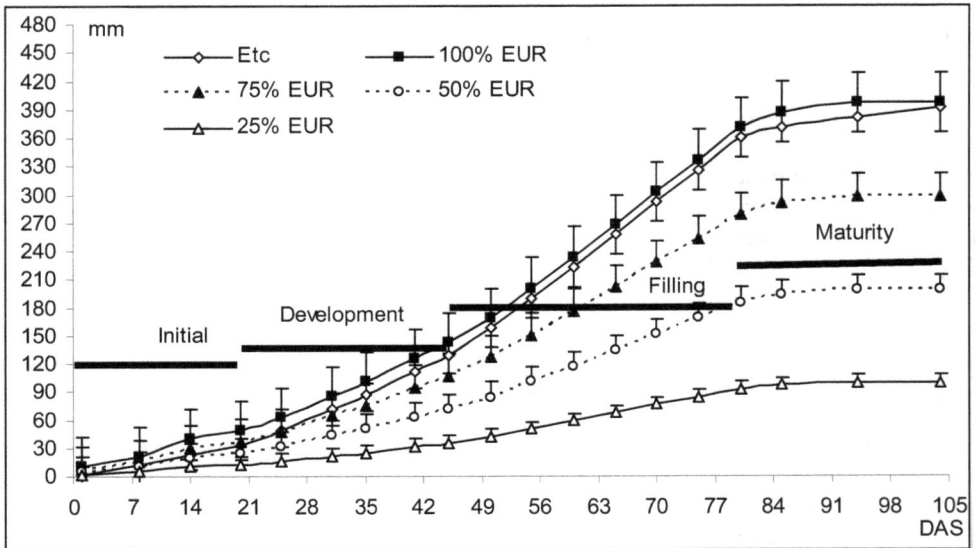

Fig. 4. Crop evapotranspiration (Etc) and cumulated water requirements varations accorading to the chickpea (Cicer arietinum L.) phonologic stages development

3.2 Individual analysis of the studied phenologic parameters

The variance analysis showed that the differences between amount irrigation are very highly significant ($P \leq 0.001$) for the early and 50% flowering dates, the early pods maturity date, the flowering and pods maturity phase durations and significant at 5% level for the 50% pods maturity. Genotypic variability is very highly significant ($P \leq 0.001$) for the early and 50% flowering dates, the early and 50% pods maturity dates, significant at 5% level for the flowering phase duration and non significant for pods maturity phase duration. The interaction (Genotype X Amount irrigation) is very highly significant ($P \leq 0.001$) for the flowering phase duration, highly significant ($P \leq 0.01$) for the early pods maturity date, significant at 5% level for the 50% flowering and 50% pods maturity dates and non significant for the early flowering date and the pods maturity phase duration. Variation coefficients vary from 4,8 to 41,1% (Table 2). These results indicate that the studied chickpea accessions present a large genotypic diversity at the level of the flowering and pods maturity dates and heir phase's duration. It appears that the chickpea flowering and pods maturity dates and the duration of these two phases are controlled by the crop water requirement. The early flowering and 50% flowering dates are inversely proportional to the amounts of irrigation. The early flowering date varied from 50,5 to 58,2 DAS; whereas the 50% flowering date varied from 61,7 to 66,5 DAS. Seedlings irrigated with 100% and 75% of EUR amount irrigation presented an early flowering; whereas those having received 50% and 25% of EUR amount irrigation expressed a late flowering (Table 3).

The abiotic stresses, in particular, drought and thermal, delay the spring chickpea flowering phase (Silim, and Saxena, 1993). Whereas Anbessa et al., (2006) noticed that early flowering is a key factor in the formation and maturation of pods before the occurrence of these abiotic stresses. Hughes et al., (1987) announced that the exposure of the culture to the final dryness shortens its biological cycle and delays its flowering. Ellis et al., (1994) indicated that high temperatures, higher than 38 °C, delay considerably the chickpea flowering. Day temperatures recorded during the flowering phase did not exceed 30 °C (Fig. 1). This reveals that the chickpea culture did not undergo thermal stress. On the other hand, the delay flowering date of the treatments irrigated with low irrigation doses, particularly, 50% and 25% of EUR amount irrigation, is allotted to the crop water requirement.

Variation source	df	EFlDt (DAS)	FlDt (DAS)	FlDr (days)	EMtDt (DAS)	MtDt (DAS)	MtDr (days)
Amount irrigation (AI)	3	261***	104***	121.2***	111.4***	59.5*	139.6***
Genotypes (G)	7	205***	109***	43.6*	56.6***	87.5***	21.6ns
Bloc	2	104ns	76.6*	17.4ns	39.2ns	42.3ns	21.3ns
AI * G	21	57.8ns	36.2*	52.6***	33.9**	44.9*	11ns
Error	62	37,7	20,3	20,5	16,313	22,2	11,5
Variation coefficient (%)	-	11,2	7,1	24,0	4,8	5,6	41,1

***: significant at 1‰ level; **: significant at 1% level; *: significant at 5% level; ns: not significant

Table 2. Variance analyzes and F tests of the chickpea (*Cicer arietinum* L.) genotypes flowering and maturity parameters

Amount irrigation	EFlDt (DAS)	FlDt (DAS)	FlDr (days)	EMtDt (DAS)	MtDt (DAS)	MtDr (days)
100% EUR	54.9a	62.6b	20.2a	83.5a	84.2a	9.5a
75% EUR	50.5b	61.7b	21.6a	80.7b	82.6a	10.9a
50% EUR	56.3a	64.1ab	16.9b	84.8a	85.5a	5.5b
25% EUR	58.2a	66.5a	17.3b	85.7a	86.2a	7.1b

Table 3. Mean comparisons (Newman-Student and Keuls test at 5% level) of the chickpea (*Cicer arietinum* L.) genotypes flowering and maturity parameters according to amounts irrigation

First flowers appearance date of the chickpea genotypes varies from 48,5 to 58,5 DAS; whereas the 50% flowering date varies from 59,5 to 67,8 DAS (Table 4). Chickpea genotypes, having received 75% EUR amount irrigation underwent drought stress 63 DAS; whereas those having received 50% and 25% EUR amount irrigation, have undergoes drought stress before even the flowering phase (fig. 3). Genotypes Kasseb and FLIP96-114C appear characterized by an early flowering; whereas Bochra, Nayer, Béja1 and ILC3279 have a late flowering. Genotypes Amdoun1 and FLIP88-42C have an intermediate flowering (Table 4). Kumar and Abbo (2001) have reported that time to flowering plays a central role in determining the adaptation and productivity of the chickpea genotypes in short growing environments.

Morizet et al., (1984) showed that genotypic variability for the drought tolerance appears only if the drought stress proceeded during the flowering phase. An early stress does not induce, necessarily, a distinction between the drought tolerant and sensitive genotypes.

Other work of Ouattar et al., (1987) concluded that sifting period for drought stress tolerance could extend until the grain development phase.

Genotypes	EFlDt (DAS)	FlDt (DAS)	FlDr (days)	EMtDt (DAS)	MtDt (DAS)	MtDr (days)
Béja I	**16.5b**	58.5a	64.5abc	84.9ab	85.6abc	**7.1a**
Amdoun I	**22.3a**	52.9ab	62.6bc	85.3ab	87.ab	8.5a
Nayer	18.1ab	**58.7a**	**67.8a**	**87.1a**	**87.7a**	7.8a
Kasseb	20.4ab	**48.5b**	59.5c	81.1b	81.6bc	7.6a
Bochra	19.3ab	59.3a	67.6a	84.1ab	87.4a	**11.4a**
FLIP 96-114 C	20.2ab	50.5b	**61bc**	82b	83abc	7.4a
FLIP 88-42 C	17.7ab	53.8ab	62bc	**80.9b**	**80.7c**	8.5a
ILC 3279	17.6ab	57.5a	64.9ab	84.1ab	83.9abc	7.7a

The values of the same column accompanied by the same letter are not significantly different at 5% level
The values in fat from the same column are extreme values

Table 4. Mean comparisons (Newman-Student and Keuls test at 5% level) of the chickpea (*Cicer arietinum* L.) genotypes flowering and maturity parameters

First flowers appearance date varies, simultaneously, according to the chickpea genotypes and crop water requirement from 39 to 69 DAS (Table 5). Mean comparisons showed that there are three interfered homogeneous groups. The genotype Kasseb presented the earliest flowering date, 39 DAS, with 75% of EUR amount irrigation. On the other hand Béja I formed its first flowers 69 DAS with 25% of EUR amount irrigation (Table 5).

According to Richa, and Singh, (2001), the appearance of the first flowers depends on several factors such as varietals precocity, the sowing date and density and the farming techniques. Singh et al., (1995) indicated that, on the basis of a collection consist of 4165 chickpea genotypes evaluated under drought conditions, they could select only 19 drought tolerant accessions characterized by an early flowering.

The 50% flowering date varies according to the amount irrigation and chickpea genotypes from 54,9 to 73,7 DAS. Mean comparisons showed that there are three interfered homogeneous groups. The earliest flowering date is produced at 55 DAS, by FLIP96-114C with 50% of EUR amount irrigation; while the latest flowering is produced at 74 DAS by Bochra under the same amount irrigation (Table 5). Singh et al., (1995) found that the flowering date of six kabuli chickpea genotypes, led in rainfed conditions, varied from 48 to 54 DAS. Berger et al., (2006) stated that the early chickpea genotypes flowering date varies from 51 to 69 DAS; whereas that of the late genotypes varies from 60 to 93 DAS. Physiological chickpea studies confirm the flowering period importance for the sifting of drought tolerant genotypes (Tollenaar, 1989). Other phenological studies indicated that the chickpea biological cycle and flowering durations are determined by the response of the genotype to the day length, the temperature and photoperiod rise. Subbarao et al., (1995) announced that, chickpea flowering date is the most important component of adaptation to the abiotic stresses such as water deficit and high temperatures. In the semi-arid zones, leguminous flowering date has a great adaptive value for the dryness. It determines the ground water use efficiency for the seeds filling (Or et al., 1999). Saxena et al., (1993)

Amount irrigation	Genotypes	EFlDt (DAS)	FlDt (DAS)	FlDr (days)	EMtDt (DAS)	MtDt (DAS)	MtDr (days)
100% EUR	Béja I	54.7abc	61.3abc	14.3bc	81ab	82.4ab	7ab
	Amdoun I	52.3abc	62.1abc	24.7ab	88.7ab	88.6ab	7.7ab
	Nayer	57.7abc	67.6abc	19.3abc	**90.7a**	91.2ab	8.7ab
	Kasseb	49.7abc	58.6bc	23.3abc	81ab	79.6ab	8ab
	Bochra	64.7ab	64.7abc	21abc	85.9ab	89.8ab	12.4ab
	FLIP 96-114 C	49.6abc	61.8abc	22.7abc	81.1ab	82.4ab	9.2ab
	FLIP 88-42 C	52abc	60.9abc	18.7abc	80.2ab	80.9ab	11.1ab
	ILC 3279	58.7ab	63.9abc	17.3abc	79.3ab	78.3ab	12.3ab
75% EUR	Béja I	52.7abc	60.9abc	20.3abc	**77.7b**	**77b**	9.3ab
	Amdoun I	49.7abc	59.3abc	23.3abc	**77.7b**	81.1ab	10.3ab
	Nayer	60.3ab	66.9abc	15bc	87ab	89ab	9ab
	Kasseb	**39c**	59.7abc	**30a**	**77.7b**	82.1ab	13ab
	Bochra	46.7bc	62.6abc	24.7ab	81ab	84.2ab	**16.7a**
	FLIP 96-114 C	49.6abc	61.8abc	25ab	82.8ab	85.4ab	9.5ab
	FLIP 88-42 C	51.3abc	60.4abc	18.7abc	79.3ab	79.6ab	11.7ab
	ILC 3279	54.3abc	62.3abc	16.7abc	82.8ab	82.8ab	7.5ab
50% EUR	Béja I	57.7abc	63abc	21.7abc	**91.7a**	**92.8a**	4.3b
	Amdoun I	54.3abc	63.7abc	22abc	85.8ab	88.1ab	8.2ab
	Nayer	62ab	73.1ab	19abc	86.9ab	86ab	5.4b
	Kasseb	55.7abc	60abc	11bc	83ab	82.2ab	4b
	Bochra	64ab	**73.7a**	17abc	85.9ab	90.9ab	8.1ab
	FLIP 96-114 C	47.3bc	**54.9c**	12bc	80.2ab	79.7ab	**2.8b**
	FLIP 88-42 C	52.3abc	59bc	15.3bc	79.3ab	78.2ab	5.7b
	ILC 3279	57.3abc	65.5abc	17.3abc	85.8ab	85.9ab	5.5b
25% EUR	Béja I	**69a**	72.9ab	**9.7c**	89.1ab	90.2ab	7.8ab
	Amdoun I	55.3abc	65.2abc	19.3abc	89.1ab	90.2ab	7.8ab
	Nayer	55abc	63.7abc	19abc	83.ab	84.6ab	8.3ab
	Kasseb	49.6abc	59.5abc	17.3abc	82.8ab	82.6ab	5.5b
	Bochra	62ab	69.5ab	14.3bc	83.ab	84.6ab	8.3ab
	FLIP 96-114 C	55.3abc	65.5abc	21abc	83.ab	84.6ab	8.3ab
	FLIP 88-42C	59.6ab	67.8abc	19abc	84.8ab	84.2ab	5.5b
	ILC 3279	59.6ab	67.8abc	19abc	88.5ab	88.6ab	5.5b

The values of the same column accompanied by the same letter are not significantly different at 5% level
The values in fat from the same column are extreme values

Table 5. Mean comparisons (Newman-Student and Keuls test at 5% level) of the chickpea (*Cicer arietinum* L.) flowering and maturity parameters according to the interaction (Amount irrigation X Genotype)

concluded that dryness escape resistance is the seedling ability to finish its biological cycle before the exhaustion of the soil water reserves. According to Malhotra and Saxena, (2002), early flowering remains the main component of the chickpea water stress avoidance. This mechanism was largely used, especially through the selection of genotypes for an early flowering. Moreover, genotypes with early flowering are characterized by a high grain yield (Berger et al., 2004); whereas genotypes with late flowering, having suffered the final drought stress, are characterized by poor yield (Thomas et al., 1996). Actually, the delay of chickpea flowering induced by drought stress, increases the potential of drought stress avoidance and generates a reduction of the duration between flowering and pods formation (Berger et al., 2006). On the other hand, Siddique and Khan, (1996) concluded that chickpea genotypes selection with early flowering does not involve necessarily an increase in the production. However, the combination of an early flowering and grain yield improvement alleles were proven at desi chickpea genotypes. Rajin et al., (2003) noticed that chickpea phenologic phases depend on accumulated thermal time. To flower, chickpea genotypes need thermal durations varying from 623 to 808 °C/day. According to the amounts irrigation, flowering phase duration varied from 16,9 to 21,6 days. With amounts irrigation 100%, 75%, 50% and 25% of EUR, flowering phase duration are similar two by two. It appears proportional to amounts irrigation. They are long with amounts 100% and 75% of EUR with respective values 20,2 and 21,6 days and short values with the amounts 50% and 25% of EUR with respective values 16,9 and 17,3 days (Table 3). Flowering phase duration of the chickpea genotypes varied from 16,5 to 22,3 days. Flowering phase of the genotype Béja1 is shortest; whereas that of Amdoun1 is longest. The other genotypes have intermediate flowering durations (Table 4). According to Richa, and Singh, (2001), flowering phase duration varies, according to genotypes, from 30 to 45 days. The early cultivars spread out their flowering phase duration and delay their pods formation period (Abdelguerfi-Laouar et al., 2001).   The interaction (Genotype X Amount irrigation) showed that the chickpea genotypes flowering period varied from 9,7 to 30 days. Mean comparisons revealed three interfered homogeneous groups. The longest flowering phase duration is 30 days and is accomplished at 75% of EUR amount irrigation by the genotype Kasseb; whereas, the shortest duration is 9,7 days and is recorded at 25% of EUR amount irrigation by the genotype Béja1 (Table 5). Bonfil and Pinthus, (1995) indicated that the chickpea flowering phase duration is a determining factor of the grain yield. Or et al., (1999) noted that the long flowering period, controlled by early flowering alleles, can increase the grain yield. In fact, cultivars with early flowering enter in hasty fructification and achieve their filling pods before the final dryness advent (Abdelguerfi-Laouar and Al, 2001).

Early flowering date and 50% flowering date are inversely proportional to the flowering phase duration. High amounts irrigation, 100% and 75% of EUR, cause an early flowering over a long duration. Conversely, limited amounts irrigation, 50% and 25% of EUR, delay the flowering phase and shorten its duration (Fig. 5A). Favorable water conditions incited plants to increase their capacities to flower enough early in the season. On the other hand, under severe drought stress conditions plants find difficulties to producing flowers even limited number. Early pods maturity and 50% pods maturity dates depend on the amounts of irrigation and vary respectively from 80,7 to 85,7 and 82,6 to 86,2 DAS (Table 3). Early date pods maturity mean comparison revealed two homogeneous groups. The first is composed of 100%, 50% and 25% of EUR amount irrigation which generated of similar and late pods maturities. The second is consisting of the amount irrigation 75% of EUR which generated an early pods maturity (Table 3). Mean comparison of 50% pods maturity dates showed only one homogeneous group which shows that all amounts of irrigation have similar effects on 50% pods maturity date (Table 3). Pods maturity date appears inversely proportional to amounts of irrigation. With the amounts 100 and 75% of EUR, 50% pods maturity is hasty, whereas under limited water doses it is, relatively, late (Table 3). These results are in conformity with those obtained by Khanna-Chopra and Sinha, (1987) and Silim, and Saxena, (1993) which noticed moisture supply during the growing period had a strong influence on phenology, that pods maturity date is prolonged by complementary irrigation and reduced by dryness.

Early pods maturity date varies, according to the chickpea genotypes from 80,9 to 87,1 DAS. Mean comparison revealed two interfered homogeneous groups. The first is formed of genotypes Béja1, Amdoun1, Kasseb, Bochra, FLIP96-114C, FLIP88-42C and ILC3279. The second is formed of Béja1, Amdoun1, Nayer, Bochra and ILC 3279 (Table 4).

Chickpea genotypes 50% pods maturity date varies from 80,7 to 87,7 DAS. Mean comparison showed three interfered homogeneous groups. The first group is consisting of Béja1, Amdoun1, Nayer, Bochra, FLIP 96-114 C and ILC 3279. The second is composed of genotypes Béja1, Amdoun1, Kasseb, FLIP 96-114 C and ILC 3279. The thread is consisting of Béja1, Kasseb, FLIP 96-114 C FLIP 88-42 C and ILC 3279 (Table 4). Siddique, et al., (2001) reported that drought avoidance and/or tolerance were observed for the some species (*C. arietinum* and *L. satius*) in the form of delayed senescence and maturity.

Chickpea 50% pods maturity depends jointly on the vegetable material and amounts irrigation. It varies from 77 to 92,8 DAS. Mean comparison showed two interfered homogeneous groups (Table 5). Silim, and Saxena, (1993) reported that, in the Mediterranean basin, chickpea pods maturity date of the spring culture varies from 85 to 101 DAS. However, this culture suffers from thermal and drought stress during flowering, seeds filling and pods maturity phases (Singh et al., 1995). According to Singh et al., (1994), early pods maturity is significantly associated with the dryness tolerance. Other authors claimed that, in the dry zones, escape to drought stress could appear through the early flowering and pods maturity (Berger et al., 2006). Gentinetta et al., (1986) noticed the possibility of sifting for drought stress tolerance during the physiological seeds maturity phase.

Chickpea pods maturity phase duration is proportional to the amounts of irrigation and varies from 5,5 to 10,9 days. With the amounts of irrigation 100 and 75% of EUR, pods maturity phase durations are lengthen with similar respective values 9,5 and 10,9 days. On the contrary, with the amounts 50 and 25% of EUR, they are shortened with respective similar values 5,5 and 7,1 days (Table 3). Maturity duration was extended by high moisture supply and reduced by drought. Irrigation extended reproductive growth duration (Silim, and Saxena, 1993).

Chickpea genotypes pods maturity phase duration varies from 7,1 to 11,4 days. Mean comparison showed that chickpea genotypes have similar pods maturity phase durations (Table 4). Chickpea 50% pods maturity date is inversely proportional to pods maturity phase duration (Fig. 5B). With the amounts irrigation not stressful, in fact 100% and 75% of EUR, 50% pods maturity is hastened and its phase duration is lengthened. On the other hand, limited amounts irrigation, 50% and 25% of EUR, delay the physiological 50% pods maturity and reduce its duration (Fig. 5B). It appears that, under not limited water conditions, the plant tends to take easily its water requirements. Vegetative development and pods filling phases are shortened in aid of pods maturity phase duration which is lengthened. It may be that pods are sufficiently water gorged and would need enough time to release it. Conversely, under drought stress conditions, vegetative development and pods filling phases are lengthened with the detriment of the pods maturity phase duration which is shortened. With the water scarcity, the plant will spend more time to be able to achieve its vegetative development and pods filling phases. As pods are less water gorged, they will be desiccated more quickly.

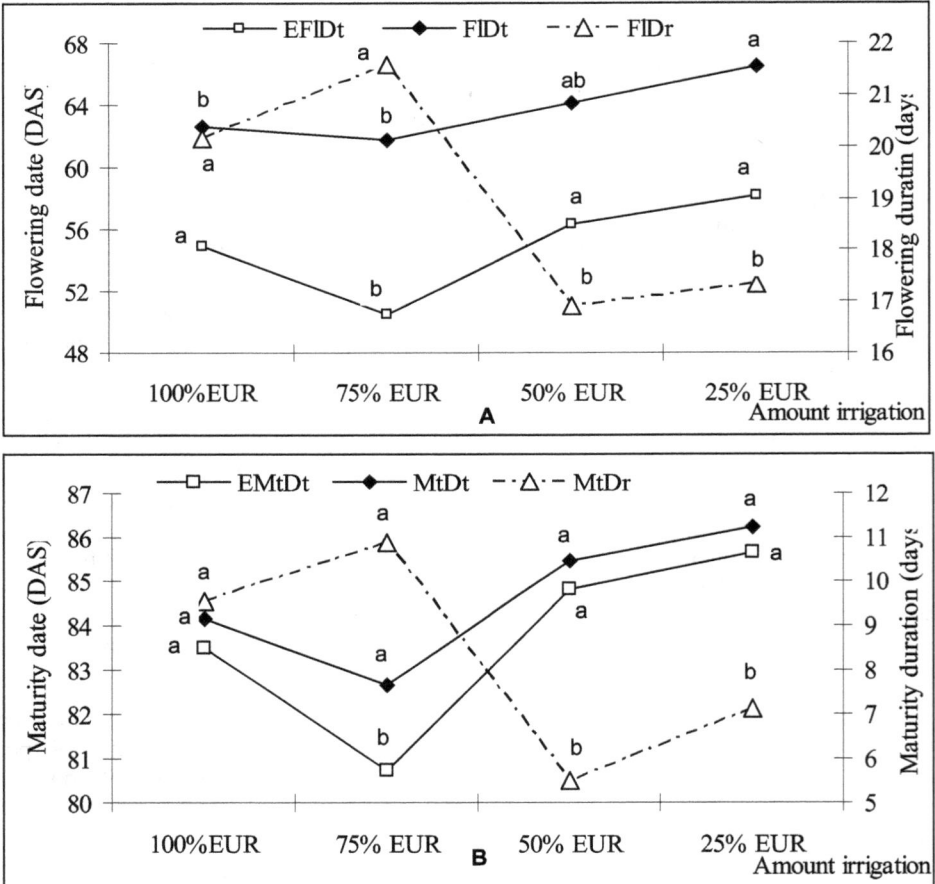

Fig. 5. Comparisons (Student-Newman and Keuls test at 5%) of (A) the flowering dates and durations; (B) maturity dates and durations of the chickpea (Cicer arietinum L) cultures according to the amounts of irrigation

4. Conclusion

Chickpea culture did not suffer from thermal stress and the critical temperature was exceeded only during the pods maturity phase. Water requirement for this culture is evaluated to 392 mm. Amounts of irrigation 50% and 25% of the EUR induced severe drought stress.

Chickpea flowering and pods maturity dates and durations are controlled by the crop water requirement. The amount irrigation 75% of the EUR induced the hastened flowering and maturity dates with longest durations. Furthermore, according to the amounts of irrigation, flowering and maturity dates were inversely proportional to their durations. The amounts of irrigation 100% and 75% of the EUR hasten flowering and maturity dates and enlarge their durations; while the amounts 50% and 25% of the EUR delay flowering and maturity dates and shorten their durations.

5. References

AAC (Agriculture et Agroalimentaire Canada) 2006. Pois chiche: Situation et perspectives. *Le Bulletin bimensuel* ; Volume 19; Numéro 13; 4 pages.

Abdelguerfi-Laouar, M. ; Bouzid L. ; Zine, F. ; Hamdi, N. ; Bouzid, H. & Zidouni, F. 2001. Evaluation de quelques cultivars locaux de pois chiche dans la région de BEJAIA ; *Recherche Agronomique*, Revue semestrielle, Institut National de la Recherche Agronomique d'Algrie ; 9: 31-42.

Allen, G.; L. Pereiral, D. Races et M. Smith, 1998. Crop evapotranspiration guidelines for computing crop water requirements. In: FAO (Eds) Irrigation and drainage, paper n° 24, 1998, 56 pp; http:www.fao.org/docrep/X0490E/x0490eob.htm, (08/12/07)

Anbessa, Y.; Warkentin, T. Vandenberg, A. & Ball, R. 2006. Inheritance of time to flowering in chickpea in a Short-Season temperate environment Journal of Heredity 97(1): 55–61

Ben Mechlia, N. 1998: Manuel de Formation: Application des Données Climatiques à la Planification et à la Gestion Efficace de l'Irrigation. Projet INAT - CGRE. Mise au point d'un système d'irrigation; 193 pages.

Berger J.D. ; Ali M.; Basu P.S.; Chaudhary B.D. ; Chaturvedi S.K.; Deshmukh P.S.; Dharmaraj, P.S. ; Dwivedi, S.K. ; Gangadhar, G.C. ; Gaur, P.M. ; Kumar, J. ; Pannu, R.K. ; Siddique, K.H.M. ; Singh, D.N. ; Singh, D.P. ; Singh, S.J. ; Turner, N.C. ; Yadava, H.S. & Yadav, S.S. 2006. Genotype by environment studies demonstrate the critical role of phenology in adaptation of chickpea (*Cicer arietinum L.*) to high and low yielding environments of India; *Field Crops Research*; 98: 230–244.

Berger, J.D.; Turner, N.C.; Siddique, K.H.M.; Knights, E.J.; Brinsmead, R.B.; Mock, I.; Edmondson, C. & Khan, T.N. 2004. Genotypes by environment studies across Australia reveal the importance of phenology for chickpea (*Cicer arietinum L.*) improvement. *Aust. J. Agric. Res.* 55:1-14.

Bonfil, D.J., & Pinthus, M.J. 1995. Response of chickpea to nitrogen, and a comparison of the factors affecting chickpea seed yield with that affecting wheat grain yield. Exp. Agric. 31:39-47.

DGPA, (Direction Générale de la Production Agricole) 2006. Rapport annuel de suivi des emblavures, Direction des grandes cultures, Direction générale de la production Agricole. Ministère de l'agriculture, de l'environnement et des ressources hydrauliques.

DGPA, (Direction Générale de la Production Agricole) 2008. Rapport annuel de suivi des emblavures, Direction des grandes cultures, Direction générale de la production

Agricole. Ministère de l'agriculture, de l'environnement et des ressources hydrauliques.

Doorenbos, J. & Pruitt, W.O. 1977. Crop water requirements. Irrigation and Drainage Paper No. 24, (rev.) FAO, Rome, Italy; 144 pages.

Ellis, R.H.; Lawn, R.J.; Summerfield, R.J.; Qi, A.; Roberts, E.H.; Chays, P.M.; Brouwer, J.B.; Rose, J.L.; Yeates, S.J. & Sandover, S. 1994. Towards the reliable prediction of time to flowering in six annual crops: V. Chickpea (*Cicer arietinum* L). *Exp Agric* 30: 271–282.

Gentinetta, E.; Cepi, D.; Leporic, G. ; Motto, M. & Salamini, F. 1986. A major gene for delayed senescence in maize. Pattern of photosynthate accumulation and inheritance. *Plant Breeding* 97: 193–206.

Golezani, G.; Dalil, B.; Muhammadi-Nasabi, A. D. & Zehtab-Salmasi, S. 2008. The Response of Chickpea Cultivars to Field Water Deficit. *Not. Bot. Hort. Agrobot. Cluj* 36 (1):25-28

Gumber RK & Sarvjeet S, 1996. Genetics of flowering time in chickpea: a preliminary report. Crop Improv 23:295–296.

Hughes, G.; Keatinge, J.D.H.; Cooper, P.J.M. & Dee, N.F. 1987. Solar-radiation interception and utilization by chickpea (*Cicer arietinum* L.) Crops in Northern Syria. *J. Agric. Sci.* 108: 419-424.

Khanna-Chopra, R. & Sinha, S.K. 1987. Chickpea: physiological aspects of growth and yield. *In:* The Chickpea. 409 pages; CAB International, (Eds.Saxena, M.C., Singh, K.B.), Wallingford, Oxon, UK; pages: 163-190.

Kumar J & van Rheenen, H.A. 2000. A major gene for time of flowering in chickpea. J Hered 91:67–68.

Kumar, J. & Abbo, S. 2001.Genetics of flowering time in chickpea and its bearing on productivity in semiarid environments. *Agronomy,* 72: 107-138.

Malhotra, R.S. & Saxena, M.C. 2002. Strategies for Overcoming Drought Stress in Chickpea. Caravan, ICARDA n°17 ; 3 pages.

Malhotra, R.S. & Johansen, C. 1996. Germplasm Program Legumes, Annual Report for 1996, International Center Agricultural Research Dry Areas (ICARDA), Aleppo, Syria; pages: 229.

McVicar, R.; Pearse, P.: Panchuk, K.; Warkentin, T.; Banniza, S.; Brenzil, C. Hartley, S.; Harris, C.; Yasinowski, J. & Goodwillie, D. 2007. Chickpea. www.agr.gov.sk.ca/Production. visité le 14/3/2008.

Morizet, J. ; Tribo, A.M. & Pollacsek, M. 1984. Résistance à la sécheresse chez le maïs : quelques mécanismes impliqués. *Physiologie et Production du Maïs.* INRA; France ; pages: 167–174.

Nayyar, H.; Singh, S.; Kaur, S.; Kumar, S. & Upadhyaya, H. D. 2006, Differential sensitivity of macrocarpa and microcarpa types of chickpea (*Cicer arietinum* L.) to water stress: association of contrasting stress response with oxidative injury. *J. Integrative Plant Biol.* 48, 1318-1329.

Or, E.; Hovav, R. & Abbo, S. 1999. A Major Gene for Flowering Time in Chickpea. *Crop Sci.* 39: 315-322.

Ouattar, S.; Jones, R.J.; Crookston, R.K. & Kajeiou, M. 1987. Effect of drought on water relation of developing maize kernels. *Crop Sci.* 27: 730–735.

Rajin, A. M.; McKenzie, B.A. & Hill, G.D. 2003. Phenology and growth response to irrigation and sowing date of kabuli chickpea (*Cicer arietinum* L.) in a cool temperature sub humid climate. *Crops and Soils* (Abstract); 141: 3-4.

Richa J. & Singh, N.P. 2001. Plant Regeneration from NaCl Tolerant Callus/Cell Lines of Chickpea International Chickpea and pigeonpea *Newsletter*; N°8; ICRISAT International Crops Research Institute for the Semi-Arid Tropics; Patancheru 502 324, Andhra Pradesh; 73 pages.

Roberts E.H.; Hadley, P. & Summerfield, R.J. 1985. Effects of temperature and photoperiod on flowering in chickpeas (*Cicer arietinum* L.). Ann Bot 55:881–892.

Saxena, M.C. 1987. Agronomy of chickpea. *In*: The Chickpea; M.C. Saxena and K.B. Singh (Editors), CAB International, Wallingford, pages: 207-232.

Saxena, N.; Johansen, P.C.; Saxena, M.C. & Silim, S.N. 1993. Breeding for stress tolerance in cool season food legumes. *In*: K.B. Singh & Saxena, M.C. (Eds.) John Weley and Sons Chichester UK; pages: 245 – 270.

Siddique, K.H.M., & Khan, T.N. 1996. Early-flowering and high yielding chickpea lines from ICRISAT ready for release in Western Australia. Int. Chickpea Pigeonpea Newsl. 3: 22–24.

Siddique, K.H.M.; Regan, K.L.; Tennant, D. & Thomson, B.D. 2001. Water use and water use efficiency of cool season grain legumes in low rainfall Mediterranean-type environments. European Journal of Agronomy 15 (2001) 267–280

Silim, S.N. & Saxena, M.C.1993. Adaptation of spring-sown chickpea to the Mediterranean basin. I. Response to moisture supply, *Field Crops Research,* 34: 121-136.

Singh, K.B.; Malhotra, R.S.; Saxena, M.C. & Bejiga, G. 1995a. Analysis of a decade of winter/spring chickpea. In: Germplasm program legumes; Annual report; International Center for Agricultural Research in the Dry Areas (ICARDA); P.O. Box 5466, Aleppo, Syria; 210 pages.

Singh, K.B.; Malhotra, R.S. Halila, M.H. Knights, E.J. & Verma, M.M. 1994. Current status and future strategy in breeding chickpea for resistance to biotic and abiotic stresses. *Euphytica* 73: 137-149.

Slama, F. 1998. Cultures industrielles et légumineuses à graines. (Ed. Centre de diffusion Universitaire Tunisie, en Arabe) ; pages: 300.

Soltner, D. 1981. Les bases de la production végétales : Le sol - Le climat-La plante; Phytotechnie Générale; Tome I: Le sol ; 10ème Eds; 456p.

Subbarao, G.V. ; Johansen, C.; Slinkard, A.E. ; R.C., Rao, R.C. N. Saxena, N.P. & Chauhan, Y.S. 1995. Strategies and scope for improving drought resistance in grain legumes. *Critical Reviews in Plant Sciences* 14: 469-523.

Summerfield R.J. & Roberts E.H. 1988. Photothermal regulation of flowering in pea, lentil, faba bean and chickpea. In: World crops: cool season food legumes (Summerfield RJ, ed.). Dordecht, The Netherlands: Kluwer Academic Publishers; 911–922.

Summerfield, R.J.; Minchen, F.R. Rberts, E.H. & Hadley, P. 1981. Adaptation to contrasting aerial environments in chickpea (*Cicer arietinum* L.) . *Trop. Agric. (Trinidad)*, 58:97-113.

Thomas, H.; Dalton, S.J.; Evans, C.; Chorlton, K.H. & Thomas, I.D. 1996. Evaluating drought resistance in Germplasm of meadow fescue. *Euphytica* 92: 401-411.

Tollenaar, M. 1989. Response of dry matter accumulation in maize to temperature. *Crop Sci.* 29: 1275-1279.

USDA, 1951. Soil Survey Manual. U.S.Department of Agriculture, Edition Washington D.C; 503 pages.

Wery, J. 1990. Adaptation to frost and drought stress in chickpea and implications in plant breeding. *In:* Saxena M.C; Cubero J.I. and Wery (Eds), Present status and future prospects of chickpea crop production and improvement in the Mediterranean countries, Options Méditerranéennes - Série Séminaires - n° 9 - CIHEAM, Paris. pages: 77-85.

Wery, J.; Silim, S.N.; Knights, E.J.; Malhotra, R.S. & Cousin, E. 1994. Screening techniques and sources of tolerance to extremes of moisture and air temperature in cool season food legumes. *Euphytica*; 73: 73-83.

3

Water Productivity and Fruit Quality in Deficit Drip Irrigated Citrus Orchards

Ana Quiñones, Carolina Polo-Folgado, Ubaldo Chi-Bacab,
Belén Martínez-Alcántara and Francisco Legaz
Instituto Valenciano de Investigaciones Agrarias, Moncada (Valencia)
Spain

1. Introduction

Citrus is one of the most relevant crops worldwide with a yearly average production of $90 \cdot 10^6$ Mg in the last decade. In Mediterranean countries, citrus is the second largest fruit crop after apples, in the European Union (EU). Spain is the leading producer in the area with nearly 60 % of tonnes produced in the whole of the EU (Ollier et al., 2009). In Spain, citrus orchards cover around $300 \cdot 10^3$ ha ($6 \cdot 10^6$ Mg) of which up to 60 % is located in the Comunidad Valenciana (CV). This area has remained more or less constant in this respect since 1990 (MARM, 2010). The Comunidad Valenciana is the most important region, not only in acreage but also with respect to its long tradition of citrus farming.

In this area, like in many regions of the world, the lack of water or lack of good water is a growing concern for the development of relevant agriculture since water is the most limiting factor for crop production. Moreover, climatic conditions are characterized by low rainfall (400-600 mm year-1) and irregular spatial and temporal distribution. On the other hand, the world's population has undergone an exponential growth, which has led to soaring food demand and, therefore, high natural-resource exploitation. For example, in Spain, irrigated land had risen up to 3.421.304 ha in 2009 (MARM, 2010).

Therefore, improved water use efficiency (WUE) or water productiviy (WP), using different strategies, is a key concept to solve this water scarcity. So nowadays, efforts are being focussed on developing not only alternative irrigation methods but also new water management methods in order to reduce water dosages while maintaining maximum tree growth, without significantly affecting yield.

2. Options for improving irrigation efficiency

Wallace and Batchelor (1997) showed the main options for improving WUE in different categories, engineering, agronomy, management and institutional improvements. Although it is not possible to discuss all the options listed in detail by these authors, three of the options are of particular interest.

Concerning engineering improvements, there are several irrigation systems to water crops that can reduce application losses and improve application uniformity.

In flood irrigation, a large amount of water is directed to the field and flows over the ground among the crops. In regions where water is abundant, flood irrigation is the cheapest irrigation method and this low tech irrigation method is commonly used in developing countries. On the contrary, localized irrigation is a system where water is distributed under low pressure through a piped network, in a pre-determined pattern, and applied as a small discharge to each plant or adjacent to it. Regarding irrigation systems, in the citrus area of Spain, about 69 % of citrus orchards are irrigated under fertigation, mainly with drip irrigation, and the remaining by flood irrigation (MARM, 2010). Similar percentages are found in other citrus areas where localized (drip or mini-sprinklers) irrigation systems are mainly used for citrus and other tree crops (olives and deciduous trees), while sprinkler irrigation is dominant for fodder crops and some vegetables.

Depending on the different localized irrigation system fertigation can be performed on surface or subsurface drip, spray, micro-jet and micro-sprinkler. These different techniques can be used, both in annual crops or fruit trees, according to soil type and the different characteristics of agricultural area. This versatility has led to a rapid expansion of fertigation in the world cultivated areas. The advantages of fertigation are listed below (Burt et al ,1998).

- High water and nutrient use efficiency as a consequence of coupling fertilizer timing to the plant requirements and, therefore, minimized fertilizer/nutrient loss due to localized application and reduced leaching.
- Reduced energy cost by the saving of labour and machinery and the efficient use of the costly chemicals to be applied.
- Minimized soil erosion by avoiding heavy equipment traffic through the field to apply fertilizers.

Moreover, fertigation allows safe use of recycled or saline water. Boman et al., (2005) affirmed that irrigation scheduling is a key factor in managing salinity. Increasing irrigation frequency and applying water exceeding the crop requirement are recommended to leach the salts and minimize their concentration in the root zone. Fertigation also reduces the risk of diseases since foliage remains dry. Schumann et al., (2009) observed a significant reduction in the number infected trees of citrus greening disease or citrus canker by optimising daily and nutrient levels for trees. Moreover, frequent and small water split with fertigation technique leads to a shallow and compact root system in comparison with a wide and deeper root system in flood irrigated trees (Sne, 2006), enhances N uptake efficiency by the fibrous roots and contributes to lower leaching below the root zone (Quiñones et al., 2007).

In drip irrigation systems, subsurface drip irrigation (SDrI) has been part of modern agriculture. Current commercial and grower interest levels indicate that future use of SDrI systems will continue to increase. SDrI applies water below the soil surface, using buried drip tapes (ASAE, 2001). SDrI uses buried lateral pipelines and emitters to apply water directly to the plant root zone. Laterals are placed deep enough to avoid damage by normal tillage operations, but sufficiently shallow so that water is redistributed in the active crop root zone by capillarity. SDrI systems must be compatible with the total farming and cultural systems being used.

SDrI requires the highest level of management of all microirrigation systems to avoid remedial maintenance. A poorly designed SDrI system is much less forgiving than an improperly designed surface drip system. Deficiencies and water distribution problems are difficult and expensive to remedy. Lamm and Camp (2007) present an excellent, detailed review of SDrI.

These systems require safeguards and special operational procedures to prevent plugging and facilitate maintenance, but they also have numerous advantages. These were:

- The top of the soil surface remains dry, limiting surface evaporation to the rate of vapour diffusion transport and preventing salt accumulations on the surface.
- The use of a very high irrigation frequency (several times per day) that matches actual crop water use will result in a constant wetted soil volume and a net upward hydraulic gradient, which minimizes leaching.
- Supplying water and nutrients directly to the root zone allows root uptake to be more efficient if irrigation and fertilization schedules are appropriate.
- Soil crusts, which may impede infiltration and cause ponding and runoff, are bypassed so that surface infiltration variability becomes insignificant.

Under proper management, appropriately designed and managed SDI irrigation systems offer several other advantages to growers (Devasirvatham, 2009) because of their potential for:

- Maintaining access to fields with tillage, planting, spray and harvest equipment that is not restricted by irrigation.
- Obtaining better weed suppression with minimal chemicals because there is less seed germination with dry soil surfaces.
- Efficiently and safely applying labelled plant-systemic pesticides and soil fumigants for improved disease and pest control.
- Reducing surface wetting often reduces fungal disease incidence (e.g., molds, mildews) by maintaining dryer plant surfaces and lower air humidity within the plant canopy.
- Reducing pesticide exposures for workers when chemicals are applied below the soil surface.
- Implementing minimum tillage, permanent beds, and multiple cropping systems (Bucks et al., 1981), although much of the necessary equipment modifications and farming techniques have yet to be developed; and minimizing flow-rate sensitivity to temperature fluctuations because emitters are buffered by the soil.

Phene et al. (1992) and Phene (1995) listed several drawbacks, including potentially high initial system costs, potential rodent damage, the fact salt may accumulate between drip lines and soil surface, low upward water movement in coarse-textured soils, high potential for emitter plugging, and insufficient technical knowledge requiring dissemination, and hands-on experience by growers and researchers. In addition, fertility management becomes more critical with SDI because roots tend to grow deeper than with surface drip systems and some surface applied nutrients may not be sufficiently available (Phene, 1995).

Improved WUE can also be affected by water regimes. Although WUE frequently decreases under water deficit conditions (García Tejero et al. 2011), in areas with significant water scarcity, like in the east of Spain, it is possible to increase efficiency under different irrigation management methods based on deficit-irrigation (DI) programmes (Bonet et al. 2010). These DI strategies are defined as a practise where the total water provided for the plant (irrigation plus effective rainfall) is below to the crop's water needs (García-Tejero, 2010) in order to reduce ETc, and hence save water, while simultaneously minimizing or eliminating negative impacts of stress on fruit yield or quality. This approach differs from season-long stress in that the deficit irrigation is restricted to stress-tolerant periods. Essentially, there are two methods to achieve DI management of a crop, reducing the amount of water supplied or increasing the period between irrigation cycles.

Regarding reducing the quantity of water applied, Sustained DI (SDI) and Regulated DI (RDI) are the most widely used practices in several tree crops. SDI is when a reduced percentage of ETc is applied throughout the irrigation season without considering its phenological period or the accumulated water stress. Regulated deficit irrigation (RDI) is based on supplying some 100% of ETc when the crop is less tolerant to water stress and a reduced percentage for the rest of the season. The water stress tolerated by the crops is closely related to crop phenology and, therefore, a detailed knowledge on tree physiology is crucial for successful use of RDI. Deficit irrigation strategies are not recommended for young orchards, since conditions must be favoured under which trees reach maturity as soon as possible.

Concerning the methods based on increasing periods between irrigation events, Low frequency DI (LFDI) is when the soil is left to dry until the readily available water is consumed; then the soil is irrigated to field capacity and left to dry again. Under this strategy the crop is kept below a certain water stress threshold value (García-Tejero, 2010).

Options for enhancing irrigation efficiency in the agronomic category are related to crop management. Different strategies to improve rainfall use or reduce evaporation can be performed. In this respect, it is important to note that precipitation is not a reliable water source, but can contribute to some degree towards water needs. Therefore, usable rainfall or effective precipitation, which is the portion of total precipitation retained by soil available for plants, must be calculated in the water plant requirements.

The research reported here summarizes the results obtained in terms of the response of yield, fruit quality and nutritional tree status in citrus orchards under two irrigation systems, three deficit irrigation strategies and two effective precipitation values during five consecutive seasons (2006-2010). Also, the benefits of each irrigation strategy were estimated in terms of agricultural water productivity (WPagr).

3. Design factors

The different experimental plots were located in Puzol, Valencia (latitude 39° 34´ N; longitude 00° 24´ W, elevation 25 m) in the East of Spain, in commercial orchards of clementine cv. Nules mandarin adult trees (*Citrus clementine* Hort. Tanaka x *C. reticulata* Blanco) grafted onto citrange Carrizo rootstock [*C. sinensis* L. (Osb.) x *Poncirus trifoliata* L.(Raff.)], planted at a spacing of 3.5 m x 5.6 m (i.e. 510 trees ha⁻¹). This variety, grafted onto Carrizo citrange is highly representative in the study area, this being the most widely used rootstock in citrus orchards in the Comunidad Valenciana area (MARM, 2010).

3.1 Soil and water characteristics

The trees were grown on Cambic Arenosol soil (62.6% sand, 19.2% silt, 18.2% clay; pH 8.2; organic matter content 1.03% and a bulk density of 1.6 kg m⁻³) with low water holding capacity (16%).

The irrigation water had an average electrical conductivity of 2.8 mS cm^{-1}, containing an annual average of 272, 212 and 100 mg L^{-1} of NO$_3^-$, Ca^{2+} and Mg^{2+}, respectively.

3.2 Climatic conditions

The general climatic conditions for the experimental sites is Mediterranean dry, with an average potential evapotranspiration (ET0) close to 1500 mm yr^{-1}, and annual rainfall between 250 and 500 mm yr^{-1}, with a high monthly variability distributed mainly from October to May. The thermal range is broad, with mild temperatures in winter, rarely below 0 °C and severe conditions in summer, with temperatures in many cases exceeding 40 °C. These environmental conditions promote an average annual water deficit of around 1000 mm.

3.3 Fertilization program

The nutrient-fertilizer rate and seasonal distribution for citrus plants was calculated for a 3.10 m canopy diameter in citrus trees grown under drip irrigation. Nitrogen (N) requirements were 400 g N tree^{-1} year^{-1} based on Legaz and Primo-Millo (1988), of which an average value of 58 and 49 % were supplied as potassium nitrate in 50 and 100 % canopy area coefficients of effective precipitation in UEP (use of effective precipitation) treatments, respectively. These coefficients were arbitrarily chosen. The N-remainder was provided by typical irrigation water in the Mediterranean area, with 272 ppm of nitrate concentration, as described above. The quantity of N contributed by the irrigation water was calculated using the formula described by Martinez et al., (2002). Phosphorus and potassium fertilizer demand was 120 g P$_2$O$_5$ tree-year^{-1} applied as phosphoric acid (48% P$_2$O$_5$) and 475 g K$_2$O tree-year^{-1} applied as potassium nitrate (44% K$_2$O equivalent). The basic iron needs per tree were distributed throughout the growing cycle in a similar way for N. Foliar spray treatments of zinc (Zn) and manganese (Mn) were applied as organic commercial fertilizer at 0.5% weight(w)/volume(v) (Zn: 6.6% w/w and Mn: 4.8% w/w) to correct deficiencies.

3.4 Irrigation scheduling

The amount of water applied to each tree was equivalent to the total seasonal crop evapotranspiration (ETc) calculated using the formula described by Aboukhaled et al. (1982).

$$ETc \left(mm\right) = \frac{ETo \left(mm\right)}{Kc}$$

Where ETo is the reference crop evapo-transpiration under standard conditions and Kc is the crop coefficient (Table 1). This coefficient (Kc per month) accounts for crop-specific effects on overall crop water requirements and is a function of canopy size and leaf properties. The ETo values were determined using the Penman-Monteith approach (Allen et al. 1998) using hourly data collected by an automated weather station situated near the orchard The values obtained were 1108, 1041, 972, 1043 and 1092 mm yr^{-1} in 2006, 2007, 2008, 2009 and 2010 respectively. The Kc values were based on guidelines provided by Castel and Buj (1994). Irrigation water requirements were met by the effective rainfall (≥3 mm and ≤ 45 mm which resulted in soil water saturation) of the entire year plus irrigation water for the three years of the assay, respectively). The annual rainfall was 315, 516, 463, 472 and 392 mm yr^{-1} in 2006 to 2010, respectively.

There is scarcely any information about the use of rainfall by crops, possibly due to the difficulty of evaluation. In this research study, rainfall was recorded as the mean of three rain gauges placed in different parts of the orchard. Irrigation water requirements covered by effective precipitation (UPe) were calculated according to the following expression:

$$UPe \ (m^3 \ tree^{-1}) = CA \ (m^2) \ x \ Pe \ (L \ m^{-2}) \ x \ F \ x \ 0.001 \ (L \ m^{-3})$$

Where CA is canopy area of the tree at the beginning of each growth cycle, Pe is effective precipitation corresponding to rainfall greater than 3 mm (lower values are not utilizable by the plant) and less than 33 mm (which saturate the soil profile and water percolates through soil to groundwater), F is the potential factor for effective precipitation use (0.5 or 1).

Month	Jan	Feb	Mar	Apr	May	Jun	Jul	Aug	Sep	Oct	Nov	Dec
Factor	0.97	0.96	0.97	0.91	0.81	0.91	1.00	1.16	1.09	1.24	1.07	0.93
Kc per month	0.505	0.500	0.505	0.474	0.422	0.474	0.521	0.604	0.567	0.646	0.557	0.484

Table 1. Month crop coefficient (Kc) = Factor x Kc mean (0.521)

Trees were surface and subsurface drip-irrigated, through eight pressure-compensating emitters (4 L h^{-1} each) per tree, placed every 88 cm in two drip lines, and at a depth of 30 cm in subsurface drip irrigated trees, both located within 100 cm of the tree trunk and producing a 33% wetted area (Keller and Karmelli, 1974). Moreover, plants were fertirrigated from 0 to 3 times per week, according to evapotranspiration demand and effective rainfall.

3.5 Experimental design

The assay treatments consisted of two irrigation systems, three regulated deficit irrigation practices and two effective precipitation coefficients. The combinations of these factors resulted in twelve treatments distributed in a randomised complete block design and with three replicates each, and fifty trees per plot (Table 2).

Irrigation treatments were subsurface drip irrigation (SDrI) and drip irrigation (DrI). Regulated deficit irrigation practices were:

i.   Control fully irrigated trees where irrigation scheduling was based on the standard FAO approach replacing crop as described above. Therefore, 100 % of crop evapotranspiration (ETc) was covered during the whole year (100 % treatments)
ii. Standard regulated deficit irrigation (RDI$_{70}$) where water was applied at 70% of ETc during July (at the beginning of fruit growth) to the end of October (post-harvest). During the rest of the season water was applied at 100% of ETc.
iii. Alternate regulated deficit irrigation (RDI$_{100-40}$) where water was applied at 100-40 % alternate irrigation events of ETc during the same period explained above. Similarly, during the rest of the season water was applied at 100% of ETc.

To calculate the portion of total precipitation used by the plants (use of effective precipitation); two coefficients of effective precipitation (UEP) were arbitrarily employed, corresponding to 50 and 100 % % canopy area (% CA).

Treatments	Irrigation system[1]	ETc [%][2]	UPe [%CA][3]	Tree Treament[1]
DrI$_{100-50}$	DrI	100	50	54
SDrI$_{100-50}$	SDrI	100	50	51
DrI$_{70-50}$	DrI	100-70	50	51
SDrI$_{70-50}$	SDrI	100-70	50	54
DrI$_{100/40-50}$	DrI	100-100/40	50	54
SDrI$_{100/40-50}$	SDrI	100-100/40	50	51
DrI$_{100-100}$	DrI	100	100	51
SDrI$_{100-100}$	SDrI	100	100	54
DrI$_{70-100}$	DrI	100-70	100	54
SDrI$_{70-100}$	SDrI	100-70	100	51
DrI$_{100/40-1000}$	DrI	100-100/40	100	51
SDrI$_{100/40-100}$	SDrI	100-100/40	100	54

Table 2. Treatments performed during 2006-2009. [1]:Irrigation system, DrI: Drip irrigation, SDrI: Subsurface Drip irrigation; [2]:Regulated deficit irrigation, 100, 70 and 100-40 % of crop evapotranspiration (ETc). [3]:Use of effective precipitation

3.6 Sample collection and measurements

Spring-flush leaves from non-fruiting shoots (around 10 leaves per tree) were randomly sampled in November, from around the canopy. Then, leaves were frozen in liquid-N_2 and freeze-dried (lyophilised). Samples were ground with a water-refrigerated mill, then sieved through a 0.3 mm mesh sieve and stored at -20 °C for further analysis, no more than one month later.

Macro and micronutrient concentration was measured to test nutritional status of the tree and quantify annual nutrient requirements of a crop. Total nitrogen content of spring flush leaves was determined using an Elemental Analyser (NC2500 Thermo Finnigan, Bremen, Germany). Other macronutrients were measured by simultaneous ICP emission spectrometry (iCAP-AES 6000, Thermo Scientific. Cambridge, United Kingdom). Results were expressed as a percentage of dry weight (DW).

In November of each year (19th, 17th and 22nd of November in the 1st, 2nd and 3rd year of the assay), which is the commercial harvest period, the yields of all replicates of each treatment were weighed and a representative sample of forty fruits per replication (5 fruits per tree from 8 trees per replication) was collected at random, weighed and internal fruit quality (including fruit weight, fruit diameter, peel thickness, peel and juice weight, total soluble-solids content, total acidity and colour index) was measured. The number of fruits was calculated using the ratio of yield to average weight of individual fruit. Fruit weight, peel and juice content were determined gravimetrically. Equatorial fruit diameter of samples was measured for each fruit using a digital calliper (Mitutoyo CD-15D, Japan) and the average value of the sample was calculated. Total soluble solids (TSS) content of juice (°Brix) was measured using a digital refractometer (Atago PR-101 Alfa, Tokyo, Japan) and total acidity (TA) was assessed by titration with 0.1 N NaOH, using phenolphthalein as indicator. The ratio between TSS and TA, commonly called maturity index, was calculated. Colour index (CI) was measured taking three readings around the equatorial surface of each fruit using a Minolta Chroma Meter CR-300 (Minolta Camera Co. Ltd., Osaka, Japan). The results are given in the Hunter Lab Colour Scale. This system is based on L, a and b measurements. The L values represent light from zero (black) to 100 (white). The a values change from $-a$ (greenness) to $+a$ (redness), while the b value is from $-b$ (blueness) to $+b$ (yellowness). From L, a and b values, a ratio of 1000·a/L·b is calculated to give the citrus colour index (Ladaniya, 2008).

In addition, impact of different irrigation scheduling on agricultural water productivity (WP) was evaluated, taking into account the water applied, through a ratio between the crop yield (kg), and the total water applied:

$$WP \left(kg\ m^{-3} \right) = \frac{Yield}{irrigation\ +\ rain}$$

3.7 Statistical analyses

Data are summarized in tables as means from three replicates ± standard errors. All data were statistically analysed using PROC ANOVA (SAS version 9, SAS Institute, Cary, NC, USA) and least significant difference multiple range-tests were used to identify differences among the means of the parameters examined. Significance was considered at $P < 0.05$.

4. Water saving irrigation response to water management strategies

The annual volumes of ETc covered by irrigation water and effective rainfall are shown in Table 3 for both irrigation systems (DrI and SdrI). The percentages of water irrigation savings due to the contribution of UPe and RDI strategies are also presented.

Regarding effective precipitation use percentage (50 or 100 % CA), the reduction in irrigation water applied was significantly higher when 100 % canopy area (an average over 36 %) was considered than for 50 % canopy area (20 %) in all the years tested. Moreover, significant differences between years were observed, with higher water saving in 2007, 2008 and 2009 (an average over 30 %) than in 2006 (20 %), due to increased rainfall volume during those years.

Concerning regulated deficit irrigation (RDI) strategies, the reduction of 100 to 70 (or 100-40 % ETc alternate irrigation event) in the percentage of ETc covered by irrigation water allowed a significant water irrigation saving of up to 16 %, without significant differences between deficit water management. In this parameter, significant differences among years were also recorded, with a lower irrigation saving in 2008 corresponding to the lower ETc. The coefficient used when calculating rainfall water use did not affect this variable.

The overall saving of irrigation water due to both UEP and RDI strategies maintained a similar trend to that described for the previous variable. The greatest savings were obtained using a coefficient of 100% CA with RDI management in 2007, achieving a 66 % reduction in the irrigation water supplied of (DrI$_{70-100}$ /SDrI$_{70-100}$ and DrI$_{100/40-100}$ /SDrI$_{100/40-100}$ treatments)

Year	Treatments	ETc covered by (m³ ha-year⁻¹)			Reduction in irrigation water by (%)		
		UPe	IW	UPe+IW	UEP	RDI	UEP + RDI
2006	DrI_{100-50}/ $SDrI_{100-50}$	853	4928	5781	14,76	0,00	14,76
	DrI_{70-50}/ $SDrI_{70-50}$	853	3903	4756	14,76	17,73	32,49
	$DrI_{100/40-50}$/$SDrI_{100/40-50}$	853	3898	4751	14,76	17,82	32,57
	$DrI_{100-100}$/ $SDrI_{100-100}$	1485	4256	5741	25,87	0,00	25,87
	DrI_{70-100} /$SDrI_{70-100}$	1485	3282	4767	25,87	16,97	42,83
	$DrI_{100/40-100}$ /$SDrI_{100/40-100}$	1485	3279	4764	25,87	17,02	42,88
2007	DrI_{100-50}/ $SDrI_{100-50}$	1230	4204	5434	22,64	0,00	22,64
	DrI_{70-50}/ $SDrI_{70-50}$	1230	3245	4475	22,64	17,65	40,28
	$DrI_{100/40-50}$/$SDrI_{100/40-50}$	1230	3219	4449	22,64	18,13	40,76
	$DrI_{100-100}$/ $SDrI_{100-100}$	2337	3011	5348	43,70	0,00	43,70
	DrI_{70-100} /$SDrI_{70-100}$	2337	1995	4332	43,70	19,00	62,70
	$DrI_{100/40-100}$ /$SDrI_{100/40-100}$	2337	2008	4345	43,70	18,75	62,45
2008	DrI_{100-50}/ $SDrI_{100-50}$	1141	3788	4929	23,15	0,00	23,15
	DrI_{70-50}/ $SDrI_{70-50}$	1141	3313	4454	23,15	9,64	32,79
	$DrI_{100/40-50}$/$SDrI_{100/40-50}$	1141	3332	4473	23,15	9,25	32,40
	$DrI_{100-100}$/ $SDrI_{100-100}$	2030	3244	5274	38,49	0,00	38,49
	DrI_{70-100} /$SDrI_{70-100}$	2030	2689	4719	38,49	10,52	49,01
	$DrI_{100/40-100}$ /$SDrI_{100/40-100}$	2030	2712	4742	38,49	10,09	48,58
2009	DrI_{100-50}/ $SDrI_{100-50}$	1136	4266	5402	21,03	0,00	21,03
	DrI_{70-50}/ $SDrI_{70-50}$	1136	3307	4443	21,03	17,75	38,78
	$DrI_{100/40-50}$/$SDrI_{100/40-50}$	1136	3292	4428	21,03	18,03	39,06
	$DrI_{100-100}$/ $SDrI_{100-100}$	2214	3350	5564	39,79	0,00	39,79
	DrI_{70-100} /$SDrI_{70-100}$	2214	2511	4725	39,79	15,08	54,87
	$DrI_{100/40-100}$ /$SDrI_{100/40-100}$	2214	2507	4721	39,79	15,15	54,94
2010	DrI_{100-50}/ $SDrI_{100-50}$	1072	4950	6022	17,80	0,00	17,80
	DrI_{70-50}/ $SDrI_{70-50}$	1072	3848	4920	17,80	18,30	36,10
	$DrI_{100/40-50}$/$SDrI_{100/40-50}$	1072	3849	4921	17,80	18,28	36,08
	$DrI_{100-100}$/ $SDrI_{100-100}$	2143	3836	5979	35,84	0,00	35,84
	DrI_{70-100} /$SDrI_{70-100}$	2143	2703	4846	35,84	18,95	54,79
	$DrI_{100/40-100}$ /$SDrI_{100/40-100}$	2143	2693	4836	35,84	19,12	54,96
ANOVA[5]	RDI[6]					***	***
	UEP[7]				***	NS	***
	Year				*	***	***
	RDI x UEP					NS	NS
	RDI x Year					NS	NS
	UEPxYear				NS	NS	NS

Table 3. Water irrigation saving response to different irrigation strategies. UPe: volume of rainfall water available to the root system. IW: irrigation water applied. [5]ANOVA: Significant effects of different irrigation strategies are given at P>0,05 (NS, not significant), P≤0,05 (*), P≤0,01 (**),P≤0,001 (***). [6]RDI: Regulated deficit irrigation (Control, 70 and 100-40 % ETc). [7]UEP: coefficient use in effective precipitation (50 and 100 % CA).

By using regulated deficit irrigation (RDI) strategies, savings in irrigation water were recorded in peach orchards without reducing yield (Mitchell and Chalmers, 1982). Citrus has also been studied under deficit irrigation strategies. Thus, deficit irrigation treatments compared with the control, drip irrigated by six pressure compensated emitters per tree, allowed seasonal water savings of between 12 and 18% (Velez et al., 2007). Similarly, experiments with RDI have been successful in citrus (Domingo et al. 1996; González-Altozano and Castel, 1999; Goldhamer and Salinas, 2000). Similar results were reported for almond (Goldhamer et al., 2000), apple (Ebel et al., 1995), apricot (Ruiz-Sánchez et al., 2000), pear (Mitchell et al., 1989), pistachio (Goldhamer and Beede, 2004), wine grape vines (Bravdo and Naor, 1996; McCarthy et al., 2002), and olive (Moriana et al., 2003; Fernández et al., 2006). Accordingly, Fereres and Soriano (2007) published a comprehensive review on the use of deficit irrigation techniques to reduce water use in agriculture. However, there is no available information on the different UEP factor effects on water irrigation savings.

5. Impact of water irrigation techniques on tree nutritional status

Regarding macronutrient and micronutrient concentration in the spring flush leaves, significant differences were observed resulting from seasonality in most of the nutrients analysed (Table 4). However, every year, values were within the range considered optimal according to the standards described by Emblenton et al. (1973) and Legaz and Primo-Millo (1988). Only, Mg concentration showed slightly higher values due to the high concentration of this element in the irrigation water.

Regarding different factor effects (IS, RDI and UEP), foliar concentrations did not differ significantly between treatments. However, several authors found a higher foliar concentration in trees under SDrI than that obtained in DrI (Chartzoulakis and Bertaki, 2001). This indicates that this irrigation system improved nutrient absorption.

6. Fruit yield and WP$_{agr}$ response to irrigation strategies

Yield, expressed in kg, and fruit number per tree, fruit weight and others fruit quality parameters are shown in Tables 5 to 7. Season was observed to exert a significant effect on yield and fruit number parameters per tree (Table 1), with an evident alternate bearing pattern of trees during the assay, with years of low production and fruit number and high fruit weight ('off year') followed by years of high yield ('on year'). Similar results were also observed by other authors. Accordingly, El-Otmani et al. (2004) and Quiñones et al. (2011), among others, concluded that 'Nules' clementine mandarin is a cultivar with an alternate-bearing pattern and poor fruit-set with a large number of small-sized fruits during the 'on year' crop mainly.

According to the irrigation system, this factor significantly affected the yield, number of fruits per tree and percentage of fruits in the first category (> 78 mm). These variables were significantly lower under drip irrigation (DrI) than in subsurface irrigated trees (SDrI). In this regard, conflicting results have been found in the literature. A broad range of yield increases have been observed under SDrI when compared to surface, sprinkler, and even surface drip irrigation systems ranging from small to up to over 100% differences. Velez et al. (2007) in clementina de Nules mandarin subjected to different deficit irrigation observed, in two years of assay, that the deficit irrigation applied did not significantly reduce yield, nor average fruit weight compared with the control treatment. Besides there were no significant differences in fruit distribution by commercial sizes. The number of fruit harvested did not also vary between treatments, indicating that there were not carry over effects of the deficit irrigation applied. Research into SDrI has also been reported on crops including alfalfa (Oron et al., 1989; Bui and Osgood, 1990), asparagus (Sterret et al., 1990), cabbage and zucchini (Rubeiz et al., 1989), cantaloupe (Phene et al., 1987), cotton (Hutmacher et al., 1995) and tomatoes (Bogle et al., 1989), potatoes (Bisconer, 1987). Most yield increases have been attributed to better fertilization, better water management, improved water distribution uniformities, and improved disease and pest control. Grattan et al. (1988) cited better weed control as the major factor in the yield increases observed in their study. However, Yazar et al. (2002) obtained similar yield results for both irrigation methods in cotton. Brilay et al. (2003) also found a similar yield and fruit size in peach trees irrigated under drip and subsurface drip irrigation. In annual crops, like melon, surface drip irrigated plants yielded a higher percentage of 'first' category fruit along with greater equatorial diameter fruit compared to other treatments (Antunez et al. 2011).

Regarding deficit irrigation strategies, water stress significantly affected fruit weight and the percentage of fruit with a high calliper (up to 58 mm). Control trees with a 100 % ETc covered by irrigation water had significantly higher fruit weight and calliper than those irrigated under deficit regimes. Furthermore, these variables significantly increased with UEP when the latter rose from 50 to 100 % CA. In citrus, Ginestar and Castel (1996) also observed a decrease in production, although not significant, as the amount of irrigation water was reduced in Nules clementine. In almond, Goldhamer et al. (2006) analysed the impact of three different water stress timing patterns. The most successful stress timing pattern in terms of yield (considering fruit size and load) was the pattern that imposed sustained deficit irrigation by applying water at a given percentage of full ETc throughout the season. Furthermore, Romero et al. (2004) analysed the influence of several RDI strategies under subsurface and surface drip irrigation. Thus, RDI, with severe irrigation deprivation during kernel-filling (20% ETc) and a post-harvest recovery at 75% ETc or up to 50% ETc under subsurface drip irrigation, may be adequate in almonds under semiarid conditions, saving a significant amount of irrigation water. Deficit irrigation effects on yield and vegetative development have also been analysed for drip irrigated olives. Thus, some authors (Tognetti et al., 2006) determined that water availability might affect fruit weight before flowering or during the early stages of fruit growth rather than later in the summer season. Thus, irrigation of olive trees with drip systems from the beginning of pit hardening may be recommendable. Comparing different treatments, deficit irrigation during the whole summer resulted in improved plant water relations with respect to other watering regimes, while severe RDI differentiated treatments only slightly from rain fed plants (Tognetti et al., 2005).

A joint assessment of the effects of different irrigation strategies based exclusively on yield and water savings is difficult because crop response depends not only on irrigation, but also on climate, soil, cultivar, age, etc. In this sense, WP enables comparisons to be made incorporating all the data, thus establishing the most effective irrigation strategy (Garcia-Tejero et al., 2011).

In this assay, the different strategies significantly affected WP values. Regarding the irrigation system, SDrI resulted in significantly higher water efficiency than that obtained with DrI. This result could be due to the fact that subsurface drip systems may further

Year	Treatments	N	P	K	Ca	Mg	S	Na	B	Fe	Mn	Zn	Cu
2007	$DrI_{100\text{-}50}$	2.02	0.109	0.47	3.51	0.45	0.254	0.059	58.3	62.3	22.4	24.4	4.13
	$SDrI_{100\text{-}50}$	2.04	0.119	0.50	3.75	0.46	0.267	0.053	59.2	63.9	25.4	26.2	4.80
	$DrI_{70\text{-}50}$	2.02	0.106	0.48	3.28	0.42	0.236	0.048	53.7	48.1	22.4	25.2	4.57
	$SDrI_{70\text{-}50}$	2.02	0.112	0.44	3.23	0.44	0.251	0.051	53.6	50.3	21.6	23.6	4.53
	$DrI_{100/40\text{-}50}$	2.03	0.103	0.46	4.04	0.44	0.242	0.054	71.4	62.1	24.3	27.2	5.37
	$SDrI_{100/40\text{-}50}$	2.04	0.114	0.47	4.33	0.49	0.282	0.080	58.3	71.2	25.1	28.5	4.87
	$DrI_{100\text{-}100}$	2.00	0.096	0.43	3.48	0.40	0.218	0.044	60.3	59.5	23.3	26.2	5.27
	$SDrI_{100\text{-}100}$	2.02	0.106	0.49	3.67	0.44	0.250	0.049	63.0	61.7	24.3	28.1	5.43
	$DrI_{70\text{-}100}$	1.99	0.095	0.42	3.33	0.43	0.226	0.055	49.5	54.3	20.3	23.8	4.10
	$SDrI_{70\text{-}100}$	2.08	0.114	0.50	3.31	0.45	0.259	0.061	52.7	60.3	24.5	25.0	4.50
	$DrI_{100/40\text{-}100}$	1.99	0.091	0.41	2.80	0.41	0.215	0.058	48.7	43.6	22.3	23.8	3.87
	$SDrI_{100/40\text{-}100}$	2.08	0.104	0.46	3.57	0.41	0.237	0.045	56.5	56.2	23.1	23.2	4.33
2008	$DrI_{100\text{-}50}$	2.18	0.120	0.56	4.43	0.53	0.274	0.072	55.1	53.5	48.4	35.1	7.77
	$SDrI_{100\text{-}50}$	2.13	0.113	0.50	4.60	0.50	0.269	0.047	49.7	71.5	56.2	37.7	9.33
	$DrI_{70\text{-}50}$	2.18	0.124	0.58	4.64	0.54	0.286	0.080	57.8	59.7	58.3	40.1	8.13
	$SDrI_{70\text{-}50}$	2.15	0.113	0.53	4.38	0.48	0.256	0.059	57.7	75.3	52.7	36.9	9.50
	$DrI_{100/40\text{-}50}$	2.24	0.126	0.55	4.33	0.54	0.290	0.067	62.3	117.5	53.1	40.4	10.70
	$SDrI_{100/40\text{-}50}$	2.30	0.131	0.61	4.57	0.53	0.297	0.079	76.5	52.6	47.6	33.4	7.57
	$DrI_{100\text{-}100}$	2.32	0.134	0.64	4.46	0.52	0.298	0.070	47.3	49.0	45.2	29.8	6.43
	$SDrI_{100\text{-}100}$	2.37	0.129	0.55	4.43	0.53	0.292	0.061	99.2	90.1	53.0	39.1	8.77
	$DrI_{70\text{-}100}$	2.23	0.128	0.53	4.41	0.58	0.283	0.071	58.0	65.7	45.1	30.7	7.80
	$SDrI_{70\text{-}100}$	2.24	0.127	0.56	4.66	0.54	0.286	0.071	48.3	49.7	48.3	33.6	8.70
	$DrI_{100/40\text{-}100}$	2.23	0.130	0.61	4.56	0.56	0.299	0.085	41.7	24.1	40.6	30.9	7.17
	$SDrI_{100/40\text{-}100}$	2.22	0.130	0.62	4.53	0.59	0.294	0.071	55.8	41.2	40.6	27.4	10.43
2009	$DrI_{100\text{-}50}$	2.25	0.118	0.83	3.52	0.50	0.237	0.062	67.3	43.9	28.6	23.7	5.93
	$SDrI_{100\text{-}50}$	2.22	0.119	0.77	3.84	0.49	0.240	0.061	62.0	44.2	29.8	26.9	5.27
	$DrI_{70\text{-}50}$	2.23	0.106	0.73	3.36	0.47	0.211	0.059	52.3	48.3	27.6	24.9	5.50
	$SDrI_{70\text{-}50}$	2.25	0.107	0.74	3.36	0.45	0.214	0.050	52.9	38.9	27.2	24.5	5.87
	$DrI_{100/40\text{-}50}$	2.27	0.112	0.81	3.59	0.52	0.228	0.060	71.4	44.3	28.5	25.8	5.33
	$SDrI_{100/40\text{-}50}$	2.25	0.108	0.71	3.57	0.48	0.222	0.058	53.0	42.6	26.4	22.5	5.13
	$DrI_{100\text{-}100}$	2.25	0.116	0.75	3.58	0.50	0.244	0.056	58.4	37.0	29.6	26.2	5.43
	$SDrI_{100\text{-}100}$	2.28	0.109	0.70	3.54	0.46	0.237	0.053	69.1	47.4	32.8	32.2	6.10
	$DrI_{70\text{-}100}$	2.27	0.122	0.85	3.46	0.53	0.242	0.070	65.2	45.4	29.3	25.1	5.80
	$SDrI_{70\text{-}100}$	2.22	0.115	0.86	3.53	0.48	0.232	0.060	81.2	49.1	31.1	30.4	5.97
	$DrI_{100/40\text{-}100}$	2.20	0.112	0.86	3.31	0.48	0.211	0.067	59.1	40.3	30.1	29.1	6.03
	$SDrI_{100/40\text{-}100}$	2.27	0.114	0.75	3.49	0.48	0.235	0.058	60.2	41.8	30.6	27.1	6.77
2010	$DrI_{100\text{-}50}$	2.44	0.105	0.83	3.28	0.43	0.222	0.056	84.5	58.0	33.37	25.3	5.43
	$SDrI_{100\text{-}50}$	2.42	0.102	0.86	3.30	0.43	0.224	0.054	86.0	55.6	33.5	24.8	5.03
	$DrI_{70\text{-}50}$	2.40	0.102	0.96	3.39	0.45	0.214	0.052	58.7	58.9	32.7	30.4	5.63
	$SDrI_{70\text{-}50}$	2.41	0.108	1.04	3.43	0.45	0.234	0.055	67.4	53.5	31.8	27.5	5.33
	$DrI_{100/40\text{-}50}$	2.43	0.101	0.97	3.34	0.44	0.221	0.057	82.9	57.9	31.0	24.4	4.57
	$SDrI_{100/40\text{-}50}$	2.44	0.108	0.82	3.49	0.46	0.250	0.061	74.2	69.4	33.1	25.9	5.40
	$DrI_{100\text{-}100}$	2.40	0.112	1.08	3.64	0.47	0.236	0.062	72.3	73.9	33.8	32.5	5.60
	$SDrI_{100\text{-}100}$	2.43	0.105	0.83	3.37	0.45	0.229	0.048	69.4	67.9	32.0	29.7	5.80
	$DrI_{70\text{-}100}$	2.39	0.092	0.87	3.14	0.43	0.201	0.058	65.0	54.2	31.4	28.0	4.67
	$SDrI_{70\text{-}100}$	2.40	0.108	1.04	3.49	0.47	0.238	0.063	73.1	59.2	31.3	19.7	5.53
	$DrI_{100/40\text{-}100}$	2.42	0.096	0.91	3.22	0.45	0.208	0.059	66.9	58.4	32.9	24.5	5.40
	$SDrI_{100/40\text{-}100}$	2.44	0.102	0.96	3.28	0.45	0.224	0.057	54.5	56.9	31.6	24.7	5.43

ANOVA[1]	IS[2]	NS	NS	NS	NS	NS	NS	NS	NS	NS	NS	NS	NS
	RDI[3]	NS	NS	NS	NS	NS	NS	NS	NS	NS	NS	NS	NS
	UEP[4]	NS	NS	NS	NS	NS	NS	NS	NS	NS	NS	NS	NS
	Year	***	***	***	***	***	NS	***	**	***	***	***	***
	IS x RDI	NS	NS	NS	NS	NS	NS	NS	NS	NS	NS	NS	NS
	IS x UEP	NS	NS	NS	NS	NS	NS	NS	NS	NS	NS	NS	NS
	IS x Year	NS	NS	NS	NS	NS	NS	NS	NS	NS	NS	NS	NS
	RDI x UEP	NS	NS	NS	NS	NS	NS	NS	NS	NS	NS	NS	NS
	RDI x Year	NS	NS	NS	NS	NS	NS	NS	NS	NS	NS	NS	NS
	UEP x Year	NS	NS	NS	NS	NS	NS	NS	NS	NS	NS	NS	NS

Table 4. Effect of irrigation strategies on macronutrient concentration (% dry weight) and micronutrient (ppm) concentration. [1]ANOVA: Significant effects of different irrigation strategies are given at P>0,05 (NS, not significant), P≤0,05 (*), P≤0,01 (**),P≤0,001 (***). [2]IS: irrigation system (SDrI and DrI); [3]RDI: Regulated deficit irrigation (Control, 70 and 100-40 % ETc). [4]UEP: coefficient use in effective precipitation (50 and 100 % CA).

improve irrigation and fertilizer use efficiency because water and nutrients are applied directly to the root zone (Camp, 1998). Boss (1985) also obtained less WP in drip irrigated trees (microjets) than trees irrigated by subsurface drip. However, Bryla et al. (2003) did not observe significant differences in this variable between peach trees irrigated by surface and subsurface drip.

Concerning the effect of deficit irrigation, water use significantly increased by increasing water stress. García-Tejero et al. (2011) affirmed that WP was strongly influenced by the irrigation strategy employed at different phenological stages, rather than the amount of water in orange trees subjected to different deficit irrigation regimes. In this sense, the best results were registered when stress was applied at fruit maturity. However, the most restrictive treatment during fruit growth had a descending WP, registering values below even the fully irrigated treatment. Treatments in which water stress was applied at flowering and maturity showed similar WP values. Clearly, WP depends not only on the total water applied but also on when it is applied. Dissimilar results were found by Ibrahim and Abd El-Samad (2009) for pomegranate trees irrigated at 70%, 50% and 30% of available soil water. WP diminishing in trees with high deficit irrigation regimes (46-52 % and 2-6% of tree water needs).

The factor used to calculate the volume of rainwater available to the root system also affected the WP variable. Thus, efficiency values were higher in trees that theoretically used 100% of CA than in those that used 50 %. As indicated, there is no information available on the different UEP factor effects on water irrigation savings.

7. Effect of irrigation management on fruit quality parameters

With regard to fruit quality (Table 7), the studied factors and their interactions did not significantly affect peel thickness, or the percentage of pulp or juice. Only UEP significantly affected juice percentage with higher values in 50 % CA irrigated trees. This result could be due to the high volume of water applied in these plants. In other studies on grapefruit (Cruse et al., 1982), orange (Castel and Buj, 1990), Satsuma mandarin (Salustiano, Rabe and Peng, 1998) and Nules mandarin (Velez et al., 2007) significant differences due to differential irrigation doses were not found for these parameters either.

Year	Treatments	Yield (kg tree^{-1})	Fruit weight (g)	N° fruit tree^{-1}	WP (kg m^{-3})
2007	DrI$_{100-50}$	62.5	107.1	597	6.45
	SDrI$_{100-50}$	66.6	110.8	610	6.90
	DrI$_{70-50}$	68.3	107.6	651	6.95
	SDrI$_{70-50}$	74.0	110.2	679	7.55
	DrI$_{100/40-50}$	72.3	106.1	700	9.10
	SDrI$_{100/40-50}$	73.0	112.3	662	9.20
	DrI$_{100-100}$	69.4	109.4	647	8.50
	SDrI$_{100-100}$	76.3	110.7	701	9.35
	DrI$_{70-100}$	59.3	99.7	596	7.50
	SDrI$_{70-100}$	73.8	105.1	707	9.30
	DrI$_{100/40-100}$	66.0	107.3	616	8.10
	SDrI$_{100/40-100}$	74.2	108.8	686	9.15
2008	DrI$_{100-50}$	26.0	103.2	252	4.09
	SDrI$_{100-50}$	38.7	94.1	411	6.44
	DrI$_{70-50}$	39.7	88.5	450	7.13
	SDrI$_{70-50}$	38.2	97.3	392	6.87
	DrI$_{100/40-50}$	28.6	92.6	306	5.10
	SDrI$_{100/40-50}$	37.6	89.6	426	6.72
	DrI$_{100-100}$	39.1	107.1	365	7.18
	SDrI$_{100-100}$	41.1	100.7	418	7.53
	DrI$_{70-100}$	41.8	94.9	422	9.24
	SDrI$_{70-100}$	45.6	102.4	445	10.10
	DrI$_{100/40-100}$	35.2	95.3	364	7.72
	SDrI$_{100/40-100}$	47.4	97.0	486	10.39
2009	DrI$_{100-50}$	39.7	112.6	353	4.63
	SDrI$_{100-50}$	48.4	110.4	442	5.64
	DrI$_{70-50}$	48.6	99.2	484	6.78
	SDrI$_{70-50}$	50.0	94.7	528	6.98
	DrI$_{100/40-50}$	33.0	95.1	350	4.63
	SDrI$_{100/40-50}$	42.4	100.4	422	5.95
	DrI$_{100-100}$	44.5	111.0	402	4.94
	SDrI$_{100-100}$	53.0	103.7	521	5.89
	DrI$_{70-100}$	35.9	107.4	334	4.84
	SDrI$_{70-100}$	47.9	104.9	482	6.46
	DrI$_{100/40-100}$	30.0	101.5	296	4.05
	SDrI$_{100/40-100}$	46.3	105.7	439	6.26
2010	DrI$_{100-50}$	31.1	96.0	324	3.8
	SDrI$_{100-50}$	32.8	95.9	343	4.0
	DrI$_{70-50}$	31.9	92.0	347	4.9
	SDrI$_{70-50}$	28.8	89.6	326	4.5
	DrI$_{100/40-50}$	28.8	693.5	314	4.5
	SDrI$_{100/40-50}$	26.1	91.8	289	4.1
	DrI$_{100-100}$	30.9	102.9	300	4.8
	SDrI$_{100-100}$	35.5	95.9	372	5.5
	DrI$_{70-100}$	21.8	103.1	210	4.8
	SDrI$_{70-100}$	33.4	100.2	335	7.4
	DrI$_{100/40-100}$	21.8	93.9	234	4.8
	SDrI$_{100/40-100}$	34.3	94.2	361	7.6

ANOVA[1]	IS[2]	***	NS	***	***
	RDI[3]	NS	*	NS	*
	UEP[4]	NS	*	NS	***
	Year	***	***	***	***
	IS x RDI	NS	NS	NS	NS
	IS x UEP	NS	NS	NS	NS
	IS x Year	NS	NS	NS	NS
	RDI x UEP	NS	NS	NS	NS
	RDI x Year	NS	NS	NS	NS
	UEP x Year	NS	NS	NS	NS

Table 5. Effect of irrigation strategies on yield and fruit parameters and water use efficiency. [1]ANOVA: Significant effects of different irrigation strategies are given at $P>0,05$ (NS, not significant), $P≤0,05$ (*), $P≤0,01$ (**), $P≤0,001$ (***). [2]IS: irrigation system (SDrI and DrI); [3]RDI: Regulated deficit irrigation (Control, 70 and 100-40 % ETc). [4]UEP: coefficient use in effective precipitation (50 and 100 % CA).

As for the other quality parameters analysed, both IS and RDI and their interaction significantly affected total acidity (TA), total soluble solids (TSS) and maturity index (IM). In drip irrigated trees, control trees showed higher TA and IM than that recorded for the juice of fruits from deficit irrigated plants (70 or alternate irrigation events 40-100 % ETc). However, under subsurface drip irrigation an opposite trend was observed. Regarding TSS, higher values were recorded in drip irrigated trees and under water stress conditions. Similarly to what occurs in SDrI, Ginestar and Castel (1996) only detected an increase (albeit insignificant) in acidity on decreasing water doses in drip irrigated trees of Nules clementine. Other researchers (Cruse et al., 1982; Koo and Smajstrla, 1985, Castel and Buj, 1990; Eliades, 1994, Peng and Rabe, 1998) described a similar pattern to that described in this assay. Velez et al. (2007) observed that fruit quality parameters were slightly altered by the deficit irrigation applied. Thus, fruit from the deficit irrigated trees was more acidic and was not sweeter than that from the control trees in the first year of the assay, but fruit from water stress trees had significantly higher Brix and similar acidity in the second year. On the other hand, in both years, the MI was not significantly altered by the water restrictions applied. Pérez-Sarmiento et al. (2010) also found that TSS values were increased significantly by RDI treatment, whereas TA was equal in control and deficit treatments, and therefore the TSS/TA ratio increased significantly in the RDI treatment. Thus, fruits from RDI treatment can be considered of high quality since TSS increased without affecting acidity (Scandella et al., 1997).

RDI has been used successfully, maintaining yield and fruit quality, including higher values of total soluble solids, tritratable acidity in many fruit species (Ebel et al., 1995; López et al., 2008), citrus species (Sánchez-Blanco et al., 1989; Castel and Buj, 1990; Domingo et al., 1996; González-Altozano and Castel, 1999; Goldhamer and Salinas, 2000), apricot trees (Ruiz-Sanchez et al., 2000), nut species (Romero et al., 2004), wine grape vines (Bravdo and Naor, 1996; McCarthy et al., 2002) and olives (Moriana et al., 2003).

The analysis of fruit peel colour (CI) showed that only IS affected this variable, with greener fruit corresponding to the DrI treatments than those obtained for SDrI. An opposite pattern was obtained in apricot fruit indices (Pérez-Sarmiento et al., 2010). Fruits from RDI treated trees showed higher CI values. The increase in this parameter in apricot fruits from RDI plants can be associated to a reduction in carotenoid accumulation, attributed to oxidation

Year	Treatments	Calliper	> 78 mm	78-67	67-58	58-50	< 50
2007	DrI_{100-50}	62.5	0.3	14.41	43.9	36.7	4,7
	$SDrI_{100-50}$	66.6	0.0	9.5	41.7	44.3	4,5
	DrI_{70-50}	68.3	0.0	3.8	40.3	45.3	10,6
	$SDrI_{70-50}$	74.0	0.0	5.1	47.3	41.6	6,0
	$DrI_{100/40-50}$	72.3	0.0	11.7	46.7	38.0	3,6
	$SDrI_{100/40-50}$	73.0	0.3	8.7	50.0	35.5	5,5
	$DrI_{100-100}$	69.4	0.3	11.4	50.5	35.3	2,5
	$SDrI_{100-100}$	76.3	0.0	9.7	55.5	29.6	5,2
	DrI_{70-100}	59.3	0.0	9.0	49.8	35.1	6,1
	$SDrI_{70-100}$	73.8	0.6	12.0	52.5	30.2	4,7
	$DrI_{100/40-100}$	66.0	0.0	11.7	53.0	31.7	3,6
	$SDrI_{100/40-100}$	74.2	0.6	7.2	53.4	35.9	2,9
2008	DrI_{100-50}	61.9	0.0	18.9	64.6	16.1	0,4
	$SDrI_{100-50}$	61.8	0.5	15.6	66.5	17.1	0,3
	DrI_{70-50}	59.7	0.0	9.0	63.6	26.1	1,3
	$SDrI_{70-50}$	60.9	0.0	15.3	66.7	17.1	0,9
	$DrI_{100/40-50}$	60.7	0.0	14.2	61.7	23.8	0,3
	$SDrI_{100/40-50}$	60.7	0.0	22.1	56.2	21.4	0,3
	$DrI_{100-100}$	62.7	0.0	25.2	61.1	13.4	0,3
	$SDrI_{100-100}$	62.1	0.9	23.6	56.6	15.2	3,7
	DrI_{70-100}	60.9	0.5	15.6	62.7	20.3	0,9
	$SDrI_{70-100}$	62.3	0.0	24.0	62.0	13.0	1,0
	$DrI_{100/40-100}$	60.4	0.0	13.7	60.3	25.2	0,8
	$SDrI_{100/40-100}$	61.2	0.0	14.4	65.4	19.9	0,3
2009	DrI_{100-50}	59.8	0.6	20.6	47.2	29.2	2,4
	$SDrI_{100-50}$	58.8	0.0	14.1	46.2	36.9	2,8
	DrI_{70-50}	57.1	0.0	6.4	47.1	39.7	6,8
	$SDrI_{70-50}$	58.3	0.0	7.8	54.1	34.5	3,6
	$DrI_{100/40-50}$	59.4	0.0	18.2	50.0	29.6	2,2
	$SDrI_{100/40-50}$	59.4	0.6	12.8	54.5	28.7	3,4
	$DrI_{100-100}$	60.0	0.0	17.1	50.0	27.4	5,5
	$SDrI_{100-100}$	59.8	0.0	14.8	59.4	23.0	2,8
	DrI_{70-100}	59.2	0.0	12.2	54.9	29.0	3,9
	$SDrI_{70-100}$	60.5	1.1	17.0	55.6	24.2	2,1
	$DrI_{100/40-100}$	60.2	0.0	17.2	56.7	24.1	2,0
	$SDrI_{100/40-100}$	59.6	1.1	10.9	57.4	28.7	1,9
2010	DrI_{100-50}	59.2	0.0	8.1	47.1	39.3	5,5
	$SDrI_{100-50}$	59.0	0.6	6.3	47.96	43.9	1,2
	DrI_{70-50}	58.3	0.0	3.8	46.8	46.2	3,2
	$SDrI_{70-50}$	59.7	0.3	10.8	46.7	40.8	1,4
	$DrI_{100/40-50}$	58.9	0.0	6.8	49.5	40.1	3,6
	$SDrI_{100/40-50}$	58.4	0.0	4.2	44.1	49.8	1,9
	$DrI_{100-100}$	60.4	0.0	10.5	57.0	31.4	1,1
	$SDrI_{100-100}$	59.9	0.0	9.0	51.4	37.6	2,0
	DrI_{70-100}	60.1	0.0	9.6	54.1	33.6	2,7
	$SDrI_{70-100}$	59.9	0.0	7.4	54.4	36.6	1,6
	$DrI_{100/40-100}$	59.7	0.0	11.3	47.8	37.4	3,5
	$SDrI_{100/40-100}$	59.2	0.3	8.6	49.3	38.5	3,3

ANOVA[1]							
	IS[2]	NS	*	NS	NS	NS	NS
	RDI[3]	*	NS	**	NS	NS	NS
	UEP[4]	**	NS	*	***	***	NS
	Year	***	NS	***	***	***	***
	IS x RDI	NS	NS	NS	NS	NS	NS
	IS x UEP	NS	NS	NS	NS	NS	NS
	IS x Year	NS	NS	NS	NS	NS	NS
	RDRI x UEP	NS	NS	NS	NS	NS	NS
	RDRI x Year	NS	NS	NS	NS	NS	NS
	UEP x Year	NS	NS	NS	NS	NS	NS

Table 6. Effect of irrigation strategies on fruit calliper and percentage of fruit in each commercial calliper. [1]ANOVA: Significant effects of different irrigation strategies are given at P>0,05 (NS, not significant), P≤0,05 (*), P≤0,01 (**),P≤0,001 (***). [2]IS: irrigation system (SDrI and DrI); [3]RDI: Regulated deficit irrigation (Control, 70 and 100-40 % ETc). [4]UEP: coefficient use in effective precipitation (50 and 100 % CA).

Year	Treatments	Peel thickness	% Peel	% Juice	TA[5] (° Brix)	TSS[6]	IM[7]	CI[8]
2007	DrI$_{100-50}$	2.95	52.4	47.6	0.88	13.2	15,0	8.8
	SDrI$_{100-50}$	2.85	51.7	48.3	0.89	12.9	14,5	8.5
	DrI$_{70-50}$	3.85	50.8	49.2	0.94	13.1	13,9	8.0
	SDrI$_{70-50}$	3.40	50.9	49.1	0.91	13.1	14,4	8.3
	DrI$_{100/40-50}$	3.00	51.6	48.4	0.89	13.2	14,8	8.3
	SDrI$_{100/40-50}$	2.85	50.4	49.6	0.89	12.9	14,5	9.4
	DrI$_{100-100}$	2.95	52.4	47.6	0.97	13.3	13,7	9.1
	SDrI$_{100-100}$	2.90	51.4	48.6	0.92	13.2	14,3	8.5
	DrI$_{70-100}$	2.75	50.7	49.3	1.04	14.1	13,6	9.0
	SDrI$_{70-100}$	2.80	50.8	49.2	0.91	13.3	14,6	9.8
	DrI$_{100/40-100}$	2.90	52.0	48	0.99	13.8	13,9	8.9
	SDrI$_{100/40-100}$	2.85	51.4	48.6	0.91	13.5	14,8	9.4
2008	DrI$_{100-50}$	3.50	49.1	50.9	0.90	13.6	15,1	9.6
	SDrI$_{100-50}$	3.37	49.1	50.9	0.97	13.2	13,6	9.3
	DrI$_{70-50}$	3.37	48.8	51.2	0.90	14.0	15,6	9.6
	SDrI$_{70-50}$	3.47	50.9	49.1	0.87	13.4	15,4	9.4
	DrI$_{100/40-50}$	3.57	51.5	48.5	0.93	14.1	15,2	8.5
	SDrI$_{100/40-50}$	3.43	48.7	51.3	0.93	13.9	14,9	9.3
	DrI$_{100-100}$	3.40	49.4	50.6	0.87	12.6	14,5	8.3
	SDrI$_{100-100}$	3.40	49.7	50.3	0.87	12.7	14,6	9.1
	DrI$_{70-100}$	3.43	49.5	50.5	0.93	13.1	14,1	9.2
	SDrI$_{70-100}$	3.53	50.0	50	0.90	13.2	14,7	11.0
	DrI$_{100/40-100}$	3.43	49.7	50.3	1.00	13.7	13,7	9.8
	SDrI$_{100/40-100}$	3.37	48.6	51.4	0.80	13.2	16,5	9.2
2009	DrI$_{100-50}$	3.29	51.9	48.1	1.02	12.7	12,5	7.8
	SDrI$_{100-50}$	3.15	51.7	48.3	1.00	124	12,4	10.4
	DrI$_{70-50}$	3.21	50.9	49.1	1.02	13.1	12,8	7.9
	SDrI$_{70-50}$	3.04	50.0	50	1.04	12.6	12,1	8.0
	DrI$_{100/40-50}$	3.04	51.2	48.8	1.15	13.0	11,3	6.8
	SDrI$_{100/40-50}$	3.24	51.6	48.4	1.09	12.8	11,7	6.4
	DrI$_{100-100}$	3.20	53.2	46.8	1.04	12.7	12,2	7.0
	SDrI$_{100-100}$	2.98	51.4	48.6	1.11	12.7	11,4	6.8

	DrI$_{70-100}$	3.27	53.0	47	1.10	12.6	11,5	7.8
	SDrI$_{70-100}$	3.18	51.8	48.2	1.02	12.6	12,4	9.0
	DrI$_{100/40-100}$	3.12	52.9	47.1	1.13	12.6	11,2	7.0
	SDrI$_{100/40-100}$	3.01	52.0	48	0.96	12.9	13,4	9.5
2010	DrI$_{100-50}$	2.84	49.6	50.4	1.16	12.9	11,1	
	SDrI$_{100-50}$	2.94	52.0	48	1.16	13.0	11,2	
	DrI$_{70-50}$	2.93	51.9	48.1	1.33	14.2	10,7	
	SDrI$_{70-50}$	2.75	50.9	49.1	1.16	13.0	11,2	
	DrI$_{100/40-50}$	3.00	51.0	49	1.33	14.0	10,5	
	SDrI$_{100/40-50}$	2.93	51.3	48.7	1.15	13.0	11,3	
	DrI$_{100-100}$	2.93	48.3	51.7	1.19	13.0	10,9	
	SDrI$_{100-100}$	2.76	52.3	47.7	1.18	13.1	11,1	
	DrI$_{70-100}$	3.12	52.1	47.9	1.28	13.6	10,6	
	SDrI$_{70-100}$	3.07	54.8	45.2	1.20	13.5	11,3	
	DrI$_{100/40-100}$	3.18	53.1	46.9	1.29	14.0	10,9	
	SDrI$_{100/40-100}$	2.94	53.7	46.3	1.23	13.5	11,0	
ANOVA[1]	IS[2]	NS	NS	NS	***	***	*	*
	RDI[3]	NS	NS	NS	NS	***	NS	NS
	UEP[4]	NS	NS	*	NS	NS	NS	NS
	Year	***	***	***	***	***	***	***
	IS x RDI	NS	NS	NS	**	NS	**	NS
	IS x UEP	NS	NS	NS	NS	NS	NS	NS
	IS x Year	NS	NS	NS	NS	NS	NS	NS
	RDRI x UEP	NS	NS	NS	NS	NS	NS	NS
	RDRI x Year	NS	NS	NS	NS	NS	NS	NS
	UEP x Year	NS	NS	NS	NS	NS	NS	NS

Table 7. Effect of irrigation strategies on fruit quality parameters. [1]ANOVA: Significant effects of different irrigation strategies are given at P>0,05 (NS, not significant), P≤0,05 (*), P≤0,01 (**),P≤0,001 (***). [2]IS: irrigation system (SDrI and DrI); [3]RDI: Regulated deficit irrigation (Control, 70 and 100-40 % ETc). [4]UEP: coefficient use in effective precipitation (50 and 100 % CA).. [4]UEP: coefficient use in effective precipitation (50 and 100 % CA). [5]TA: total acidity. [6]TSS: total soluble solids. [7]MI: maturity index, ratio between TSS/TA. [8]CI: colour index.

by exposure to light (Ruiz et al., 2005). This exposure to light in fruits from the RDI treatment is related to a significant reduction in the vegetative growth of the trees during fruit development, implying a high exposure of fruits to the light. Similar trends were observed in peach fruits under RDI (Gelly et al., 2003; Buendía et al., 2008).

Significant differences were observed resulting from seasonality in all the fruit quality variables analysed (Table 7). However, values were within the range considered optimal according to the standards established by González-Sicilia (1968).

8. Conclusions and future research

Efficient irrigation systems management at the farm level appears to be a very important factor in irrigated agriculture and, given the competition for water resources with other sectors, is a key issue in terms of the economic and environmental sustainability of agriculture. In general, surface and pressurized irrigation systems can attain a reasonable level of efficiency, when they are well designed, adequately operated and appropriately selected for specific conditions.

Subsurface drip irrigated leads to higher fruit production and water use efficiency and regulated deficit irrigation also provides savings in irrigation water without reducing yield. Thus, both subsurface drip systems and water stress at certain phenological stages of the crop have been demonstrated as a useful tool to improve irrigation management and maintain sustainable production levels at the field scale under arid and semi-arid conditions. Moreover, highly stressful deficit irrigation should not be applied during flowering or fruit-growth periods, in order to ensure yield.

In addition, it is important to emphasize the importance on the use of effective rainfall in reducing the volumes of water applied, without affecting either production or fruit quality.

Other potential strategies for future use, such as partial root-zone, drying by irrigating half of the root-zone while the other half is kept under dry soil, alternating irrigation from one half to the other every 2-3 weeks, low-frequency deficit irrigation, or higher water stress all appear to be promising techniques. Further research is needed to analyse the effects of these strategies on yield, nutritional status, fruit quality and water irrigation savings.

9. References

Aboukhaled, A.A.; Alfaro, A. & Smith, M. (1982). Lysimeters. *FAO Irrigation and Drainage.* paper N° 39, FAO (Food and Agriculture Organization of the United Nations), Rome.

Allen, R.G.; Pereira, L.S.; Raes, D. & Smith, M. (1998). Crop evapotranspiration (guidelines for computing crop water requirements). *FAO Irrigation and Drainage.*, paper N° 56, FAO (Food and Agriculture Organization of the United Nations), Rome, Italy

Antunez, A.; Martínez, J.P.; Alfaro, C. & Alé, M. (2011). Impact of surface and subsurface drip irrigation on yield and quality of "Honey Dew" Melon. *Acta Horticulturae*, (ISHS), 889, pp. 417-422

ASAE. (2001) ASAE Standard S526.2, JAN01, In: *Soil and Water Terminology*, ASAE, St.Joseph, Michigan USA

Bogle, C.R.; Hartz, T.K. & Nuñez. C. (1989). Comparison of subsurface trickle and furrow irrigation on platic-mulched and bare-soil for tomato production. *J. Am. Soc. Hort. Sci.*, 114(1), pp. 40-43

Bonet, L.; Ferrer, P.; Castel, J.R. & Intrigliolo, D.S. (2010). Soil capacitance sensors and stem dendrometers. Useful tools for irrigation scheduling of commercial orchards?. *Spanish Journal of Agricultural Research* 8(S2), pp. S52-S65

Boman, B.J.; Zekri, M. & Stover, E. (2005). Managing salinity in citrus. *HortTechnol.* 15(1), pp. 108-113

Boss, M.G. (1985). Summary of ICID definitions of irrigation efficiency. *Intl. Comm. Irr. And Drainage Bull.* 34, pp. 28-31

Bravdo, B. & Naor, A. (1996). Effect of water regime on productivity and quality of fruit and wine. *Acta Horticulturae*, 427, pp. 15–26

Bryla, D.R.; Trout, T.J. & Ayars, J.E. (2003) Growth and production of young peach trees irrigated by furrow, microjet, surface drip, or subsurface drip systems. *HortScience*, 38(6), pp.1112-1116

Bucks, D.A.; Erie, L.J.; French, O.F.; Nakayama, F.S. & Pew, W.D. (1981). Subsurface trickle irrigation management with multiple cropping. *Transactions of the ASAE* 24(6), pp. 1482-1489

Buendía, B.; Allende, A., Nicolás, E.; Alarcón, J.J. & Gil, M.I. (2008). Effect of regulated deficit irrigation and crop load on the antioxidant compounds of peaches. *J Agric Food Chem*, 56, pp. 3601-3608

Bui, W.& Osgood, R.V. 1(990). Subsurface irrigation trial for alfalfa in Hawaii. In: *Proceedings. Third Nat'l Irrigation Symposium*, pp. 658-660. St. Joseph, Mich.: ASAE.

Burt, C.; O'Connor, K. & Ruehr, T. (1998). Fertigation. Irrigation Training and Research Center, California Polytechnic State Univ. San Luis Obispo, CA.

Camp, C.R. (1998). Subsurface drip irrigation: A review. *Trans. Amer. Soc Agr. Eng*, 41, pp. 1353-1367

Castel, J.R. & Buj. A. (1990). Response of Salustiana oranges to high frequency deficit irrigation. *Irrg. Sci.*, 11, pp. 121-127

Castel, J.R. & Buj, A. (1994). Growth and evapotranspiration of young, drip irrigated Clementine trees, *Proceedings of International Congress of Citriculture*. Seventh meeting of the International Society of Citriculture, 8-13 March 1992 Acireale, Italy 2, pp. 651-656

Chartzoulakis, K. & Bertaki, M. (2001). Towards Sustainable Water Use on Mediterranean Islands: Addressing Conflicting Demands and Varying Hydrological, Social and Economic Conditions WORK PACKAGE 2 In: *Investigation of irrigation methods – Recommendations*. Deliverable D14 and D22. Project No EVK1 – CT – 2001 – 00092. Funded by the European Commission

Cruse, R.; Wiegand, C.L. & Swason. W.A. (1982). The effect of rainfall and irrigation management on citrus juice quality in Texas. *J. Am. Soc. Hortic. Sci.*, 107, 767-770.

Devasirvatham, V. (2009). A review of Subsurface Drip Irrigation in Vegetable Production. In: *Irrigation Matters Series* no 03/09. Cooperative Research Centre for Irrigation Futures. IF Technologies Pty. Ltd. Australia

Domingo, R.; Ruiz-Sánchez, M.C.; Sánchez-Blanco, N.J. & Torrecillas, A. (1996). Water relations, growth and yield of Fino lemon trees under regulated deficit irrigation. *Irrigation Science* 16, pp. 115–123

Ebel, R.C.; Proebsting, E.L. & Evans, R.G. (1995). Deficit irrigation to control vegetative growth in apple and monitoring fruit growth to schedule irrigation. *HortScience*, 30, pp. 1229–1232

Eliades, G. (1994). Response of grapefruit to different amounts of water apllied by drippers and minisprinklers. *Acta Horticulturae*, 365, pp. 129-146

El-Otmani, M.; Ait-Oubahou, A.; El-Hassainate, F.; Kaanane, A. & Lovatt, C.J. (2004). Effect of Gibberellic acid, urea and KNO$_3$ on yield and on composition and nutritional quality of clementine mandarin fruit juice. *Acta Horticulturae (ISHS)*, 2, pp. 149-157

Emblenton, T.W.; Reitz, H.J. & Jones, W.W. (1973). Citrus fertilization. In: *Citrus Industry.* Reuther W (Ed) 3, pp. 122-181

Fereres, E. & Soriano, M.A. (2007). Deficit irrigation for reducing agricultural water use. *Journal of Experimental Botany*, 58, pp. 147–159

Fernández, J.E.; Díaz-Espejo, A.; Infante, J.M.; Durán, P.; Palomo, M.J.; Chamorro, V.; Girón, I.F. & Villagarcía, L. (2006). Water relations and gas exchange in olive trees under regulated deficit irrigation and partial rootzone drying. *Plant and Soil*, 284, pp. 273-291

García-Tejero, I. (2010). Deficit irrigation for sustainable Citrus cultivation in Guadalquivir river basin. Ph. D. Thesis. Universidad de Sevilla, Spain 285 pp.

García-Tejero, I.; Durán Zuazo, V.H.; Jiménez Bocanegra, J.A. & Muriel Fernández, J.L. (2011). Improved water-use efficiency by déficit irrigation programmes: Implications for saving water in citrus orchards. *Scientia Horticulturae* 128, pp. 274-282

Gelly, M.; Recasens, I.; Mata, M.; Arbones, A.; Rufat, J.; Girona, J. & Marsal, J. (2003). Effects of water deficit during stage II of peach fruit development and postharvest on fruit quality and ethylene production. *J Hort Sci Biotechnol*, 78, pp. 324-330

Ginestar, C. & Castel, J.R. (1996). Responses of young Clementine citrus trees to water stress during Different phenological periods. *Journal of Horticultural Science*, 71, pp. 551–559

Goldhamer, D.A. & Beede, R.H. (2004). Regulated deficit irrigation effects on yield, nut quality and water-use efficiency of mature pistachio trees. *Journal of Horticultural Science and Biotechnology*, 79, pp. 538–545

Goldhamer, D.A. & Salinas, M. (2000). Evaluation of regulated deficit irrigation on mature orange trees grown under high evaporative demand. In: *Proceedings of the International Society of Citriculture*, IX Congress. Orlando, FL: ISC, pp. 227–231

Goldhamer, D.A. & Viveros, M. (2000). Effects of preharvest irrigation cut off durations and postharvest water deprivation on almond tree performance. *Irrigation Science*, 19, pp. 125-131

Goldhamer, D.A.; Viveros, M. & Salinas, M. (2006). Regulated deficit irrigation in almonds: effects of variations in applied water and stress timing on yield and yield components. *Irrigation Science*, 24, pp. 101–114

González-Altozano, P. & Castel, J.R. (1999). Regulated deficit irrigation in Clementina de Nules' citrus trees. I. Yield and fruit quality effects. *Journal of Horticultural Science and Biotechnology* 74, pp. 706–713

González-Sicilia, E. (1968). In: *El cultivo de los agrios*. 3rd. edición.Bello (Ed). Valencia

Grattan, S.R.; Schwankl, L. J. & Lanini, W.T. (1988). Weed control by subsurface drip irrigation. *Calif. Agric.* 42(3), pp. 22-24

Hutmacher, R.B.; Phene, C.J.; Davis, K.R.; Vail, S.S.; Kerby, T.A.; Peters, M.; Hawk, C.; Keeley, A.M.; Clark, D.A.; Ballard, D. & Hudson, N. (1995). Evapotranspiration, fertility management for subsurface drip Acala and Pima cotton. In: *Proc. 5th Int'l Microirrigation Congress*, pp. 147-154. St. Joseph, Mich.: ASAE

Ibrahim, A.M. & Abd El-Samad, G.A. (2009). Effect of different irrigation regimes and partial substitution of N-Mineral by organic manures on water use, growth and productivity of Pomegranate Trees. *European Journal of Scientific Research,* 38(2), pp. 199-218. ISSN 1450-216X

Keller, J. & Karmelli, D. (1974). Trickle irrigation design parameters. *Transactions of the ASAE* 17, pp. 678-684

Koo, R.C.J. & Smajstrla. A.G. (1985). Effects of trickle irrigation and fertigation on fruit production and juice porduction and juice quality of 'Valencia' orange. Proc. Fla. State Hort. Soc., 97, pp. 8-10

Ladaniya, M. (2008). Fruit quality control, evaluation and analysis. In Citrus fruit: biology, technology and evaluation, pp. 475-500. (Ed Elsevier Academic Press., San Diego CA, USA

Lamm, F.R. & Camp, C.R. (2007). Subsurface drip irrigation. Microirrigation for crop production design, operation and management, Lamm, F.R., Ayars, J.E., and Nakayama, F.S. (Eds.) Elsevier, pp: 473-551

Legaz, F. & Primo-Millo, E. (1988). In: Guidelines for Citrus fertilization. Technical Report, Department of the Agriculture, Fish and Food of the Valencian Government, Valencia, Spain, No 5-88 (In Spanish)

López, G;, Rabones, A.; del Campo, J.; Mata, M.; Vallverdú, X.; Girona, J. & Marsal, J. (2008). Response of peach trees to regulated deficit irrigation during stage II of fruit development and summer pruning. *Spanish Journal of Agricultural Research,* 6, pp. 479-491

McCarthy, M.G.; Loveys, B.R.; Dry, P.R. & Stoll, M. (2002). Regulated deficit irrigation and partial rootzone drying as irrigation management techniques for grapevines. In: *Deficit irrigation practices,* FAO Water Reports No. 22. Rome, Italy: FAO, pp. 79–87

MARM. (2010). Ministry of Environment, rural and marine. Agricultural Report

Martínez, J.M.; Bañuls, J.; Quiñones, A.; Martín, B.; Primo-Millo, E. & Legaz, F. (2002). Fate and transformations of 15N labelled nitrogen applied in spring to Citrus trees. *Journal of Horticultural Science & Biotechnology* 77, pp. 361-367

Mitchell, P.D. & Chalmers, D.J. (1982). The effect of reduced water supply on peach tree growth and yields. *Journal of the American Society of Horticultural Science* 107, pp. 853–856

Mitchell, P.D.; van den Ende, B.; Jerie, P.H. & Chalmers, D.J. (1989). Response of 'Bartlett' pear to withholding irrigation, regulated deficit irrigation, and tree spacing. *Journal of the American Society of Horticultural Science,* 114, pp. 15–19

Moriana, A.; Orgaz, F.; Pastor, M. & Fereres, E. (2003). Yield responses of mature olive orchard to water deficits. *Journal of the American Society for Horticultural Science,* 123, pp. 425–431

Ollier, C; Cardoso, F. & DrInu, M. (2009). Summary results of the EU-27 orchard survey In: *Statistics in Focus, Agriculture and Fisheries.* Eurostat. European Commission.

Oron, G.; DeMalach, Y. & Hoffman, Z. (1989). Subsurface trickle irrigation of alfalfa with treated wastewater. In: *Progress Report.* Israel: Ben Gurion Univ., Institute of Desert Research.

Peng Y.H. & Rabe. E. (1998). Effect of differing irrigations regimes on fruit quality, yield, fruit size and net CO_2 assimiliation on "Mihbowase" satsuma. *J. Hort. Sci. Biotecnology*, 73(2), pp. 229-234

Phene, C.J. (1995). The sustainability and potential of subsurface drip irrigation. In: *Proc. 5th Int'l Microirrigation Congress*, pp. 359-367. St. Joseph, Mich. ASAE.

Phene, C.J.; Davis, K.R. & McCormick, R.L. (1987). Evapotranspiration and irrigation scheduling of drip irrigated cantaloupes. ASAE Paper No. 87-2526. St. Joseph, Mich.: ASAE

Phene, C.J.; Hutmacher, R.B. & Ayars, J.E. (1992). Subsurface drip irrigation: Realizing the full potential. In: *Proc. of Conference on Subsurface Drip Irrigation*, pp. 137-158. CATI Publication 921001. Fresno, Calif.: California State Univ. USA

Pérez-Sarmiento F.; Alcobendas, R; Mounzer, O; Alarcón, J. & Nicolás, E. (2010). Effects of regulated deficit irrigation on physiology and fruit quality in apricot trees. *Spanish Journal of Agricultural Research*, 8(S2), pp. S86-S94

Romero, P.; Botia, P. & Garcia, F. (2004). Effects of regulated deficit irrigation under subsurface drip irrigation conditions on water relations of mature almond trees. *Plant and Soil*, 260, pp. 155-168

Ruiz-Sánchez, M.C.; Torrecillas, A.; Pérez-Pastor, A. & Domingo, R. (2000). Regulated deficit irrigation in apricot trees. *Acta Horticulturae*, 537, pp. 759–766.

Sánchez-Blanco, M.J.; Torrecillas, A.; León, A. & del Amor, F. (1989). The effect of different irrigation treatments on yield and quality of Verna lemon. *Plant and Soil*, 120, pp. 299-302

Scandella, D.; Kraeutler, E. &V´enien, S. (1997). Anticiper la qualit´e gustative des p^eches et nectarines. *Infos CTIFL129*, pp. 16–19 (In French).

Schumann, A.W.; Syvertsen, J.P. & Morgan, K.T. (2009) Implementing Advanced Citrus Production Systems in Florida-Early Results. *Proc Fal State Hort Soc* 122, pp. 108-113

Sne, M. (2006). Micro irrigation in arid and semi-arid regions. In Kulkarni SA (Ed.) Guidelines for planning and design. In: *International Commission on Irrigation and Drainage*. New Delhi, India

Sterret, S.B.; Ross, B.B.& Savage, C.P. (1990). Establishment and yield of asparagus as influenced by planting and irrigation method. *J. American Soc. Hort. Sci.*, 115(1), pp. 29-33

Quiñones, A.; Martínez-Alcántara, B. & Legaz, F. (2007) Influence of irrigation system and fertilization management on seasonal Distribution of N in the soil profile and on N-uptake by citrus trees. *Agri Ecosyst Environ* 122, pp. 399-409

Quiñones, A.; Martínez-Alcántara, B.; San-Francisco, S.; García-Mina, J.M. & Legaz, F. (2011), Methyl Xanthine as a potential alternative to Gibberellic acid in enhancing fruit set and quality in clementine citrus trees in Spain. *Experimental Agricultural*, 47(1), pp. 159–171, doi:10.1017/S0014479710000906.

Tognetti, R.; d'Andria, R.; Morelli, G. & Alvino, A. (2005). The effect of deficit irrigation on seasonal variations of plant water use in *Olea europaea* L. *Plant and Soil*, 273, pp. 139-155

Tognetti, R.; d'Andria, R.; Lavini, A. & Morelli. G. (2006). The effect of deficit irrigation on crop yield and vegetative development of Olea europaea L. (cvs. *Frantoio and Leccino*). *European Journal of Agronomy*, 25, 356-364

Velez, J.E.; Intrigliolo, D.S. & Castel, J.R. (2007) Scheduling deficit irrigation of citrus trees with maximum daily trunk shrinkage agricultural water management 90, pp. 197-204

Wallace, J.S. & Batchelor, C.H. (1997). Managing water resources for crop production, Philos. Trans. R. Soc. London Ser. B. 352, pp. 937–947

Yazar, A.; Sezen, S.M. & Sesveren S. (2002). LEPA and trickle irrigation of cotton in the Southeast Anatolia. Project (GAP) area in Turkey. *Agricultural Water Management*, 54, pp. 189-203

4

Crop Evapotranspiration and Water Use Efficiency

Bergson Guedes Bezerra
National Institute of Semi Arid (INSA)
Brazil

1. Introduction

The efficient use of water in any sector of human activity has become an increasingly important need in our daily lives, especially in arid and / or semi-arid regions where water resources have become increasingly scarce. In irrigated agriculture this concern becomes more relevant because globally, water for agriculture is the primary user of diverted water, reaching a proportion that exceeds 70–80% of the total water resources in the arid and semi-arid zones (Fereres & Soriano, 2007).

Currently, world food production from irrigated agricultural represents >40% of total, with this coming form only about 17% of the total land area devoted to food production. However, these percentages tend to increase due to increased human population and climate change, because there is widespread agreement that increasing anthropogenic climate change will exacerbate the present shortages of water, and are likely to increase drought (Intergovernmental Panel on Climate Change [IPCC], 2007). According to Perry et al. (2008) as a consequence of climate change, some areas will receive higher rainfall but most of the currently water-scarce regions will become drier and warmer. These two changes will exacerbate scarcity: reduced rainfall means less flow in rivers; higher temperatures mean increased evaporation and water consumption of natural water demand for agriculture use. On the other hand, the increase in population will result in a greater demand for food. Thus, the competition for water intensifies worldwide, water for food production must be used more efficiently (Steduto et al., 2007). Another concern with water use for irrigated agriculture is the question of sustainability because food production tends to rely increasingly on irrigation. In developing countries, agriculture continues to be an important economic sector as it constitutes a significant contribution to national incomes and economic growth and provides livelihood support for 60–80% of the population (Hussain et al., 2007).

One way to achieve greater water use efficiency in irrigation is switching from the less efficient flood or furrow system to more efficient systems such as microirrigation or to adopt irrigation strategies, such as deficit irrigation, in order to maximize crop yield and/or minimize water losses. According to Perry et al. (2009) switching from flood or furrow to low-pressure sprinkler systems reduces water use by an estimated 30%, while switching to drip irrigation typically cuts water use by half. In addition to having a direct relationship to the total water used, irrigation systems also have a bearing on the crop yield. Cetin & Bilgel (2002) evaluated the effect of the irrigation system on cotton yield and concluded that drip irrigation produced 21% more seed-cotton than with a furrow system and 30% more than with a sprinkler system.

The most fundamental requirement of scheduling irrigation is the determination of crop evapotranspiration, ETc. According to Allen et al. (1998), evapotranspiration is not easy to measure, because specific devices and accurate measurements of various physical parameters or the soil water balance in lysimeters are required to determine evapotranspiration. The two-step crop coefficient (Kc) versus reference evapotranspiration (ET$_0$) method is a practical and reliable technique for estimating ETc, and it is being widely used. Besides the accuracy and reliability, the advantage of this method is related to the fact that is inexpensive, requiring only meteorological data to estimate ET$_0$ which is then multiplied by a crop coefficient to represent the relative rate of ETc under a specific condition (Allen et al., 1998). Additionally, the knowledge of the Kc for each specific crop growth stage is necessary. This inexpensive method makes it popular, accessible and vastly applied by the small farmers which have restricted financial resources. However, several methods which measure evapotranspiration indirectly have been proposed, such as the micrometeorological methods. All these methods present advantages, disadvantages and limitations but generally provide reasonable accuracy.

This chapter presents a review on evapotranspiration and its importance for agricultural water management. This review will focus on the concepts and main methods of estimating crop evapotranspiration. The strategies for improving water use efficiency such as through deficit irrigation and partial root-zone irrigation beyond of irrigation performance indicators will also be discussed.

2. Crop evapotranspiration – Concepts

The crop evapotranspiration is defined like the water transferred to atmosphere by plant transpiration and surface evaporation. The evapotranspiration term was proposed by Thornthwaite (1948) to conceptualize the process of plant transpiration and surface evaporation which occurs simultaneously and no easy way of distinguish them. The evapotranspiration process occurs naturally only if there is inflow of energy in the system, from the sun, atmosphere, or both, and is controlled by the rate of energy in the form of water vapor that spreads from the surface of the Earth (Tucci and Beltrame, 2009). This transfer takes place physically in the forms of molecular and turbulent diffusion.

According to Allen et al. (1998), evaporation is the process whereby liquid water is converted to water vapor (vaporization) and removed from the evaporating surface (vapor removal). Water evaporates from a variety of surfaces, such as lakes, rivers, pavements, soils and wet vegetation. Transpiration, in turn, is the transfer of water from plants through their aerial parts. Water transfer from plant to atmosphere occurs mainly through the stomata through which they pass more than 90% water transpired.

The evapotranspiration can be derived from a range of measurement systems including lysimeters, eddy covariance, Bowen ratio, water balance (gravimetric, neutron meter, other soil water sensing), sap flow, scintillometer and even satellite-based remote sensing and direct modeling (Allen et al., 2011). The micrometeorological method of Bowen Ratio Energy Balance (BREB) has been widely applied due its relative simplicity, practicality, robustness and accuracy. Recently, the Eddy Covariance method has been increasingly applied mainly after the sensors have reduced costs.

However, the most popular method for crop evapotranspiration estimates is the crop coefficient (K_c), which is defined like the ratio between crop and reference evapotranspiration. The K_c method has the advantage of being inexpensive, since it requires only daily weather data to estimate the reference evapotranspiration which is multiplied by a Kc dimensionless value. The Kc value depends of crop growth stage.

2.1 Crop coefficient (K_c)

The K_c concept was introduced by Jensen (1968) and is widely discussed and refined by the Food and Agricultural Organization (FAO) in its Bulletin-56 (Irrigation and Drainage Paper, Allen et al., 1998). In the crop coefficient approach the crop evapotranspiration is calculated by multiplying the reference evapotranspiration, ET_0 (mm d^{-1}), by a crop coefficient, Kc (dimensionless) according to Equation 01.

$$ET = K_c.ET_0 \qquad (1)$$

According to Allen et al. (1998) the effects of the various weather conditions are incorporated into the ET_0 estimate. Thus, ET_0 represents an index of climate demand and K_c varies predominately with the specific crop characteristics and only to a limited extends with climate. The FAO-56 report K_c values for the initial, middle and end growth stages, K_c-ini, K_c-mid and K_c-end, respectively, for many crops. However, the K_c values presented in FAO-56 (Table 12) are expected for a sub-humid climate with average daily minimum relative humidity (RH_{min}) values of about 45% and calm to moderate wind speed (u_2) averaging 2 m s^{-1}. For humid, arid and semiarid climates it has been suggested corrections to their values according to equations proposed in FAO-56 (Allen et al., 1998). However, the use of these values can contribute to ETc estimates which are substantially different from actual ETc (Hunsaker et al., 2003), because it has been demonstrated that K_c-ini, K_c-mid and K_c-end values experimentally determined differ from those values listed in the FAO-56. Farahani et al. (2008) compared the cotton evapotranspiration obtained based on FAO-56 Kc and Kc obtained experimentally and found differences ranging from 10 to 33% in three years of observation, in a Mediterrenean environments. Thus, to the accurate application of this methodology, it is necessary to obtain the K_c curve values experimentally, to represent the local weather and water management conditions. Allen et al. (1998) suggest accurate evapotranspiration observed experimentally during years multiples. The Kc values has been experimentally obtained using the evapotranspiration derived from micrometeorologicals methods such as Bowen Ratio Energy Balance (BREB) (Inmam-Bamber & McGlinchey, 2003; Hou et al., 2010; Bezerra et al., 2010)

2.2 Bowen Ratio Energy Balance (BREB)

The crop evapotranspiration is estimated by use of the BREB technique from latent heat flux density which is derived from energy balance equation (Equation 2). Neglicting the advection effects, energy stored in the canopy, and photosysthetic energy flux, the energy balance can be writted as follows:

$$Rn = LE + H + G \tag{2}$$

where: Rn is the net radiation flux density, LE is latent heat flux density, H is sensible heat flux density, and G is soil heat flux density.

Use of the BREB concept (β = H/LE $\rightarrow$ H = β.LE) (Bowen, 1926) enables solving the energy balance and the latent heat flux density can be written, as:

$$LE = \frac{Rn - G}{1 + \beta} \tag{3}$$

The crop evapotranspiration is estimates dividing the Equation 3 by latente heat of vaporização (L = 2.501 MJ kg^{-1}). The Bowen ratio (β) is calculated assuming equality between the turbulent exchange coefficients for heat and water vapor, according to equation following:

$$\beta = \gamma . \Delta T / \Delta e \tag{4}$$

where γ is psichometric constant, ΔT and Δe are the gradients of air temperature and vapor pressure above canopy, respectively.

The application of BREB method requires horizontal advection to be negligible when compared to the magnitude of the vertical fluxes. In this case, the closure of the energy balance equation for an imaginary plane located above the canopy must be satisfied (Figuerola & Berliner, 2006). The disregard of advection effects should be one relevant worry mainly in regions which presenting advection natural events coming to regional circulation. In this case, the ETc has been understimated when compared to lysimeter measurements ranging from 5 to 20% (Blad & Rosenberg, 1974; Gavilán & Berengena, 2007). However, if the advection is originated from local circulation, the effects can be compensated taking some precautions in the instalation procedure. On precautions is to establish an equilibrium boundary layer (EBL). This equilibrium can be achieved providing uniform fetch of sufficient distance from boundary field in the predominant wind direction (Allen et al., 2011). Another caution is to stablish sufficient elevation above the canopy to avoid the roughness sublayer. This elevation is quite varied because the height where the energy closure is calculated may vary because it depends on the distance to the border of the field, humidity conditions of the soil, plant density and height, and also, the energetic and dynamic conditions of the flow field (Allen et al., 2011; Figuerola & Berliner, 2006). According to Steduto & Hsiao (1998) this technique must be used with caution since it does not reproduce the turbulent nature of the evapotranspiration process.

Another relevant limitation was detected on the data colected during nigh-time period and periods during precipitation and irrigation events. According to Perez et al. (1999) the data observed in this periods must be rejected, which corresponds to 40% of the total data.

For application of Bowen ratio technique is necessary accomplish measurements of net radiation and soil heat flux. Additional measures of the air temperatures of the dry and wet bulbs at two levels above canopy. The air temperature wet bulb are used to calculated water vapor pressure.

A accuracy of ETc provided by BREB method depends of accuracy and representativeness to measurements of net radiation (Rn) and soil heat flux (G) (Allen et al., 2011). This dependence is considered a disadvantage of the BREB method mainly related to representativeness of soil heat flux measurements, which is not easy to get due to ground cover provides by crops is not always heterogeneous. Another relevant factor is the difference of scales of Rn and G measurement. While G measurements are representative for a specific location of the field Rn measurements are originate from several hundred meters upwind.

On the other hand Allen et al. (2011) lists some potential advantages of BREB method, which are:

- non-destructive, direct sampling of the turbulent boundary layer;
- no aerodynamic data are required;
- simple measurement of temperature and vapor pressure at two heights;
- can measure ET over both potential and non-potential surfaces;
- gradient-based fluxes are averaged over a medium sized area (200–100,000 m²);
- automated.

Additionally it's relative simplicity of method made it widely applied (Bezerra et al., 2010, Gavilán & Berengena, 2007, Hou et al., 2010, Silva et al., 2007, Steduto & Hsiao, 1998, Todd et al., 2000).

2.3 Eddy covariance method

The Eddy Covariance method is one of the most direct, defensible ways to measure and calculate turbulent fluxes within the atmospheric boundary layer (Burba & Anderson, 2007). Flux measurements using the eddy-covariance method are a direct measuring method without any applications of empirical constants (Folken, 2008, Lee et al., 2004). However, the method is mathematically complex and requires significant care to set up and process data. The main challenge of the method for a non-expert is the complexity of system design, implementation, and processing of the large volume of data (Arya, 2001; Burba & Anderson, 2007; Stull, 1998). According to Allen et al. (2011), the concept of eddy covariance draws on the statistical covariance (correlation) between vertical fluxes of vapor or sensible heat within upward and downward legs of turbulent eddies.

The eddy covariance method assumes that all atmospheric entities show short-period fluctuations about their longer term mean value (Oke, 1978). Still according to Oke (1978) this is the result of turbulence which causes eddies to move continually around carrying with them their properties derived elsewhere. Therefore the value of an entity variable in time (s) consists of its mean value ($\bar{s}$), and a fluctuating part (s'), according ilustrate the Fig. 1.

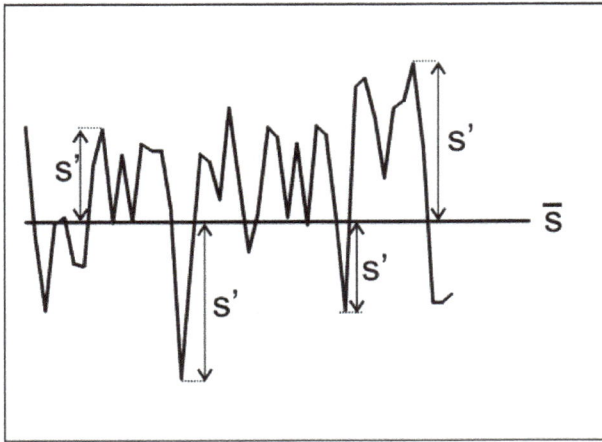

Fig. 1. Instantaneous values of turbulent variable is obtained by its mean ($\bar{s}$) and flutuing values (s')

Thus, its instantaneous value is obtained from following equation, also known as Reynold's decomposition (Arya, 2001; Folken, 2008; Oke, 1978,):

$$s = \bar{s} + s' \tag{5}$$

where the overbar indicates a time-averaged property and the prime signifies instantaneous deviation from the mean.

The air flow over an agricultural ecosystem can be understood as a horizontal flow of numerous rotating eddies, according to Fig. 2. Each eddy has three-dimensional components, including vertical movement of the air (Burba & Anderson, 2007).

Fig. 2. Air flow over an agricultural ecossystem (Source: Burba & Anderson, 2007)

From these consideration the eddy covariance method is based in the covariance between the properties contained by, and therefore transported by a eddy, which are its mass (which by considering unit volume is given by its density, ρ), its vertical velocity (w) and the volumetric content of any entity it possesses (s). Thus, the turbulent flux of any variable (momentum flux, latent and sensible heat fluxes and CO_2 flux) is equal to product of mean values of variables transported by eddies, according to to following equation:

$$F = \overline{\rho_a . w. s} \tag{6}$$

Applyind the Reynold's decomposition to break into means and deviations:

$$F = \overline{(\overline{\rho_a} + \rho'_a)(\overline{w} + w')(\overline{s} + s')} \tag{7}$$

Opening the parentheses:

$$F = \overline{(\overline{\rho_a}\overline{w}\overline{s} + \overline{\rho_a}\overline{w}s' + \overline{\rho_a}w'\overline{s} + \overline{\rho_a}w's' + \rho'_a\overline{w}\overline{s} + \rho'_a\overline{w}s' + \rho'_a w'\overline{s} + \rho'_a w's')} \tag{8}$$

Although Equation 8 looks quite complex Oke (1978) outlines some considerations which allow simplify it. These considerations are sequentially presented:

1. First, all terms involving a single primed quantity are eliminated because by definition the average of all their fluctuations equals zero (i.e. we lose the second, third and fifth terms);
2. Second, we may neglect terms involving fluctuations of ρ_a since air density is considered to be virtually constant in the lower atmosphere (i.e. we lose the sixth, seventh and eighth terms);
3. Third, if observations are restricted to uniform terrain without areas of preferred vertical motion (i.e. no 'hotspots' or standing waves) we may neglect terms containing the mean vertical velocity (i.e. we lose the first term).

With these assumptions the Equation 7 is reduced to fourth term, but dropping the bar over the ρ_a because it is considered to be a constant.

$$F = \rho_a \overline{w's'} \tag{9}$$

where $\overline{w's'}$ is the covariance between vertical wind speed and between any variable which depends of user interest. In the specific case of evapotranspiration the covariance is between vertical wind speed (w) and specific humidity (q), according to following equation:

$$ET = \rho_a \overline{w'q'} = \frac{0.622}{P} \rho_a \overline{w'e'} \tag{10}$$

where P is atmospheric pressure, $\overline{w'e'}$ is covariance between vertical wind speed and actual vapor pressure.

The covariance indicates the degree of common relationship between the two variables, "a" and "b" (Stull, 1998; Wilks, 2006). In statistics, the covariance between two variables "a' and "b" is defined as:

$$\text{covar}(a, b) = \frac{1}{N} \sum_{i=0}^{N-1} (a - \overline{a}).(b - \overline{b}) \tag{11}$$

Applying the Reynold's decomposition appears;

$$\text{covar}(a, b) = \frac{1}{N}\sum_{i=0}^{N-1}(\bar{a} + a' - \bar{a}).(\bar{b} + b' - \bar{b}) \tag{12}$$

Thus, we can show, that:

$$\text{covar}(a, b) = \frac{1}{N}\sum_{i=0}^{N-1}a'b' = \overline{a'b'} \tag{13}$$

The data processing is mathematically complex. Several mathematical operations and assumptions, including Reynolds decomposition, are involved in getting from physically complete equations of the turbulent flow to practical equations for computing "eddy flux" (Burba & Anderson, 2007). This mathematical operations use existing methodologies for the control and certification of data quality, such as crosswind correction of sonic anemometer (if not already implemented in the software of sensor), coordinates transformations, spectral corrections, conversions of sonic temperature fluctuations in the actual temperature fluctuations and corrections to the scalar densities of the water vapor flux density based Webb et al. (1980) and described in details by Lee et al. (2004). Several software programs to process eddy covariances and derive quantities such as heat, momentum, and gaseous fluxes. Currently (2011) there are several software programs to process eddy covariances and derive quantities such as heat, momentum, and gaseous fluxes. Examples include EdiRe, ECpack, TK2, Alteddy, EddyPro and EddySof. The eddy covariance method requires high speed measurement of T, w, and and q for evapotranspiration estimates. According to Allen et al. (2011) usually at frequencies of 5-20 Hz (5-20 times per second) using quick response sensors, but 10 Hz is common.

3. Water use efficiency

The increase in human population has caused increased demand for food. On the other hand, the shortage of drinking water in arid and / or semi-arid and is becoming increasing its use efficiently is becoming increasingly necessary. According to Perry et al. (2009) the competition for scarce water resources is already widely evident, from Murray Darling basin in Australia to rivers of the middle East, southern Africa and Americas, and from the aquiffers of northern India, to the Maghreb and the Ogallala in central of United States. The cause of much of the shortage of drinkable water has been the predatory exploitation of natural resources and the almost total absence of effective public policies for water resources management, especially in developing countries. When water use is not regulated or controlled, the imbalance between supply and demand is evident, and occurs as a consequence failling water tables, drying estuaries, inadequate to lower riparian and damaged aquatic ecosystems (Perry et al., 2009). Thus, water use efficiency whether in any human activity, domestic, industrial or agricultural, has become a necessity. The optimization of water use in irrigation has significant relevance in this context because it accounts for approximately 50% of the total world food production. Thus is currently is the main user of water worldwide, reaching a proportion that exceeds 80% of the total available in arid and semiarid. Another worrying fact is that the increase in population coupled with impacts of global climate change pose to global food security under threat (Strzepek & Boehlert, 2010). This threat comes from the increased demand for irrigated agriculture which increases in the same proportion of other sectors demand such as domestic and industrial.

When water supplies are limiting, the farmer's goal should be to maximize net income per unit water used rather than per land unit (Fereres & Soriano, 2007). Thus, producing more with relatively less water has become a challenge for irrigation sector (Kassam et al., 2007, Fereres & Soriano, 2007). The water productivity (WP) reflects this challenge by exposing the relationship between the net benefits of agriculture, forestry, fishing and / or livestock and the amount of water consumed to produce these benefits (Kassam et al., 2007, Molden, 1997, Steduto et al. 2007). In other words, WP represents the fresh crops (in kg ha^{-1}) produced per unit of water applied or consumed (in m^3 ha^{-1}) (Molden, 1997; Teixeira et al., 2009), according to equations followings:

$$WP_{ET} = \frac{Y(kg\ ha^{-1})}{ET(m^3 ha^{-1})} \tag{14}$$

$$WP_I = \frac{Y\ (kg\ ha^{-1})}{I\ (m^3 ha^{-1})} \tag{15}$$

where WP_{ET} is the WP calculate in terms of crop evapotranspiration, Y is the crop yield, ET is crop evapotranspiration, WP_I is WP calculated in terms of irrigation water applied and I is irrigation water applied.

Some authors (i.e., Droogers & Kite, 1999) recommend analyze the WP in terms of crop evapotranspiration (ET) because this indicator also includes non-irrigation water, such as rainfall, capillary rise, and soil moisture changes. However, according to Oweis et al. (2011), WP_{ET} is more a biological indicator while the WP_I is influenced by the performace irrigation system and the degree of water losses beyond transpiration.

So, the challenge of irrigation sector is producing more with relatively less water implies in increasing water productivity. Several strategies have been widely used at irrigation management for increase WP. The partial root zone irrigation (PRI) and deficit irrigation (DI) are the most used.

PRI is an irrigation practice with which only part of the rootzone is wetted through proper irrigation design and management while the rest of the root system is left in drying soil (Mavi & Tupper, 2004, Tang et al., 2010, Zhang et al., 2001). The dried and wetted side is irrigated by shift periodically according to the rate of soil drying and crop water consumption (Kang & Zhang, 2004, Tang et al., 2010). According to Zhang et al. (2001), this practice is predicted to reduce plant water consumption and maintain the biomass production according to two theoretical backgrounds. Firstly, fully irrigated plants usually have widely opened stomata. A small narrowing of the stomatal opening may reduce water loss substantially with little effect on the photosynthesis. Secondly part of the root system in drying soil can respond to the drying by sending a root sourced signal to the shoots where stomata may be inhibited so that water loss is reduced. In the field, however, this prediction may not be materialized because stomatal control is only part of the transpirational resistance. Because prolonged exposure to drying soil may cause anatomical changes in the roots, such as suberization of the epidermis, collapse of the cortex, and loss of succulent it is necessary to alternatively irrigate the different part of roots system so that the plants could be succulent enough to sense soil drying and produce root-sourced signal to regulate the opening of leaf stomata (Kang and Zhang, 2004; Zhang et al., 2001).

Other strategy which is most used for increase WP is the deficit irrigation (DI). DI is the application of water below ET requirement, i.e., the scheduling irrigation derived as fraction from full irrigation. In other words, DI is the application of only a predetermined percentage of calculated potential plant water use (Morison et al., 2008). DI is also mentioned in literature as regulated deficit irrigation (RDI). Thus, according to Fereres & Soriano (2007) to quantify the level of DI it is first necessary to define the full. It has been experimentally established that the DI translates increase of WP (Zwart & Bastiaanssen, 2004). Thus, has become an important strategy in the maintenance of agricultural production in arid and semi arid zones due of declining water resources in these areas. Besides the improvement of irrigation efficiency, the costs reduction and environmental benefits are potential virtues of DI practice. According to Fereres & Soriano (2007) there are several reasons for the increase in WP under DI. One of these reasons is the relationship between yield and irrigation water for a crop. Small irrigation amounts increase crop ET, more or less linearly up to a point where the relationship becomes curvilinear because part of the water applied is not used in ET and is lost. In this point yield reaches its maximum value and additional amounts of irrigation do not increase it any further. Still according to Fereres & Soriano (2007), the location of that point is not easily defined and thus, when water is not limited or is cheap, irrigation is applied in excess to avoid the risk of a yield penalty. These points are called I_W and I_M and indicate the point beyond which the water productivity of irrigation starts to decrease, and the point beyond which yield does not increase any further with additional water application, respectively (Fereres & Soriano, 2007). For investigate the reasons presented by Fereres & Soriano (2007) was simulated the effect of DI in the water balance components and yield of irrigated cotton crop in Brazilian Semiarid using the SWAP model. The SWAP model was calibrated and validated from data collected in two experimental campaigns carried out at the Rio Grande do Norte state, Northeast of Brazil. The procedure of SWAP model calibration and validation are described minutely in Bezerra (2011). The Table 1 shows the variables observed, measurement frequency, method and finality of each variable. The experimental area are located in west region of Rio Grande do Norte state. The soil texture of experimental area is sandy-clay-loam, according to USDA classification. The SWAP is a physical based, detailed agro-hydrological model that simulates vertical transport of water, solutes and heat in the saturated-unsaturated zone in relation to crop growth.

A first version of the SWAP model was developed by Feddes et al. (1978) with continuous development since. The version used for this study is SWAP 3.2 and is described by Kroes et al. (2008). According to Droogers et al. (2010) the SWAP requires various data as input which can be divided into state variables, boundary conditions (model forcing) and calibration/validation data. The most important state variables are related to soil and crop characteristics. The soil characteristics were often described by van Genuchten-Mualem (VGM) parameters which called hydraulic functions. The growth and yield of cotton crop were simulated using the detailed crop growth module which is based on the World Food Studies (WOFOST) model (Supit et al., 1994). The detailed crop growth is based on the incoming photosynthetically active radiation absorbed by crop canopy and photosynthetic characteristics of leaves, and accounts for water and salt stress of the crop (van Dam et al., 2008).

Data	Method	Frequency	Purpose
Meteorological data	Meteorology station	Daily	Input
Crop evapotranspiration	Bowen Ratio Energy Balance	Daily	Validation
Soil moisture	Probe of soil moisture profile	Two times for week	Calibration
Crop development stage, i.e. emergence, anthesis, maturity and harvest	Field observation	Once	Input
Leaf area	Leaf area meter	5 – 6 times	Input
Plant height	Field observation	5 – 6 times	Input
Dry matter portioning	Field observation and drying in oven	5 – 6 times	Input
Soil texture	Granulometric method and USDA classification	Once	Input
VGM parameters	Gravimetric method	Once	Input
Soil saturated conductivity	Porchet method	Once	Input
Irrigation depth	From crop coefficient	Weekly	Input
Irrigation date	Field observation	After each irrigation	Input
Crop Yield	Field observation	Once (harvest)	Validation

Table 1. Summary of data field observation for calibration and validation of the SWAP model

The SWAP model was calibrated for full irrigation condition. The scheduling of full irrigation was defined weekly using the crop coefficient method. The crop was irrigated using sprinkler irrigation system.

The irrigation depth of each treatment simulated was scheduled as full irrigation fraction, according showed in Table 2. Still in the Table 2 are showed the irrigation depth for each treatment. The irrigation frequency was same for all treatments.

The simulated water balance components for all treatments are presented in Table 3. These values correspond to mean of two study years.

Treatment	Water level (% of full irrigation)	Irrigation amount (mm)		
		2008	2009	Mean
DI_{40}	40	357.2	353.6	355.4
DI_{60}	60	535.8	530.5	533.2
DI_{75}	75	670.5	663.0	666.8
FI	100	894.0	884.0	889.0
FI_{130}	130	1161.0	1149.2	1155.1

Table 2. Irrigations treatments and its irrigation amount

The DI effects are evidenced in several water balance components. For example, the relative water use (RWU) (i.e., ratio between actual and potential transpiration) is lower than one, evidencing the water stress, which cotton crop has been submitted. However, in the full and excessive irrigation treatments (FI and FI_{130}, respectively), the relative water use lower than one is caused for water loss for drainage. These losses are evidenced by drainage at 100 cm (D) which increased as the applied irrigation has also increased. The water loss ranged 13.8 mm (DI_{40}) to 352 mm (FI_{130}). In the treatments under DI regimes the water loss by drainage was lower than 6% of irrigation water applied, while that in full and excessive irrigation treatments (FI and FI_{130}, respectively) the water loss by drainage was 14.9% and 31.1%, respectively.

Treatment	ΔW (mm)	Water balance component (mm)					
		I	D	T_P	T_A	E_S	RWU
DI_{40}	-9.5	355	13.8↓	256.1	136.2	208.8	0.53
DI_{60}	-25.4	533	31.2↓	373.1	264.4	226.1	0.71
DI_{75}	-25.6	674	34.2↓	466.3	389.4	235.6	0.84
FI	16.8	899	133.8↓	538.0	510.2	224.3	0.95
FI_{130}	18.9	1154	358.8↓	549.0	537.2	223.3	0.98

Table 3. Mean water balance component simulated by SWAP. ΔW = change water stored, I = irrigation, D = drainage at 100 cm or bottom flux, T_P = potential transpiration, T_A = actual transpiration, E_S = soil evaporation, RWU = relative water use = ratio between actual and potential transpiration and ↓ = downward flux.

The water stored was reduced at all deficit irrigation treatments, according to change water stored (ΔW), exhibiting of shortages of water. In the full and excessive irrigation treatments, the water stored increased and in this case can also be considered a loss. The soil evaporation, such as drainage and change water storage, also shows the evidences of deficit irrigation effect. In the DI_{40} treatment the soil evaporation corresponds to an amount in excess of 60% of ET. This ratio decreases with the irrigation increase in such way that decreases to 29% in the FI_{130} treatment. This decrease occurs due to increase crop growth as irrigation increases.

The greatest crop growth in line with the irrigation increase is evidenced by leaf area index (LAI) showed in Fig. 3. Usually the leaf area and consequently the LAI of cotton crop, increase with increasing soil moisture. The lower crop growth provides lower ground cover which favors water loss by soil evaporation mainly if sprinkler irrigation system was frequently used. The frequent use of sprinkler irrigation system causes surface wetting intense. Thus, when the crop does not provide full surface coverage, soil evaporation losses are inevitable (López-Urrea et al., 2009, Cavero et al., 2009).

The means values of irrigation (I), yield (Y), and evapotranspiration (ET) of cotton crop for all treatments simulated by SWAP are showed in Table 4. The increments in the irrigation depth implied in the increments in ET values, while increments of irrigation depth did not implied in increments in the yield values, corroborating with described by Fereres & Soriano (2007). Note that crop yield increased as the irrigation increased from treatments under DI regimes to full irrigation condition. In the excessive irrigation depth treatment the yield decreased and presented values lower than presents to DI_{75}. The yield loss verified in the treatment under excessive irrigation is attributed to excessive vegetative growth that causes reduction in the total number of bolls by plants which occurs due to increased competitiveness for assimilated available. This yield loss can also be attributed to the appearance of diseases and nematodes in the roots of the plant due to excessive soil moisture in the root zone.

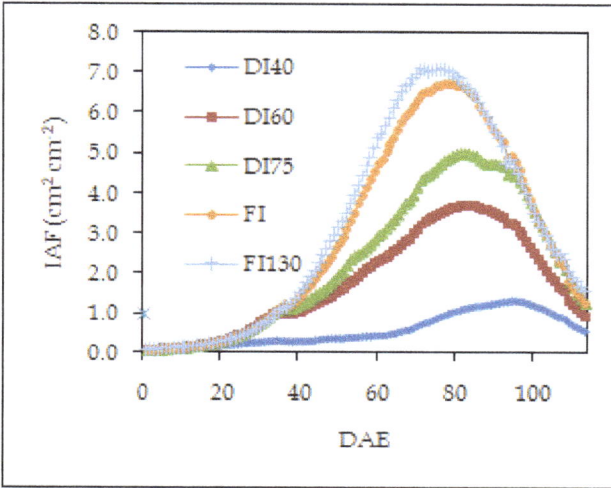

Fig. 3. IAF curves of all treatments simulated by SWAP

	I (mm)	Y (kg ha⁻¹)	ET (mm)
DI_{40}	356	1367	345
DI_{60}	533	2167	491
DI_{75}	666	3336	625
FI	899	3517	734
FI_{130}	1154	3317	761

Table 4. Irrigation depth, yield and evapotranspiration of cotton for each treatment simulated by SWAP model

The relationships between evapotranspiration and irrigation and between yield and irrigation are showed in Fig. 4 (top). The Fig. 4 (bottom) shows the relationship between WP_{ET} and irrigation and between WP_I and irrigation.

The curves of these relations expose the reasons pointed by Fereres & Soriano (2007) which guide the DI practices. As showed in Fig. 4 the points I_M and I_W do not was located in same treatment. The crop yield was increasing from DI_{40} and reached its maximum value at the FI treatment. In the FI_{130} treatment the yield decreased in relation to FI treatment and present yield similar to DI_{75}. The ET, in turn increased in all treatment, corroborating with Fereres & Soriano (2007), i.e., additional irrigation amounts causes increase of crop ET. From FI to FI_{130} crop ET increased, unlike yield. So in the FI treatment was identified the I_M point. The I_W point was located in the DI_{75}. In this treatment the performance indicators of

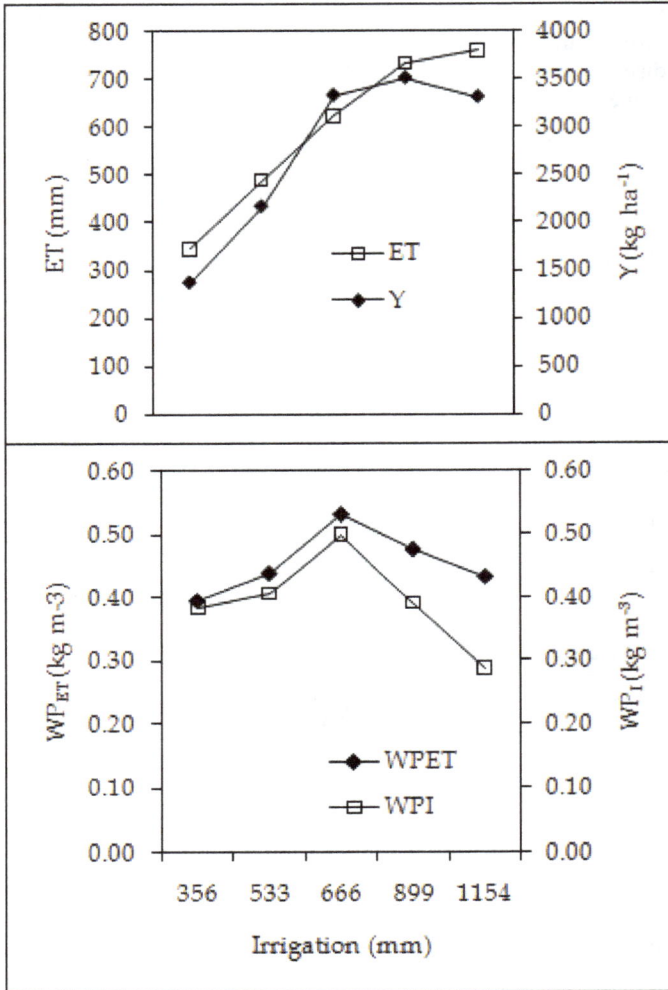

Fig. 4. Relationships between ET and irrigation and between yield and irrigation (upper) and WP_{ET} and WP_I curves of cotton crop in Brazilian semi arid

irrigations WP_{ET} and WP_I achieve its maximum values. The average WP_{ET} varied from 0.395 to 0.535 kg m^{-3} in both years (Table 5). WP_{ET} for the DI_{75} treatment was the largest, while for DI_{40} it was the smallest in both years. The WP_{ET} values increased with increasing water stress. However, from FI_{75} WP_{ET} decreased in such way that in FI_{130} excessive water treatment WP_{ET} value was equal to DI_{60}. Note that WP_I, unlike other studies (Dağdelen et al., 2009, Du et al., 2008, Ibragimov et al., 2007, Singh et al., 2010), was always less than WP_{ET}. This behavior could be attributed to groundwater depth of Brazilian semi arid which becomes impossible capillary rise. So, the water consumed by plants is restricting to irrigation water supplied.

	WP_{ET} (kg m^{-3})			WP_I (kg m^{-3})		
	2008	2009	Mean	2008	2009	Mean
DI_{40}	0.39	0.40	0.395	0.37	0.40	0.385
DI_{60}	0.43	0.45	0.440	0.37	0.44	0.405
DI_{75}	0.53	0.54	0.535	0.46	0.54	0.500
FI	0.48	0.48	0.480	0.38	0.41	0.395
FI_{130}	0.44	0.44	0.440	0.28	0.29	0.285

Table 5. Water Productivity values for cotton crop simulated by SWAP

These WP_{ET} values of cotton in semi arid lands of Brazil according to SWAP simulations were lower than most those of other studies in different regions (Table 6). This should be attributed to used irrigation system because in general the studies which used irrigation system less efficient, for example flood and furrow (Jalota et al., 2006, Saranga et al., 1998, and Singh et al., 2010) presented the worst performance.

Source	Irrigation system	WP_{ET} (kg m^{-3})	WP_I (kg m^{-3})
In this study	Sprinkler	0,39 – 0,54	0,38 – 0,50
Dağdelen et al. (2006)	Furrow	0,61 – 0,72	0,77 – 1,40
Dağdelen et al. (2009)	Drip	0,77 – 0,96	0,82 – 1,44
Du et al. (2008)	Drip	0,52 – 0,79	1,07 – 1,51
Ibragimov et al. (2007)	Drip	0,63 – 0,88	0,82 – 1,12
Ibragimov et al. (2007)	Furrow	0,46 – 0,50	0,55 – 0,62
Jalota et al. (2006)	Flood	0,26 – 0,31	0,25 – 0,87
Karam et al. (2006)	Drip	0,80 – 1,30	-
Saranga et al. (1998)	Furrow	0,22 – 0,35	-
Singh et al. (2010)	Drip	0,39 – 0,42	0,54 – 0,65
Tang et al. (2010)	Furrow	0,54 – 0,76	-
Ünlü et al. (2007)	Furrow	0,19 – 0,53	0,11 – 0,81
Yazar et al. (2002)	Drip	0,50 – 0,74	0,60 - 0,81
Yazar et al. (2002)	LEPA	0,55 – 0,68	0,58 – 0,78

Table 6. Comparison of WP_{ET} and WP_I values with other studies

4. Conclusions

The current global scenario on disputes over scarce water resources associated with climate change highlights the urgent need for the efficient water use in any human activity. In irrigated agriculture this need is even more relevant because more than 40% of total world food production comes from agricultural lands under irrigation. The water use efficiency in irrigated agriculture can be improved through the adoption of strategies, such as deficit irrigation (DI). It has been established that this can result in increased water productivity (WP). The use of DI as tool for agriculture water management was evaluated through simulations of the SWAP model. The results indicate that with DI it was able to increase WP of cotton crop in conditions of semi arid lands in Brazil, conforming to what has been reported in the literature. However, DI do not result in an increase of cotton yield per area used, but increased the relation between yield and consumed water. According to Fereres &

Soriano (2007) when water for irrigation is limited the goal should be to maximize net income per unit water used rather than per land unit. Thus, the DI strategy meets this criteria establishing water economy and achieving efficiency.

For the case of cotton crop in semi arid lands of Brazil, the treatment which presented superior performance was the DI_{75} treatment. In this treatment only 5% of irrigation water supplied was percolated, while in the full irrigation and excessive water (FI and FI_{130}) the percolated volume was more than 10%, resulting in substantial water losses and thus efficiency. Finally, based on the results DI practices it is recommended as an important tool for optimization in agriculture water management.

5. References

Allen, G.R., Pereira, S.L., Raes, D., Smith, M. (1998). Crop evapotranspiration: Guidelines for computing crop water requerements. *Food and Agrocultural Organization of the United Nations*, FAO-56. Rome. 300p.

Allen, R.G., Pereira, L.S., Howell, T.A., Jensen, M.E. (2011). Evapotranspiration information reporting: I. Factors governing measurement accuracy. *Agricultural Water Management*, Vol.98, No.6, (April 2011), pp. 899-920, ISSN 0378-3774,

Arya, S.P. (2001). *Introduction to Micrometeorology*, Academic Press, ISBN 0-12-059354-8, San Diego, United States.

Bezerra, B.G. (2011). *Avaliação da eficiência de uso da água no algodoeiro irrigado na Chapada do Apodi segundo simulações do modelo SWAP* (Evaluation of the water use efficiency for irrigated cotton crop at the Apodi Plateau second simulation of SWAP model). PhD thesis, Federal University of Campina Grande, Academy Unit of Atmospheric Science, Campina Grande, Brazil.

Bezerra, J.R.C., Azevedo, P.V., Silva, B.B., Dias, J.M. (2010). Evapotranspiração e coeficiente de cultivo do algodoeiro BRS-200 Marrom, Irrigado. *Revista Brasileira de Engenharia Agrícola e Ambiental*, Vol.14, No.6, (June 2010), pp. 625-632, ISSN 1807-1929.

Blad, B.L., Rosenberg, N.J. (1974). Lysimetric calibration of the Bowen Ratio-Energy Balance method for evapotranspiration estimation in the Central Great Plains. *Journal of Applied Meteorology*, Vol.13, No.2, (March 1974), pp. 227-236, ISSN 0894-8763

Bowen, I.S. (1926). The ratio of heat losses byu conduction and by evaporation from any water surface. *Physics Review*, Vol.27, No.6, (March 1926), pp. 779-787, ISSN 1063-651X.

Burba, G., Anderson, D. (2007). *A Brief Practical Guide to Eddy Covariance Flux Measurements: Principles and workflows examples for scientific and industries applications*, Li-Cor Biosciences, Retrieved from <https://www.licor.com/env/applications/eddy _covariance/book.jsp>.

Cavero, J., Medina, E.T., Puig, M., Martínez-Cob, A. (2009). Sprinkler irrigation changes maize canopy microclimate and crop water status, transpiration, and temperature. *Agronomy Journal*, Vol.101, No.4, (July 2009), pp. 854-864, ISSN 1435-0645.

Cetin, O., Bilgel, L. (2002). Effects of different irrigation methods on shedding and yield of cotton. *Agricultural Water Management*, Vol.54, No.1, 4, (March 2002), pp. 1-15, ISSN 0378-3774.

Dağdelen, N., Başal, H., Yilmaz, E., Gürbüz, T., Akçay, S. (2009). Different drip irrigation regimes affect cotton yield, water use efficiency and fiber quality in western

Turkey. *Agricultural Water Management*, Vol.96, No.1, (January 2009), pp. 111-120, ISSN 0168-1923.

Dağdelen, N., Yilmaz, E., Sezgin, F., Gürbüz, T. (2006). Water-yield relation and water use efficiency of cotton (Gossypium hirsutum L.) and second crop corn (Zea mays L.) in western Turkey. *Agricultural Water Management*, Vol.82, No.1-2, (april 2006), pp.63-85, ISSN 0168-1923.

Droogers, P., Immerzeel, W.W., Lorite, I.J. (2010). Estimating actual irrigation application by remotely sensed evapotranspiration observations. *Agricultural Water Management*, Vol.97, No.9, (September 2010), pp. 1351-1359, ISSN 0168-1923.

Droogers, P., Kite, G. (1999). Water productivity form integrated basin modeling. *Irrigation and Drainage System*, Vol.13, No.3, (September 1999), pp. 275–290, ISSN 1573-0654.

Du, T., Kang, S., Zhang, J., Li, F. (2008). Water use and yield responses of cotton to alternate partial root-zone drip irrigation in the arid area of north-west China. *Irrigation Science*, Vol.26, No.2, (January 2008), pp. 147-159, ISSN 0342-7188.

Farahani, H.J., Oweis, T.Y., Izzi, G. (2008). Crop coefficient for drip-irrigated cotton in a Mediterranean environment. *Irrigation Science*, Vol.26, (July 2008), pp. 375-383, ISSN 0342-7188.

Feddes, R.A., Kowalik, P.J., Zaradny, H. (1978). *Simulation of Field Water Use and Crop Yield*. Simulation Monographs. Pudoc, Wageningen, The Netherlands.

Fereres, F., Soriano, M.A. (2007). Deficit irrigation for reducing agricultural water use. *Journal of Experimental Botany*, Vol.58, No.2, (January 2007), pp. 147-159, ISSN 0022-0957.

Figureola, P.I., Berliner, P.R. (2006). Characteristics of the surface layer above a row in the presence of local advection. *Atmósfera*, Vol.19, No.2, (April 2006), pp. 75-108, ISSN 0187-6236.

Folken, T. (2008). *Micrometeorology*, Spring-Verlag, ISBN 978-3-540-74665-2, Berlin, Germany.

Gavilán, P., Berengena, J. (2007). Accuracy of the Bowen ratio-energy balance method for measuring latent heat flux in a semiarid advective environment. *Irrigation Science*, Vol.25, No.2, (June 2006), pp. 127-140, ISSN 0342-7188.

Hou, L.G., Xiao, H.L., Si, J.H., Zhou, M.X., Yang, Y.G. (2010). Evapotranspiration and crop coefficient of Populus euphratica Olivi forest during the growin season in the extreme arid region northwest China. *Agricultural Water Management*, Vol.97, No.2, (February 2010), pp. 351-356, ISSN 0168-1923.

Hussain, I., Turral, H., Molden, D., Ahmad, M. (2007). Measuring and enhancing the value of agriculture water in irrigated river basins. *Irrigation Science*, Vol.25, No.3, (February 2007), pp. 263-282, ISSN 0342-7188.

Ibragimov, N., Evett, S.R., Esanbekov, Y., Kamilov, B.S., Mirzaev, L., Lamers, J.P.A. (2007). Water use efficiency of irrigated cotton in Uzbekistan under drip and furrow irrigation. *Agricultural Water Management*, Vol.90, No.1-2, (may 2007), pp. 112-120, ISSN 0168-1923.

Inman-Bamber, N.G., McGlinchey, M.G. (2003). Crop coefficients and water-use estimates for sugarcane based on long-term Bowen ratio energy balance measurements. *Field Crops Research*, Vol.83, (August 2003), pp. 125-138, ISSN 0378-4290.

IPCC, 2007 Climate change 2007: the physical science basis. Contribution of Working Group I to the Fourth Assessment Report of the Intergovernmental Panel on Climate Change (eds S. D. Solomon, M. Qin, M. Manning, Z. Chen, M. Marquis, K. B.

Averyt, M. Tignor & H. L. Miller). Cambridge, UK and New York, USA: Cambridge University Press.

Jalota, S.K., Sood, A., Chahal, G.B.S., Choudhury, B.U. (2006). Crop water productivity of cotton (Gossypium hirsutum L.) – wheat (Triticum aestivum L.) system as influenced by deficit irrigation. soil texture and precipitation. *Agricultural Water Management*, Vol.84, No.1-2, (July 20067), pp. 137-146, ISSN 0168-1923.

Kang, S.Z., Zhang, J.H. (2004). Controlled alternate partial root-zone irrigation: its physiological consequences and impact on water use efficiency. *Journal of Experimental Botany*, Vol.55, No.407, (November 2004), pp.2437–2446, ISSN 0022-0957.

Karam, F., Lahoud, R., Masaad, R., Daccache, A., Mounzer, O., Rouphael, Y. (2006). Water use and lint yield response of drip irrigated cotton to the length of irrigation season. *Agricultural Water Management*, Vol.85, No.3, (October 2006), pp. 287-295.

Kassam, A.H., Molden, D., Fereres, E., Doorenbos, J. (2007). Water productivity: science and practice – introduction. *Irrigation Science*, Vol.25, No.3, (February, 2007), pp. 185-188, ISSN 0342-7188.

Kroes, J.G., van Dam, J.C., Groenendijk, P., Hendricks, R.F.A., Jacobs, C.M.J., 2008. SWAP Version 3.2. theory Description and User Manual. Alterra-Report 1649, Alterra, Research Institute, Wageningen, The Netherlands, 262 pp.

Lee, X., Massman, W. and Law, B.E. (2004). *Handbook of micrometeorology. A guide for surface flux measurement and analysis*. Kluwer Academic Press, ISBN 1-4020-2265-4, Dordrecht, The Netherlands.

López-Urrea, R., Montoro, A., González-Piqueras, J., López-Fuster, P., Fereres, E., 2009. Water use of spring wheat to rise water productivity. *Agricultural Water Management*, Vol.96, No.9, (September 2009), pp. 1305-1310, ISSN 0168-1923.

Mavi, H.S., Tupper, G. J. (2004). *Agrometeorology: principles and application of climate studies in agriculture*, Food Products Press, ISBN 1-56022-972-1, New York, United States.

Molden, D. (1997). *Accounting for water use and productivity. SWIM Paper 1*, International Irrigation Management Institute (IIMI), ISBM 91-9090-349 X, Colombo, Sri Lanka.

Morison, J.I.L., Baker, N.R., Mullineaux, P.M., Davies, W.J. (2008). Improving water use in crop production. *Philosophical Transactions of the Royal Society B*, Vol.363, (February 2008), pp. 639-658, ISSN 1471-2970.

North, G.B., Nobel, P.S., 1991. Changes in hydraulic conductivity and anatomy caused by drying and rewetting roots of Agave deserti (Agavaceae). *American Journal of Botany*, Vol.78, p.906–915, ISSN

Oke, T.R. (1978). *Boundary Layer Climates*, Routledge, ISBN 0-203-71545-4, Abindon, United Kingdon.

Oweis, t.Y., Farahani, H.J., Hachum, A.Y. (2011). Evapotranspiration and water use of full and deficit irrigation cotton in the Mediterranean environment in southern Syria. *Agricultural Water Management*, Vol.98, No.8, (May 2011), pp.1239-1248, ISSN 0168-1923.

Perez, P.J., Castelvi, F., Ibañez, M., Rosell, J.I. (1999). Assessment of reliability of Bowen ratio method for partitioning fluxes. *Agricultural and Forest Meteorology*, Vol.97, No.3, (November 1999), pp. 141-150, ISSN 0168-1923.

Perry, C., Steduto, P., Allen, R.G., Burt, C.M. (2009). Increasing productivity in irrigated agriculture: Agronomic constraints and hydrological realities. *Agricultural Water Management*, Vol.96, No.11, (November 2009), pp. 1517-1524, ISSN 0378-3774.

Saranga, Y., Flash, I., Yakir, D. (1998). Variation in water-use efficiency and its relation to carbon isotope ratio in cotton. *Crop Science*, Vol.38, No.3, (March 1998), pp. 782–787, ISSN 0011-183X.

Silva, V.P.R., Azevedo, P.V., Silva, B.B. (2007). Surface energy fluxes and evapotranspiratio of a mango orchard grown in a semiarid environment. *Agronmy Journal*, Vol.99, (November 2007), pp. 1391-1396, ISSN 1435-0645.

Singh, Y., Rao, S.S., Regar, P.L. (2010). Deficit irrigation and nitrogen effects on seed cotton yield. water productivity and yield response factor in shallow soils of semi-arid environment. *Agricultural Water Management*, Vol.97, No.7, (July 2010), pp. 965-970, ISSN 0168-1923.

Steduto, P., Hsiao, T.C. (1998). Maize canopies under two soil water regimes: IV. Validity of Bowen ratio–energy balance technique for measuring water vapor and carbon dioxide fluxes at 5-min intervals, *Agricultural and Forest Meteorology*, Vol.89, No.3-4, (February 1998), pp.215-228, ISSN 0168-1923.

Steduto, P., Hsiao, T.C., Fereres, E. (2007). On the conservative behavior of biomass water productivity. *Irrigation Science*, Vol.25, No.3, (February, 2007), pp. 189-207, ISSN 0342-7188.

Strzepek, K., Boehlert, B. (2010). Competition for water for the food system. *Philosophical Transactions of the Royal Society B*, Vol.365, (September 2010), pp. 2927-2940, ISSN 1471-2970.

Stull, R.B. (1998). *An Introduction to Boundary Layer Meteorology*, Kluwer Academy Publishers, ISBN 90-277-2769-4, Dordrecht, The Netherlands.

Supit, I., Hooyer, A.A., van Diepen, C.A. (1994). System description of the WOFOST 6.0, crop simulation model implemented in CGMS, vol. 1: Theory and algorithms. EUR publication 15956, Agricultural series, Luxembourg.

Tang, L-S., Li, Y., Zhang, J. (2010). Partial rootzone irrigation increases water use efficiency, maintains yield and enhances economic porfit of cotton in arid area. *Agricultural Water Management*, Vol.95, No.10, (October 2010), pp. 1527-1533, ISSN 0168-1923.

Teixeira, A.H.C., Bastiaanssen, W.G.M., Ahmad, M.D., Bos, M.G. (2009). Reviewing SEBAL input parameters for assessing evapotranspiration and water productivity for the Low-Middle São Francisco River basin, Brazil Part B: Application to the regional scale. *Agricultural and Forest Meteorology*, Vol.149, No.3-4, (March 2009), pp. 477-490, ISSN 0168-1923.

Thornthwaite, C.W. (1948). An approach toward a rational classification of climate. *Geographical Review*, Vol.38, No.1, (January 1948), pp. 55-94, ISSN 0016-7428.

Todd, R.W., Evett, S.R., Howell, T.A. (2000). The bowen ratio-energy balance method for estimated latent heat flux of irrigated alfalfa evaluated ina asemi-arid advective environment. *Agricultural and Forest Meteorology*, Vol.103, No.4, (July 2000), pp. 335-348, ISSN 0168-1923.

Tucci, C.E.M. & Beltrame, L.F.S. (2009). Evaporação e evapotranspiração, In: *Hidrologia: Ciência e Aplicação*, C.E.M. Tucci, (Ed.), 253-287, UFRGS Editora, ISBN 978-85-7025-924-0, Porto Alegre, Brasil.

Ünlü, M., Kanber, R., Onderm, S., Metin, S., Diker, K., Ozekici, B., Oylu, M. (2007). Cotton yields under different furrow irrigation management techniques in the Southeastern Anatolia Project (GAP) area. Turkey. *Irrigation Science*, Vol.26, No.1, (October 2007), pp. 35-48, ISSN 0342-7188.

van Dam, J.C., Groenendijk, P., Hendriks, R.F.A., Kroes, J.G. (2008). Advances of modeling water flow in variably saturated soils with SWAP. *Vadose Zone Journal*, Vol.7, No.2, (May 2008), pp.640–653, ISSN 1539-1663.

Webb, E.K., Pearman, G.I., Leuning, R. (1980). Correction of flux measurements for density effects due to heat and water vapour transfer. *Quarterly Journal of the Royal Meteorology Society*, Vol.106, No.447, (January 1980), pp. 85-100, ISSN 0035-9009.

Wilks, D.S. (2006). *Statistical methods in the atmospheric science*, Academic Press, ISBN 978-0-12-751966-1, San Diego, United States.

Yazar, A., Sezen, S. M., Sesveren, S. (2002). LEPA and trickle irrigation of cotton in the Southeast Anatolia Project (GAP) area in Turkey. *Agricultural Water Management*, Vol.54, No.3, (Aril 2002), pp.189-203, ISSN 0168-1923.

Zhang, J.H., Jia, W.S., Kang, S.Z. (2001). Partial rootzone irrigation: its physiological consequences and impact on water use efficiency. *Acta Botanica Boreali-occidentali Sinica*, Vol.21, No.2, (March 2001), pp. 191-197, ISSN 1000-4025.

Zwart, S.J., Bastiaanssen, W.G.M. (2004). Review of measured crop water productivity values for irrigated wheat, rice, cotton and maize. *Agricultural Water Management*, Vol.69, No.2, (September 2004), pp. 115-133, ISSN 0168-1923.

5

Strategies for Improving Water Productivity and Quality of Agricultural Crops in an Era of Climate Change

Zorica Jovanovic and Radmila Stikic
University of Belgrade - Faculty of Agriculture
Serbia

1. Introduction

Climate change is one of the most serious problems facing the world today. The recent Intergovernmental Panel on Climate Change reports confirmed that climate change will have a significant impact on global surface temperature. The projections of IPCC are that the rise of the mean temperature will be as high as 6.4°C by 2100, while the concentration of CO_2 will be 1.3 times higher than it was 20 years ago. Furthermore, the number of extreme events, including heat waves, storms and flooding will increase (IPCC, 2007).

Climate change scenarios for Europe are that global warming resulting from anthropogenic greenhouse gas emissions (mainly carbon dioxide and methane) will lead to substantial temperature increases in Northern Europe during winter and in Southern Europe during summer. The especially vulnerable for future European summer climate would be the countries in South-East European and Mediterranean areas. Predictions of different scenarios of climate are that due to the expected increase in temperature and decrease in precipitation, drought would start earlier and last longer in these comparing to other European areas (Beniston et al., 2007). Some scenarios also predict the higher incidence of heat waves and extreme temperature in South East than in Central Europe (Hirschi et al., 2011).

Agriculture is highly sensitive to climate change and especially, to drought. The increase in temperature can increase duration of the crop growing season in regions with a relatively cool spring and shortened the season in regions where high summer temperature already limits production. Therefore, in the areas of water scarcity, the irrigation is necessary for successful agricultural production. Currently, due to the climate change impacts many countries are faced with the increased competitions for water resources between different sectors (agriculture, industry or domestic consumption). The clean freshwater becoming a limited resource and its use for crop irrigation is in competition with the demand for household consumption, as well as with the need to protect the aquatic ecosystems. Therefore, the challenge is to minimize the use of water for irrigation. Another problem is that water in many countries is seriously contaminated with either inorganic or organic pollutants, mainly from intensive animal production and urban areas. Uncontrolled use of contaminated waters (chemically or microbiologically) could have serious environmental and health implications. It is obvious that saving clean water, increasing agricultural productivity per unit of water ("more crop per drop") and producing safe food are becoming of strategic importance for many countries (Luquet et al., 2005).

The aim of chapter is to provide an overview of some of the current challenges and opportunities to minimize the problem of agricultural production under water scarcity. The focus will be on the two approaches: use of the deficit irrigation methods and use of genotypes with increased drought resistance and water productivity. Furthermore, the problems of the use contamination of water for irrigation will be briefly reviewed, as well the novel technologies by which low water quality could be used to improve water productivity and to ensure food safety and quality. The special emphases will be on the current efforts to create genotypes resistant to drought and thus to reduce the existing gap between potential crop yield and crop yield in drought conditions.

2. Climate change impacts on agriculture

Climate change models for Europe highlight a particularly worrying trend in terms of rising temperatures and decreasing precipitation. The mean annual precipitation will increase in Western and Northern Europe (from 5 to 15%) and decrease in Central, Eastern and Mediterranean Europe (from 0 to 20%), while the change in seasonal precipitation will vary substantially from season to season and across different regions. Besides the projected increase in the yearly maximal temperature, it is also expected a large increase in yearly minimum temperature across most of Europe. The increase of minimum year temperature in many areas is connected with an increase of temperature during winter period (Kjellström et al., 2007). Due to the effects of the summer temperature increase and reduced precipitation, the number of extreme events (heat waves, drought, storms) will also increase.

Although agricultural production is highly dependent on climate factors, the climate change is expected to affect agriculture very differently in different parts of the world. Furthermore, the climate change could produce positive or negative effects on agriculture depending on the region. The final effects on crop productivity and food safety depend on current climatic and soil conditions, the direction of change and the availability of resources and infrastructure in specific region or country to cope with predicted change in specific (Parry et al., 2004).

Increase in greenhouse gases can affect agriculture directly (primarily by increasing photosynthesis at higher CO_2) or indirectly *via* effects on climate (primarily temperature or precipitation). Of special importance is the increase of CO_2 concentration. Over the last century, atmospheric concentrations of carbon dioxide increased from a pre-industrial value of 278 parts per million to 379 parts per million in 2005. Most of the increase in carbon dioxide comes from burning of fossil fuels such as oil, coal and natural gas, and from deforestation. It is also certain that the accumulation of CO_2 and other greenhouse gases will cause a further increase in mean global temperature (IPCC, 2007). As a consequence of increased photosynthesis at elevated CO_2, dry matter production of C_3 plants is expected to increase more than in C_4 plants. C_3 plants are those that use the C_3 carbon fixation pathway in photosynthesis in which the CO_2 is first fixed into organic compounds containing three carbon atoms, while C_4 plants use C_4 carbon fixation pathway for producing compounds containing four C atoms. Reducing stomatal opening and thus transpiration, high CO_2 can have another direct effect on plants, and it could be expected an increase in water use efficiency (WUE) of C_3 and C_4 plants. Kimball et al. (2002) reported several experiments in controlled, semi-controlled, and open-field conditions, which have shown that a doubling of atmospheric CO_2 from 330 to 660 ppm may increase the productivity of C_3 species by an average of 33% at optimal growing conditions. The effects of CO_2 and other greenhouse gases depend also on their interaction with other environmental factors, especially drought. Results for potato showed that elevated CO_2 can only partially alleviate long-term whole plant responses to water stress (Fleisher et al., 2008). These results pointed out that the CO_2 "fertilization" effect cannot totally compensate for the negative effects of other environmental stresses.

In general, the effects of global change on Europe are likely to increase productivity of agricultural plants, because increasing CO_2 concentration will directly increase resource use efficiencies of plants, and warming will give more favorable conditions for plant production in Northern Europe. The sensitivity of Europe agriculture to climate change, especially drought, has a distinct north-south gradient and many studies indicating that Southern Europe will be more severely affected than Northern Europe. The particularly vulnerable to agricultural drought in Europe are Mediterranean and South-East Europe regions. In a lot of the countries of these regions, economic development is heavily dependent upon growth in the agriculture and, therefore, the climate change impacts on agriculture could have significant social consequences (EEA, 2008).

In Northern Europe increases in productivity and expansion of suitable cropping areas are expected to dominate, whereas disadvantages from increases in water shortage and drought will dominate in Southern Europe. The increased crop productivity in Northern Europe will be also caused by lengthened growing season, decreasing cold effects on crop growth and extension of the frost-free period. On the contrary, the expected decrease in productivity in Southern Europe will be the consequence of the shortening of the growing period, with subsequent negative effects on grain filling (Iglesias et al., 2009).

Climate change effects may reinforce the current trends of intensification of agriculture in Northern and Western Europe and extensification in the Mediterranean and South Eastern parts of Europe (Olesen & Bindi, 2002). Furthermore, the area of some of the crop's cultivations in Northern Europe will be expanding, especially for cereals. According to Alcamo et al. (2007) it is expected that increase in wheat yield related to climate change will be from +2 to +9% by 2020 and from +10 to +30% by 2080, while for sugar beet yield increase will be in the range from +14 to +20% until the 2050s. Study for Southern Europe predicted a general yield decreases (e.g., legumes -30 to + 5%; sunflower -12 to +3% and tuber crops -14 to +7% by 2050) as well as increases in water demand (e.g., for maize +2 to +4% and potato +6 to +10% by 2050). The same study showed that the impacts on autumn sown crops are more geographically variable; yield is expected to decrease in most southern areas, and increase in northern or cooler areas (e.g., wheat: +3 to +4% by 2020, -8 to +22% by 2050, -15 to +32% by 2080). Furthermore, predictions are that by 2050 energy crops (e.g., oilseeds such as rape oilseed and sunflower), starch crops (e.g., potatoes), cereals (e.g., barley) and solid biofuel crops (such as sorghum and Miscanthus) will show a northward expansion in potential cropping area, but a reduction in Southern Europe (Alcamo et al., 2007).

2.1 Drought effects

The increase of temperature or drought may have a significant impact on plant growth and productivity. At the whole plant and the crop level, the important repercussions of high temperature or drought stresses are mediated by their effects on plant phenology, phasic development, growth, carbon assimilation, assimilate partitioning and plant reproduction processes. These major effects account for the most of the variation in crop yield caused by drought stress. However, there is a large variability in stress sensitivity at different periods during the life cycle of a given plant or during an increase in stress duration and severity. For crop plants, stresses during the generative phase can have been much more dramatic effects on plant yield than the stress during the vegetative phase (Craufurd & Wheeler, 2009). Table 1 presents an overview of the growth stages that are most sensitive to drought in different agricultural crops.

Crop	Stage of development
Field crops	
Maize	flowering and grain filling
Wheat	flowering more than yield formation
Rice	head development and flowering
Soybean	flowering and yield formation
Sunflower	flowering more than yield formation
Sugarbeet	first month after emergence
Cotton	flowering and boll formation
Sugarcane	tillering and stem elongation
Tobacco	period of rapid growth
Vegetables	
Lettuce	head development
Pea	flowering and yield formation
Bean	flowering and pod filling
Carrots	root enlargement
Cabbage	head development and ripening
Cucumbers	flowering and fruit development
Potato	tuber initiation and enlargement
Tomato	flowering, fruit setting and enlargement
Pepper	flowering and fruit development
Fruit tree and grape	
Grape	vegetative period and flowering
Citrus	flowering and fruit setting
Olive	flowering and yield formation

Table 1. Stages of plant development that are the most sensitive to drought.

2.2 Water scarcity and contamination

Water is essential for high and stable yield of agricultural plants and in many areas modern farming would be impossible without irrigation. However, only a small proportion of the world cultivated area is equipped for irrigation. According to FAO (2003), more than 80% of global agricultural land is rain-fed and in these regions, crop productivity depends solely on sufficient precipitation to meet evaporative demand and associated soil moisture distribution.

Furthermore, most of the climatic scenarios predicted that climate change will have a range of impacts on water resources. A simulation study done by Eitzinger et al. (2003) predicted that groundwater recharge will be reduced in Central and Eastern Europe. Although there is still a considerable range of uncertainty related to changes in climate variability in future climate scenarios for these regions, the study showed that summer crops will be very vulnerable and dependent on soil water reserves, as the soil water or higher groundwater tables during the winter period cannot be utilized as much as by winter crops and evapotranspiration losses during summer due to higher temperatures could increase significantly.

The Mediterranean and South East European regions are especially vulnerable to water scarcity. They are faced with increased competitions for water resources between different sectors (agriculture, industry or domestic consumption). Climate change projection for the Mediterranean area is a gradual increase of temperature and lower rainfall by the end of this century. Moreover, increased average temperatures will be coupled to an increase in extreme events frequency and magnitude as heat waves.

Investigations of impact of warming climate on the phenology of typical Mediterranean crops indicated an earlier development of crops and a reduction of the length of growing season for winter and summer annual crops, grapevine and olive tree (Moriondo & Bindi, 2007). These responses may allow some crops to escape summer drought stress (e.g., winter crops). However, at the same time the climate change will increase the frequency of extreme climate events during the most sensitive phenological stages and without irrigation this will reduce the final crop yield quantity and quality.

Although drought in South East European region is shorter than in Mediterranean, its impact on agricultural production in South East European region (Serbia, Bosnia and Herzegovina, continental part of Bulgaria, Croatia, Montenegro and Albania) could be also very serious. During summer period growth and productivity of a lot of agricultural plants are in the most sensitive phase to drought, and therefore, the reduction of yield could be significant. In accordance to current agricultural drought effects and prediction of increasing agricultural drought in the future, farmers are forced to irrigate crops. However, the maximal use of water for irrigation usually occurs during 2 or 3 summer months that is a significantly shorter period for crop irrigation than in Mediterranean region, where sometimes the irrigation period is more than six months. Therefore, the mitigation of drought by irrigation is economically more profitable in Mediterranean than in South East European climate conditions.

Together with water scarcity current problem in many areas is also contamination of water resources. In the most of European countries` water for irrigation is abstracted from surface water. The surface water resource may be recipients of treated wastewater and may be polluted from other anthropogenic activities or natural sources (Vinten et al., 2004). The problem is very serious because about 10% of crops are irrigated with untreated wastewater (Anon, 2003) and currently the potential for contamination via irrigation water is further increased worldwide.

Water for irrigation could be contaminated microbiologically or/and chemically. The pathogens, organic and inorganic chemical compounds in wastewater, can induce health risks for workers and consumers, exposed *via* the direct or indirect contact with such waters during field work and ingestion of fresh and processed food (Peralta-Videa et al., 2009).

The wide spectrum of pathogenic organisms in low quality water poses the most immediate and direct risk to public health. The most frequent microbiological contaminants in water are faecal microorganisms, including disease-causing pathogens like *Salmonella, Campylobacter, Shigella*, enteric viruses, protozoan parasites and helminth parasites (Steele & Odumeru, 2004). The potential risk is transport of pathogenic from water for irrigation to soil or crops. Moreover, the edible portions of a plant can become contaminated by contaminant uptake from soil by the root system and subsequent transport of the pathogen inside the plant. Therefore, irrigation with water contaminated with bacteria can be the starting point of a water–soil–plant–food contamination pathway (Battilani et al., 2010).

Increasing trend of consumption of fresh fruit and vegetables present also a risk factor for infection with enteric pathogens such as *Salmonella* and *Escherichia coli* O157 (Heaton & Jones, 2008). Routes of contamination with enteropathogens may vary. Usually they include application of organic wastes to agricultural land or contamination of irrigation waters with faecal material. If the crops are irrigated with wastewater an increased incidence of enteropathogens in different fruit and vegetables will happen (Steele & Odemeru 2004).

Pathogen survival will depend on the different environments conditions associated with the method of irrigation, e.g. surface irrigation like furrow and sprinkler irrigation, exposure to high temperatures, desiccation and UV-light factors which all lead to a faster die-off of pathogens on the soil surface. Studies on plant nursery irrigation (Lubello et al., 2004) have shown that tertiary treatment technologies like filtration and peracetic acid need to be added to primary and secondary treated wastewater to eliminate the risk posed by waterborne pathogens.

Results of Enriquez et al. (2003) showed that the use of the subsurface drip line could delay the movement of pathogens to the surface and inhibit the further impact on the above ground product. To test the hypothesis that subsurface application of urban wastewater could provide the potato safe for consumption, Forslund et al. (2010) compared different irrigation techniques (sprinkler, furrow and subsurface drip irrigation) for using treated urban wastewater, canal water and tap water. These results showed no significant number of *E. coli* in soil and potato tubers during irrigation. They also pointed out that soil could be a very effective filter barrier for pathogenic in ensuring food safety.

The use of low quality water for irrigation may also introduce hazardous heavy metals into the food chain (Behbahaninia et al., 2009). Heavy metals are dangerous as they tend to accumulate in living organisms faster than they are metabolized or excreted (Järup, 2003). Water filters designed to protect irrigation systems, offers no barriers against heavy metals contamination, except for the fraction of metals bound or trapped into the suspended solids. Several techniques are applicable to remove heavy metals from contaminated water. Heavy metal removal device (HMR) is based on heavy metal adsorption to granular ferric hydroxide (GFH) and HMR application is recommended if severe heavy metal pollution occurs in the irrigation water source, which cannot be sufficiently treated by the gravel filter (Battilani et al., 2010).

As the result of EU FP6 project SAFIR (www.safir4eu.org) new decentralized water treatment devices (prototypes) were developed to allow a safe direct or indirect reuse of wastewater produced by small communities/industries or the use of polluted surface water. The testing was done of a small-scale compact pressurized membrane bioreactor and a modular field treatment system that include commercial gravel filters and heavy-metal specific adsorption materials. These results indicated that decentralised compact pressurised membrane biobooster (MBR) could remove up almost all *Escherichia coli* and total coliforms. MBR from inlet flow also removed arsenic, cadmium, chromium, copper and lead. The field treatment system (FTS) also proved to be effective against faecal contamination when applied with its complete set up including UV treatment. FTS removed arsenic, cadmium, copper, chromium, lead and zinc (Battilani et al., 2010). Using new technology Surdyk et al. (2010) investigated the transfer of heavy metals from low quality surface water to the soil and potato plants in a Serbian field study during 2007 and 2008 seasons. These results indicated that after passing water through the FTS no significant impact of the irrigation water on potato heavy metal accumulation could be detected.

In general, the use of low quality water for irrigation of agricultural plants as a substitute for groundwater and surface water can only be accepted if the health of farm workers and consumers of irrigated produce can be ensured.

3. Agricultural strategies for adaptation to climate change

To avoid or at least reduce negative effects of drought, several agronomic adaptation strategies have been suggested, including both short-term adjustments and long-term adaptations. The short-term adjustments include efforts to optimize production without major system changes. Most of them are already available to farmers and communities. Examples of short-term adjustments include use of varieties/species with increased resistance to heat shock and drought, introducing new crops, changes in sowing dates (Olesen et al., 2007) and fertilizer use, improvement and modification of irrigation techniques (amount, timing or technology), other different soil or crop managements as mulching, crop rotation, intercropping, skip rows, protected cropping (Davies et al., 2011). They are autonomous in the sense that no other sectors (e.g., policy, research, etc.) are needed in their development and implementation.

Long-term adaptations refer to major structural changes to overcome adversity caused by climate change (Bates et al., 2008). This involves changes of land use that result from the farmer's response to the differential response of crops to climate change. The changes in land allocation may also be used to stabilize production, substitution of crops with high inter-annual yield variability (e.g., wheat) by crops with lower productivity but more stable yields (e.g., pasture). Other examples of long-term adaptations include breeding of crop varieties and new land management techniques to conserve water or increase irrigation use efficiencies and more drastic changes in farming systems (including land abandonment). They are planned actions, and they should be focused on developing new infrastructure, policies, and institutions that support, facilitate, co-ordinate and maximize the benefits of new management and land-use arrangements.

In this chapter, we will be focused only on two strategies for saving water in agriculture: improvement of irrigation techniques and breeding and use of stress drought and heat stress resistant genotypes.

4. Water saving irrigation strategies

Under current and predicted climate conditions of drought and scarce water supply, the challenge for agricultural production is to increase water productivity (ratio between yield and amount of water used for irrigation) and to sustain or even increase crop yield. Therefore, considerable emphasis in the research is placed on crop physiology and crop management for dry conditions with the aim to increase crop water use efficiency (Chaves et al., 2002; Morison et al., 2008).

Another approach is to improve the irrigation management. Many results confirmed that the deficit irrigation strategy has the potential to save water for irrigation and optimize water productivity in agriculture. The term *deficit irrigation* describes an irrigation scheduling strategy that allows a plant's water status to decrease to the certain point of drought stress.

Currently, two deficit irrigation methods are in use: regulated deficit irrigation and partial root-zone drying (FAO, 2002). Both methods are based on the understanding of the physiological responses of plants to water supply and water deficit, especially the perception and transduction of root-to-shoot drought signals (Chaves et al., 2002; Morison et al., 2008; Stikic et al. 2010).

4.1 Regulated deficit irrigation (RDI)

Regulated deficit irrigation (RDI/DI) is a method that irrigates the entire root zone with an amount of water less than the potential evapotranspiration during whole or specific periods of the crop cycle (English & Raja, 1996). The principle of the RDI technique is that plant sensitivity to drought is not constant during the growing season and that intermittent water deficit during specific periods of ontogenesis may increase water savings and improve yield quality (Loveys et al., 2004).

The key to the RDI strategy is the timing of the water deficit and the degree of the deficit applied to the plants. To avoid the possible reducing effect of RDI on yield, the monitoring of soil water status is required in order to maintain a plant water regime within a certain degree of drought stress that could not limit yield.

Implementing RDI could also be difficult where there is a high water table or deep soil with a high water holding capacity. However, if RDI is managed carefully, the negative impact on yield could be avoided. Results for numerous field crops (maize, wheat, soybean, sunflower), tree crops and grapevine showed that optimal RDI managing might increase water productivity or yield quality, maintain or even increase farmers' profits (reviewed by Fereres & Soriano, 2007).

4.2 Partial root-zone drying (PRD)

Regulated deficit irrigation is a method where water application is manipulated over time, while partial root–zone drying (PRD) is a method where water is manipulated over space. PRD is designed to maintain half of the root system in a dry or drying state, while the other half is irrigated. The treatment is then reversed, allowing the previously well-watered side of the root system to dry down while fully irrigating the previously dry side.

The principle behind PRD is that irrigating part of the root system keeps the leaves hydrated and in a favorable plant water status, while drying on the other part of the root system promote synthesis and transport of so-called chemical signals (particularly plant hormone abscisic acid) from roots to the shoot *via* the xylem to induce a physiological response (Dodd et al., 2006). The frequency of the switch is determined according to soil type, genotypes or other factors such as rainfall and temperature and in most of the published data the PRD cycle includes 10 to 15 days (Davies et al., 2000).

Effects of PRD on plant physiology are different from RDI because wet roots under PRD sustain shoot and fruit turgor that are important for plant growth. The drying roots in the PRD produce the sufficient amount of the chemical signals to maintain a physiological response to water stress. Triggering partial stomatal closure under PRD irrigation prevent excessive water loss and also the metabolic inhibition of CO_2 assimilation, that otherwise would occur in extensively development of drought stress (Chaves et al., 2002; Costa et al., 2007).

PRD may be applied by different techniques in the field depending on the cultivated crops or soil condition. PRD irrigation (alternate or fixed) could be done by subsurface or surface drip lines, furrow, micro-sprinkler or vertical soil profile methods (Kang & Zhang, 2004).

Figure 1 shows the scheme of the full irrigation (FI) and partial root-zone drying drip line installation in the potato field experiment. This experiment was a part of research activity in EU FP6 project SAFIR (www.safir4eu.org). For PRD irrigation, the subsurface drip system was applied, which consisted of two parallel bundled lines, each with 60 cm distance between emitters, but displaced to give 30 cm distance. In this way emitter from one line irrigated one side of the root, and emitters from the other line irrigated another side of the root system (Jensen et al., 2010; Jovanovic et al., 2008, 2010).

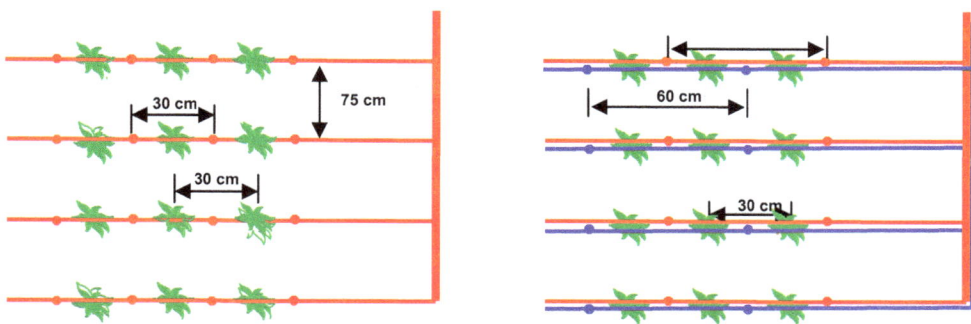

Fig. 1. Scheme of FI (left) and PRD drip lines (right) installation in potato experimental field.

Partial root drying method is applied in a wide range of different crops (Kirda et al., 2007; Sepaskhah & Ahmadi, 2010). Some of PRD experiments applied to the different agricultural plants are presented in Table 2. Comprehensive data sets from the most of these field and glasshouse studies have shown that under PRD irrigation, water may be reduced by approximately 30-50% without significant yield reduction and in some cases with an improved yield quality. An important mechanism of plant response to PRD, in addition to increase WUE or yield quality, may be the promotion of root growth and increase of root biomass. Enhanced root growth will increase the plants' ability to explore a greater soil volume potentially increasing soil water and nutrient acquisition.

Crops	Species	References
Field crops	maize	Kang et al. (2000); Li et al. (2010)
	wheat	Sepaskhah & Hosseini (2008)
	sunflower	Metin Sezen et al. (2010)
	sugar beet	Sepaskhah & Kamgar-Haghighi (1997)
Vegetables	potato	Ahmadi et al. (2010a, 2010b); Jovanovic et al. (2010); Liu et al. (2006); Saeed et al. (2008); Shahnazari et al. (2007)
	tomato	Davies et al (2000); Kirda et al. (2004); Mingo et al. (2004); Zegbe et al. (2006); Zegbe-Dominguez et al. (2004)
	beans	Genocoglan et al. (2006); Wakrim et al. (2005)
Fruit tree and grape	grape	de Souza et al. (2003); dos Santos et al. (2003); de la Hera et al. (2007)
	apple	Leib et al. (2006); Zegbe & Serna-Péreza (2011)
	pear	Kang et al. (2002); O'Connell & Goodwin (2007)
	peach	Goldhammer et al. (2002)
	olive	Centritto et al. (2005); Wahbi et al. (2005)
	citrus	Hutton & Loweys (2011)
	almond	Egea et al. (2010)

Table 2. PRD experiments with different agricultural plants.

Table 3 presents some of our recent results in an experiment with tomato cultivar *Amati* grown under PRD and FI in commercial polytunnel conditions. These results, similarly to the others published showed that with the PRD method is possible to increase WUE and save water for irrigation, without statistically significant reduction of tomato yield. Furthermore, in our experiment the antioxidative activity was significantly increased in tomato fruits under PRD compared to the fruits of control plants. This improvement of PRD fruit quality could be also beneficial from the aspect of health-promoting value of tomato fruits.

PRD irrigation method has been also successfully trailed with potato (Table 2). Recently, we conducted potato PRD field trials with cultivar *Liseta* and with the aim to compare "static" PRD management approach with "dynamic" system, when amounts of water irrigated in PRD were changed according to the plant growth phases by increasing water saving during later robust growth stages (Jovanovic et al., 2010). In the 2007 season PRD plants received

70% of fully irrigated (FI), and in 2008 year 70% of PRD was replaced by 50% in the last 3 weeks of the irrigation period in order to further save water. Comparison of the effects of PRD and FI irrigation technologies did not show significant differences in yield in investigated seasons. However, water use efficiency of PRD plants compared to FI was significantly bigger in 2008 season when "dynamic" PRD was applied than in 2007 years when "static" approach of PRD irrigation was used (by 14%). Tuber quality data showed in both seasons a significant increase in antioxidant activity and in starch content in the tubers of PRD plants comparing to FI tubers. Table 4 present some of our data from potato PRD experiments (Jovanovic et al., 2008, 2010).

Water treatment	Yield (t ha^{-1})	WUE (kg FW m^{-3})	TSS (°Brix)	TA (citric acid µmol g^{-1} FW)	AA (µmol TEAC 100g^{-1} FW)
FI	48.71	34.90^A	5.10	19.60	33.33^A
PRD	43.41	56.02^B	5.10	19.90	50.87^B

Table 3. Treatments means of yield, water use efficiency (WUE), fruit quality (total soluble solids - TSS, titrable acidity - TA and antioxidant activity - AA) in fully irrigated tomato (FI) and tomato under partial root-zone drying (PRD). Different letters show significant differences at 95% level for comparison between irrigation treatments.

Water treatment	Yield (t ha^{-1})	IWUE (kg ha^{-1} mm^{-1})	N (%)	Starch (% FW)	AA (µmol TE 100 g^{-1} FW)
FI $_{-2007}$	45.31^{AB}	241.00^A	2.18^A	13.72^A	19.92^A
PRD$_{-2007}$	41.78^A	334.27^B	2.45^B	15.02^{BC}	22.63^B
FI$_{-2008}$	53.19^C	236.40^A	2.25^A	13.45^{AB}	19.13^A
PRD$_{-2008}$	50.46^{BC}	380.14^C	2.68^B	15.76^C	22.81^B

Table 4. Treatments means of yield, irrigation water use efficiency (IWUE), tuber quality (%N, starch content and antioxidant activity - AA) of fully irrigated potato (FI) and potato under partial root-zone drying (PRD) during 2007 and 2008. Different letters show significant differences at 95% level for comparison between irrigation treatments and investigated seasons.

Furthermore, our potato results indicated that PRD treatment could improve the allocation of N from the shoot to tuber at final harvest and increase the N-use efficiency (Jovanovic et al., 2008, 2010). Similarly, Shahnazari et al. (2008) results also confirmed that PRD treatment may improve soil nitrogen availability during the late phases of potato growing season indicating a higher N mineralization. In general, our results indicate that "dynamic" PRD approach could be a more promising strategy for saving water for potato irrigation than the classical "static" approach.

4.3 Choice of RDI or PRD irrigation

Although RDI and PRD methods functioning differently, some of their main effects are similar. Both methods limit vegetative vigour and improve water use efficiency or water productivity (Kriedmann & Goodwin, 2003). Reduction of vegetative vigour is desirable characteristics for many crops in drought regions. Excessive vegetative vigor is a major problem for many fruit crops, since the use of assimilates in leaf growth restricts fruit set and development, and may cause shading and more fungal diseases (Morison et al., 2008). Reduction of vegetative growth may also induce a change of assimilate partitioning and source/sink relationships. The photosynthetically active tissue of mature leaves is an active source of assimilate for sink tissues, such as flowers, fruits, or roots. Among sink organs, fruits or tubers are defined as a high priority in the context of competition for assimilates between alternative sinks. Davies et al. (2000) results pointed out that reduction of carbo-hydrate strength (side shoots) in PRD-treated tomato plants resulted in a relative increase in the sink strength of tomato fruit such as carbohydrate previously partitioned towards the side shoots is redirected towards the fruit.

Both irrigation methods significantly increase WUE and may save 30 to 50% of water for irrigation depending on crops, soil or climatic conditions. Their effects also depend on the crop phenological stage and on the severity of stress that is imposed to the crops. For example, in Mediterranean or South-East European conditions, it is common to apply water deficit during the final phases of grape development to avoid water stress during the ripening stage, whereas in Australia the common practice is to apply less water early in the season with the aim to control berry size (McCarthy et al., 2002).

The potential reduction of yield is the main problem in the use of RDI, although this depends on the timing of application and degree of stress imposed by RDI. According to Kriedmann & Goodwin (2003) soil type is also an issue with regulated deficit irrigation. Sandy loams dry and re-wet more readily than clay soils, and are generally easier to manage. Although a clay soil has theoretically a greater range of plant-available moisture, root growth under RDI in this type of soil can be slower and water extraction by root smaller than in sandy loam soil. According to Fereres & Soriano (2007) to quantify the level of RDI it is first necessary to define the full crop ET requirements and then, adjustment of timing of irrigation with permanent control of management is necessary.

Many results showed that PRD may be a more beneficial technique than RDI, particularly in terms of lesser risk of yield reduction, especially during heat waves. Under both PRD and RDI treatments, stomatal conductance was reduced, but PRD plants due to wet side of the roots remained less stressed and pre-dawn water potential values are higher than in RDI plants (Kriedmann & Goodwin, 2003). Beneficial effects of PRD comparing to RDI are also in the increase in root growth and development (Mingo et al, 2004), quality of fruits and better control of vegetative growth and assimilate partitioning (Costa et al., 2007). Increased yield quality in many different crops could minimize the negative effects of PRD on the yield quantity in some experiments (Kang & Zhang, 2004).

For PRD irrigation scheduling less emphasis should be on evaporative indicators of the irrigation requirement and more emphasis on direct measurement of root-zone soil water content to drive both duration of irrigation, and timing of the switch from drying to re-wetting (Kriedmann & Goodwin, 2003). A key factor of PRD irrigation scheduling is re-watering of the dry side. During PRD irrigation, water must be switched regularly from one side of the root to the other to keep roots in dry soil alive and fully functional and sustain the supply of root signals. The time of switching required could present significant difficulty in operating PRD irrigation. This is one of the mean reasons that Sadras (2009) in his meta-analyses challenge the beneficial effects of PRD technique. He concluded that substantial improvement in water use efficiency can be achieved by closely monitored RDI, without the complexity and additional cost of PRD. Furthermore, PRD method is more costly than RDI because it requires installation of two drip lines.

Usually in the most applied PRD systems the switching is based on soil water depletion or stomatal reactions. Zhang & Davies (1990) suggested that the early wilting of older leaves may indicate the right time for irrigation. Recently, a novel model for prediction of the switching side was developed and is based on accumulation of xylem ABA in potato (Liu et al., 2008). The model was further improved and finally implemented into modified the agro-ecological model DAISY which simulate the mechanisms underlying the water saving effects of PRD irrigation (Plauborg et al., 2010).

It is difficult to recommend RDI or PRD for irrigation. Successful application and choice between RDI or PRD depends on different factors, including the irrigated crops, outputs of crop growing (increase WUE or yield quality, sustained yield etc.) and severity, timing and duration of the stress imposed to plants in specific agro-climatic conditions. Both methods require high management skills and the knowledge of crop response to drought stress (FAO, 2002). Recently, Jensen et al. (2010) suggested a new RDI and PRD irrigation guidelines for tomato and potato based on EU FP6 project SAFIR field experiments conducted under different climatic conditions (www.safir4eu.org). For these vegetables grown in the field conditions full irrigation is needed until the crops are well established, and then RDI and PRD should start. To avoid the yield decrease, water saving irrigation should start in potato after the end of tuber initiation, while in tomato after the first trusses were developed. After these periods 30% water saving can be applied, while finally during the last 2 weeks before harvest water saving could be increased to 50%.

In general, it could be expected that the successful implementation of deficit irrigation strategies can lead to greater economic gains for farmers, especially in the water scarcity areas, where there is not enough water for irrigation or in the areas where price of water is high.

5. Plant resistance to drought

Adaptation measures to mitigate the reduction of yield induced by drought besides the increase in crop water productivity includes the production and use of drought resistant genotypes. Additional opportunities for new cultivars also include changes in phenology or enhanced responses to elevated CO_2.

The prerequisite to produce resistant genotypes is a better understanding of the plant response and adaptation to drought stress, the improvement of phenotyping, the selection of key-genes involved in the resistance to drought and the evaluation of the impact of resistance on crop yield and quality. These are very difficult tasks because reactions of plants to drought is the complex phenomenon where the plant response depends on species or genotypes, the type, duration or intensity of drought and on phenological stage in which drought stress is experienced (Chaves et al., 2003).

According to the classic definition of Levitt (1980) plant resistance to drought stress can be divided to three main strategies namely escape, avoidance and tolerance. The plants "escapers" exhibit a rapid phenological development and thus are able to complete their life cycle before the water deficit occurs. This is associated with the plant's ability to store reserves in some organs and to mobilize them for yield production (Chaves et al., 2003). A short life cycle and maximal use of resources are particularly advantageous in environments with terminal drought stress or where physical or chemical barriers inhibit root growth (Blum, 1998).

Drought avoidance (DA) refers to the plant's ability to retain a relatively high level of hydration under water stress and involves two components: maximizing water uptake and minimizing water loss (Blum, 1998). Maximizing of water uptake can be achieved by increasing root growth, root thickness, root depth and mass (Price et al., 2002). Water loss can be minimized by closing stomata, through reduced absorption of radiation by leaf rolling, decreasing canopy area by reducing growth and shedding of older leaves. In selection and phenotyping of potato, these DA traits are often used as criteria Schafleitner (2009).

Drought tolerance (DT) response is defined as the capacity of plants to maintain functional growth under low resources (water and minerals). Drought causes the reduction in water potential of the cell, as a result of solute concentration gradients and osmosis, and leads to the loss of cell turgor. Furthermore, the reduction of available water, induces also a reduction in nutrients, especially nitrogen. Some plants have the ability to tolerate dehydration or maintain turgor pressure through an osmotic adjustment *via* the active accumulation of solutes called osmoprotectants (amino acids, sugar alcohols, polyols and quaternary ammonium and tertiary sulfonium compounds), ABA content or by an increase of antioxidative and/or other defense mechanisms (Reddy et al., 2004).

All drought resistance strategies are not mutually exclusive and plants may combine a range of different response types for optimal reaction to drought. In most temperate climates, dehydration tolerance is the only relevant mechanism but in more severe conditions, such as in southern Australia and other Mediterranean climates, a combination of different mechanisms can be achieved (Berger et al., 2010).

According to Munns et al. (2010) strategies for water use that confers drought tolerance can be quite different for annual and perennial species, and for dry land versus irrigated agriculture. For annual crops such as wheat and barley in semi-arid environments, with mild winters and hot summers, one successful strategy is a fast rate of development, and a short time to flowering and grain maturity, allowing the available water to be used by the plant before it is lost from the soil as the temperature increases. Another is to choose a slow-developing cultivar and sow early. Perennial species can employ a conservative strategy, minimizing the use of water to avoid the risk of leaf dehydration, and resuming a fast growth rate when the rainy season returns. According to the same authors, the sensitive growth response to drought would be beneficial in rain-fed conditions, while the less sensitive response for crops growing in irrigated land. Selecting genotypes with diverse responses to a decrease in soil water potential would provide an option to growers in different environments (Munns et al., 2010).

5.1 Breeding for drought resistance

A major goal in plant breeding is the production of crops with increased tolerance to abiotic stress. While natural selection has favored mechanisms for adaptation to stress conditions, the breeding efforts have directed selection towards increasing the economic yield of cultivated species, hence, stress-adaptive mechanisms have been lost in the elite gene pool of our current crop plants. The special problem is that the genetic pressure imposed on crop plants throughout early domestication, and modern plant-breeding has severely eroded the allelic variation of genes originally found in the wild, making crop species increasingly susceptible to diseases, pests and environmental stresses (Tanksley & McCouch, 1997).

The complexity of drought tolerance mechanisms explains the slow progress in breeding for drought conditions. Breeding for drought tolerance is further complicated by the fact that several types of abiotic stress (as high temperature or high irradiance, water and nutrient deficiency) in the field conditions can influence plants simultaneously and activate different molecular mechanisms.

Retrospective studies have demonstrated that selection of plants characterized by high yield potential and high yield stability has frequently led to yield improvements under both favorable and stress conditions (Cattivelli et al., 2008). Rizza et al. (2004) tested in rain/fed and irrigated conditions 89 barley genotypes representing a sample of the germplasm grown in Europe. Eight of them showed the best yield in both irrigated and rain-fed conditions. Now, further progress will depend on the introduction of traits in high yielding genotypes that are able to improve stress tolerance to multiple stress factors without detrimental effects on yield potential.

According to Zamir (2001) development of exotic genetic libraries consisting of marker-defined genomic regions taken from wild species and introgressed to the background of elite crop varieties will provide a resource for the discovery and characterization of genes that underlie traits of agricultural value. Using this approach Gur & Zamir (2004) were able to demonstrate that introgressed tomato lines carrying three independent yield-promoting genomic regions produced significantly higher yield than then control lines grown under drought conditions.

Concerning drought resistance strategies, the improvement was done in the breeding for the drought escape mechanism and for earlier flowering due to the relative simple screening traits which were on the control of only few genes (Ludlow & Muchow, 1990). Earliness is an effective breeding strategy for enhancing yield stability in Mediterranean environments where wheat and barley are exposed to terminal drought stress. In this condition shortening crop duration, a typical escape strategy, can be useful in synchronizing the crop cycle with the most favorable environmental conditions (Cattivelly et al., 2008).

Drought tolerance is a quantitative trait, with complex phenotype and genetic control. Therefore, the molecular approaches in crop improvement must be linked with suitable phenotyping protocols at all stages, such as the screening of germplasm collections, mutant libraries, mapping populations, transgenic lines and breeding materials and the design of OMICS and quantitative trait loci (QTLs) experiments (Salekdeh et al., 2009). However, despite the increasing knowledge on the mechanisms involved in plant response to stress, the advancement of high-throughput OMICS technologies (refers to the comprehensive analyses of plants ending in the suffix-omics such as genomics, proteomics and metabolomics) to screen large numbers of genes induced by drought mechanisms to regulate plant traits and also the increasing development of marker assisted selection in many crop species, the improvement of breeding to drought has been relatively modest.

Cattivelli et al. (2008) suggested that further breeding progress requires the introduction of traits that reduce the gap between yield potential and actual yield in drought-prone environments. To achieve these three main approaches can now be exploited: (1) plant physiology has provided new insights and developed new tools to understand the complex network of drought-related traits, (2) molecular genetics has discovered many QTLs affecting yield under drought or the expression of drought tolerance-related traits, (3) molecular biology has provided genes useful either as candidate sequences to dissect QTLs or for a transgenic approach.

Although there is evidence for a lot of physiological traits associated with the tolerance to drought (Table 5), the success in trait-based approaches considering the drought avoidance and drought tolerance mechanisms is not big. Table 5 presents some of these traits in different plants.

Traits	Plants	References
Plant growth and phenological phases (early or late flowering, extended crop duration, anthesis-silking interval, grain number, leaf growth, stay-green)	wheat, maize, sorghum, barley	Borrell et al. (2000); Edmeades et al. (1999); Rajcan & Tollenaar (1999); Richards (2006); Siddique et al. (1990); Slafer et al. (2005); Tardieu & Tuberosa (2010)
Photosynthesis (gas exchange, activities of key-enzymes, chlorophyll fluorescence)	grapevine, durum wheat	Chaves et al. (2002); Yousfi et al. (2010)
Assimilate partitioning and stem carbohydrates utilization	wheat, rice	Blum (1988); Kumar et al. (2006); Slafer et al. (2005)
Root growth and hydraulic properties	wheat, barley, oat	Hoad et al. (2001); Richards (2006)
Water status, osmotic adjustment, stomatal opening and related traits (leaf and canopy temperature, different spectral indices)	wheat, barley, maize, soybean	Chen et al. (2005); Morgan (2000); Munns et al. (2010)
Water use efficiency (WUE), carbon isotope discrimination	wheat, sunflower	Lambrides et al. (2004); Rebetzke et al. (2002); Siddique et al. (1990)

Table 5. Physiological traits associated with tolerance to drought in different agricultural plants.

Most of the physiological traits that impact on response to environmental stress require detailed, sophisticated and usually expensive techniques to phenotype plants, and can be applied only to a very limited number of genotypes (Sinclair, 2011). Plant resistance is usually assessed on the short term experiments in controlled conditions and many of the investigated traits are more appropriate for plant survival rather than maintaining plant productivity. Therefore, there is a need do develop the new phenotyping methods and platforms that will allow to screen available genetic resources and to monitor in situ the plant response to drought in the field conditions. Very efficient and promising are new non-imaging technologies as thermal infrared, near infrared, RGB visible or fluorescence that enable the dissection of plant responses to drought into a series of component traits (Berger et al., 2010; Munns et al., 2010).

As traits maximizing productivity normally expressed in the absence of stress can still sustain a significant yield improvement under mild/moderate stress, yield is therefore, a suitable target for breeding. Salekdeh et al. (2009) in his review paper presented a conceptual framework for drought phenotyping based on expressing yield as the product of 3 components: water use (WU), water use efficiency (WUE) and harvest index (HI). They suggested that such a phenotyping is also relevant for molecular biologists and geneticist working on grain crops. Furthermore, they identified protocols that address each of these factors, described their key features and illustrated their integration with different molecular approaches.

Quantitative trait locus (QTL) mapping provides a means to dissect complex traits, such as drought tolerance, into their components, each of which is controlled by QTLs. Molecular marker-supported genotypic information at the identified QTLs then enables quick and accurate accumulation of desirable alleles in plant breeding programmes. Plant tolerance to abiotic stress is mediated by complex traits that are sustained by multiple genetic factors with large QTLs-by-environment interactions. Due to these features, the practical application of marker-assisted selection for stress-related QTLs has proven difficult (Francia et al., 2005). The development of molecular marker technologies will help to identify a particular chromosomal location for genes regulating specific traits. The coincidence of loci for yield with the loci for the investigated trait will help in identifying if investigated trait is significant for drought resistance.

Genes connected to the drought could be those which encode an enzyme or other proteins. Many genes related to drought have been isolated and characterized in the last two decades in a variety of crop species. However, a lot of them was investigated in controlled conditions and not often proved in the field conditions. Therefore, it is difficult to exploit their expression and function for breeding processes. According to Cattivelli et al. (2008) the isolation of gene ERECTA that regulates transpiration efficiency in Arabidopsis and the transcriptional analysis of wheat genotypes with contrasting transpiration efficiency, is an example that demonstrated future approach for successful breeding. Significant progress in breeding for drought resistance will be achieved by integration of traditional breeding with physiology and genomics.

6. Conclusion

Agricultural production is highly dependent upon environmental variables, and it is expected that the climate change, especially drought, extreme temperature and water scarcity, will have significant effects on the food production and safety in many regions of the world. To address these challenges, the effort should be intensified to save water resources and to increase agricultural productivity per unit of water ("more crop per drop"). Better crop management and irrigation practice, deficit irrigation techniques and techniques for use of waste water for irrigation, will moderate the impact of climate change on water resources. However, the more efficient use of available water resources alone without growing of drought resistant crops could not have a significant long-term impact on reducing the impact of drought on agricultural production. Therefore, the more effort must be made in the future to produce crops able to deliver increased yields under drought conditions. In order to achieve this goal the focus should be in multidisciplinary approach, that integrates knowledge and research in the areas of crop physiology, genetic and molecular biology with the state-of-the-art breeding technologies.

7. Acknowledgment

This study was supported by EU Commission (FP6 project SAFIR) and Serbian Ministry of Education and Science (project TR 31005).

8. References

Ahmadi, S.H.; Andersen, M.N.; Plauborg, F.; Poulsen, R.T.; Jensen, C.R.; Sepaskhah, A.R. & Hansen, S. (2010a). Effects of irrigation strategies and soils on field-grown potatoes: Gas exchange and xylem [ABA]. *Agricultural Water Management*, Vol.97, No.10, (October 2010), pp. 1486-1494, ISSN 0378-3774

Ahmadi, S.H.; Andersen, M.N.; Plauborg, F.; Poulsen, R.T.; Jensen, C.R.; Sepaskhah, A.R. & Hansen, S. (2010b). Effects of irrigation strategies and soils on field grown potatoes: Yield and water productivity. *Agricultural Water Management*, Vol.97, No.11, (November 2010), pp. 1923-1930, ISSN 0378-3774

Alcamo, J.; Moreno J.M.; Nováky, B.; Bindi, M.; Corobov, R.; Devoy, R.J.N.; Giannakopoulos, C.; Martin, E.; Olesen, J.E. & Shvidenko, A. (2007). *Europe. Climate Change 2007: Impacts, Adaptation and Vulnerability. Contribution of Working Group II to the Fourth Assessment Report of the Intergovernmental Panel on Climate Change*, M.L. Parry, O.F. Canziani, J.P. Palutikof, P.J. van der Linden & C.E. Hanson, (Ed.), 541-580, Cambridge University Press, ISBN 978 0521 88010-7, Cambridge, UK

Anon (2003). *Water for People, Water for Life*: Executive Summary, United Nations World Water Development Report 003. Paris, France: UNESCO Publ., Available from http://unes doc.unesco.org/images/0012/001295/129556e.pdf.

Bates, B.C.; Kundzewicz, Z.W.; Wu, S. & Palutikof, J.P. (2008). *Climate Change and Water*, Technical Paper of the Intergovernmental Panel on Climate Change, 210, IPCC Secretariat, ISBN 978-92-9169-123-4, Geneva

Battilani, A.; Steiner, M.; Andersen, M.; Bak, S.N.; Lorenzen, J.; Schweitzer, A.; Dalsgaard, A.; Forslund, A.; Gola, S.; Klopmann, W.; Plauborg, F. & Andersen, M.N. (2010). Decentralised water and wastewater treatment technologies to produce functional water for irrigation. *Agricultural Water Management*, Vol.98, No.3, (December 2010), pp. 385–402, ISSN 0378-3774

Behbahaninia, A.; Mirbagheri, S.A.; Khorasani, N.; Nouri, J. & Javid, A.H. (2009). Heavy metal contamination of municipal effluent in soil and plants. *Journal of Food, Agriculture and Environment*, Vol.7, No.3-4, (July-October 2009), pp. 851-856, ISSN 1459-0263

Beniston, M.; Stephenson, D.B.; Christensen, O.B.; Ferro, C.A.T.; Frei, C.; Goyette, S.; Halsnaes, K.; Holt, T.; Jylhä, K.; Koffi, B.; Palutikof, J.; Schöll, R.; Semmler, T. & Woth, K. (2007). Future extreme events in European climate: an exploration of regional climate model projections. *Climatic Change*, Vol.81, Suppl.1, (May 2007), pp. 71-95, ISSN 0165-0009

Berger, B.; Parent, B. & Tester, M. (2010). High-throughput shoot imaging to study drought responses. *Journal of Experimental Botany*, Vol.61, No.13, (August 2010), pp. 3519–3528, ISSN 1460-2431

Blum, A. (1998). Improving wheat grain filling under stress by stem reserve mobilization. *Euphytica*, Vol.100, No.1-3, (April 1998), pp. 77–83, ISSN 1573-5060

Borrell, A.K.; Hammer, G.L. & Henzell, R.G. (2000). Does maintaining green leaf area in sorghum improve yield under drought? II. Dry matter production and yield. *Crop Science*, Vol.40, No.4, (July-August 2000), pp. 1037–1048, ISSN 1435-0653

Cattivelli, L.; Rizza, F.; Badeck, F.W.; Mazzucotelli, E.; Mastrangelo, A.M.; Francia, E.; Mare, C.; Tondelli, A. & Stanca, A.M. (2008). Drought tolerance improvement in crop plants: an integrated view from breeding to genomics. *Field Crops Research*, Vol.105, No.1-2, (January 2008), pp. 1–14, ISSN 0378-4290

Centritto, M.; Wahbi, S.; Serraj, R. & Chaves, M.M. (2005). Effects of partial rootzone drying (PRD) on adult olive tree (*Olea europaea*) in field conditions under arid climate. II. Photosynthetic responses. *Agriculture, Ecosystems & Environment*, Vol.106, No.2-3, (April 2005), pp. 303–311, ISSN 0167-8809

Chaves, M.M.; Pereira, J.S.; Maroco, J.P.; Rodrigues, M.L.; Ricardo, C.P.P.; Osorio, M.L.; Carvalho, I.; Faria, T. & Pinheiro, C. (2002). How plants cope with water stress in the field: photosynthesis and growth. *Annals of Botany*, Vol.89, No.7, (June 2002), pp. 907–916, ISSN 1095-8290

Chaves, M.M.; Pereira, J.S. & Maroco, J. (2003). Understanding plant response to drought — from genes to the whole plant. *Functional Plant Biology*, Vol.30, No.3, pp. 239–264, ISSN 1445-4416

Chen, D.Y.; Huang, J.F. & Jackson, T.J. (2005). Vegetation water content estimation for corn and soybeans using spectral indices derived from MODIS near- and short-wave infrared bands. *Remote Sensing and Environment*, Vol.98, No.2-3, (October 2005), pp. 225-236, ISSN 0034-4257

Costa, J.M.; Ortuno, M.F. & Chaves, M.M. (2007). Deficit irrigation as a strategy to save water: physiology and potential application to horticulture. *Journal of Integrative Plant Biology*, Vol.49, No.10, (October 2007), pp. 1421–1434, ISSN 1744-7909

Craufurd, P.Q. & Wheeler, T.R. (2009). Climate change and the flowering time of annual crops. *Journal of Experimental Botany*, Vol.60, No.9, (July 2009), pp. 2529-2539, ISSN 1460-2431

Davies, W.J.; Bacon, M.A.; Thompson, D.S.; Sobeigh, W. & Rodriguez, L.G. (2000). Regulation of leaf and fruit growth in plants in drying soil: exploitation of the plant's chemical signalling system and hydraulic architecture to increase the efficiency of water use in agriculture. *Journal of Experimental Botany*, Vol.51, No.350, (September 2000), pp. 1617-1626, ISSN 1460-2431

Davies, W.J.; Zhang, J.; Yang, J. & Dodd, I.C. (2011). Novel crop science to improve yield and resource use efficiency in water-limited agriculture. *Journal of Agricultural Science*, Vol.149, Suppl. S1, (February 2011), pp. 123-131, ISSN 0021-8596

De Souza, C.R.; Maroco, J.P.; Santos, T.; Rodrigues, M.L.; Lopes, C.; Pereira, J.S. & Chaves, M.M. (2003). Partial rootzone-drying: Regulation of stomatal aperture and carbon assimilation in field grown grapevines (*Vitis vinifera* cv. 'Moscatel'). *Functional Plant Biology*, Vol.30, No.6, pp. 653–662, ISSN 1445-4416

Dodd, I.C.; Theobald, J.C.; Bacon, M.A. & Davies, W.J. (2006). Alternation of wet and dry sides during partial rootzone drying irrigation alters root-to-shoot signalling of abscisic acid. *Functional Plant Biology*, Vol.33, No.12, pp. 1081–1089, ISSN 1445-4416

dos Santos, T.P.; Lopes, C.M.; Rodrigues, M.; Claudia, L.; de Souza, R.; Maroco, J.P.; Pereira, J.S.; Silva, J.R. & Chaves, M.M. (2003). Partial rootzone drying: effects on growth and fruit quality of field-grown grapevines (*Vitis vinifera*). *Functional Plant Biology*, Vol.30, No.6, pp. 663–671, ISSN 1445-4416

de la Hera, M.L.; Romero, P.; Gomez-Plaza, E. & Martinez, A. (2007). Is partial root-zone drying an effective irrigation technique to improve water use efficiency and fruit quality in field-grown wine grapes under semiarid conditions? *Agricultural Water Management*, Vol.87, No.3, (February 2007), pp. 261-274, ISSN 0378-3774

EEA (2008). *Impacts of Europe's changing climate – 2008 indicator-based assessment*, EEA Report No 4/2008, 242, EEA, European Communities, ISSN 1725-9177, Copenhagen

Eitzinger, J.; Stastna, M.; Zalud, Z. & Dubrovsky, M. (2003). A simulation study of the effect of soil water balance and water stress in winter wheat production under different climate change scenarios. *Agricultural Water Management*, Vol.61, No.3, (July 2003), pp. 195-217, ISSN 0378-3774

Edmeades, G.O.; Bolan˜os, J.; Chapman, S.C.; Lafitte, H.R. & Banziger, M. (1999). Selection improves drought tolerance in tropical maize populations: I. Gains in biomass, grain yield and harvest index. *Crop Science*, Vol.39, No.5, pp. 1306–1315, ISSN 1435-0653

Egea, G.; Nortes, P.A.; Gonzalez-Real, M.M.; Baille, A. & Domingo, R. (2010). Agronomic response and water productivity of almond trees under contrasted deficit irrigation regimes. *Agricultural Water Management*, Vol.97, No.1, (January 2010), pp. 171–181. ISSN 0378-3774

English, M.J. & Raja, S.N. (1996). Perspectives on deficit irrigation. *Agricultural Water Management*, Vol.32, No.1, (November 1996), pp. 1-14, ISSN 0378-3774

Enriquez, C.; Alum, A.; Suarez-Rey, E.M.; Choi, C.Y.; Oron, G. & Gerba, C.P. (2003). Bacteriophages MS2 and PRD1 in turfgrass by subsurface drip irrigation. *Journal of Environmental Engineering*, Vol.129, No.8, (September 2003), pp. 852–857, ISSN 0733-9372

FAO (2002). *Deficit Irrigation Practices. Water Reports Publication. 22*, 101, FAO, ISSN 1020-1203, Rome

FAO (2003). *World Agriculture Towards 2015/2030*. Available from http://www.fao.org/documents/show_cdr.asp?url_file=/docrep/004/y3557e/y3 557e00.htm.

Fereres, E. & Soriano, M.A. (2007). Deficit irrigation for reducing agricultural water use. *Journal of Experimental Botany*, Vol.58, No.2, (January 2007), pp. 147-159, ISSN 1460-2431

Fleisher, D.H.; Timlin, D.J. & Reddy, V.R. (2008). Interactive effects of carbon dioxide and water stress on potato canopy growth and development. *Agronomy Journal*, Vol.100, No.3, (May 2008), pp. 711-719, ISSN 1435-0645

Forslund, A.; Ensink, J.H.J.; Battilani, A.; Kljujev, I.; Gola, S.; Raicevic, V.; Jovanovic, Z.; Stikic, R.; Sandei, L.; Fletcher, T. & Dalsgaard, A. (2010). Faecal contamination and hygiene aspect associated with the use of treated wastewater and canal water for irrigation of potatoes (*Solanum tuberosum* L.). *Agricultural Water Management*, Vo.98, No.3, (December 2010), pp. 440-450, ISSN 0378-3774

Francia, E.; Tacconi, G.; Crosatti, C.; Barabaschi, D.; Bulgarelli, D.; Dall'Aglio, E. & Valè, G. (2005). Marker assisted selection in crop plants. *Plant Cell, Tissue and Organ Culture*, Vol.82, No.3, (September 2005), pp. 317–342, ISSN 0167-6857

Gencoglan, C.; Altunbey, H. & Gencoglan, S. (2006). Response of green bean (*P.vulgaris* L.) to subsurface drip irrigation and partial rootzone drying irrigation. *Agricultural Water Management*, Vol.84, No.3, (August 2006), pp. 274-280, ISSN 0378-3774

Goldhammer, D.A.; Salinas, M.; Crisosto, C.; Day, K.R.; Soler, M. & Moriana, A. (2002). Effects of regulated deficit irrigation and partial root zone drying on late harvest peach tree performance. *Acta Horticulturae*, No.592, (November 2002), pp. 343–350, ISSN 0567-7572

Gur, A. & Zamir, D. (2004). Unused natural variation can lift yield barriers in plant breeding. *PLoS Biology*, Vol.2, No.10, (October 2004), pp. 1610–1615, ISSN 1544-9173

Heaton, J.C. & Jones, K. (2008). Microbial contamination of fruit and vegetables and the behaviour of enteropathogens in the phyllosphere: a review. *Journal of Applied Microbiology*, Vol.104, No.3, (March 2008), pp. 613–626, ISSN 1364-5072

Hirschi, M.; Seneviratne, S.I.; Alexandrov, V.; Boberg, F.; Boroneant, C.; Christensen, O.B.; Formayer, H.; Orlowsky, B. & Stepanek, P. (2011). Observational evidence for soil-moisture impact on hot extremes in southeastern Europe. *Nature Geoscience*, Vol.4, No.1, (January 2011), pp. 17-21, ISSN 1752-0894

Hoad, S.P.; Russell, G.; Lucas, M.E. & Bingham, I.J. (2001). The management of wheat, barley and oat root systems. *Advances of Agronomy*, Vol.74, pp. 193–246, ISBN 978-0-12-000792-9

Hutton, R. & Loveys, B.R. (2011). A partial root zone drying irrigation strategy for citrus - ffects on water use efficiency and fruit characteristics. *Agricultural Water Management*, Vol.98, No.10, (August 2011), pp. 1485-1496, ISSN 0378-3774

Iglesias, A.; Garrote, L.; Quiroga, S. & Moneo, M. (2009). *Impacts of climate change in agriculture in Europe. PESETA-Agriculture study*, 59, Office for Official Publications of the European Communities, ISSN 1018-5593, Luxembourg

IPCC, (2007). *Climate Change 2007: Impacts, Adaptation, and Vulnerability*. Contribution of Working Group II to the Third Assessment Report of the Intergovernmental Panel on Climate Change. M.L. Parry, O.F. Canziani, J.P. Palutikof, P.J. van der Linden & C.E. Hanson, (Ed.), 976, Cambridge University Press, ISBN 978 0521 88010-7, Cambridge

Järup, L. (2003). Hazards of heavy metal contamination. *British Medical Bulletin*, Vol.68, No.1, (December 2003), pp. 167-182, ISSN 1471-8391

Jensen, C.R.; Battilani, A.; Plauborg, F.; Psarras, G.; Chartzoulakis, K.; Janowiak, F.; Stikic, R.; Jovanovic, Z.; Li, G.; Qi, X.; Liu, F.; Jacobsen, S.E. & Andersen, MN. (2010). Deficit irrigation based on drought tolerance and root signalling in potatoes and tomatoes. *Agricultural Water Management*, Vol.98, No.3, (December 2010), pp. 403-413, ISSN 0378-3774

Jovanovic, Z.; Brocic, Z. & Stikic, R. (2008). Effects of Partial Root Drying on Nitrogen Distribution in Potato. *X European Congress of the European Society for Agronomy*,

Italian Journal of Agronomy, Vol.3, No.3, Suppl. (July-September 2008), pp. 337-338, ISSN 1125-4718, Bologna, Italy, September 15-19, 2008

Jovanovic, Z.; Stikic, R.; Vucelic-Radovic, B.; Paukovic, M.; Brocic, Z.; Matovic, G.; Rovcanin, S. & Mojevic, M. (2010). Partial root zone drying increases WUE, N and antioxidant content in field potatoes. *European Journal of Agronomy*, Vol.33, No.2, (August 2010), pp. 124-131, ISSN 1161-0301

Kang, S.Z.; Liang, Z.S.; Pan, Y.H.; Shi, P.Z. & Zhang, J.H. (2000). Alternate furrow irrigation for maize production in an arid area. *Agricultural Water Management*, Vol.45, No.3, (August 2000), pp. 267-274, ISSN 0378-3774

Kang, S.; Hu, X.; Goodwin, I. & Jerie, P. (2002). Soil water distribution, water use, and yield response to partial root zone drying under shallow groundwater table condition in a pear orchard. *Scientia Horticulturae*, Vol.92, No.3-4, (February 2002), pp. 277–291, ISSN 0304-4238

Kang, S.Z. & Zhang, J.H. (2004). Controlled alternate partial root-zone irrigation: its physiological consequences and impact on water use efficiency. *Journal of Experimental Botany*, Vol.55, No.407, (November 2004), pp. 2437-2446, ISSN 1460-2431

Kimball, B.A.; Kobayashi, K. & Bindi, M. (2002). Responses of agricultural crops to free-air CO_2 enrichment. *Advances in Agronomy*, Vol.77, pp. 293-368, ISBN 978-0-12-000795-0

Kirda, C.; Cetin, M.; Dasgan, Y.; Topcu, S.; Kaman, H.; Ekici, B.; Derici, M.R. & Ozguven, A.I. (2004). Yield response of greenhouse-grown tomato to partial root drying and conventional deficit irrigation. *Agricultural Water Management*, Vol.69, No.3, (October 2004), pp. 191-201, ISSN 0378-3774

Kirda, C.; Topcu, S.; Cetin, M.; Dasgan, H.Y.; Kaman, H.; Topaloglu, F.; Derici, M.R. & Ekici, M.R. (2007). Prospects of partial root zone irrigation for increasing irrigationwater use efficiency of major crops in the Mediterranean region. *Annals of Applied Biology*, Vol.150, No.3, (June 2007), pp. 281–291, ISSN 0003-4746

Kjellström, E.; Bärring, L.; Jacob, D.; Jones, R. & Lenderink, G. (2007). Modelling daily temperature extremes: recent climate and future changes over Europe. *Climatic Change*, Vol.81, Suppl.1, (May 2007), pp. 249-265, ISSN 1573-1480

Kriedmann, P.E. & Goodwin, I. (2003). *Irrigation insights No. 3, Regulated deficit irrigation and partial rootzone drying*, Land and Water Australia, ISBN 0642 76089 6, Canberra

Kumar, R.; Sarawgi, A.K.; Ramos, C.; Amarante, S.T.; Ismail, A.M. & Wade, L.J. (2006). Partitioning of dry matter during drought stress in rainfed lowland rice. *Field Crops Research*, Vol.96, No.2-3, (April 2006), pp. 221-231, ISSN 0378-4290

Leib, B.G.; Caspari, H.W.; Redulla, C.A.; Andrews, P.K. & Jabro, J.J. (2006). Partial rootzone drying and deficit irrigation of 'Fuji' apples in a semiarid climate. *Irrigation Science*, Vol.24, No.2, (October 2005), pp. 85–99, ISSN 1432-1319

Lambrides, C.J.; Chapman, S.C. & Shorter, R. (2004). Genetic variation for carbon isotope discrimination in sunflower: Association with transpiration efficiency and evidence for cytoplasmic inheritance. *Crop Science*, Vol.44, No.5, (September-October 2004), pp. 1642–1653, ISSN 1435-0653

Lewitt, J. (1980). *Responses of plants to environmental stresses*, Academic Press, ISBN 0124455026, New York

Li, F.; Wei, C.; Zhang, F.; Zhang, J; Nong, M. & Kang, S. (2010). Water-use efficiency and physiological responses of maize under partial root-zone irrigation. *Agricultural Water Management*, Vol.97, No.8, (August 2010), pp. 1156–1164, ISSN 0378-3774

Liu, F.; Shahnazari, A.; Andersen, M.N.; Jacobsen, S.E. & Jensen, C.R. (2006). Physiological responses of potato (*Solanum tuberosum* L.) to partial root-zone drying: ABA signalling, leaf gas exchange, and water use efficiency. *Journal of Experimental Botany*, Vol.57, No.14, (November 2006), pp. 3727-3735, ISSN 1460-2431

Liu, F.; Song, R.; Zhang, X.; Shahnazari, A.; Andersen, M.N.; Plauborg, F.; Jacobsen, S.E. & Jennsen, C.R. (2008). Measurement and modelling of ABA signalling in potato (*Solanum tuberosum* L.) during partial root-zone drying. *Environmental and Experimental Botany*, Vol.63, No.1-3, (May 2008), pp. 385-391, ISSN 0098-8472

Loveys, B.; Stoll, M. & Davies, W.J. (2004). Physiological approaches to enhance water use efficiency in agriculture: exploiting plant signalling in novel irrigation practice. In: *Water use efficiency in plant biology*, M. Bacon, (Ed.), 113–141, Blackwell Publishing, ISBN 1-4051-1434-7, Oxford

Lubello, C.; Gori, R.; Nicese, F.P. & Ferrini, F. (2004). Municipal-treated wastewater reuse for plant nurseries irrigation. *Water Research*, Vol.38, No.12, (July 2004), pp. 2939–2947, ISSN 0043-1354

Ludlow, M.M. & Muchow, R.C. (1990). A critical evaluation of traits for improving crop yields in water-limited environments. *Advances in Agronomy*, Vol.43, pp. 107–153, ISBN 9780120007431

Luquet, D.; Vidal, A.; Smith, M. & Dauzat, J. (2005). 'More crop per drop': how to make it acceptable for farmers? *Agricultural Water Management*, Vol.76, No.2, (August 2005), pp. 108-119, ISSN 0378-3774

McCarthy, M.G.; Loveys, B.R. & Dry, P.R. (2002). Regulated deficit irrigation and partial rootzone drying as irrigation management techniques for grapevines. In: *Deficit Irrigation Practices. Water Reports Publication. 22*, 79–87, FAO, ISSN 1020-1203, Rome

Metin Sezen, S.; Yazar, A. & Tekin, S. (2010). Effects of partial root zone drying and deficit irrigation on yield and oil quality of sunflower in a Mediterranean environment. *Irrigation and Drainage*, Published online 23 Dec. 2010, Wiley Online Library (wileyonlinelibrary.com) DOI: 10.1002/ird.607

Mingo, D.M.; Theobald, J.C.; Bacon, M.A.; Davies, W.J. & Dodd, I.C. (2004). Biomass allocation in tomato (*Lycopersicum esculentum*) plants grown under partial rootzone drying: Enhancement of root growth. *Functional Plant Biology*, Vol.31, No.10, pp. 971–978, ISSN 1445-4416

Morgan, J.M. (2000). Increases in grain yield of wheat by breeding for an osmoregulation gene: relationship to water supply and evaporative demand. *Australian Journal of Agricultural Research*, Vol.51, pp. 971–978, ISSN 1836-5795

Moriondo, M. & Bindi, M. (2007). Impact of climate change on the phenology of typical Mediterranean crops. *Italian Journal of Agrometeorology*, No.3, (October 2007), pp. 5-12, ISSN 2038-5625

Morison, J.I.L.; Baker, N.R.; Mullineaux, P.M. & Davies, W.J. (2008). Improving water use in crop production. *Philosophical Transactions of the Royal Society B: Biological Sciences*, Vol.363, No.1491, (February 2008), pp. 639-658, ISSN 1471-2970

Munns, R.; James, R.A.; Sirault, X.R.R.; Furbank, R.T. & Jones, H.G. (2010). New phenotyping methods for screening wheat and barley for beneficial responses to water deficit. *Journal of Experimental Botany*, Vol.61, No.13, (August 2010), pp. 3499–3507, ISSN 1460-2431

O'Connell, M.G. & Goodwin, I. (2007). Water stress and reduced fruit size in micro-irrigated pear trees under deficit partial rootzone drying. *Australian Journal of Agricultural Research*, Vol.58, No.7, pp. 670–679, ISSN 1836-5795

Olesen, J.E. & Bindi, M. (2002). Consequences of climate change for European agricultural productivity, land use and policy. *European Journal of Agronomy*, Vol.16, No.4, (June 2002), pp. 239-262, ISSN 1161-0301

Olesen, J.E.; R. Carter, R.; Díaz-Ambrona, C.H.; Fronzek, S.; Heidmann, T.; Hickler, T.; Holt, T.; Mínguez, M.I.; Morales, P.; Palutikof, J.; Quemada, M.; Ruiz-Ramos, M.; Rubæk, G.; Sau, F.; Smith, B. & Sykes, M. (2007). Uncertainties in projected impacts of climate change on European agriculture and terrestrial ecosystems based on scenarios from regional climate models. *Climatic Change*, Vol.81, Suppl.1, (May 2007), pp. 123-143, ISSN 1573-1480

Parry, M.; Rosenzweig, C.A.; Iglesias, M.; Livermore, M. & Fisher, G. (2004). Effects of climate change on global food production under SRES emissions and socioeconomic scenarios. *Global Environmental Change*, Vol.14, No.1, (April 2004), pp. 53–67, ISSN 0959-3780

Peralta-Videa, J.R.; Lopez, M.L.; Narayan, M.; Saupe, G. & Gardea-Torresdey, J. (2009). The biochemistry of environmental heavy metal uptake by plants: Implications for the food chain. *The International Journal of Biochemistry and Cell Biology*, Vol.41, No.8-9, (August-September 2009), pp. 1665-1677, ISSN 1357-2725

Plauborg, F.; Abrahamsen, P.; Gjettermann, B.; Mollerup, M.; Iversen, B.V.; Liu, F.; Andersen, M.N. & Hansen, S. (2010). Modelling of root ABA synthesis, stomatal conductance, transpiration and potato production under water saving irrigation regimes. *Agricultural Water Management*, Vol.98, No.3, (December 2010), pp. 425-439, ISSN 0378-3774

Price, A.H.; Steele, K.A.; Moore, B.J. & Jones, R.G.W. (2002). Upland rice grown in soil-filled chambers and exposed to contrasting water-deficit regimes: II. Mapping quantitative trait loci for root morphology and distribution. *Field Crop Research*, Vol 76, No.1, (June 2002), pp. 25-43, ISSN 0378-4290

Rajcan, I. & Tollenaar, M. (1999). Source-sink ratio and leaf senescence in maize. I. Dry matter accumulation and partitioning during the grain-filling period. *Field Crops Research*, Vol.60, No.3, (February 1999), pp. 245–253, ISSN 0378-4290

Rebetzke, G.J.; Condon, A.G.; Richards, R.A. & Farquhar, G.D. (2002). Selection for reduced carbon isotope discrimination increases aerial biomass and grain yield of rainfed bread wheat. *Crop Science*, Vol.42, No.3, (May-June 2002), pp. 739–745, ISSN 1435-0653

Reddy, A.R..; Chaitanya, K.V. & Vivekanandan, M. (2004). Drought-induced responses of photosynthesis and antioxidant metabolism in higher plants. *Journal of Plant Physiology*, Vol.161, No.11, (November 2004), pp. 1189–1202, ISSN 0176-1617

Richards, R.A. (2006). Physiological traits used in the breeding of new cultivars for water-scarce environments. *Agricultural Water Management*, Vol.80, No.1-3, (February 2006), pp. 197–211, ISSN 0378-3774

Rizza, F.; Badeck, F.W.; Cattivelli, L.; Li Destri, O.; Di Fonzo, N. & Stanca, A.M. (2004). Use of a water stress index to identify barley genotypes adapted to rainfed and irrigated conditions. *Crop Science*, Vol.44, No.6, (November-December 2004), pp. 2127–2137, ISSN 1435-0653

Sadras, V.O. (2009). Does partial root-zone drying improve irrigation water productivity in the field? A meta-analysis. *Irrigation Science*, Vol.27, No.3, (December 2008), pp. 183-190, ISSN 1432-1319

Salekdeh, G.H.; Reynolds, M.; Bennett, J. & Boyer, J. (2009). Conceptual framework for drought phenotyping during molecular breeding. *Trends in Plant Science*, Vol.14, No.8, (August 2009), pp. 488-496, ISSN 1360-1385

Saeed, H.; Grove, I.G.; Kettlewell, P.S. & Hall, N.W. (2008). Potential of partial root zone drying as an alternative irrigation technique for potatoes (*Solanum tuberosum*). *Annals of Applied Botany*, Vol.152, pp. 71-80, ISSN 0003-4746

Sepaskhah, A.R. & Kamgar-Haghighi, A.A. (1997). Water use and yields of sugarbeet grown under every-other furrow irrigation with different irrigation intervals. *Agricultural Water Management*, Vol.34, No.1, (July 1997), pp. 71-79, ISSN 0378-3774

Sepaskhah, A.R. & Hosseini, S.N. (2008). Effects of alternate furrow irrigation and nitrogen application rates on winter wheat (*Triticum aestivum* L.) yield, water- and nitrogen-use efficiencies. *Plant Production Science*, Vol.1, No.2, (April 2008), pp. 250-259, ISSN 1349-1008

Sepaskhah, A.R. & Ahmadi, S.H. (2010). A review on partial root-zone drying irrigation. *International Journal of Plant Production*, Vol.4, No.4, (October 2010), pp. 241-258, ISSN 1735-8043

Schafleitner, R. (2009). Growing more potatoes with less water. *Tropical Plant Biology*, Vol.2, No.3-4, (December 2009), pp. 111-121, ISSN 1935-9756

Shahnazari, A.; Liu, F.; Andersen, M.N.; Jacobsen, S.E. & Jensen, C.R. (2007). Effects of partial root zone drying (PRD) on yield, tuber size and water use efficiency in potato under field conditions. *Field Crop Research*, Vol.100, No.1 (January 2007), pp. 117-124. ISSN 0378-4290

Shahnazari, A.; Ahmadi, S.H.; Laerke, P.E.; Liu, F.; Plauborg, F.; Jacobsen, S.E.; Jensen, C.R. & Andersen, MN. (2008). Nitrogen dynamics in the soil – plant system under deficit and partial root–zone drying irrigation strategies in potatoes. *European Journal of Agronomy*, Vol.28, No.2, (February 2008), pp. 65-73, ISSN 1161-0301

Siddique, K.H.M.; Tennant, D.; Perry, M.W. & Belford, R.K. (1990). Water use and water use efficiency of old and modern wheat cultivars in a Mediterranean-type environment. *Australian Journal of Agricultural Research*, Vol.41, No.3, pp. 431–447, ISSN 0004-9403

Sinclair, T.R. (2011). Challenges in breeding for yield increase for drought. *Trends in Plant Science*, Vol.16, No.6, (June 2011), pp.289-293, ISSN 1360-1385

Slafer, G.A.; Araus, J.L.; Royo, C. & Del Moral, L.F.G. (2005). Promising ecophysiological traits for genetic improvement of cereal yields in Mediterranean environments. *Annals of Applied Biology*, Vol.146, No.1, (January 2005), pp. 61–70, ISSN 1744-7348

Steele, M. & Odemeru, J. (2004). Irrigation water as a source of foodborne pathogens on fruit and vegetables. *Journal of Food Protection*, Vol.67, No.12, (December 2004), pp. 2839–2849, ISSN 0362-028X.

Stikic, R.; Savic, S.; Jovanovic, Z.; Jacobsen, S.E.; Liu, F. & Jensen, C.R. (2010). Deficit irrigation strategies: use of stress physiology knowledge to increase water use efficency in   tomato and potato, In: *Horticulture in 21st Century*, A.N. Sampson, (Ed.), 161-178, Nova Science, ISBN 978-1-61668-582-9, NewYork

Surdyk, N.; Cary, L.; Blagojevic, S.; Jovanovic, Z.; Stikic, R.; Vucelic-Radovic, B.; Zarkovic, B.; Sandei, L.; Pettenati, M. & Kloppmann, W. (2010). Impact of irrigation with treated low quality water on the heavy metal contents of a soil-crop system in Serbia. *Agricultural Water Management*, Vol.98, No.3, (December 2010), pp. 451-457, ISSN 0378-3774

Tanksley, S.D. & McCouch, S.R. (1997). Seed banks and molecular maps: Unlocking genetic potential from the wild. *Science*, Vol.277, No.5329, (August 1997), pp. 1063-106, ISSN 0036-8075

Tardieu, F. & Tuberosa, R. (2010). Dissection and modelling of abiotic stress tolerance in plants. *Current Opinion in Plant Biology*, Vol.13, No.2, (April 2010), pp. 206-212, ISSN 1369-5266

Vinten, A.J.A.; Lewis, D.R.; McGechan, M.; Duncan, A.; Aitken, M.; Hill, C. & Crawford, C. (2004). Predicting the effect of livestock inputs of E.coli on microbiological compliance of bathing waters. *Water Research*, Vol.8, No.14-15, (August-September 2004), pp. 3215–3224, ISSN 0043-1354

Wahbi, S.; Wakrim, R.; Aganchich, B.; Tahi, H. & Serraj, R. (2005). Effects of partial rootzone drying (PRD) on adult olive tree (*Olea europaea*) in field conditions under arid climate I. Physiological and agronomic responses. *Agriculture, Ecosystems & Environment*, Vol.106, No.1-2, (January 2005), pp. 289-301, ISSN 0167-8809

Wakrim, R.; Wahbi, S.; Tahi, H.; Aganchich, B. & Serraj, R. (2005). Comparative effects of partial root drying (PRD) and regulated deficit irrigation (RDI) on water relations and water use efficiency in common bean (*Phaseolus vulgaris* L.). *Agriculture, Ecosystems & Environment*, Vol.106, No.1-2, (January 2005), pp. 275-287, ISSN 0167-8809

Yousfi, S.; Serret, M.D.; Voltas, J. & Araus, J.L. (2010). Effect of salinity and water stress during the reproductive stage on growth, ion concentrations, $\Delta^{13}C$, and $\delta^{15}N$ of durum wheat and related amphiploids. *Journal of Experimental Botany*, Vol.61, No.13, (August 2010), pp. 3529–3542, ISSN 1460-2431

Zamir, D. (2001). Improving plant breeding with exotic genetic libraries. *Nature Review Genetics*, Vol.2, No.12, (December 2001), pp. 983–989, ISSN 1471-0056

Zegbe, J.A.; Behboudian, M.H. & Clothier, B.E. (2006). Yield and fruit quality in processing tomato under partial rootzone drying. *European Journal of Horticultural Science*, Vol.71, No.6, pp. 252–258, ISSN 1611-4426

Zegbe, J.A. & Serna-Péreza, A. (2011). Partial rootzone drying maintains fruit quality of 'Golden Delicious' apples at harvest and postharvest. *Scientia Horticulturae*, Vol.127, No.3, (January 2011), pp. 455–459, ISSN 0304-4238

Zegbe-Domínguez, J.A.; Behboudian, M.H. & Clothier, B.E. (2004). Partial rootzone drying is a feasible option for irrigation processing tomatoes. *Agricultural Water Management*, Vol.68, No.3, (July 2004), pp. 195-206, ISSN 0378-3774

Zhang, J. & Davies, W.J. (1990). Changes in the concentration of ABA in the xylem sap as a function of changing soil water status can account for changes in leaf conductance and growth. *Plant Cell Environment*, Vol.13, No.3, (April 1990), pp. 277-285, ISSN 1365-3040

6

A Review on Creating Drought Tolerant Crop Varieties

Ramesh Thatikunta
Acharya N. G. Ranga Agricultural University
College of Agriculture, Rajendranagar
India

1. Introduction

Sustainability and the overall management of water resources has been the one of the greatest challenges of the century. The world population has crossed the six billion mark. Based on the proportion of young people in the developing countries, the requirement for water would continue to increase significantly during the next few decades. This places enormous demand on the world's limited fresh water supply (Bharucha, 2005). United Nations estimates a total amount of water present on earth to be about 1400 M km³. Precipitation annually amounts to 1,10, 000 km³. Global current withdrawal amounts to 4500 km³, which exceeds the availability of approximately 4200 km³. By the year 2030, a global deficit of 40 per cent has been forecasted. At the same time, biotic and abiotic stresses which serve as barriers to achieve high crop yields tends to be on the increase. Such challenges can be met only by paying adequate attention to sustainable agriculture (Swaminathan and Jana,1992).

Available water resources for crop production include rainfall, canal water and ground water. Precipitation acts as a primary source of water for crops in command areas. Surface and ground water systems of irrigation supplement precipitation. Ground water through dug wells and bore wells supplement irrigation water requirement. Conjunctive use of surface and ground water can result in optimum utilization of water resources. Fresh water for agriculture around the world has become scarce, thereby threatens the productivity of crop lands. Causes for the increase in water scarcity have been diverse and location specific. These include falling ground water table, chemical pollution, malfunctioning of irrigation systems and increased competition from other sectors such as urban and industrial users. Farmers often experience total crop failure because of lack of water. Most of the rainfall received in rainfed areas show either erratic distribution or shortage at any particular stage of the crop leading to drought. To help the fate of water scarce farmers and to ensure global food security, every drop of water counts. Farmers do not invest enough in inputs to increase the crop production in such areas (Bouman and Aureus, 2009). Scarcity of irrigation water calls for well planned long term strategy for sustainable water resources management. Modernization of existing irrigation systems with better operation and maintenance involves participatory irrigation management (PIM), which could play an important role in irrigation management. PIM includes participation of users – the farmers – in planning, design, construction, operation, maintenance, financing, decision rules, monitoring and evaluation of irrigation system. For efficient use of limited irrigation water, precise knowledge of crop water requirements has become essential prerequisite (Reddy, 1999).

Water forms a major structural component and constitutes to 90 per cent of vegetative biomass. Only a small fraction is absorbed by plants and used in photosynthesis. As much as 99 per cent escapes as vapour by the process of transpiration. Accurate crop water requirement data has therefore become essential in irrigated agriculture. Methods of estimating crop water requirements depend on the desired level of accuracy, availability of equipment and technical know- how. The current review focuses on three objectives i.e., to explore the interactions and issues involved in extrapolating water use efficiency (WUE), second to outline the methods adopted to improve the efficiency in water use by soil and crop and third to spell out the success stories of effectiveness of water use.

2. Historical perspective

In rainfed areas, many traditional cultivars tolerant to drought have evolved because of continuous farmer selection pressure in drought environments over time. Besides, there also exists, a wide variety of indigenous plants that thrive with minimal or no irrigation, fertilizer or pest control. Ideally these native land races that thrive on natural rainfall do best in an ecosystem. There are many native Hawaiian plants that are less thirsty.

Farmers in their fields, mainly have to deal with technologies that reduce the non productive outflows (percolation, seepage and evaporation) while transpirational flows are maintained by plants. This can be practiced during land preparation, and during the actual crop growth period. Sandy soil drains well. The loose structure makes it easy to dig, warm up quickly after the winter months, aiding plant growth. At the same time less organic matter and minimum ability to retain water, offers plants less water and nutrients needed for healthy growth. The addition of compost greatly enhances the ability of sandy soil to retain enough water. Mulch placed on top of the beds helps to reduce the water loss. Such soil based inputs form the basis to improve crop productivity. Realizing the importance of soil and water conservation, farmers used to protect their lands especially rainfed by construction of long lines of bunds made of stones and or earth. Small tanks were construnted to retain and store rain water. Soil erosion control studies were initiated by German soil scientist Wollny between 1877 and 1895. Ten research centres were established in USA during the period from 1928 to 1933. A major stimulus for water requirement studies have been from the development of irrigation facilities in the western United States in the middle of nineteenth century and which expanded during the later part of that century (Jensen, 1968).

In the last century the increase in yield was mainly due to three reasons. First of this rise was due to better nutrition. Second was due to the introduction of highly productive pasture legumes which provided rich source of nitrogen to the following crops. Third and most spectacular was the control of endemic root diseases which gave farmers the confidence to apply more nitrogen which resulted in large responses in yield (Passioura, 2004).

3. Reduction in evapotranspiration

Attempt to control the supply of water to crops must be based on the understanding of variable state of water in the soil and its cyclic movement into, within and out of root zone.

The cycle of water in the field and its uptake by plants has been governed by evaporation (E), transpiration (T) which are collectively known as evapotranspiration (ET) and ET along with metabolic needs of the plant has been referred to as consumptive use (CU). Evaporation referred to as a cooling process includes latent heat of vapourization (585 cal g^{-1} of evaporated water). Transpiration on the other hand, leaves the body of a living plant and reaches the atmosphere as water vapour. However, a major difference exists between transpiration and evaporation. Transpiration essentially confines itself to day light hours and the rate depends on growth periods of plants. Evaporation, on the other hand continues all through the day and night with different rates. Factors that play an important role in evapotranspiration include physical (vapour pressure deficit, wind speed, quality of water), plant (rooting characteristics, distribution of stomata, roughness of leaf) and soil factors (available soil moisture, exposed soil surface). Potential yields are realized upon adequate supply of water through out the crop growing season and when ET is expected to be at its maximum. An ideal irrigation schedule for optimum yield indicates the correct timing (when to irrigate). Other factors of importance are soil and topography. In poorly permeable soils, percolation losses will be low and surface losses relatively high. In very permeable soils, percolation losses will be high and surface losses relatively low. With increase in slope, the risk of excessive infiltration losses become less, but the surface losses increase. Run off from the soil surface may be substantial during heavy rain, but much of the rain becomes run on in lower parts of the field. The net result is a little net loss from the field as a whole as there occurs greater vertical drainage from the low lying areas.

In principle the rate of water uptake by plants depends on the ability of roots to absorb water from soil with which they are in contact, as well as on the ability of soil to supply and transmit water towards roots at a rate sufficient to meet transpiration and growth requirements. These variables in turn depend on plant, soil and weather conditions. Cooper *et al*, (1983) gave a technique to calculate seasonal evaporation which ranged from about 60-160 mm.

Loss of water varies with seasonal weather and as well with the rate of development of leaf canopy and the presence of leaf mulches such as from stubbles of the previous crop. Impact of canopy size on evaporative loss from the soil can be manipulated by varying the crops nitrogen supply. Overall WUE values were not necessarily higher at luxurious nitrogen supply and ranged from 11-20 kg ha^{-1} mm^{-1}, as the crop could use too much of water during its vegetative stage and run out of water during grain filling period.

Evaporative losses of water from soil surface, can be minimized by the crop if there exists a rapid early development of leaf area which enhances the productive flow of water through the plants. Plant coverage of soil can also improve the carbon dioxide uptake. However trade off of too much water use during vegetative growth may leave little for grain filling.

The method of irrigation also contributes to irrigation losses. Application of the least amount of water required to bring the root zone depth to field capacity has been considered as efficient irrigation. The major control of irrigation rests in the hands of irrigator in the field. Quality of irrigation water assumes importance in arid climates where high rates of evaporation contribute to increased salt concentration in soil and drainage water. Application of saline or brackish water may hinder crop growth directly and also cause soil degradation. In a season application of 1000 mm of medium quality water introduces five tones of salt to one hectare. In all such instances proper leaching and drainage become an integral part of farming (Reddy, 1999).

Water plays a central role in the metabolism of plants. Loss of water from plant canopies by transpiration sets up a chain of reactions to replace the water lost. To design a successful irrigation system it becomes imperative to know the plant rooting characteristics, moisture extraction pattern and moisture sensitive stages.

Root systems in the field are seldom uniform with depth. Rooting depth of annual field crops on deep well drained soils range from 0.30 to 2.0 m. In general, root zone depth of crops reduces by 25 to 35 per cent on clayey soils and increased by 25-35 per cent on sandy soils. Usual moisture extraction shows that, about 40 per cent of the extracted moisture comes from upper quarter of the root zone while, 30, 20 and 10 per cent from lower quarters. Low frequency irrigation leads to depleting soil moisture from lower quarter of the root zone depth. Based on rooting depth field crops have been classified into shallow (rice, onion, cabbage, cauliflower, potatoes), medium (barley, wheat, castor, tobacco, chillies, peas, tomato) and deep rooted (cotton, maize, sorghum, pearl millet, sugarcane, soybean).

Optimum soil moisture for plant growth varies with the stage of crop growth. Certain periods of crop growth and development are more sensitive to soil moisture stress compared with others (panicle initiation, heading, flowering for rice, heading and flowering for pearl millet, flowering, seed formation for soybean and commencement of fruit set for tomato). Inadequate water supply during sensitive periods will irrevocably reduce the yield. Provision of adequate water and along with other management practices or at other growth stages would not help in recovering of the lost yield.

Lack of rainfall either through insufficient irrigation or rainfall causes drought stress in plants wherein too little water is available in a suitable thermodynamic state (Larcher, 2003). Rice crop needs twice as much of water, than many other crops to produce good yields. For production of one kg of rough rice 2500 liters of water is needed, of which, 1400 liters are used up in evapotranspiration and 1100 liters are lost in seepage and deep percolation (Bouman and Aureus, 2009). To grow crops like rice with special water requirements, lack of water at one critical plant growth stage or another can decrease productivity. Besides, erratic distribution and shortage particularly at flowering and again at grain filling seriously limits productivity depending on the growing duration of the crops or varieties. Such of those losses need to be quantified in various crops and efforts made to reduce them without any decrease in the yield. These aspects need to be justified in crops and situations.

4. Available technologies

Annual rainfall in several parts of drylands may be sufficient to raise one or two crops. In such areas high intensity storms and erratic rainfall leads to runoff and erosion. The effective rainfall is less than 50 per cent and seldom exceeds and reaches 65 per cent. In all such instances soil management practices need to be tailored to store and conserve as much rainfall as possible by reducing runoff and increasing storage capacity of soil profile. A number of simple technologies (tillage, fallowing, mulching, contour bunding) have been developed to prevent or reduce water losses and to increase water intake.

Farmers who practice alternate wetting and drying (AWD) method, use the field tube technology to monitor the underground water. When the water level fell below 15-20 cm below the surface of the soil, field was again flooded. The technology reduced the amount of water by a quarter and more importantly it did not reduce the crop yields (Bouman and Aureus, 2009).

Over the seasons water escapes the roots of annual crops and accumulates in deep sub soil and becomes accessible to the roots of deep rooted perennial agricultural crops like Lucerne grown for two to three years. Such crops prove valuable when they are sown at the start of the drought. Lucerne also reduces the risk of environmental damage such as dry land salinity and eutrophication of discharge areas (Ridley et al., 2001).

Semi dwarf varieties dominate wheat cultivation throughout the world. The dwarfism in these arises from insensitivity to gibberellic acid (GA) conferred by genes Rht 1 and Rht2 which exhibit short coleoptiles. Seed sown deeply i.e., > 60 cm dies as the leaf does not emerge from soil. Here agronomy and breeding strongly affect crop canopy development. The aspect assumes importance when the seed is directly sown on untilled soil in dry lands where soil structural problems arise. Rht8 a candidate gene enables emergence from sowing depths as great as 120 mm but yet provides adequate dwarfing of the canopy (Rebetzke et al., 1999).

Intensified efforts of International Rice Research Institute (IRRI) to cope with this looming water shortage, has lead to development of drought tolerant cultivars that use less water (aerobic rice). Reduction in length of growing period by two weeks, ahead of the previously used varieties of shorter duration allowed crop to be harvested, three times instead of two times a year. Such varieties can be planted even in dry season without any fear of loss. Once again, with the onset of rain the same crop which showed wilting has recovered. Thus it has been revealed that the crop could experience drought at any stage and withstand drought even at the reproductive stage, when crop suffers the maximum. The drought tolerance trait enables broadcasting of seeds in lieu of transplanting whereby saving a lot of money on labourers to plant the seedlings. The success of adoption of the variety was more attributed to farmer to farmer influence and support from the local government. Farmers were also pleased with plant traits like its ability to tolerate a month long drought, recovery from stress, early maturity and good eating quality.

The "stay green" character in case of sorghum arises from the positive feed back in nitrogen acquisition. Plants in this case maintain nitrogen in their leaves during grain filling and fix more carbon, which in turn enables roots to continue extracting soil nitrogen, so that the system is self reinforcing (Borrell et al., 2001). "Haying off" in which the crop senesces prematurely and its yield responds negatively to nitrogen can result from the phenomenon that when ever plant takes up nitrogen from either fertilizer or mineralization of organic matter, grows vigourously, utilizes too much of water before flowering, sets large number of flowers for which the plant can not mobilize carbohydrates from current photosynthesis or from remobilized sources and results in low yields.

Several earlier studies reported low selection efficiency for grain yield under drought stress. Much of the initial efforts focused on improvements of secondary traits such as root architecture, leaf water potential, panicle water potential, osmotic adjustment and relative water content. Recent studies have proved that these traits rarely have higher broad sense heritability (H) than grain yield under drought stress and are often not highly correlated with grain yield (Kumar et al., 2008).

Plant breeders have produced a large number of cultivars that flower close to optimal time in a given environment. Global warming in the coming decades could alter the time of flowering. Breeders are now in a position to tune the phenology even without being consciously aware that they are doing so.

In the face of the troubling reality, to help farmers to cope with the water scarce situations several approaches / technologies have been developed across the globe. Concerted efforts are needed through an interdisciplinary approach, to define criteria for stress assessment and to evolve breeding approaches for incorporating stress tolerance in crops (Paroda, 1986). Some of these methodologies include those of IRRI which works with national agricultural and extension system (NARES) for the evaluation of newly developed breeding lines, India – IRRI collaborative project and drought breeding network (DBN) that identifies the promising entries for drought prone ecosystem. Participatory varietal selection (PVS) testing and evaluation aims to develop prototype aerobic rice production systems for water scarce environments. Variety identification committee (VIC) recommends release of varieties to the central sub committee on crop standards Notification and Release of Varieties (Reyes, 2009).

In the recent times new concepts were initiated and experiments were conducted on direct selection for grain yield under drought stress wherein high yield potential under irrigated situation was combined with good yield under drought. Drought Physiologist involved in dissecting the mechanisms of drought tolerance and its genetic variation in rice, says that combining high yield potential and drought tolerance through direct selection for grain yield is one of the right approaches for developing drought-tolerant lines. In addition, new molecular tools like marker-assisted selection can aid in crop improvement. Previously scientists who worked on to improve the traits thought that drought tolerance was related to traits like leaf rolling, rooting depth etc. They believed that yield under drought could be increased by improving these secondary traits. Their efforts resulted in limited success. In this context, it is noteworthy to state that a successful recurrent selection for increased grain yield in drought stressed tropical maize was associated with a decrease in root mass (Bolanos et al., 1993).

5. Success stories

Biotic and abiotic stresses which become a part of climate change show uncertainties that can be met only by paying adequate attention to the biological software essential for sustainable agriculture. Identification of plant attributes that are important for such farming conditions forms an efficient plant breeding strategy. On the other hand conservation of genetic diversity under the conditions that promote desirable combinations of plant attributes in individual genotypes and enhance their frequency acts as an effective strategy to reduce genetic vulnerability (Swaminathan and Jana, 1992).

Conventional breeding for drought tolerance revolved around to develop populations with large genetic variation by phenotyping populations. Extensive efforts made in the last ten years to tackle the most complex and recalcitrant abiotic stress i.e., drought, improved rice yields in rainfed drought prone ecosystems. Direct selection for grain yield assessed under low land reproductive stage stress involved crosses between tolerant parents and lines with high yield potential, followed by direct selection of progeny in replicated trials for yield under optimal conditions and managed stress. In all such instances creation of stress levels that reduce the yield by 65 to 85% relative to unstressed controls only proved worthwhile. Stress was created by use of reliable tensiometers in trails, by use of number of drought susceptible and tolerant checks at repeated intervals, monitoring leaf rolling and leaf drying, proper monitoring of water table depth in low land trials. Irrigation by over head sprinklers were used in evaluation of upland stress. These served as tools that can assist breeders in selection. Three genotypes viz., IR 42253-61-1-1-2-3, JGL 384 and Badshah Bhog significantly out yielded the two mega varieties IR64 and Swarna. Yield advantage in stress selected lines was mainly due to high harvest index (HI) under stress (Kumar *et al.*, 2008). IRRI has come out with two drought tolerant rice breeding lines and recommended their release. IR74371-70-1-1 (Sahbhagi dhan) in India and its sister line IR74371-54-1-1 (5411) in the Philippines showed wide adaptability and have major impacts on the life and sustainability of farmers (Reyes, 2009).

Stable isotopes of carbon and oxygen have been used extensively as an interdisciplinary approach to save and utilize the available water by crops. The work on stable isotopes has led to breeding programme that involved carbon isotope discrimination (CID). The isotopic analysis of plant tissues enabled selection for intrinsically transpiration efficient plants. The technique resulted in production of wheat cultivar "Drysdale" now in commercial use in Australia. This is one of the very few examples of physiological analysis leading directly to a new cultivar. The variety has a promise to increase water limited yields by ten percent above the widely grown cultivar – Hartog. The breeding line was selected for intrinsically higher transpiration efficiency. As expected the trait has greater impact when the rainfall has been low (Passioura, 2004).

6. Novel approaches

Revolutionary changes in cropping patterns are expected when availability of irrigation water extends to five to six months in a year. Farmers can take up one or more additional crops after the harvest of main crop in the same land. Allotment of 33 and 67 per cent of area for wet cultivation in kharif in black soils and ill drained and water logged areas appears to be the ideal pattern for localization. Where ever single distributory or pipe outlet was available, pattern of irrigation to be followed would be wet or light irrigation with cultivation of less water intensive high value crops. The criteria for selection of main and second crops would be that main crop has to be grown in wet or monsoon season and second crop to be grown in dry season. For multiple cropping to be successful, a combination of fine and dry weather with sufficient irrigation and adequate drainage becomes essential.

The traditional methods (like diallel analysis, generation mean analysis, factorial design etc.) were used to investigate the genetic control of quantitative traits. These techniques however, do not provide information on the chromosome regions governing the naturally occurring variability in WUE and on the genetic causes like linkage or pleiotropy and their association with yield. Roots were observed to be positively associated with yield under drought stress and the region RG939-RG214 on chromosome 4 strongly affected root traits and yield. It is notable that successful recurrent selection for increased grain yield in drought stressed maize was associated with decrease in root mass. Whereas in crops which have limited capacity to adjust osmotically, water use and yield stabilization under drought with stored moisture available in deeper soil layers was recovered in genotypes with selection for faster growing and deeper roots (Bolonos *et al.*, 1993). The difficulty in studying roots was circumvented by growing plants in pots and / or chambers filled with soil as is the case with rice (Wade *et al.*, 2000).

There has been a considerable debate in recent years on the potential impact of biotechnologies in agriculture. The impact of biotechnology in overcoming hunger may have to wait till the next millennium. The tools of biotechnology can help raising the productivity of major crops through an increase in total dry matter production which can then be partitioned in a way favourable to the economic part. Enhancing biomass production and its conversion into energy have important applications like as in cardamom, oil palm through tissue culture methodologies (Swaminathan, 1992).

For improving crop performance in terms of WUE and other traits associated with yield through marker assisted selection (MAS), it has become essential to map the quantitative trait loci (QTLs). The next logical step is to identify suitable candidate genes accounting for QTL effects, validate their roles and proceed with manipulation with the gene itself. Identification of genes and elucidation of their role can be facilitated through high throughput profiling of "omics" approach which includes transcriptomics, proteomics and metabolomics. These new approaches provide opportunities and challenges to analyze the changes in concerted expression of thousands of genes to evaluate the responses to water deficit (Tuberosa, 2004). Some of the constraints can be partially overcome through the identification of QTLs. The dissection of quantitative trait has been possible through the production of experimental population consisting of either 100-200 F_2 plants, F_3 families, recombinant inbred lines (RIL), double haploids derived from a cross between two inbred lines differing for a trait of interest. Molecular distance of about 20 cM becomes essential to detect the presence of QTL having a major effect on the phenotype. Model species like arabidopsis need 40-50 markers and over 300 needed in large genome species like durum and bread wheat to detect QTLs. Logarithm of odds ratio (LOD) ($\log_{10}$ of the ratio between the chance of a real QTL measured at a position divided by the chance of having similar effect with no QTL present can be set to a score of > 2.5 in the mapping population under study. Such detection also avoids false positive QTLs (i.e., declaring the presence of a ghost QTL when the actual QTL is absent). The technique of QTL analysis also enables with precision the genetic basis of trait association merely by looking for co location of the genetic map of the corresponding QTL and comparing the effect of the trait. The differing mapping populations profiled with a set of restriction fragment length polymorphism (RFLP) and simple sequence repeat (SSR) markers has made it possible for detailed comparison between the QTLs. For example in rice, the mapping populations most extensively used for root traits and other WUE traits has been derived from Bala x Azucena (Price and Tomos, 1997). It becomes orthwhile to mention here that QTLs for flowering time has already been cloned in rice (Kojima *et al.*, 2002) and the positional cloning of *Vgt1* is well advanced in wheat (Salvvi *et al.*, 2002).

Now through molecular approaches, major QTLs for grain yield under drought have been identified in the background of popular mega varieties. The aim is to pyramid these QTLs and enhance the yield. Physiological and molecular mechanisms like Marker Assisted Selection (MAS) related to drought tolerance in some of the QTLs have been studied in rice and other crops as well and varieties have been in pipeline for release. From the information reviewed it becomes evident that genetic variability in WUE is an interaction of multitude of quantitatively inherited morpho-physiological traits whose effect on yield varies considerably according to the prevailing environment. Undermining such traits and understanding their cause effect relationships simplifies the complexity of a trait and makes it amenable for direct selection by a breeder (Tuberosa, 2004).

7. Summary

There has been an increase in the demand for more food from the natural resources to feed the growing population. At the same time, the primary concern has been to improve the efficiency of soil and plant water use to bring about productivity. Among the traits more attention needs to be paid to improve intrinsic and / or seasonal WUE, root architecture and photosynthetic efficiency.

In the past both empirical and analytical breeding have contributed to the improvement of seasonal WUE and yield of crops under both well watered and water limited conditions. This has been possible not only due to better partitioning but also due to increased biomass production. The latter has been possible by increased extraction of water from soil rather than by increased intrinsic WUE. Dyrsdale has been an exception as its yield advantage and selection is based on Δ ^{13}C studies.

New initiatives integrating population biology with socio economics are required for sustainable management and utilization of worlds rapidly diminishing biological wealth. It must be realized that past approaches to achieving quick fix genetic advance are no longer appropriate. Interdisciplinary coordinated nature of work is needed and has resulted in development of less thirsty varieties, though, much hope depends on use of conventional breeding techniques with the available new technologies that include use of markers to bring about sustainability.

The challenge would be to best and most effectively integrate the materials developed by conventional breeding and the information generated through the new innovative techniques and approaches.

8. References

Bharucha Erach. 2005. Text book of environmental studies for undergraduate courses. University press, New Delhi, pp 13-49.

Bolanos, J., Edmeades, G and Martinez, L. 1993. Eight cycles of selections for drought tolerance in lowland tropical maize. III. Response in drought adaptive physiological and morphological traits. Field Crops Research, 31: 269-286.

Borrell, A., Hammer G and Van Oosterom, E. 2001. 2001. Stay green: A consequence of balance between supply and demand for nitrogen during grain filling? Annals of Applied Biology, 138: 91-95.

Bouman, B and Aureus, A. 2009. Every drop counts. Rice Today. IRRI Publishers, pp 16-18.

Cooper, P. J. M., Keatinge, J. D. H and Hughes, G. 1983. Crop evapotranspiration – a technique for calculation of its components by field measurements. Field Crops Research, 7: 299-312.

Jensen, M. E. 1968. Water consumption by agricultural plants. In: (Kozlowski, T. T. Ed.). Water deficit and plant growth, Volume II. Plant water consumption and response. Academic Press. New York. 1-22.

Kojima, S., Takahashi, Y., Kobayashi, Y., Monna, L., Saski, T., Araki, T and Yano, M. 2002. *Hd3a*, a rice orthologue of the Arabidopsis FT gene promotes transition to flowering downstream of *Hd1* under short day conditions. Plant and Cell Physiology, 43: 1096-1105.

Kumar, A., Jerome Bernier., Satish Verulkar., Lafitte, H. R. and Atlin, G.N. 2008. Breeding for drought tolerance: Direct selection for yield, response to selection and use of drought tolerant donors in upland and low land – adapted populations. Field Crops Research, 107: 221-231.

Larcher Walter. 2003. Physiological plant ecology. (Larcher Walter. Ed.). Ecophysiology and stress physiology of functional groups. Springer. 401-416.

Paroda, R. S.1986. Breeding approaches for drought resistance in crop plants. In: (Chopra, V. L. and Paroda, R. S. Ed.). Approaches for incorporating drought and salinity resistance in crop plants. Oxford and IBH Publishing Co Pvt. Ltd. New Delhi. 87-107.

Passioura John. 2004. Water use efficiency in farmers fields. In: (Mark A Bacon. Ed.) Water use efficiency in plant biology. Blackwell Pub. CRC press. U. K. 302-318.

Price, A. H and Tomos, A. D.1997. Genetic dissection of root growth in rice (Oryza sativa L.) II. Mapping quantitative trait loci using molecular markers. Theoretical and Applied Genetics, 95: 143-152.

Rebetzke, G. J., Richards, R. A., Fischer, V. M and Mickelson, B. J. 1999. Breeding long coleoptiles, reduced height wheats. Euphytica, 106, 159-168

Reddy S, R. 1999. Irrigation water management. In Principles of Agronomy (Reddy S.R. Ed.), Kalyani publishers. New Delhi. 346-520.

Reyes, C Lanle. 2009. Making rice less thirsty. Rice Today, 12-15.

Ridley, A, M., Christy, B., Dunin, F. X., Haines, P. J., Wilson, K. F and Ellington, A. 2001. Lucerne in crop rotations on the reverine plains 1. The soil water balance. Australian Journal of Agricultural Research, 52: 263-277.

Salvi, S., Tuberosa, R., Chiapparino, E., Maccarerri, M., Veillet, S., Van Beuningen, L., Issac, P., Edwards, K and Phillips, R, L. 2002. Towards positional cloning of Vgt1, a QTL controlling the transition from the vegetative the reproductive phase in maize. Plant Molecular Biology, 48: 601-613.

Swaminathan, M. S. 1992. Biodiversity and biotechnology : Biodiversity implications for global food security (Swaminathan, M, S. and Jana, S. Ed.) . Macmillan India limited, Madras. 264-277.

Swaminathan, M. S. and S Jana. 1992. Introduction In : Biodiversity implications for global food security (Swaminathan, M, S. and Jana, S. Ed.). Macmillan India ltd. Madras. 1-7.

Tuberosa, R. 2004. Molecular approaches to unravel the genetic basis of water use efficiency. In : (Mark A Bacon. Ed.). Water use efficiency in plant biology. Blackwell publishers. CRC press. U. K. 228-282.

Wade, L. J., Kamoshita, A., Yamauchi, A. and Azhiri-Sigari, T. 2000. Genotypic variation in response of rainfed low land rice to drought and rewatering. I. Growth and water use. Plant Production Science, 3, 173-179.

7

Drought Stress and the Need for Drought Stress Sensing in a World of Global Climate Change

Rita Linke
Society for the Advancement of Plant Sciences
Austria

1. Introduction

Water scarcity imposes huge reductions in crop yield and is one of the greatest limitations to crop expansion outside present-day agricultural areas. Because the scenarios for global environmental change suggest a future increase in aridity and in the frequency of extreme events in many areas of the earth (Schär, 2006), maintaining crop yields under adverse environmental conditions is probably the major challenge facing modern agriculture. Nowadays, approximately 70% of the global available water is used in agriculture and 40% of the world food is produced in irrigated soils (Somerville & Briscoe, 2001). During the next 25 years, world population is expected to increase by about 2.5 billion people expecting food requirements in the developing world to double by 2025. An efficient use of water is therefore needed for the conservation of this limited resource (Somerville & Briscoe, 2001).

2. Effects of drought stress on plant physiology

Drought is a meteorological term which is commonly defined as a period without significant rainfall. Drought stress in plants generally occurs when the water available in the soil is reduced and atmospheric conditions further cause a continuous loss by transpiration and evaporation (Jaleel, 2009). Responses of plants to drought stress however are complex, involving adaptive changes and/or deleterious effects. Strategies to cope with drought stress normally involve a mixture of stress avoidance and tolerance mechanisms that vary with plant genotype (Chaves, 2002). Plant growth is accomplished through cell division, cell enlargement and differentiation, and involves genetic, physiological, ecological and morphological events and their complex interactions. Many yield-determining physiological processes in plants respond to water stress. Yield integrates several of these processes in a complex way. Thus, it is difficult to interpret how plants accumulate, combine and display the ever-changing and indefinite physiological processes over the entire life cycle. For water stress, severity, duration and timing of stress, as well as responses of plants after stress removal, and interaction between stress and other factors are extremely important (Plaut, 2003).

The continuity of water columns from soil pores through the plant to leaf cells, linked to the evaporative flux, is known as the soil–plant–atmosphere continuum (SPAC). Maintenance of this hydraulic system is needed to ensure a continuous water supply to leaves. The higher the capacity to provide such supply, the faster the leaf expansion (Nardini & Salleo, 2002) and the higher the potential for carbon gain (Sperry, 2000; Tyree, 2003), as has been observed for different life forms (Brodribb *et al.*, 2005), species (Brodribb & Field, 2000; Sack *et al.*, 2003), and genotypes (Sangsing *et al.*, 2004; Maseda, 2006). Under conditions of drought stress, however, genetic variations in leaf area growth, leaf area duration and/or leaf photosynthesis might become very important (Richards, 2000).

Abiotic stress occurring during canopy development will modify many of the canopy characteristics compared with a well watered crop. Leaves are often smaller creating a more erectophile canopy than when unstressed (Araus, 1986). In addition, there may be fewer tillers in grain crops, and this, together with the lower leaf area reduces the leaf area index (LAI), which is defined as the one sided green leaf area per unit ground area. Finally, due to accelerated senescence, shorter green area duration may reduce the potential for assimilation (Araus, 2002) and therewith crop yield, which is dependent on seed filling duration (Egli & Crafts-Brandner, 1996) as well as on leaf area duration (Geisler, 1983; De Costa, 1997). Nonetheless, there is a wide consensus that the reproductive growth stage is the most sensitive to water deficit. It is also recognized that drought stress at the reproductive stage is the most prevalent problem in rainfed drought prone agriculture, at least simply because in most rainfed ecosystems the crop season's rains diminish towards flowering and harvest time (Blum, 2009). In summary, prevailing drought reduces plant growth and development, leading to hampered flower production and grain filling and thus smaller and fewer grains. The reduction in grain filling occurs mainly due to a reduction in the assimilate partitioning and activities of sucrose and starch synthesis enzymes (Farooq, 2009).

On plant level, depletion of soil water reserve causes a variety of symptoms, with timescales ranging from a few minutes (wilting, stomatal closure), to weeks (change in leaf growth, senescence) or months (decrease in total biomass or yield; Tardieu, 1996). Increased senescence rates are regularly observed in plants subjected to water deficit in the field. They already occur at relatively moderate leaf water potentials, and begin in older leaves located in the lowest layer of the canopy (Tardieu, 1996). The shedding of older leaves also contributes to water saving and can be viewed as a recycling program within the plant, allowing the reallocation of nutrients stored in older leaves to the stem or younger leaves (Chaves, 2003).

While the hydration states of different tissues are very sensitive to the magnitude of the hydraulic conductance, the direct physical control of transpiration itself resides almost entirely in environmental conditions of temperature and humidity and the stomatal regulation of gas-phase diffusion between leaf-air spaces and the atmosphere (van den Honert, 1948 in Comstock, 2002). Stomata must regulate transpiration in a way that sufficient carbon is gained while leaf water potential (ψ_w) is prevented from becoming too negative and the break-down of the plants hydraulic system is avoided (Tyree & Sperry, 1988; Jones & Sutherland, 1991; Schultz & Matthews, 1997). A decrease in stomatal conductance can correlate with a declining ψ_w during soil drying, but can also occur before any measurable change in ψ_w is recorded (Gollan, Turner & Schulze, 1985; Trejo & Davies 1991; in Schultz, 2003). However, the relationship between stomatal closure and the lowering of plant water potential varies between different life forms and species as well as between plants of the different photosynthetic types (i.e. C_3 and C_4 and CAM plants).

Depending on how narrowly plants control their ψ_w, homeohydric plants are further classified as either isohydric or anisohydric (Maseda & Fernandez, 2006). A perfectly isohydric plant would close stomata, reducing transpiration as needed to maintain pre-drought leaf water status, whereas a perfectly anisohydric plant would keep stomata comparatively more open, reducing ψ_w just enough to maintain pre-drought leaf transpiration (Maseda & Fernandez, 2006). Maize for example, is an isohydric plant and therefore shows a less negative ψ_w during drought periods compared to wheat which is anisohydric (Henson et al., 1989; Tardieu, 1998). The distinction between isohydric and anisohydric plants, however, is often a matter of degree, and most plants operate under a relatively well-buffered range of ψ_w (Maseda & Fernandez, 2006).

Owing to reduced leaf water potential under conditions of low soil water content, leaf osmotic potential is reduced due to the simple effect of solute accumulation. However, if during the course of cellular water loss, solutes are actively accumulated, osmotic potential would be reduced beyond the rate dictated by the mere effect of concentration. Such accumulation of solutes during the development of water deficit is termed osmotic adjustment or osmoregulation (Zhang, 1999 and references therein). In general, osmotic adjustment (OA) is achieved by absorbing ions (e.g., K^+, Na^+, Ca^{2+}, Mg^{2+}, Cl^-, NO_3^-, SO_4^{2-}, and HPO_4^-) or by accumulating organic solutes (e.g. free amino acids, sugar alcohols, quaternary ammonium compounds and sugars). As a consequence, the osmotic potential of the cell is lowered, which in turn, attracts water into the cell and, thereby, tends to maintain its turgor (Morgan, 1984; Serraj & Sinclair, 2002). The accumulation of such compounds can protect cell membranes, proteins and metabolic machinery, which helps to preserve subcellular structure from damage as a result of cell dehydration (Serraj & Sinclair, 2002 and references therein).

Accumulation of solutes in roots leads to a lowering of the osmotic potential of the root, which maintains the driving force for extracting soil water under water deficit conditions (Wright et al., 1983). An increasing number of reports provide evidence on the association between high rate of osmotic adjustment and sustained yield or biomass under water-limited conditions across different cultivars of crop plants. Since osmotic adjustment helps to maintain higher leaf relative water content (RWC) at low leaf water potential, it is evident that OA helps to sustain growth while the plant is meeting transpirational demand by reducing its leaf water potential (Blum, 2005).

At the leaf level, drought stress is generally characterized by a reduction of water content, diminished leaf water potential and turgor loss, closure of stomata and a decrease in cell enlargement and growth, whereby cell enlargement is more strongly inhibited than cell division. Further, various physiological and biochemical processes, such as photosynthesis, respiration, translocation, ion uptake, carbohydrates, nutrient metabolism and growth promoters (Jaleel et al., 2009; for review see Farooq et al., 2009) are affected. Severe water stress may finally result in the arrest of photosynthesis, disturbance of metabolism and at last the death of the plant (Jaleel et al., 2009).

In higher plants, water loss and CO_2 uptake are tightly regulated by stomata. Under continuously changing environmental conditions stomata optimise gas exchange between the interior of the leaf and the surrounding atmosphere. Stomata close in response either to a decline in leaf turgor and/or water potential, or to a low-humidity atmosphere (Maroco et *al.*, 1997). As a rule, stomatal responses are more closely linked to soil moisture content than to leaf water status suggesting that stomata are responding to chemical signals (e.g. abscisic acid, ABA) produced by dehydrating roots whilst leaf water status is kept constant (Gowing *et al.*, 1990; Davies & Zang, 1991; Chaves, 2002; Yordanov, 2003). Under mild to moderate stress the reduction of leaf conductance (g_L) helps to avoid excessive water loss and provides higher water use efficiency to the plant (e.g. Lawlor, 1995; Cornic & Massacci, 1996; Lawlor, 2002; Flexas & Medrano, 2002). However, stomatal conductance not only regulates the efflux of water vapour by the leaf but also controls the influx of CO_2 into the leaf. As reviewed by Cornic (1994), stomatal closure is mainly responsible for the decline of net photosynthetic rates in C_3 plants subjected to moderate drought stress. Nevertheless, under more severe conditions of stress internal CO_2 concentration (C_i) frequently increases indicating the predominance of non-stomatal limitations to photosynthesis (Lawlor, 1995; Brodribb, 1996; Flexas & Medrano, 2002). To the non-stomatal mechanisms, under prolonged or severe soil drought, belong changes in chlorophyll synthesis, functional and structural changes in chloroplasts and also disturbances in accumulation and distribution of assimilation products (Medrano *et al.*, 2002). With this respect, processes like photophosphorylation (Haveaux *et al.*, 1987; Meyer & de Kouchkovsky, 1992), ribulose-1,5-bisphosphate (RuBP) regeneration (Gimenez *et al.*, 1992) and RubisCO activity (Castrillo & Calcagno, 1989; Medrano *et al.*, 1997; Medrano, 2002) are impaired under drought.

3. Effects of drought stress on plant physiology under conditions of elevated CO_2 concentration and temperature

Rising concentrations of atmospheric carbon dioxide (CO_2) contribute to global warming and thus to changes in both precipitation and evapotranspiration (Kruijt, 2008). Climatic shifts in both mean and variability could threat ecosystem functions and human welfare (Tubiello, 2002). Current research confirms that, while crops would respond positively to elevated CO_2 concentrations ($[CO_2]$) in the absence of climate change, the associated impacts of higher temperatures, altered patterns of precipitation, and possibly increased frequency of extreme events, such as drought and floods, will likely combine to depress yields and increase production risks in many regions of the world. These will widen the gap between rich and poor countries further. A consensus has emerged that developing countries are more vulnerable to climate change than developed countries, because of the predominance of agriculture in their economies, the scarcity of capital for adaptation measures, their warmer baseline climates, and their heightened exposure to extreme events. Thus, climate change may have particularly serious consequences in the developing world, where about 800 million people are currently undernourished (Tubiello & Fischer, 2007; and references therein).

Except for regions where irrigation is employed or dewfall is significant, precipitation is the source of almost all soil moisture. Plants extract almost all their moisture from the soil (a small amount may be absorbed through the surface of wet plant leaves). Any change in timing and/or quantity of precipitation will affect soil moisture supply and crop yield. At times of low precipitation, soil moisture may be insufficient to meet the evaporative demand imposed by the atmosphere. Plant leaves will lose turgor and stomata close to prevent further dehydration of the plant (see section 2). The entry of CO_2, in the leaf is inhibited and photosynthesis, crop growth and yield are reduced (Brown, 1997).

Despite numerous works, up to date quantitative information on the impact of changes in precipitation, temperature and atmospheric CO_2 concentration on the soil and water resources is still required. Such information is needed principally at the watershed level where most of the processes underlying landscape functioning act and which is the scale at which decision making is taken (Chaplot, 2007).

The consequences of global change on plant biomass production and water use are manifold. Such multi-factor interactions are difficult to predict since the different effects might intensify each other, nullify each other or even change the sign of the overall change.

It is well documented that atmospheric CO_2 enrichment typically enhances photosynthesis (Long, 1991; Loreto & Centritto, 2004) mainly due the repression of photorespiration and because of an increased substrate supply (Poorter & Navas, 2002). The increases will be, based on the Farquhar *et al.* (1980) model of leaf photosynthesis, larger at higher temperatures whereas the benefit from CO_2 enrichment will be little when temperature is low (<15°C; Wolfe *et al.*, 1998). Further, growth rates will increase followed by enhanced biomass accumulation (Saralabai, 1997; Fuhrer, 2003) in a wide range of plants. One of the major consequences of this phenomenon is an increased production of edible biomass (Cure & Acock, 1986; Kimball, 1993) as well as an increased allocation of assimilated carbon to the roots (Arp, 1991; Fuhrer, 2003). The relative increase in biomass and yield under elevated [CO_2], however, is expected to be largest if all the other growth parameters remain constant (Amthor, 2001) which is not likely since general circulation models of the atmosphere predict increases in global temperature in the range of 1.5-3°C (and perhaps as much as 5°C) by the end of the 21st century (Lawlor & Mitchell, 1991; IPCC, 2007).

In determinate annual species warmer temperature accelerates ontogenetic development resulting in a (substantial) shortening of the growth period. This in turn leads to less time for carbon fixation and biomass accumulation (Morison & Lawlor, 1999). However, since seed yield is directly related to the seed filling duration (Egli & Crafts-Brandner, 1996) as well as to the leaf area duration (Geisler, 1983; De Costa, 1997), a combined increase of CO_2 concentration and temperature does not necessarily translate into improved yield, especially when other factors like water and nutrient supply are limiting (Amthor, 2001).

Different lines of evidence further indicate that growth in elevated [CO_2] leads to a change in the sink-source balance of the plant. In this context, carbohydrate accumulation in the source leaves is expected if the rate of photosynthesis exceeds the capacity of the sinks to utilize the photosynthates for growth. Therefore, the repeatedly observed variability in the response to CO_2 in different species, developmental stages or environmental conditions can be explained in terms of differing sink strength of the plants. Further, some of the morphological changes seen under elevated [CO_2] can be explained by an increased supply of photosynthates which 'forces' the development of new sinks (Stitt, 1991; Bowes, 1993). For this reason, it has been supposed that plants with a smaller sink size or capacity to develop new or alternative sinks will acclimate to higher levels of [CO_2] by decreasing their photosynthetic capacity to adjust the assimilate production to the demand.

Referring to this, plant species developing N_2 fixing nodules might present a special case. The increased availability of carbohydrates might enhance the development of nodules since they represent a considerable sink for carbohydrates. Higher atmospheric [CO_2] could therefore have complex indirect effects on growth and photosynthetic rates in plants which develop nodules, because increased nodule development will increase the supply of organic nitrogen to the plant. This might be one explanation for the positive response of soybeans to elevated $[CO_2]$, and could be of considerable significance under natural conditions (Stitt, 1991). In other plant species, however, one of the most prominent, but not always observed, consequences of atmospheric CO_2 enrichment that has been found is decreased foliar N concentration. In a review of 378 observations obtained from 75 published studies Cotrufo (1998) found that 82% of the experiments related to this subject reported a reduction in plant N concentration under conditions of atmospheric CO_2 enrichment, with a mean concentration reduction for all studies of 14% on a plant dry weight basis whereby C_3 plants showed a mean decrease of 16% and C_4 and N-fixing plants of 7%. These decreases in foliar leaf N concentration are, amongst other factors, due to dilution effects. Higher growth rates in elevated $[CO_2]$ will lead to an increased demand for mineral nutrients. The acceleration of growth and the increased biomass production in elevated $[CO_2]$ may further change the nutrient status in the plant (Stitt, 1991). In agricultural situations where man has the capacity to alter the growing environment in a number of different ways, it has been demonstrated by Rogers (1996) and Kimball (1993) that the provision of high levels of nitrogen fertilizer to the soil has the capacity to offset the reduced foliage nitrogen concentration caused by higher levels of CO_2. As Rogers *et al.* (1996) have described it, "the widely reported reduction in leaf or shoot N concentration in response to elevated CO_2 is highly dependent on nitrogen supply and virtually disappears when N is freely available to the roots".

Alongside with an increase of photosynthetic rates, the reduction of stomatal conductance and consecutively transpiration rates are commonly observed under conditions of elevated atmospheric CO_2 concentrations. However, the response to water stress is variable, in part because, although high $[CO_2]$ reduces transpiration per unit leaf area, it often increases the total leaf area per plant (Cure & Acock, 1986; Allen, 1990). Whether elevated CO_2 reduces evapotranspiration therefore depends on the effects of elevated $[CO_2]$ on leaf area index (LAI) as well as on stomatal conductance. No savings in water can be expected in canopies where elevated CO_2 stimulates the increase in LAI relatively more than it decreases stomatal conductance (Drake, 1997). Particularly, in C_3 species such as wheat or cotton (Kang *et al.*, 2002) which are more responsive to increasing $[CO_2]$ compared to C_4 plants this might become effective (Fuhrer, 2003). Further, in canopies with high LAI, leaf boundary layer and aerodynamic conductance may exert a stronger control on water vapour exchange than stomatal conductance, so that any change in stomatal conductance induced by elevated $[CO_2]$ may only marginally affect transpiration and hence, plant and stand water use (Wullschleger, 2002).

It is, however, worth to be noted that the reduced evapotranspiration could be cancelled out also by other changes caused by the increase in $[CO_2]$. At the leaf level, stomatal closure would reduce transpirational cooling and thereby increase leaf temperature (Yoshimoto, 2005). Indeed, increases of leaf temperature in the order of 1-2°C have been measured in various crops with a doubling of CO_2 concentration (Idso *et al.*, 1987). Higher leaf temperature will reduce the longevity and photosynthetic capacity of individual leaves (Kimball, 1995) but also result in larger vapour pressure deficits between the leaf and the air, thereby negating some of the positive effects of elevated $[CO_2]$ like decreased stomatal conductance (Yoshimoto, 2005). At the canopy level, the accelerated aging of leaves can shorten the growing season (Kimball, 1995) and therewith lead to a reduction in crop yield.

Finally, the decreased stomatal conductance and latent heat transfer might cause a warming of the order of 1-2°C over the continents in addition to warming from the CO_2 greenhouse effect (Sellers *et al.*, 1996).

In summary, while elevated [CO_2] alone tends to increase growth and yield of most agricultural plants (Kimball, 1983; Cure & Acock, 1986; Kimball *et al.*, 2002) as well as increases water use efficiency, warmer temperatures and changed precipitation regimes may either benefit or damage agricultural systems. Water and fertilizer application regimes will further modify crop responses to elevated CO_2 (Tubiello, 2002 and references therein). Consequently, the picture emerging from experiments at the whole plant level is rather diffuse, and this holds even more if we try to scale up CO_2 -induced growth responses from the individual to the stand level (Poorter & Navas, 2002). However, due to the increased need for food supply worldwide and the prediction of the emergence of drier regions in the world, the development of cost effective methods for early stress detection and therewith a possibility to reduce yield losses is inevitable.

4. Sensing drought stress

Irrigation is important in raising crops and achieving considerable yields in many areas of the world. It is essential especially in arid environments but is also becoming increasingly important in semi arid to humid regions due to the increased demand for food. Owing to this growing demand the supply of water available is decreasing and costs are going up (Gonzales-Dugo, 2006). During the last decades, therefore, the effects of drought stress on plant physiological traits have been intensively studied (see section 2 and 3) to develop new methods for early detection and monitoring of drought stress. This should allow developing both short and long term agro-technical measures (e.g. irrigation scheduling) and thus help avoiding substantial yield losses and at the same time reduce water consumption. With this respect, special focus was put on sensing leaf/canopy reflectance, thermal radiation as well as fluorescence emission.

4.1 Leaf spectral reflectance as a measure of plant drought stress

Remote sensing techniques have evolved rapidly during the past decades. Ecological remote sensing now encompasses a wide range of applications including vegetation mapping, land-cover change detection, disturbance monitoring and the estimation of biophysical and biochemical attributes of ecosystems (Asner *et al.*, 1998a). A lot of effort has also been made towards the use of spectral reflectance of leaves and canopies for stress detection in agricultural environments since these techniques could offer a powerful tool not only for crop stress detection but also for quantifying crop development and yield (Asner, 1998b).

Leaf reflectance is driven mainly by the chemical composition of the leaves but can vary independently of pigment concentrations due to differences in internal structure, surface characteristics (e.g. hairs, waxes) and moisture content (Blackburn, 2007). The reflectance pattern of a canopy is even more complex since it is influenced not only by the reflectance of single leaves but also by its geometry - the leaf area index, inclination and clumping of the leaves - as well as the percentage of canopy ground coverage and presence of non-leaf elements (Gao, 2000; Blackburn, 2007).

Agricultural monitoring is a process by which crop development is tracked and treatments are applied with the aim of increasing income while minimizing expenditure (Beeri, 2006). To accomplish this numerous spectral reflectance indices were developed, most of them based on simple mathematical formulas, such as ratios or differences between the reflectance at given wavelengths (Araus et al., 2001; Babar, 2006). Most of the indices based on the reflectance of single leaves aim to trace plant physiological status, i.e. plant water status or nutritional status. However, indices developed for canopy spectra can not only be used for the assessment of plant water and nitrogen status but also for the estimation of plant productivity (measurement and interpretation of absorbed photosynthetically active radiation; Ferri, 2004).

Largely as a result of interests in remote sensing, leaf reflectance has been studied intensively. Pioneering efforts in this field have been reviewed elsewhere (Myers et al., 1983; Jackson, 1986; Carter, 2001). The shape of leaf reflectance spectra is determined by the absorption of leaf pigments, mainly chlorophyll. Reflectance spectra of green leaves are characterized by a low signal in the blue region between 400 and 500nm and a high signal in the near-infrared between 750 and 800nm (Figure 1). With increasing leaf chlorophyll content the reflectance signal around 680nm decreases (Buschmann, 1993). Typical leaf reflectance spectra (500 - 2500nm) can be separated in three parts:

1. Wavelength spectrum between 500-750nm, the visible light absorbance region which is dominated by pigments (chlorophyll a and b, carotene, and xanthophyll pigments);
2. Wavelength spectrum between 750-1350nm, the near-infrared region which is affected by internal leaf structure; and
3. Wavelength spectrum between 1350-2500nm, a region influenced to some amount by leaf structure, however significantly affected by water concentration in the tissue. Strong water absorption bands occurring between 1450 and 1950nm (Myers, 1983; in Tanriverdi, 2006).

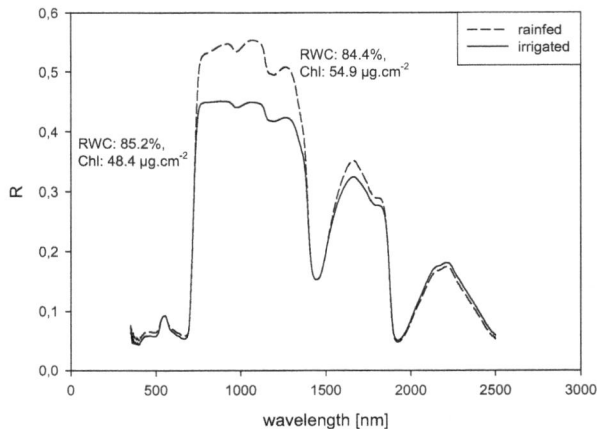

Fig. 1. Reflectance spectra of leaves from irrigated and rainfed grown *Triticum asetivum* L. Abbreviations: Chl: chlorophyll content, RWC: relative water content.

Concerning the occurrence of stress (e.g drought stress, nutrient deficiency, diseases), leaf reflectance is altered more consistently at visible wavelengths (400–720nm) than in the remainder of the incident solar spectrum (730–2500nm). These changes were found to be spectrally similar among many common stressors and vascular plant species. Increased reflectance in the far-red 690–720nm spectrum is a particularly generic response, providing an earlier or more consistent indication of stress than reflectance in other regions of the incident solar spectrum (Carter, 2001 and references therein).

Under conditions of drought stress, absorption of radiation by the leaf tends to decrease due to lower leaf water content. Although water absorbs most strongly in the wavelengths of the infrared region of the spectrum from approximately 1300 to 2500nm (Curcio & Petty, 1951), some absorption also occurs at lower wavelengths. As water is lost from a leaf reflectance increases and absorption decreases, primarily as a result of water's radiative properties (Bowman, 1981; Hunt & Rock, 1989). Even after accounting for the radiative characteristics of water, secondary effects occur. These include the influence of water content on absorption by other substances in the leaves, such as pigments. Also included as secondary are the effects of water content on wavelength-independent processes, particularly multiple reflections inside the leaf (Carter, 1991).

Moreover, drought stress not only causes leaf water content to decline but also affects physiological processes such as, for example, leaf conductance and photosynthetic rates (see section 2). Changes in pigment and nitrogen concentration of plant tissue will follow. For example, chlorophyll and RubisCO contents decline as the leaf remobilizes resources under stress conditions (Parry, *et al.*, 2002). Chlorophyll and accessory pigments absorb strongly in the visible range (Knipling, 1970). Carter & Knapp (2001) described a consistent stress induced alteration of leaf reflectance at visible wavelengths (~400–720nm) since chlorophyll is the major absorber in the leaf and the metabolic disturbance brought about by stress alters leaf chlorophyll concentrations (Knipling, 1970). Plant responses to water deficit therefore include both biochemical and morphological changes that primarily lead to acclimation and later to functional damage and the loss of plant parts (Chaves, *et al.*, 2002). However, it is worth pointing out that leaf reflectance in the visible range of plants experiencing nutrient deficiency was also found to increase since nitrogen (and magnesium) is essential in the formation of chlorophyll. As leaves become more chlorotic, reflectance increases and the reflectance peak, normally centred at about 550nm, broadens towards the red as absorption of incident light by chlorophyll decreases (Ayala-Silva, 2005). Therefore, identifying the release of a stress situation by leaf reflectance spectra alone might be difficult to a certain extent.

Concerning the detection of plant water status and nutrient deficiency numerous spectral indices have been developed. Table 1 summarizes some of the most often used indices without attempting to give an exhaustive overview of all indices appearing in literature. Further, several indices were later on modified to better fit different plant species and/or conditions in different geographical regions.

Some of the most common indices are, for example, the photochemical reflectance index (PRI), indices to determine leaf chlorophyll content, nitrogen content or indices for the estimation of leaf water content. The PRI is widely used for the estimation of photosynthetic radiation use efficiency. It was proposed based on the finding that the interconversion of xanthophyll cycle pigments in intact leaves can be detected as subtle changes in absorbance at 505-510nm (Bilger *et al.*, 1989) or the reflectance at 531nm (Gamon *et al.*, 1990). The

Index	Related to	Reference
NPQI $(R_{415}-R_{435})/(R_{415}+R_{435})$	chl, stress	Barnes *et al.*, 1992
SR R_{800}/R_{680}	chl	Birth & McVey, 1968
REIP	chl, stress	Collins, 1978
PRI $(R_{531}-R_{570})$ x $(R_{570}+R_{531})$	chl	Filella, 2004
SIPI $(R_{800}-R_{445})/(R_{800}-R_{680})$	chl	Penuelas *et al.*, 1995
PSSRa R_{800}/R_{680}	chl a	Blackburn, 1998
PSSRb R_{800}/R_{635}	chl b	Blackburn, 1998
PSSRc R_{800}/R_{500}	chl	Blackburn, 1998
PSNDa $(R_{800}-R_{680})/(R_{800}+R_{680})$	chl a	Blackburn, 1998
PSNDb $(R_{800}-R_{650})/(R_{800}+R_{650})$	chl b	Blackburn, 1998
PSNDc $(R_{800}-R_{500})/(R_{800}+R_{500})$	chl	Blackburn, 1998
(R_{675}/R_{700})	chl a	Chapelle, 1992
$R_{675}/(R_{700}$ x $R_{650})$	chl b	Chapelle, 1992
$(R_{800}-R_{700})/(R_{800}+R_{700})$	chl	Gitelson & Merzlyak, 1994
$(R_{750}-R_{800})/(R_{695}-R_{740})$ - 1	chl	Gitelson, 2003
$R_{860}/(R_{708}$ x $R_{550})$	chl a, chl tot	Datt, 1998
R_{675}/R_{700}	chl	Datt, 1998
$R_{675}/(R_{650}$ x $R_{700})$	chl	Datt, 1998
R_{760}/R_{500}	chl	Datt, 1998
R_{750}/R_{700}	chl	Datt, 1998
R_{750}/R_{550}	chl	Datt, 1998
$R_{672}/(R_{550}$ x $R_{708})$	chl a, chl tot	Datt, 1998
R_{672}/R_{708}	chl	Datt, 1998
$R_{800}-R_{550}$	chl	Buschmann & Nagel, 1993
R_{800}/R_{550}	chl	Buschmann & Nagel, 1993
PSR R_{430}/R_{680}	total pigments, chl, stress	Penuelas *et al.*, 1994
NPCI $(R_{680}-R_{430})/(R_{680}+R_{430})$	total pigments, chl, stress	Penuelas *et al.*, 1994
(Chl)RI$_{green}$ $[(R_{750-800}-R_{430-470})/(R_{520-580}-R_{440-480})]^{-1}$	chl	Gitelson, 2004
(Chl)RI$_{red\ edge}$ $[(R_{750-800}-R_{430-470})/(R_{695-740}-R_{440-480})]^{-1}$	chl	Gitelson, 2004
CRI$_{green}$ $[(R_{510})^{-1} - (R_{550}-R_{570})^{-1}]$ x $(R_{750}-R_{800})$	carotenoids	Gitelson, 2004
CRI$_{red\ edge}$ $[(R_{510})^{-1} - (R_{700}-R_{710})^{-1}]$ x $(R_{750}-R_{800})$	carotenoids	Gitelson, 2004
ARI $(R_{550})^{-1} - (R_{700})^{-1}$ or	car, anthocyanin	Gitelson, 2001
ARI $[(R_{550-570})^{-1} - (R_{700-710})^{-1}]$ x $R_{750-800}$		Gitelson, 2004
R_{1483}/R_{1650}	LWC	Yu *et al.*, 2000
R_{1100}/R_{1430}	LWC	Yu *et al.*, 2000
R_{1121}/R_{1430}	LWC	Yu *et al.*, 2000
R_{1430}/R_{1650}	RWC	Yu *et al.*, 2000
R_{1430}/R_{1850}	RWC	Yu *et al.*, 2000
R_{1483}/R_{1650}	RWC	Yu *et al.*, 2000
R_{2200}/R_{1430}	RMP	Yu *et al.*, 2000
R_{1430}/R_{1650}	RMP	Yu *et al.*, 2000
R_{1483}/R_{1430}	RMP	Yu *et al.*, 2000
R_{695}/R_{420}	stress	Carter, 1994
R_{695}/R_{760}	stress	Carter, 1994
R_{605}/R_{760}	stress	Carter, 1994
R_{710}/R_{760}	stress	Carter, 1994
R_{695}/R_{670}	stress	Carter, 1994

Table 1. Compilation of frequently used spectral indices to detect stress situations (e.g. drought stress, nutrient deficiency, etc.) in plants at the leaf level. *Abbreviations:* chl: chlorophyll, car: carotenoids, ARI: Anthocyan Reflectance Index, CRI: Carotenoids Reflectance Index, LWC: Leaf Water Content; NPCI: Normalized Difference Pigment Index, PSR: Pigment Simple Ratio, REIP: Red Edge Inflection Point, RMP: Relative Leaf Moisture Percentage on Fresh Weight Basis, SIPI: Structure Independent Pigment Index, SR: Simple Ratio.

photochemical reflectance index (PRI), incorporating reflectance at 531nm (xanthophyll cycle signal), was then defined as $[(R_{570}-R_{531})/(R_{570}+R_{531})]$ to establish a reflectance-based photosynthetic index (Gamon et al., 1997). Concerning the attempt to trace the relative and actual leaf water content (RWC and AWC) with spectral indices, a lot of effort has been made and a number of different indices have been developed for numerous crop species, amongst many others the water index (WI; R_{900}/R_{970}; Penuelas & Filella, 1998), the water band index (WBI; R_{905}/R_{980}; Davenport et al., 2000), RWC (R_{1483}/R_{1650}) or AWC (R_{1121}/R_{1430}).

Although much hyperspectral reflectance work to date has been done at the leaf scale, in situ measurements made above the canopy are becoming more widely used, driven by the need to simulate the scales involved in airborne or satellite measurements (i.e. remotely sensed imagery at the canopy scale; Strachan, 2002). Table 2 summarizes some of the indices frequently used in remote sensing.

The most known and often used index in remote sensing of green phytomass is the NDVI (normalized difference vegetation index). The NDVI is a broad-band vegetation index which has largely been employed to determine quantitative parameters of green phytomass, using wide spectral bands in the red and near infrared, generally acquired by multispectral sensors in satellites (Ferri, 2004). It was proposed by Rouse et al. (1973) and is based on the contrast between the maximum absorption in the red due to chlorophyll and the maximum reflection in the infrared caused by leaf cellular structure (Haboudane, 2004). However, despite its intensive use in remote sensing applications the NDVI has the disadvantage to saturate in cases of dense and multi-layered canopies and further shows a non linear relationship to biophysical parameters such as the leaf area index (LAI; Haboudane, 2004). Several studies reveal a saturation level of NDVI at LAI values between 2 and 3 implying that a further increase in standing biomass does not yield a further increase in NDVI (e.g. Gilabert et al. 1996; Haboudane, 2004).

However, as much as the saturation of vegetation indices at high LAI values is a problem, the influence of soil background is one at very low LAI values. To account for changes in soil optical properties, soil adjusted indices minimizing the effect of soil background were developed (Haboudane, 2004). The leading index with this respect is the Soil-Adjusted Vegetation Index (SAVI) which is less sensitive to soil reflectance at low LAI than NDVI (Huete, 1988). It is based on the linear relationship between near-infrared and visible reflectance for bare soil and therefore reduces the influence of the soil on canopy reflectance. The SAVI index was modified further several times to optimize the removal of soil background influences (Dorigo, 2007 and references therein). For a good overview of all the modified SAVI indices see e.g. Broge & Leblanc (2000).

Tracing changes in plant water status can, for example, be done by the setup of simple ratios between two wavelengths, one of which characterized by strong water absorption and a second one outside the absorption band. One of these indices is the NDWI which is given by $(R_{860} - R_{1240})/(R_{860} + R_{1240})$ and is sensitive to changes in liquid water content of vegetation canopies (Gao, 2000; Serrano, 2000). Another index commonly used to trace plant water deficits is the water index (WI;), which was developed by Penuelas et al. (1997) and is calculated as the ratio between reflectance at 900nm and 970nm (R_{900}/R_{970}). Further indices used for the estimation of plant water status from remotely sensed data would be for example the moisture stress index (MSI) which is given by the ratio of R_{1600}/R_{817} (Hunt & Rock, 1989) or the maximum difference water index (MDWI, Eitel et al., 2006).

To assess leaf chlorophyll (and leaf N) status from remotely sensed observations, spectral indices are needed that are sensitive to leaf chlorophyll concentration and minimize variations in canopy reflectance associated with background reflectance and LAI (Daughtry, 2000). Most hyperspectral ratios used for estimating leaf chlorophyll content make use of the three discrete bands describing the typical reflectance pattern of green vegetation: the reflectance peak in the green and NIR and the region of maximum absorption in the red (Dorigo, 2007). Amongst many other indices, some of the most widely used ones to measure chlorophyll (and leaf N) are the CARI (Chlorophyll Absorption Ratio Index), the MCARI (Modified Chlorophyll Absorption Ratio Index), the TVI (Triangular Vegetation Index) and the REIP (red edge inflexion point).

The Chlorophyll Absorption Ratio Index (CARI), which measures the depth of chlorophyll absorption at 670nm relative to the green reflectance peak at 550nm and the reflectance at 700nm, was developed by Kim et al. (1994) for minimizing the effects of non-photosynthetic materials on spectral estimates of absorbed photosynthetically active radiation (PAR; Daughtry; 2000, Haboudane, 2004). Subsequently, different alterations of this index were proposed (see e.g.: Daughtry et al., 2000; Haboudane et al., 2002) to make it more sensitive to chlorophyll. The MCARI was simplified from the CARI by Daughtry et al. (2000) and is given by $[(R_{700}-R_{670})-0.2(R_{700}-R_{550})]*(R_{700}-R_{670})$ (Haboudane, 2004). The TVI (Triangular Vegetation Index), however, follows a different concept. It was introduced by Broge et al. (2000) and is based on the fact that the total area of the triangle (green, red, infrared) will increase as a result of chlorophyll absorption (decrease of red reflectance) and leaf tissue abundance (increase of near-infrared reflectance; Broge & Leblanc, 2000; Haboudane, 2004).

The red-edge, finally, describes the abrupt increase in leaf reflectance at wavelengths between 680nm and 740nm which is caused by the combined effects of strong chlorophyll absorption and leaf internal scattering (Dawson, 1998). Increases in the amount of chlorophyll visible to the sensor, either through an increase in leaf chlorophyll content or Leaf Area Index, result in a broadening of a major chlorophyll absorption feature centred around 680nm. The effect is to cause a movement of the point of maximum slope, termed the red edge position (REP; Dawson, 1998). Several studies have subsequently illustrated the use of the red edge in the estimation of foliar chlorophyll content (e.g. Lamb, 2002 and references therein). To date various techniques have been developed for parameterizing the shape of the red-edge and determining the position of the red edge inflection point (REIP), including inverted Gaussian models (Miller et al., 1990), fitted high-order polynomials, linear interpolation (Guyot et al., 1988; Clevers et al., 2002) and Langrangian interpolation (Dawson & Curran, 1998b; in Dorigo, 2007). The structure of the chlorophyll red-edge might be best observed by plotting $dR/d\lambda$, the first derivative, with respect to wavelengths. A common approach for locating the red-edge wavelength is to manually or computationally locate the highest peak in the derivative spectra (Lamb, 2002 and references therein). The use of derivative spectrometry is commonly employed to resolve or enhance absorption features that might be masked by interfering background absorption (Curran et al., 1990; Filella & Penuelas, 1994). Spectral derivatives also aid in suppressing the continuum caused by other leaf biochemicals (such as lignin and secondary pigments) and canopy background effects (Elvidge 1990; Curran et al. 1991).

Index	Related to	Reference
SR R_{800}/R_{680}	chl	Birth & McVey, 1968
CARI	chl	Kim et al., 1994
MCARI $[(R_{700} - R_{670})-0.2(R_{700}-R_{550})](R_{700}-R_{670})$	chl, LAI	Daughtry et al., 2000
TCARI $3[(R_{700} - R_{670}) - 0.2(R_{700}-R_{550})(R_{700}/R_{670})]$		
NDVI $(R_{800} - R_{680}) / (R_{800} + R_{680})$	chl, LAI, Yield	Rouse et al., 1973
Green NDVI $(R_{780}-R_{550})/(R_{780}+R_{550})$	chl	Gitelson et al., 1996
DVI $R_{800}-R_{680}$	chl	Jordan, 1969
RDVI $(R_{800}-R_{670})/sqrt(R_{800}+R_{670})$	chl, LAI	Roujean, 1995
MSR $(R_{880}/R_{670}-1)/sqrt(R_{880}/R_{670}+1)$	chl, LAI	Jordan, 1969
SAVI $(1+L)(R_{801}-R_{670})/(R_{801}+R_{670}+L)$	chl, LAI	Huete, 1988
OSAVI $(1+0.16)(R_{801}-R_{670})/(R_{801}+R_{670}+0.16)$	chl, LAI	Rondeaux, 1996
TVI $0.5[120(R_{750}-R_{550})-200(R_{670}-R_{550})]$	chl	Broge & Lelanc, 2000
REIP	chl, LAI, stress	Collins, 1978
PRI $(R_{531}-R_{570})(R_{570}+R_{531})$	chl	Filella, 2004
CCCI	chl, N	Fitzgerald, 2010
CNI	chl, N	Fitzgerald, 2010
NDRE	N	Fitzgerald, 2010
WI R_{900}/R_{970}	water	Penuelas et al., 1997
NDWI $(R_{860} - R_{1240}) / (R_{860} + R_{1240})$	water	Gao, 1996
MDWI $(R_{max1500-1750} - R_{min1500-1750}) / (R_{max1500-1750} + R_{min1500-1750})$	water	Eitel et al., 2006

Table 2. Compilation of frequently used spectral indices to detect stress situations (e.g. drought stress, nutrient deficiency, etc.) in plants at the canopy level. *Abbreviations*: CARI: Chlorophyll Absorption Ratio Index, CCCI: Canopy Chlorophyll Content Index, CNI: Canopy Nitrogen Index, MCARI: Modified Chlorophyll Absorption in Reflectance Index; MDWI: Maximum Difference Water Index; NDRE: Normalized Difference Red Edge, NDWI: Normal Difference Water Index, NDVI: Normalized Difference Vegetation Index; PRI: Photochemical Reflectance Index, REIP: Red Edge Inflection Point; SAVI: Soil Adjusted Vegetation Index, SR: Simple Ratio, TCARI: Transferred Chlorophyll Absorption in Reflectance Index, TVI: Triangular Vegetation Index, WI: Water Index.

As can be seen from the many indices which were developed, intensive research has been made towards sensing the response of leaf optical characteristics to different stressors, such as for example, drought stress and nutrient deficiency, exposure to heavy metals, exposure to gaseous pollutants, UV-B radiation, ozone or increased temperature and CO_2 (Baltzer & Thomas, 2005 and references therein). However, it should be stated that most of the vegetation indices have temporal effects, which are not necessarily correlated to the temporal effects of the spectral indices. Care has to be taken to apply the right indices at the right time in the growing season. Also, it should be mentioned that these vegetation indices are inter-related (e.g. most of them influence the total yield; Zwiggelaar, 1998).

4.1.1 Case study 1 – An attempt to sense reoccurring drought stress events remotely

This study aimed to evaluate the impact of drought stress on plant physiological traits and leaf reflectance of wheat (*Triticum aestivum* L.) occurring at different phenological stages (flowering and/or grain filling). Further, the consequences of two consecutive drought events and recovery of plants after drought were investigated. The analysis of the effect of consecutive stress periods and recovery on changes in leaf reflectance has rarely been performed until now but might gain in importance considering the predicted increased frequency of drought events whereby plants could be exposed to drought repeatedly (Schär et al., 2004; Seneviratne et al., 2006; Vidale et al., 2007; IPCC, 2007).

4.1.1.1 Material and methods

Plants (*Triticum aestivum* L. cv. Xenos) were grown in 8 litre plastic pots. Simulation of seasons in the growth chamber was based upon long-time observation of temperature and relative air humidity (meteorological station: 16°29′ eastern longitude and 48°15′ northern latitude). For a detailed description of growing conditions and measurement techniques see Linke et al. (2008).

Four different treatments were set up - one control treatment and three treatments exposed to drought at different times during phenology: AC: control plants; AF: plants exposed to drought stress at flowering, recovery after anthesis; AG: plants exposed to drought stress at grain filling and AFG: plants exposed to drought stress at flowering and grain filling.

Soil moisture content of control plants was consistently held at 20-23 vol% (AC; TDR Trime, Imko Micromodultechnik GmbH, Germany). Drought stress at flowering was imposed by halving water supply 10 days before the beginning of pollen shedding resulting in a soil moisture content of ~10 vol% at flowering (AF). After flowering, plants receiving a second stress at grain filling were allowed to recover for 8 days (water supply similar to control plants) before the second stress was imposed by halving water supply again (soil moisture content during measuring period ~10 vol%; AFG). Plants receiving drought stress only at grain filling (AG) were treated similar to control plants until after flowering. Drought stress was imposed at the same time as in plants of the treatment stressed twice.

All physiological and spectral measurements were made in the mid region of the youngest fully expanded leaves at three developmental stages: vegetative growth, flowering and grain filling. Light saturated photosynthetic rates (A_{sat}) refer to measurements at growth conditions under saturating light intensities (CO_2: 350-370 $\mu mol.mol^{-1}$; light: 1000 $\mu mol.m^{-2}.s^{-1}$; CIRAS-I, PP-Systems, U.K), actual leaf conductance (g_L) was measured with a steady state porometer (PMR-4, PP-Systems; U.K.) and total chlorophyll content (Chl_{tot}) of leaves was determined with a SPAD-502 hand held chlorophyll meter (Minolta, Japan). Relative water content was calculated as RWC = ((fresh weight - dry weight)/(saturated weight - dry weight)) * 100 [%] and actual leaf water content was calculated as AWC = ((fresh weight - dry weight) / (fresh weight)) * 100 [%].

Leaf spectral reflectance was measured with a FieldSpec Pro FR in connection with a plant reflectance probe from Analytical Spectral Devices Inc., Boulder, CO. Relative difference of reflectance spectra between stress and control treatments ($\Delta R/R$) was calculated as (($R_{stress}-R_{control}$)/$R_{control}$)*100 [%].

4.1.1.2 Results and discussion

Drought stress significantly influenced plant physiological traits independently of the time of its application in phenology (Table 3). A lowering of the actual leaf conductance (g_L), as observed during all stress periods, is one of the first processes occurring under decreased soil water availability providing a higher water use efficiency to the plant (Cornic & Massacci, 1996; Lawlor & Cornic, 2002; Flexas & Medrano, 2002). Moreover, as reviewed by Cornic (1994), stomatal closure is mainly responsible for the decline in net photosynthetic rate of C_3 leaves subjected to moderate drought stress. However, at a certain stage of stress, internal CO_2 concentration (C_i) frequently increases, indicating the predominance of non-stomatal limitations to photosynthesis (Lawlor, 1995; Brodribb, 1996; Medrano et al., 2002). Reductions of light saturated photosynthetic rates (A_{sat}) in the present experiment were mainly due to stomata limitation since a significantly lower C_i was found (data not shown). Leaf reflectance (R) increased over the entire spectrum due to drought stress, a response also found elsewhere (e.g. Wooley, 1971; Penuelas & Inoue, 1999; Yu et al., 2000). However, five regions with relatively high differences were observed: 520–530nm, 570-590nm, 690-710nm, 1410–1470nm and 1880–1940nm.

Rewatering plants after the stress period at flowering allowed them to restore their physiological traits until grain filling (15 days rewatered). Relative water content (RWC) of recovered plants even exceeded that of control (+7%). Therewith, A_{sat} also recovered. Only g_L remained somewhat lower than that of control plants (Table 3). However, the results from leaf reflectance (R) did not follow this trend. The relative difference $\Delta R/R$ within the range of 1410-1470nm and 1880-1940nm remained nearly as high as during the stress period at flowering despite the 7% higher RWC of recovered plants. Within the visible range of leaf spectra $\Delta R/R$ even increased during recovery compared to the actual stress period. These results indicate that quantifying the extent of change for either leaf water content or Chl_{tot} and leaf [N] from changes in leaf R might be problematic. Especially recovery from drought could not be traced using leaf R since the differences between formerly stressed plants and control plants remained rather high despite the complete recovery of physiological traits.

The reason for the enduring differences in leaf R between fully recovered plants and control plants remains rather unclear and information on leaf R during recovery of plants after a stress period is rare in literature. However, it is assumed that secondary effects following drought stress might be involved. Drought can affect the cell structure and biochemistry (e.g.: Yordanov et al., 2000; Larcher, 2003; Read & Stokes, 2006) and is further known to influence the morphology of the leaf surface by means of changes in the content and/or composition of epicuticular waxes (Jordan et al., 1983; Johnson et al., 1983; Deng, 2005; Sehperd, 2006) or the occurrence of hairs (Foyer, 1994). Moreover, drought has the potential to accelerate ontogenetic development (Foyer, 1994; Kimball, 1995). Such alterations of leaf morphology and/or biochemical composition could not only have influenced leaf R after recovery but also have attributed to (or might be the reason for) the unexpectedly great differences in leaf R observed in plants subjected to a second stress period at grain filling. The less pronounced reaction of physiological traits to a second drought period is attributed to the preconditioning of plants already exposed to drought at flowering and/or the higher amount of green biomass (transpiring surface) of plants from the treatment stressed solely at grain filling. Plants of the treatment stressed twice were watered optimally for eight days after the drought stress event at flowering before water supply was halved again. Leaf osmotic potential remained below (more negative) that of control plants during these days providing a better initial situation concerning osmotic adjustment (data not shown) for plants already experiencing a first drought period at flowering.

		Triticum aestivum L.			
		AC	AF	AG	AFG
A_{sat}	Vegetative	21.2			
	Flowering	16.9	10.7***		
	grain filling	13.8	**12.2 n.s.**	4.4***	6.9***
g_L LS	Vegetative	84.0			
	Flowering	164.1	18.4***		
	grain filling	171.8	**116.3****	15.3***	20.3***
RWC	Vegetative	86.7			
	Flowering	83.8	74.0**		
	grain filling	76.3	**81.9***	57.1***	64.0***
AWC	Vegetative	81.2			
	Flowering	72.2	68.8**		
	grain filling	74.1	**74.8 n.s.**	68.3 n.s.	71.1**
Chl_{tot}	Vegetative	46.8			
	Flowering	55.0	59.2***		
	grain filling	48.3	**50.3****	61.7***	55.6***
Leaf [N]	Vegetative	4.3			
	Flowering	4.4	4.2**		
	grain filling	2.4	**2.3 n.s.**	1.9**	2.0**

Table 3. Summary of physiological traits of *T. aestivum*. Significance levels refer to the differences between control and stress treatments. n=5-30; n.s.: not significant, *: $p \leq 0{,}05$; **: $p \leq 0{,}01$; ***: $p \leq 0{,}001$. *Abbreviations:* AC: control; AF: drought stress at flowering, plants were recovered at grain filling; AG: drought stress at grain filling; AFG: drought stress at flowering and grain filling. A_{sat} [$\mu mol.m^{-2}.s^{-1}$], g_L: [$mmol.m^{-2}.s^{-1}$], RWC: [%], AWC: [%], Chl_{tot}: [$\mu g.cm^{-2}$]; Leaf [N]: leaf nitrogen content in % dry matter; LS: lower leaf surface. Bold values highlight performance of recovered plants (measured at grain filling).

The differences observed in $\Delta R/R$ during recovery show that no general prediction can be made concerning the potential to trace recovery from a stress situation with leaf reflectance. Apparently, different species and even cultivars respond inconsistently to drought stress with respect to their spectral signature (compare Linke *et al.*, 2008).

In contrast to changes in leaf R within the range of 1410-1470nm and 1880-1940nm, which can be attributed mainly to differences in leaf water content, the changes within the visible range are not well defined with respect to a certain stressor. As already described by Carter (1994) an increased reflectance at visible wavelengths (400-700nm) is the most consistent response to stress within the 400-2500nm range. The often made assumption that the chlorophyll content of leaves was proportional to moisture content (e.g. Tucker, 1977) may be correct for some species but cannot be generalized to different plant species and ecotypes. Variations in chlorophyll content can be caused by water stress but also by phenological status of the plant, atmospheric pollution, nutrient deficiency, toxicity, plant disease and radiation stress (Ceccato, 2001; Larcher, 2003). These findings are supported by the results from the present study where different trends for RWC, Chl_{tot} and leaf [N] were found. Due to these adverse effects of leaf [N] (decrease) and Chl_{tot} (increase) an interpretation of the increased leaf R is difficult. At least the specific cause of these differences remains uncertain. However, the increased Chl_{tot} content found might result from leaf shrinkage leading to seemingly higher chlorophyll content per unit leaf area ($\mu g.cm^{-2}$).

Finally, three spectral indices (RWC_i, AWC_i and PRI) were tested towards their ability in estimating biophysical parameters (RWC, AWC and A_{sat}). Concerning the estimation of leaf water content a better correlation was found for AWC. Unfortunately, the AWC is the less meaningful parameter since it only gives the water content as percentage of fresh weight which might vary greatly between species, phenology and environmental conditions (Larcher, 2003). The RWC, however, represents the actual leaf water content with respect to a standard measure (leaves under conditions of water saturation; Larcher, 2003) and is therefore the more appropriate indicator of plant water status. Moreover, following changes in biophysical parameters using these indices was not possible due to the different extent of changes in leaf R compared to physiological traits under drought stress at different phenological stages. From these results it is concluded that a good relationship between spectral indices and biophysical parameters does not necessarily lead to an appropriate estimation of biophysical parameters at a given phenological state and/or physiological status.

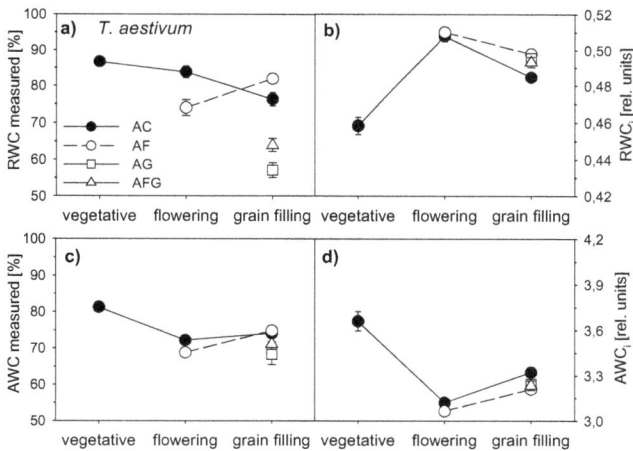

Fig. 2. Comparison of the phenological course of measured and estimated RWC and AWC of *T. aestivum*. **a)-b)** Measured and calculated relative water content; **c)-d)** Measured and calculated actual leaf water content. Legend: AC: control, AF: drought at flowering, recovered at grain filling, AG: drought at grain filling, AFG: drought at flowering and grain filling. n=6 for measured RWC and AWC, n=20-30 for estimated RWC (RWC_i) and AWC (AWC_i). Errors represent standard error.

In the here presented study, estimating leaf water content (RWC and AWC) as well as Chl_{tot} and leaf [N] from reflectance measurements gave good correlations. For tracing changes in physiological parameters during phenology and stress periods, however, the use of these indices was not promising due to false estimation of stress situations and recovery (Figure 2 and 3). An appropriate estimation appeared possible only in unstressed control plants. A good correlation between spectral indices and physiological parameters alone is therefore not necessarily sufficient for estimating physiological parameters from leaf spectra appropriately.

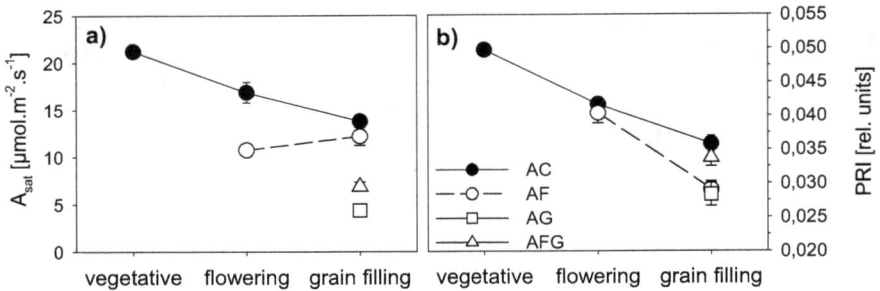

Fig. 3. Comparison of the phenological course of **a)** light saturated photosynthetic rates (A_{sat}) and **b)** photochemical reflectance index (PRI). AC: control, AF: drought at flowering, recovered at grain filling, AG: drought at grain filling, AFG: drought at flowering and grain filling. n=12 for A_{sat} and n=20-30 for PRI. Errors represent standard error.

To summarize, drought stress occurring at different phenological stages increased leaf R throughout the whole spectrum. Unfortunately, the degree to which plant physiological traits and water relations changed could not be quantified by the extent of change in leaf R, at least when drought occurred at different phenological stages. The main concern of this study, however, was to test the ability of leaf reflectance to follow recovery of physiological traits after a stress period which may be of essential importance when considering the occurrence of repeated drought events. Distinguishing between a currently occurring stress situation and an already passed one could become crucial in context with the application of spectral measurements in the field to trace stress situations and to make recommendations on fertilization or irrigation. Unfortunately, recovery from drought stress could not be traced by leaf R since the differences between formerly stressed plants and control plants remained high despite the complete recovery of physiological traits. Further investigations using different species with different leaf morphology and anatomy would be needed.

4.2 Drought stress detection by thermal infrared

Quantifying drought stress by measuring leaf/canopy temperature has become subject of intensive research within the last decades (Tanner, 1963; Wiegand, 1966). In the 1960ies researchers first used crude infrared thermometers to remotely monitor leaf temperature (Fuchs & Tanner, 1966). With the commercial availability of handheld instruments the focus moved from single leaf measurements toward the assessment of canopy temperatures.

The scientific basis for this method relies on the fact that evaporating surfaces are cooled as soon as liquid water is converted to water vapour. Therefore, the less water is available to the plant stomata will close reducing transpiration rates, lowering evaporation and therewith evaporative cooling. The result is an increase in leaf temperature. As the stress situation becomes more severe, leaf temperature (T_c) will reach the temperature of the surrounding air (T_a) and finally exceed it. Based on this, several indices have been proposed to measure the onset of plant water stress, such as the Critical Temperature Variability (CTV) (Blad et al., 1981; Clawson & Blad 1982), Temperature Stress Day (TSD) (Gardner et al., 1981), Stress Degree Day (SDD) (Idso et al., 1977; Jackson et al., 1977), or Crop Water Stress Index (CWSI) (Jackson et al., 1981; Jackson 1982).

The concept of stress degree days (SDD) relies on a daily value of the difference between T_c and T_a at the time of maximum surface temperature (approximately 1-2 hours after solar noon; Idso, 1977; Idso 1981). However, from basic considerations concerning energy balance it becomes evident that the stress degree day parameter is additionally influenced by other environmental factors like the vapour pressure deficit of the air, net radiation or wind speed (Idso, 1981). Ben-Asher (1992) further cautioned that the sensitivity of IR sensors would be insufficient to sense very small differences in temperature and it would be unable to assess short term fluctuations of transpiration. Moreover, errors might result from stomatal closure during periods of peak solar radiation (midday depression), high ambient CO_2 concentrations, or because of disease (Ehret, 2001).

All together, these findings led to the development of the crop water stress index (CWSI) an index that essentially normalizes the stress degree parameter for environmental variability (Idso, 1981). The crop water stress index (CWSI), which is derived from canopy-air temperature differences (T_c -T_a) versus the vapour pressure deficit of the air (VPD), was found to be a promising tool for quantifying crop water stress (Jackson et al., 1981; Idso & Reginato, 1982; Jackson, 1982). The calculation of CWSI relies on the establishment of two baselines: the non water stressed baseline (lower limit), which represents a fully watered crop, and the maximum stressed baseline (upper limit), which corresponds to a non-transpiring crop (stomata fully closed; Yuan et al., 2004; Erdem, 2005). The resultant values of the CWSI normally cover a range from 0 (no stress) to 1 (severe stress). The critical value signifying a reduction in transpiration of plants can be found between 0.25-0.35 (Roth et al., 2004) but the boundary threshold of CWSI indicating irrigation requirements is crop specific, depending, amongst others, on yield response to water stress.

During the last decades many researchers have used the CWSI method for irrigation management (e.g. Pinter & Reginato, 1982; Wanjura et al., 1990; Irmak et al., 2000). Due to the dependency on species, location and climate zones, quite different slope and intercept values have been established in the different studies. Beside theses factors the heterogeneity of different plant canopies has to be considered to assure that the fraction of soil background sensed plays only a minor role. This might become a challenging factor especially during early growth stages until complete canopy closure or in crop species where complete soil cover is generally not reached, since spots of soil between the plants induce higher heterogeneity and thus lead to erroneous plant temperature measures.

Another factor strongly influencing the applicability of CWSI for irrigation scheduling is the local climatic situation. The majority of studies which have successfully applied the CWSI concept were carried out in arid or semi arid regions where cloud cover plays a minor role. In contrast, under more humid conditions the validity of CWSI should be seen critical due to low vapour pressure deficit (VPD) values (with a small range) and the frequent occurrence of clouds (Roth et al., 2004). Faraj et al. (2001) emphasized that because of the strong impact of changing environmental conditions (such as VPD, net radiation and wind speed) on the performance of the lower baseline and variable canopy resistance the usefulness of CWSI for irrigation scheduling is rather limited. Moreover, Yuan *et al.* (2004) compared different CWSI approaches and concluded that, due to its large fluctuations and variations, the empirical CWSI is of little practical value for detecting crop water stress in winter wheat in China. When using the empirical approach, CWSI may even range outside of 0 – 1, leading for example to negative values as it was observed by Faraj *et al.* (2001).

Wanjura *et al.* (1995; Wanjura & Upchurch, 1996) introduced an alternative method to determine the non-water stressed baseline. They based their method on the Penman-Montheith equation and considered the surface temperature as a wet bulb temperature which can be determined when further parameters like net radiation (R_n) and the aerodynamic resistance (r_a) to heat flow between the surface and a reference level are known. The baseline, when established by the method of Idso *et al.* (1981), has to be determined experimentally which bears considerable constraints: it precludes its transfer to other regions since baselines will be site specific. They might also not be transferred to different years (or other times of the day) and they will be valid only for the same clear sky conditions (Alves, 2000). Alves & Pereira (2000), however, concluded from their studies that the infrared surface temperature of fully transpiring crops can indeed be regarded as a wet bulb temperature that can be used to calculate the surface temperatue (T_s) for non-water-stressed conditions when net radiation (R_n), aerodynamic resistance (r_a) and air temperature are known. This method has the advantage over the experimentally determined non-water stressed baseline that measurements can be made at any time of the day from sunrise to sunset, that they can be made independently of climatic conditions including cloudy conditions and finally, previous observations, to derive or to validate a baseline, are not necessary (Alves & Pereira, 2000). For a detailed derivation of the equation see Alves & Pereira (2000).

To summarize, several studies have successfully applied the CWSI concept to their regions but all these studies have in common that they were carried out in arid or semi arid regions of the world where cloud cover plays a minor role. In these climatic regions the concept provides a solid and cost effective method to schedule irrigation and reduce water consumption. It has, however, to be kept in mind that the CWSI is only an indicator of the onset of a drought event but does not give any further information about the amount of water needed to retain maximum possible yields. But this should not be a detriment to use this technique for irrigation scheduling since in many cases the irrigation amounts are limited by other factors like the irrigation system application rate, soil water intake rate or the amount of water available for irrigating crops. In such situations the knowledge about the theoretically needed amount of water is of little use due to the other restrictions (Nielson & Gardener, 1987).

4.2.1 Case study 2 – Use of CIR sensors for drought stress detection in Pannonian climate

4.2.1.1 Materials and methods

This study was carried out at Versuchsgarten Augarten, Vienna (48°13'35" N, 16°22'30" E, 164 m a.s.l.), and aimed at the short term drought stress detection by the use of thermal infrared measurements. Two areas in the size of 5 x 6 m were available for crop growing: a reference plot, where plants were irrigated ("irrigated") and a second plot where plants were only irrigated until the first leaves were fully developed and then exposed to precipitation only ("rainfed").

Triticum aestivum L. cv. Xenos was sown on 5 April 2006 at a rate of 600 seeds m^{-2}. The soil, a chernozem, was fertilized with a total amount of 120 kg N/ha (70 kg N from Nitramoncal and 50kg from KNO$_3$) with the application split in three bits according to local agricultural practice: before sowing, at tillering and at heading. Application of KNO$_3$ further supplied plants with a total of amount 60 kg K/ha.

For the measurement of plant canopy temperature cloud infrared (CIR) sensors were used. CIR-Sensors are ground-based instruments which were originally designed to infer day and night cloud cover. They operate in a 9 to 14 µm spectral range with a 12-degree FOV. In the here presented study two such sensors of which each generated two distinct output signals, the temperature of the sensor's shell and the infrared temperature of the measured body, were used. From these data and under implementation of meteorological parameters such as air temperature (T_a), vapor pressure deficit (VPD), net radiation (R_n) and aerodynamic resistance (r_a) the canopy temperature of the plant stand was calculated.

In addition, basic plant physiological parameters were determined: instantaneous leaf conductance (g_L) was measured with a steady state porometer (PMR-4, PP-Systems; U.K.), leaf water potential (Ψ_w) was determined predawn (02:30-04:00) and at noon (12:30-14:00) using a pressure Bomb (Scholander *et al.*, 1965) and leaf temperature of single leaves within the plots was measured with a handheld infrared thermometer (Raytek). Relative water content was calculated as RWC = ((fresh weight - dry weight)/(saturated weight - dry weight)) * 100 [%]. All measurements were performed on the youngest fully expanded leaf at three developmental stages (vegetative growth, flowering and grain filling). The determination of plant physiological parameters served as a reference to indicate the occurrence of drought stress in rainfed plants.

4.2.1.2 Results and discussion

Regarding the climate, the study region belongs to the northeastern part of Austria, a semi-arid area characterized by deep groundwater level and low precipitation levels. The mean annual precipitation is 577mm and the mean annual temperature is 9.9°C (Eitzinger *et al.*, 2003). The climate is therefore more humid than in other regions where CWSI is applied for irrigation management. It is further known from other studies that the calculation of CWSI is ideally performed during cloudless skies (e.g. Idso *et al.*, 1981). Since such conditions are not frequently available in the study area, the present study aimed at testing the uncertainties of the CIR-Sensor measurements and to calculate CWSI for non-ideal conditions such as cloudy sky, fetch effects and suboptimal orientation of the sensors.

An early response of plants to a lowering of soil water availability is the reduction of leaf conductance (g_L) thus avoiding excess water loss and providing higher water use efficiency to the plant (Cornic & Massacci, 1996; Lawlor, 2002; Flexas & Medrano, 2002; compare also section 2). At all observation dates, g_L of rainfed plants was lower than that of irrigated ones (Table 4) indicating poorer water supply. This further resulted in lower transpiration rates and reduced transpirational cooling. Therewith, slightly increased leaf temperatures were observed in plants of the rainfed plot throughout the whole growing season (Table 4). The differences between leaf temperature of plants from the rainfed and the irrigated plot were highest around noon, where the surface received maximum net radiation. The smallest mean differences were observed between 0:00 and 3:00 MEZ. Therefore, values of the 12:00 to 15:00 MEZ time period were used for the calculation of crop water stress index (CWSI).

time	growth stage	Ψ_w		RWC		g_L US		g_L LS		Tl$_{eaf}$	
		ir	rf	ir	rf	ir	rf	ir	rf	ir	rf
pre-d.	veget.	-1.8	-1.9	93.2	94.8						
	flow.	-2.4	-4.6***	94.4	89.4						
	grain f.	-6.8	-11.3***	82.8	83.8						
noon	veget.	-6.8	-7.8	94.4	92.2	541.4	364.2**	601.1	177.0***	18.7	22.4
	flow.	-17.2	-20.1*	77.5	82.1	632.7	523.2**	568.7	361.1**	27.7	28.6*
	grain f.	-18.3	-19.9	79.3	81.5	721.0	537.4**	609.1	328.5***	25.0	30.2***

Table 4. Summary of the results from physiological measurements. *Abbreviations*: pre-d.: pre dawn measurement; noon: noon measurement; veget.: vegetative growth, flow.: flowering; grain f.: grain filling; ir: irrigated; rf: rainfed; Ψ_w: leaf water potential [bar]; RWC: relative water content [%]; g_L: leaf conductance [mmol.m^{-2}.s^{-1}]; US: upper leaf surface; LS: lower leaf surface, Tl$_{eaf}$: leaf temperature. Significance levels refer to the differences between rainfed and irrigated plants. ***: p ≤ 0.001; **: p ≤ 0.01; *: p ≤ 0.05; n=5-30.

In addition to the measurement of stomatal conductance and leaf temperature, leaf water potential was determined. A decrease of soil water content resulted in a lowering of leaf water potential (Ψ_w, more negative values; Chaves, 1991; Cornic, 1994; Lawlor, 1995). Predawn leaf water potential ($\Psi_{w,pd}$), which gives a pretty good estimation of soil water content (Richter, 1997; Taiz & Zeiger, 2000; Lösch, 2003), did not differ greatly at vegetative growth (+8%, Table 4) between plants of the two plots. At flowering and grain filling, rainfed plants exhibited significantly lower $\Psi_{w,pd}$ (more negative; +94% and +65%, respectively) compared to irrigated plants.

For wheat, the most critical period concerning the occurrence of drought stress is the period which brackets anthesis. Irrigation recommendations are therefore given for drier periods during vegetative growth until shortly after flowering (Geisler, 1983). Thus, for testing the energy balance method to determine a non-water stressed baseline, data from June (flowering period) were used. Concerning the results obtained from infrared thermometry it could be shown that the effect of wind was rather small and, both the orientation of the CIR sensors as well as fetch effects were not very prominent in this environment during noon time (results not shown). The simulated leaf temperature for the irrigated plot was calculated by applying both the classical method after Idso (Idso *et al.*, 1981) and the energy balance based method (Alves & Pereira, 2000) for all sky conditions. The resultant coefficient of determination (r^2) between simulated and measured leaf temperature was higher for the latter method ($r^2=0.8$), for which reason the set up of a non water stressed baseline by the method based on the energy balance is to be advantaged under such climatic conditions. Figure 4 indicates a satisfying relationship between measured and calculated leaf temperature both for a single week (11.6-17.06.2006) and for almost one month (8.6.-4.7.2006) during the experiment (Table 5 gives a summary of leaf temperatures measured with a handheld IR thermometer). However, when comparing the simulated (energy balance method) and measured CWSI calculated from data recorded at noon time (12.00-15.00) throughout the whole experimental period, the result is not at all satisfying ($r^2=0.3366$).

From this study it can be concluded that one of the most detrimental factors for establishing a non water stressed baseline seems to be cloudiness, a result which was also obtained by Da Silva & Rao (2005). However, the results further suggest that influences other than cloudiness might be responsible for the poor relationship between measured and simulated CWSI. Idso (1982), for example, defined non-water-stressed baselines for 26 different species for clear sky conditions and found that these baselines were different for various phenological stages in certain crops. He further suggested that, for example, for winter wheat different baselines should be developed for pre and post head stages. Gardner *et al.* (1992) also urged that care has to be taken concerning the use of an inappropriate base line since small errors in its determination can lead to large errors in the calculation of CWSI.

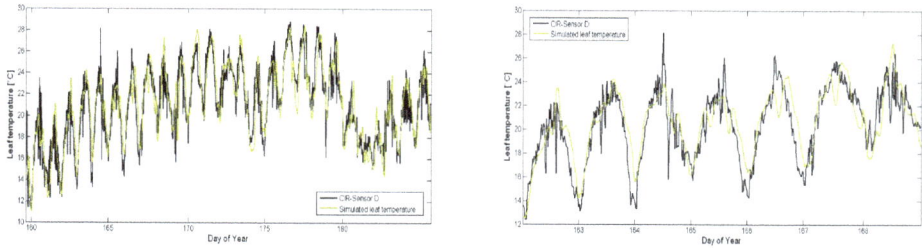

Fig. 4. Simulated (energy balance method) and measured leaf temperature for plants of the irrigated plot during a) one month (8.6.-4.7.2006) and b) one week (11.6-17.06.2006, beginning of flowering) of the experimental period.

date	treatment	time of the day	g_L (Std error)	T_{leaf} (Std error)	T_{air}
19/06/2006	rainfed	morning	531.8 (31.2)	26.8 (1.5)	26.2
		noon	582.6 (57.1)	31.0 (0.9)	29.1
		afternoon	325.4 (39.2)	26.6 (0.4)	29.2
	irrigated	morning	746.7 (88.9)	29.0 (1.1)	26.8
		noon	558.1 (41.7)	28.5 (0.5)	29.7
		afternoon	520.5 (54.3)	26.3 (0.4)	28.5
21/06/2006	rainfed	morning	606.8 (33.3)	27.1 (0.7)	28.0
		noon	538.6 (65.4)	26.1 (0.2)	28.0
		afternoon	513.8 (74.5)	28.5 (0.3)	29.7
	irrigated	morning	651.2 (22.1)	24.0 (0.3)	27.2
		noon	696.6 (72.7)	27.0 (0.4)	28.3
		afternoon	628.6 (61.1)	25.4 (0.4)	29.9

Table 5. Daytime course of leaf temperatures measured with a handheld IR thermometer (T_{leaf}), air temperature (T_{air}) and stomatal conductance (g_L) in *T. aestivum* for the 19.06.2006 and the 21.06.2006 (flowering period).

5. Conclusions

Water scarcity is an increasingly important issue in many parts of the world. Within the next centuries global climate change is expected to result in a long-term trend towards higher temperatures, greater evapotranspiration, and an increased incidence of drought in specific regions. Concurrently, the increased need for food supply worldwide and the prediction of the emergence of drier regions, demand the development of cost effective methods for early stress detection to reduce yield losses.

The first study presented herein aimed at the evaluation of drought stress, applied at different phenological stages (flowering and/or grain filling), on plant physiological traits and leaf reflectance and their subsequent recovery. An increase of leaf reflectance (R) as observed in the range from 500-600nm is mainly attributed to a lower chlorophyll or nitrogen content. However, in this study, a lower relative water content (RWC) also increased R in this range of the spectrum. A higher R would normally be attributed to a decreased chlorophyll content or nitrogen deficiency but would not be primarily addressed to a lower RWC. Results further showed that rehydrating plants recovers physiological traits but the recovery could not be traced by reflectance measurements since R remained above that of control plants. A distinction between a current stress situation and an already passed one was not possible. Fertilization commendations based on such results would be ineffective since the uptake of nitrogen by plants is strongly restricted under drought.

From the second study presented herein it can be concluded that sensing drought stress by thermal IR works well under clear sky conditions at larger scales without fetch effects and when sensor orientation is optimal. Non-ideal conditions such as small study sites and frequently changing environmental conditions (e.g. cloudiness), however, may introduce uncertainties which might be larger than the drought stress signal itself. As a result, the calculation of CWSI for this study site, characterized by changing environmental factors (e.g. cloudiness), seems not accurate enough to be used for irrigation scheduling.

6. References

Allen, L. (1990). Plant responses to rising carbon dioxide and potential interactions with air pollutants. *Journal of Environmental Quality*, Vol.19, pp.15 - 34

Alves, I. & Pereira, L. (2000). Non-water-stressed baselines for irrigation scheduling with infrared thermometers: A new approach. *Irrigation Science*, Vol.19, pp.101-106

Amthor, J. (2001). Effects of atmospheric CO_2 concentration on wheat yield: review of results from experiments using various approaches to control CO_2 concentration. *Field Crops Research*, Vol.73, No.1, pp.1-34

Araus, J.; Alegre, L.; Tapia, L. & Calafell, L. (1986). Relationship between leaf structure and gas exchange in wheat leaves at different insertion levels. *Journal of Experimental Botany* Vol.37, pp.1323-1333

Araus, J.; Casadesus, J. & Bort, J. (2001). Recent tools for the screening of physiological traits determining yield. In M.P. Reynolds, J.I. Ortiz-Monasterio, and A. Mcnab (ed.) Application of physiology in wheat breeding. CIMMYT, Mexico, D.F., pp. 59–77

Araus, J.; Slafer, G.; Reynolds, M. & Royo, C. (2002). Plant breeding and drought in C3 cereals: what should we breed for? *Annals of Botany*, Vol.89, pp.925-940

Arp, W. (1991). Effects of source-sink relations on photosynthetic acclimation to elevated CO_2. *Plant Cell and Environment*, Vol.14, pp.869-875

Asner, G.; Braswell, B. ; Schimel, D. & Wessman, C. (1998a). Ecological research needs from multi-angle remote sensing data. *Remote Sensing of Environment*, Vol.63, pp.155–165.

Asner, G. (1998b). Biophysical and biochemical sources of variability in canopy reflectance. *Remote Sensing of the Environment*, Vol.63, pp.234-253

Ayala-Silva, T. & Beyl, C. (2005). Changes in spectral reflectance of wheat leaves in response to specific macronutrient deficiency. *Advances in Space Research*, Vol.35, pp.305-317

Babar, M.; Reynolds,M.; van Ginkel, M.; Klatt, A.; Raun, W. & Stone M. (2006). Spectral Reflectance to Estimate Genetic Variation for In-Season Biomass, Leaf Chlorophyll, and Canopy Temperature in Wheat. *Crop science*, Vol.46, pp.1046-1057

Baltzer, J. & Thomas, S. (2005). Leaf optical responses to light and soil nutrient availability in temperate deciduous trees. *American Journal of Botany*, Vol.92, No.2, pp.214–223

Barnes, J.; Balaguer, L.; Manrique, E.; Elvira, S. & Davison, A. (1992). A reappraisal of the use of DMSO for the extraction and determination of chlorophylls a and b in lichens and higher plants. *Environmental and Experimental Botany*, Vol.32, No.2, pp.85– 100

Bilger, W.; Björkmann, O. & Thaya, S. (1989). Light-induced spectral absorbance changes in relation to photosynthesis and the epoxidation state of xanthophylls cycle components in cotton leaves. *Plant Physiology*, Vol.91, pp.542-551

Blackburn, G. (1998). Spectral indices for estimating photosynthetic pigment concentrations: a test using senescent tree leaves. *International Journal of remote Sensing*, Vol.19, pp.657-675

Blackburn, G. & Steele, C. (1999). Towards the remote sensing of matorral vegetation physiology: relationships between spectral reflectance, pigment, and biophysical characteristics of semiarid bushland canopies. *Remote Sensing of environment*, Vol.70, pp. 278-292

Blackburn, G. (2007). Hyperspectral remote sensing of plant pigments. *Journal of Experimental Botany*, Vol.58, No.4, pp.855-867

Blad, B.; Gardner, B.; Watts, D. & Rosenberg, N. (1981). Remote sensing of crop moisture status. *Remote Sens. O.*, Vol.3, pp.4-20

Beeri, O. & Peled, A. (2006). Spectral indices for precise agriculture monitoring. *International Journal of Remote Sensing*, Vol.27, No.10, pp.2039-2047

Ben-Asher, J.; Phene, C.; Kinarti, A. (1992). Canopy temperature to assess daily evapotranspiration and management of high frequency drip irrigation systems. *Agric. Water Manage.*, Vol.22, pp.379–390

Birth, G. & McVey, G. (1968). Measuring the color of growing turf with a reflectance spectrophotometer. *Agronomy Journal*, Vol.60, pp.640– 643

Blum, A. (2005). Drought resistance, water-use efficiency, and yield potential—are they compatible, dissonant, or mutually exclusive? *Australian Journal of Agricultural Research*, Vol.56, pp.1159–1168

Blum, A. (2009). Effective use of water (EUW) and not water-use efficiency (WUE) is the target of crop yield improvement under drought stress. *Field Crops Research*, Vol.112, pp.119–123

Bowes, G. (1993). Facing the inevitable: plants and increasing atmospheric CO_2. *Annual Review of Plant Physiology and Plant Molecular Biology*, Vol.1993, pp.309-33

Bowman, W. (1989). The relationship between leaf water status, gas exchange, and spectral reflectance in cotton leaves. *Remote Sensing of Environment*, Vol.30, pp.249–255

Brodribb, T. (1996). Dynamics of changing intercellular CO_2 concentration (Ci) during drought and determination of minimal functioning Ci. *Plant Physiology*, Vol.111, pp.179-185

Brodribb, T. & Field, T. (2000). Stem hydraulic supply is linked to leaf photosynthetic capacity: evidence from New Caledonian and Tasmanian rainforests. *Plant, Cell and Environment* Vol.23, pp.1381–1388

Brodribb, T. et al. (2005). Leaf hydraulic capacity in ferns, conifers and angiosperms: impacts on photosynthetic maxima. *New Phytologist* Vol.165, pp.839–846

Broge, N. & Leblanc, E. (2000). Comparing prediction power and stability of broadband and hyperspectral vegetation indices for estimation of green leaf area index and canopy chlorophyll density. *Remote Sens. Environ.* Vol.76, No.2, pp.156–172

Brown, R. & Rosenberg, N. (1997). Sensitivity of crop yield and water use to change in a range of climatic factors and CO_2 concentrations: a simulation study applying EPIC to the centralUSA. *Agricultural and Forest Meteorology*, Vol.83, pp.171-203

Buschmann, C. & Nagel, E. (1993). In vivo spectroscopy and internal optics of leaves as basis for remote sensing of vegetation. *International Journal of Remote Sensing*, Vol.14, No.4, pp.711 - 722

Clawson, K. & Blad, B. (1982). Infrared thermometry for scheduling irrigation of corn. *Agronomy Journal*, Vol.74, pp.311-316

Carter, G. (1991). Primary and secondary effects of water content on the spectral reflectance of leaves. *American Journal of Botany*, Vol.78, pp.916–924

Carter, G. (1993). Responses of leaf spectral reflectance to plant stress. *American Journal of Botany*, Vol.80, pp.239–243

Carter, G. (1994). Ratios of leaf reflectances in narrow wavebands as indicators of plant stress. *International Journal of Remote Sensing*, Vol.15, No.3, pp.697–703

Carter, G. & Knapp, A. (2001). Leaf optical properties in higher plants: Linking spectral characteristics to stress and chlorophyll concentration. *American Journal of Botany*, Vol.88, No.4, pp.677-684

Castrillo, M. & Calcagno, A. (1989). Effects of water stress and rewatering on ribulose-1,5-bisphosphate carboxylase activity, chlorophyll and protein contents in two cultivars of tomato. *Journal of horticultural Science*, Vol.64, pp.717-724

Ceccato, P.; Flasse, S.; Tarantola, S.; Jaquemoud, S. & Gregoire, J.-M. (2001). Detecting vegetation leaf water content using reflectance in the optical domain. *Remote Sensing of Environment*, Vol.77, pp.22-33

Chappelle, E.; Kim, M. & McMurtrey, J. III (1992). Ratio analysis of reflectance spectra (RARS): An algorithm for the remote estimation of the concentrations of chlorophyll a, chlorophyll b and carotenoids in soybean leaves. *Remote Sensing of Environment*, Vol.39, No.3, pp.239– 247

Chaplot, V. (2007). Water and soil resources response to rising levels of atmospheric CO_2 concentration and to changes in precipitation and air temperature. *Journal of Hydrology*, Vol.337, pp.159– 171

Chaves, M. (1991). Effects of water deficits on carbon assimilation. *Journal of experimental Botany*, Vol.42, pp.1-46

Chaves, M.M. et al. (2002). How plants cope with water stress in the field. Photosynthesis and Growth. *Annals of Botany*, Vol.89, pp. 907-916

Chaves, M.; Maroco, J. & Pereira, J. (2003). Understanding plant responses to drought – from genes to whole plant. *Functional Plant Biology*, Vol.30, pp.239-263

Cibula, W. & Carter, G. (1992). Identification of a far-red reflectance response to ectomycorrhizae in slash pine. *International Journal of Remote Sensing,* Vol.13, pp.925–932

Clevers, J.; de Jong, S.; Epema, G.; van der Meer, F.; Bakker,W.; Skidmore, A.; Scholte, K.; (2002). Derivation of the red edge index using the MERIS standard band setting. *Int. J. Remote Sens.* Vol.23, no.16, pp.3169–3184.

Collins, W. (1978). Remote sensing of crop type and maturity. *Photogrammetric Engineering and Remote Sensing,* Vol.44, pp.43– 55

Comstock, J. (2002). Hydraulic and chemical signalling in the control of stomatal conductance and transpiration. *Journal of Experimental Botany* Vol.53, No.367, pp.195-200

Cornic, G. (1994). Drought stress and high light effects on leaf photosynthesis. In: Photoinhibition of Photosynthesis (eds N.R. Baker and J.R. Bowyer), BIOS, Scientific Publishers, pp. 297-313

Cornic, G. & Massacci, A. (1996). Leaf photosynthesis under drought stress. In: Photosynthesis and the Environment (ed. N.R. Baker), Kluwer Academic Press, pp. 347-366

Cotrufo, M. et al. (1998). Elevated CO_2 reduces the nitrogen concentration of plant tissues. *Global Change Biology,* Vol.4, pp.43-54

Curcio, J. & Petty, C. (1951). The near infrared absorption spectrum of liquid water. *Journal of the Optical Society of America,* Vol.41, pp.302–304

Curran, P. ; Dungan, J. & Gholz, H. (1990). Exploring the relationship between reflectance red edge and chlorophyll content in slash pine. *Tree Physiology,* Vol.7, pp.33-48.

Cure, J. & Acock, B. (1986) Crop response to carbon dioxide doubling: a literature survey. *Agricultural and Forest Meteorology,* Vol.38, pp.127 - 145

Curran, P.; Dungan, J.; Macler, B. & Plummer, S. (1991). The effect of a red leaf pigment on the relationship between red edge and chlorophyll concentration. *Remote Sensing of Environment,* Vol.35, pp.69-76

Datt, B. (1998). Remote sensing of chlorophyll *a,* chlorophyll *b,* chlorophyll *a* + *b,* and total carotenoid content in eucalyptus leaves. *Remote Sensing of Environment,* Vol.66, pp.111–121

Daughtry, C.; Walthall, C.; Kim, M.; Brown de Colstoun, E. & McMurtrey III, J. (2000). Estimating corn leaf chlorophyll concentration from leaf and canopy reflectance. *Remote Sens. Environ.,* Vol.74, pp.229– 239

Davenport, J.; Lang, N. & Perry, F. (2000). Leaf spectral reflectance for early detection of misorders in model annual and perennial crops. ASA-CSSA-SSSA, 677 Segoe Road, Madison, WI 53711, USA. *Proceedings of the Fifth International Conference on Precision Agriculture.*

Davies, W. & Zhang, J. (1991). Root signals and the regulation of growth and development of plants in drying soil. *Annual Reviews of Plant Physiology,* Vol.42, pp.55-76

Dawson, T. & Curran, P. (1998). Technical note A new technique for interpolating the reflectance red edge position, *International Journal of Remote Sensing,* Vol.19, No.11, pp.2133 - 2139

Dawson, T. & Curran, P. (1998b). A new technique for interpolating the reflectance red edge position. *Int. J. Remote Sens.* Vol.19, pp.2133–2139

De Costa, W. ; *et al.* (1997). Effects of different water regimes on field-grown determinate and indeterminate faba bean (*Vicia faba* L.). II. Yield, yield components and harvest index. *Field Crops Research*, Vol.52, pp.169-178

Deng, X.-P.; Shan, L.; Inanaga, S. & inoue, M. (2005). Water-saving approaches for improving wheat production. *Journal of the Science of Food and Agriculture*, Vol.85, pp.1379-138

Dorigo, W.; Zurita-Miller, R.; de Wit, A.; Brazile, J.; Singh, R. & Schaepman, M. (2007). A review on reflective remote sensing and data assimilation techniques for enhanced agroecosystem modelling. *International Journal of Applied Earth Observation and Geoinformation*, Vol.9, pp.165–193

Drake, B. & Gonzales-Meler, M. (1997). More efficient plants: a consequence of rising atmospheric CO_2? *Annual Review of Plant Physiology and Plant Molecular Biology*, Vol.48, pp.609-639

Egli, D. & Crafts-Brandner, S. (1996). Soybean. In: *Photoassimilate distribution in plants and crops*. (Eds Zamski E and Schaffer AA) Marcel Dekker, Inc. pp. 595-624.

Ehret, D.; Lau, A.; Bittman, S.; Lin, W. & Shelford, T. (2001). Automated monitoring of greenhouse crops. *Agronomie*, Vol.21, pp.403–414

Eitel, J.; Gessler, P.; Smith, A. & Robberecht, R. (2006). Suitability of existing and novel spectral indices to remotely detect water stress in Populus spp. *Forest Ecology and Management*, Vol.229, pp.170–182

Eitzinger, J.; Stastná, M.; Zalud, Z.; & Dubrovsky, M. (2003). A simulation study of the effect of soil water balance and water stress on winter wheat production under different climate change scenarios. *Agricultural Water Management*, Vol.61, pp.195-217.

Elvidge, C. (1990). Visible and near infrared reflectance characteristics of dry plant materials. *International Journal of Remote Sensing*, Vol.11, pp.1775-1795

Erdem, Y.; Erdem, T.; Orta, A. & Okursoy, H. (2005). Irrigation scheduling for watermelon with crop water stress index (CWSI). *Journal Central European Agriculture*, Vol.6, No.4, pp.449-460

Faraj, A.; Meyer, G. & Horst, G. (2001) A crop water stress index for tall fescue (*Festuca arundinacea* Schreb.) irrigation decision-making — a traditional method. *Computers and Electronics in Agriculture*, Vol.31, pp.107–124

Farooq, M. (2009). Plant drought stress: effects, mechanisms and management. *Agronomy for Sustainable Development*, Vol.29, pp.185–212

Farquhar, G. et al. (1980). A biochemical model of photosynthetic CO_2 assimilation in leaves of C3 species. *Planta*, Vol.149, pp.78-90

Ferri, C.; Formaggio, A. & Schiavinato, M. (2004). Narrow band spectral indexes for chlorophyll determination in soybean canopies [*Glycine max* (L.) Merril]. *Braz. J. Plant Physiol.*, Vol.16, No.3, pp.131-136

Filella, I. & Penuelas, J. (1994). The red edge position and shape as indicators of plant chlorophyll content, biomass and hydric status. *International Journal of Remote Sensing*, Vol.15, pp.1459-1470

Flexas, J. & Medrano, H. (2002). Drought inhibition of photosynthesis in C3 plants: stomatal versus non-stomatal limitations revisited. *Annals of Botany* Vol.89, pp.183-189

Flexas, J.; Bota, J.; Escalona, J.; Sampol, B. & Medrano, H. (2002). Effects of drought on photosynthesis in grapevines under field conditions: an evaluation of stomatal and mesophyll limitations. *Functional Plant Biolology* Vol.29, pp.461–471

Foyer, C.; Descourvières, P. & Kunert, K. (1994). Protection against oxygen radicals: an important defence mechanism studied in transgenic plants. *Plant, Cell & Environment*, Vol.17, pp.507– 523

Fuchs, M. & Tanner, C. (1966): Infrared thermometry of vegetation. *Agronomy Journal*, Vol.58, pp.597-601

Fuhrer, J. (2003). Agroecosystem responses to combinations of elevated CO_2, ozone and global climate change. *Agriculture, Ecosystems & Environment*, Vol.97, No.1-3, pp.1-20

Gamon, J.; Field, C.; Bilger, W.; Björkmann, O.; Fredeen, A. & Penuelas, J. (1990). Remote sensing of the xanthophylls cycle and chlorophyll fluorescence in sunflower leaves and canopies. *Oecologia*, Vol.85, pp.1-7

Gamon, J.; Serrano, L. % Surfus, J. (1997). The photochemical reflectance index: an optical indicator of photosynthetic radiation use efficiency across species, functional types, and nutrient levels. *Oecologia* Vol.112, pp.492-501

Gao, X.; Huete, A.; Ni, W. & Miura, T. (2000). Optical-Biophysical relationships of vegetation spectra without background contamination. *Remote sensing of Environment*, Vol.74, pp.609-620

Gardner, B.; Blad, B.; Garrity D. & Watts, D. (1981). Relationships between crop temperature, grain yield, evapotranspiration and phonological development in two hybrids of moisture stressed sorghum. *Irrigation Science*, Vol.2, pp.213-224

Gardner, B.; Nielsen, D. & Shock, C. (1992): Infrared thermometry and the crop water stress index. I. History, theory, and baselines. *Journal of Production Agriculture*, Vol.5, No.4, pp.462–466.

Geisler, G. (1983). Ertragsphysiologie von Kulturarten des gemäßigten Klimas. Berlin und Hamburg, Paul Parey. ISBN 3-489-61010-5

Gilabert, M.; Gandia, S. & Melia, J. (1996). Analysis of spectral-biophysical relationships for a corn canopy. *Remote Sensing of Environment*, Vol.55, pp.11-20

Gimenez, C.; Mitchell, V. & Lawlor, D. (1992). Regulation of photosynthesis rate of two sunflower hybrids under water stress. Plant Physiology, Vol.98, pp.516-524

Gitelson, A. & Merzlyak, M. (1994). Quantitative estimation of chlorophyll *a* using reflectance spectra: Experiments with autumn chestnut and maple leaves. *J Photochem Photobiol*, (B) Vol.22, pp.247–252

Gitelson, A.; Kaufman, Y. & Merzlyak, M. (1996). Use of a green channel in remote sensing of global vegetation from EOS-MODIS. *Remote Sensing of Environment*, Vol.58, No.3, pp.289– 298

Gitelson, A.; Merzlyak, M. & Chivkunova, O. (2001). Optical Properties and Nondestructive Estimation of Anthocyanin Content in Plant Leaves. *Photochemistry and Photobiology*, Vol.74, No.1, pp.38–45

Gitelson, A. &Merzlyak, M. (2004). Non-destructive assessment of chlorophyll, carotenoid and anthocyan content in higher plant leaves: principles and algorithms. *Remote sensing for Agriculture and the Environment*. (S. Stamatiadis, J.M. Lynch, J.S. Schepers, eds.) Grece, Ella, 2004: 78-94

Gollan, T.; Turner, N. & Schulze, E. (1985). The response of stomata and leaf gas exchange to vapour pressure deficits and soil water content. III. In the sclerophyllous woody species *Nerium oleander. Oecologia* Vol.65, pp.356–362

Gonzales-Dugo, M.; Moran, M.; Mateos, L. & Bryant, R. (2006). Canopy temperature variability as an indicator of crop water stress severity. *Irrigation Science*, Vol.24, pp.233-240.

Gowing, D. ; Davies, W. & Jones, H. (1990). A positive root-source signal as an indicator of soil drying in apple, *Malus domestica*. *Journal of Experimental Botany* Vol.41, pp.1535-1540

Henson, I. ; Jenson, C. ; et al. (1989). Leaf gas exchange and water relations of lupins and wheat. I. Shoot responses to soil water deficits. *Australian Journal of Plant Physiology* Vol.16, pp.401-413.

Guyot, G.; Baret, F. & Major, D. (1988). High spectral resolution: determination of spectral shifts between the red and the near infrared. *Int. Arch. Photogrammetry Remote Sens.* Vol.11, pp.750-760

Haboudane, D.; Miller, J.; Tremblay, N.; Zarco-Tejada, P. & Dextraze, L. (2002). Integrated narrow-band vegetation indices for prediction of crop chlorophyll content for application to precision agriculture. *Remote Sensing of Environment*, Vol.81, pp.416-426

Haboudane, D.; Miller, J.; Pattey, E.; Zarco-Tejada, P. & Strachan, I. (2004). Hyperspectral vegetation indices and novel algorithms for predicting green LAI of crop canopies: Modeling and validation in the context of precision agriculture. *Remote Sensing of Environment*, Vol., 90, pp337-352

Haveaux, M.; Canaani, O. & Malkin, S. (1987). Inhibition of photosynthetic activities under slow water stress measured in vivo by the acoustic method. Physiologia Plantarum, Vol.70, pp.503-510

Huete, A. (1988). A soil-adjusted vegetation index (SAVI). *Remote Sens. Environ.*, Vol.25, pp.295-309

Hunt, E. & Rock, B. (1989). Detection of changes in leaf water content using near- and middle-infrared reflectance. *Remote Sensing of Environment*, Vol.30, pp.43-54

Idso, S. (1977). Remote-Sensing of crop yields. *Science*, Vol.196, No.4285, pp.19-25

Idso, S.; Jackson, R.; Pinter, J.; Reginato, R. & Hatfield, J. (1981). Normalizing the stress-degree-day parameter for environmental variability. *Agricultural Meteorology*, Vol.24, pp.45-55

Idso, S. & Reginato, R. (1982). Soil and atmosphere induced plant water stress in cotton as inferred from foliage temperatures. *Water Resource. Res.*, Vol.18, pp.1143-1148

Idso, S. (1982). Non-water-stressed baselines: a key to measuring and interpreting plant water stress. *Agriculture Meteorology*, Vol.27, pp.59-70

Idso S.B et al. (1987). Effects of atmospheric CO_2 enrichment on plant growth: the interactive role of air temperature. *Agriculture, Ecosystems & Environment*, Vol.20, No.1, pp.1-10

IPCC 2007 Climate Change (2007). The physical science basis, summary for Policymakers, Contribution of Working Group I to the Fourth Assessment Report of the Intergovernmental Panel on Climate Change, International Panel on Climate Change. WMO UNEP report.; www.ipcc.ch

Irmak, S.; Haman, D. & Bastug, R. (2000). Determination of Crop Water Stress Index for Irrigation Timing and Yield Estimation of Corn. *Agronomy Journal*, Vol.92, pp.1221-1227

Jackson, R.; Reginato, R. & Idso, S. (1977). Wheat canopy temperature: a practical tool for evaluating water requirements. *Water Resour. Res.*, Vol.13, pp. 651-656

Jackson, R.; Idso, S.; Reginato, R. & Pinter Jr, P. (1981). Canopy temperature as a crop water stress indicator. *Water Resources Research*, Vol.17, No.4, pp.1133–1138

Jackson, R. (1982). Canopy temperature and crop water stress. Advances in irrigation, 1, Academic Press, New York, 43-85. ISBN 0-12-24301-6

Jackson, R. (1986). Remote sensing of biotic and abiotic plant stress. *Annual Review of Phytopathology*, Vol.24, pp.265–287

Jaleel, C.; Manivannan P.; Wahid, A.; Farooq, M.; Somasundaram, R. and Panneerselvam, R. 2009. Drought stress in plants: a review on morphological characteristics and pigments composition. *Int. J. Agric. Biol.*, 11: 100–105

Johnson, D.; Richards, R. & Turner, N. (1983). Yield, water relations, gas exchange and surface reflectances of near-isogenic wheat lines differing in glaucousness. *Crop Science*, Vol.23, pp.318–325 72

Jones, H. & Sutherland, R. (1991). Stomatal control of xylem embolism. *Plant, Cell and Environment* Vol.11, pp.111–121

Jordan, C. (1969). Derivation of leaf area index from quality of light on the forest floor. *Ecology*, 50, 663– 666

Jordan, W.; Monk, R.; Miller, F.; Rosenow, D.; Clark, L. & Shouse, P. (1983). Environmental physiology of sorghum. I. Environmental and genetic control of epicuticular wax load. *Crop Science*, Vol.23, pp.552–558

Kang, S.; Zhang, F.; Hu, X. & Zhang, J. (2002). Benefits of CO_2 enrichment on crop plants are modified by soil water status. *Plant Soil*, Vol.238, pp.69–77

Kim, M.; Daughtry, C.; Chappelle, E.; McMurtrey III, J. & Walthall, C. (1994). The use of high spectral resolution bands for estimating absorbed photosynthetically active radiation (Apar). Proceedings of the 6th Symp. on Physical Measurements and Signatures in Remote Sensing, Jan. 17–21, 1994, Val D'Isere, France, pp. 299–306

Kimball, B. & Mauney, J. (1993). Response of cotton to varying CO_2, irrigation, and nitrogen: yield and growth. *Agronomy Journal*, Vol.85, 706 -712

Kimball, B. (1995). Productivity and water use of wheat under free air CO_2 enrichment. *Global Change Biology*, Vol.1, pp.429-442

Kimball, B.; Kobayashi, K. & Bindi, M. (2002). Responses of agricultural crops to free-air CO_2 enrichment. *Adv. Agron.*, Vol.77, pp.293-368

Knipling, E. (1970). Physical and physiological basics for the reflectance of visible and near-infrared radiation from vegetation. *Remote Sensing of Environment*, Vol1, pp.155-159

Kruijt, B.; Witte, J.-P.; Jacobs, C. & Kroon, T. (2008). Effects of rising atmospheric CO_2 on evapotranspiration and soil moisture: A practical approach for the Netherlands. *Journal of Hydrology*, Vol.349, pp.257– 267

Lamb, D.; Steyn-Ross, M.; Schaare, P.; Hanna, M.; Silvester, W. & Steyn-Ross, A. (2002). Estimating leaf nitrogen concentration in ryegrass (*Lolium spp.*) pasture using the chlorophyll red-edge: theoretical modelling and experimental observations, *International Journal of Remote Sensing*, Vol.23, No.18, pp.3619 - 3648

Larcher, W. (2003): Physiological Plant Ecology – Ecophysiology and Stress Physiology of Functional Groups. Springer (4th edition). ISBN 3-54043516-6

Lawlor, D. & Mitchell, R. (1991). The effects of increasing CO_2 on crop photosynthesis and productivity: a reviwe of field studies. *Plant Cell and Environment*, Vol.14, pp.807-818

Lawlor, D. (1995). The effects of water deficits on photosynthesis. In: Environment and plant metabolism (ed. Smirnoff). BIOS Scientific Publishers, Oxford. pp. 129-160

Lawlor, D. & Cornic, G. (2002). Photosynthetic carbon assimilation and associated metabolism in relation to water deficit in higher plants. *Plant Cell and Environment* Vol.25, pp.275-294

Linke, R.; Richter, K.; Haumann, J.; Schneider, W. & Weihs, P. (2008). Occurrence of repeated drought events: can repetitive stress situations and recovery from drought be traced with leaf reflectance? Periodicum Biologorum, Vol.110, No.3, pp.219-229

Long, S. (1991). Modification of the response of photosynthetic prodctivity to rising temperature by atmospheric CO_2 concentrations: Has its importance been underestimated? *Plant Cell and Environment*, Vol.14, pp.729-739.

Loreto, F. & Centritto, M. (2004). Photosynthesis in a changing world. *Plant Biology*, Vol.6, pp.239-241

Maroco, J.; Pereira, J. &Chaves, M. (1997). Stomatal responses to lea-to-air vapour pressure deficit in Sahelian species. *Australian Journal of Plant Physiology* Vol.24, pp.381-387

Maseda, P. & Fernandez, R. (2006). Stay wet or else: three ways in which plants can adjust hydraulically to their environment. *Journal of Experimental Botany* Vol.57, Issue15, pp.3963-3977

Medrano, H.; Parry, M., Socias, X. & Lawlor, D. (1997). Long term water stress inactivates Rubisco in subterranean clover. Annals of Applied Biology, Vol.131, pp.491-501

Medrano, H.; Escalona, J.; Bota, J.; Gulias, J. & Flexas, J. (2002). Regulation of photosynthesis in C_3 plants in response to progressive drought: stomatal conductance as a reference parameter. *Annals of Botany*, Vol.89, pp.895-905

Meyer, S. & de Kouchkovsky, Y. (1992). ATPase state and activity in thylakoids from normal and water stressed lupin. *FEBS Letters*, Vol.303, pp.233-236

Miller, J.; Hare, E. & Wu, J. (1990). Quantitative characterization of the vegetation red edge reflectance. 1: An inverted-Gaussian reflectance model. *Int. J. Remote Sens.* Vol.11, pp.1755-1773

Morison, J. & Lawlor D. (1999). Interactions between increasing CO_2 concentration and temperature on plant growth. *Plant, Cell and Environment*, Vol.22, pp.659-682

Myers, V. ; Bauer, M.; Gausman, H.; Hart, H.; Heilman, J.; MacDonald, R.; Park, A.; Ryerson, R.; Schmugge, T. & WESTIN, F. (1983). Remote sensing applications in agriculture. *In* R. N. Colwell [ed.], Manual of remote sensing, vol. 2, 2nd ed., American Society of Photogrammetry, Falls Church, Virginia, USA, pp. 2111–2228

Morgan, J. (1984). Osmoregulation and water stress in higher plants. *Annual Review of Plant Physiology* Vol.18, pp.249-257

Nardini, A. & Salleo, S. (2002). Relations between efficiency of water transport and duration of leaf growth in some deciduous and evergreen trees. *Trees–Structure and Function* Vol.16, pp.417–422

Nielsen, D. & Gardner, B. (1987). Scheduling irrigations for corn with the crop water stress index (CWSI). *Applied Agricultural Research*, Vol.2, No.5, pp.295-300

Parry, M.; Andralojc, P.; Khan, S. & Keys, A. (2002). Rubisco activity: effects of drought stress. *Annals of Botany*, Vol.89, pp.833-839

Peñuelas, J.; Filella, I.; Biel, C.; Serrano, L. & Savé, R. (1993). The reflectance at the 950-970 nm region as an indicator of plant water status, *International Journal of Remote Sensing*, Vol.14, No.10, pp.1887 - 1905

Penuelas, J.; Gamon, J.; Fredeen, A.; Merino, J. & Field, C. (1994). Reflectance indices associated with physiological changes in nitrogen- and water-limited sunflower leaves. *Remote Sensing of Environment*, Vol.48, pp.135– 146

Penuelas, J.; Baret, F. & Filella, I. (1995). Semi-empirical indices to assess carotenoids/chlorophyll a ratio from leaf spectral reflectance. *Photosynthetica*, Vol.31, No.2, pp.221– 230

Peñuelas, J. *et al.* (1997). Estimation of plant water content by the reflectance water index WI (R900/R970). *Int. J. Remote Sens*ing Vol.18, pp.2869–2875

Penuelas, J. & Filella, I. (1998). Visible and near infrared reflectance techniques for diagnosing plant physiological status. *Trends in Plant Science* Vol.3, No.4, pp.151-156

Penuelas, J. & Inoue, Y. (1999). Reflectance indices indicative of changes in water and pigment contents of peanut and wheat leaves. *Photosynthetica*, Vol36, Np.3, pp.355-360

Pinter, P. Jr. & Reginato, R. (1982). A thermal infrared technique for monitoring cotton water stress and scheduling irrigation. *Trans. ASAE*, Vol.25, pp.1651–1655.

Plaut, Z. (2003). Plant exposure to water stress during specific growth-stages, Encyclopedia of Water Science, Taylor & Francis, pp. 673– 675

Poorter, H. & Navas, M.-L. (2002). Plant growth and competition at elevated CO_2: on winners, losers and functional groups. *New Phytologist*, Vol.157, pp.175–198

Read, J. & Stokes, A. (2006). Plant biomechanics in an ecological context. *American Journal of Botany*, Vol.93, No.10, pp.1546-1565

Richards, R. (2000). Selectable traits to increase crop photosynthesis and yield of grain crops. *Journal of Experimental Botany* Vol.51, pp.447-458

Richter, H. (1997). Water relations of plants in the field: some comments on the measurement of selected parameters. *Journal of Experimental Botany*, Vol.48, pp.1–7

Rogers, G. et al. (1996). Interactions between rising CO_2 concentration and nitrogen supply in cotton. I. Growth and leaf nitrogen concentration. *Australian Journal of Plant Physiology*, Vol.23, pp.119 - 125

Roth, G. & Goyne, P., (2004). Measuring plant water status. In WATERpak, Australian Cotton CRC/CRDC (http://www.cotton.crc.org.au), p. 157-164

Roujean, J.-L. & Breon, F.-M. (1995). Estimating PAR absorbed by vegetation from bidirectional reflectance measurements. *Remote Sensing of Environment*, Vol.51, No.3, pp.375–384

Rouse, J.; Haas, R.; Schell, J. & Deering, D. (1973). Monitoring vegetation systems in the great plains with ERTS. In N. SP-351, Ed. Third ERTS symposium, vol. 1 (pp. 309 – 317). Washington:NASA

Sack, L.; Cowan, P.; Jaikumar, N. & Holbrook, N. (2003). The 'hydrology' of leaves: co-ordination of structure and function in temperate woody species. *Plant, Cell and Environment* Vol.26, pp.1343–1356

Sangsing, K.; Cochard, H.; Kasemsap, P.; Thanisawanyangkura, S.; Sangkhasila, K.; Gohet, E. & Thaler, P. (2004). Is growth performance in rubber (*Hevea brasiliensis*) clones related to xylem hydraulic efficiency? *Canadian Journal of Botany* Vol.82, pp.886–891

Saralabai, V. et al. (1997). Plant responses to high CO_2 concentration in the atmosphere. *Photosynthetica*, Vol.33, No.1, pp.7-37

Schär, C. et al. (2004). The role of increasing temperature variability in European summer heatwaves. *Nature* Vol.427, pp.332-336

Scholander, P.; *et al.* 1965. Sap pressure in vascular plants. *Science*, Vol.148, pp.339-346

Schultz, H. & Matthews, M. (1997). High vapour pressure deficit exacerbates xylem cavitation and photoinhibition in shadegrown *Piper auritum* H.B. & K. during prolonged sunflecks. I. Dynamics of plant water relations. *Oecologia* Vol.110, pp.312–319

Schultz, H. (2003). Differences in hydraulic architecture account for near isohydric and anisohydric behaviour of two field-grown *Vitis vinifera* L. cultivars during drought. *Plant, Cell and Environment Vol.*26, pp.1393–1405

Sehperd, T. & Griffiths, D. (2006). The effect of stress on plant cuticular waxes. *New Phytologist*, Vol.171, pp.469-499

Sellers P., et al. (1996). Comparison of radiative and physiological-effects of doubled atmospheric CO_2 on climate. *Science*, Vol.271, No.5254, pp.1402 - 1406

Seneviratne, S.; Lüthi, D.; Litschi, M. & Schär, C. (2006). Land–atmosphere coupling and climate change in Europe. *Nature*, Vol.443, pp.205-209

Serraj, R.; & Sinclair, T. (2002). Osmolyte accumulation: can it really help increase crop yield under drought conditions? *Plant, Cell and Environment* Vol.25, pp.333-341

Serrano, L.; Ustin, S.; Roberts, D.; Gamon, J. & Penuelas, J. (2000). Deriving Water Content of Chaparral Vegetation from AVIRIS Data. *Remote sensing of Environment*, Vol.74, pp.570-581

Silva, B. da & Rao, T. (2005). The CWSIvariations of a cotton crop in a semi-arid region of Northeast Brazil. *Journal of Arid Environments*, Vol.62, pp.649–659

Somerville, C. & Briscoe, J. (2001). Genetic engineering and water. *Science* Vol.292, p.2217

Sperry, J. (2000). Hydraulic constraints on plant gas exchange. *Agricultural and Forest Meteorology* Vol.104, pp.13–23

Strachan, I. ; Pattey, E. & Boiswert, J. (2002). Impact of nitrogen and environmental conditions on corn as detected by hyperspectral reflectance. *Remote Sensing of Environment*, Vol.80, pp.213- 224

Stitt, M. (1991). Rising CO_2 levels and their potential significance for carbon flow in photosynthetic cells. *Plant Cell and Environment*, Vol.14, pp.741-762

Taiz, L. & Zeiger, E. (2000). Physiologie der Pflanzen. Heidelberg-Berlin, Spektrum Akademischer Verlag. ISBN 3-8274-0538-6

Tanner, C. (1963). Plant temperatures, *Agronomy Journal*, Vol.55, pp.210-211

Tanriverdi, C. (2006). A Review of Remote Sensing and Vegetation Indices in Precision Farming. *KSU. Journal of Science and Engineering*, Vol.9, No.1, pp.69-76

Tardieu, F. (1996). Drought perception by plants. Do cells of droughted plants experience water stress? *Plant Growth Regulation*, Vol.20, pp.93-104

Tardieu, F. & Simmoneau, T. (1998). Variability among species of stomatal control under fluctuating soil water status and evaporative demand: modelling isohydric and anisohydric behaviours. *Journal of Experimental Botany* Vol.49, pp.419-432

Trejo, C. & Davies, W. (1991). Drought-induced closure of *Phaseolus vulgaris* L. stomata precedes leaf water deficit and any increase in xylem ABA concentration. *Journal of Experimental Botany* Vol.42, pp.1507–1515

Tubiello, F. & Ewert, F. (2002). Simulating the effects of elevated CO_2 on crops: approaches and applications for climate change. *Europ. J. Agronomy*, Vol.18, pp.57-74

Tubiello, F. & Fischer, G. (2007). Reducing climate change impacts on agriculture: Global and regional effects of mitigation, 2000–2080. *Technological Forecasting & Social Change*, Vol.74, pp.1030–1056

Tucker, C. (1977). Assymptotic nature of grass canopy spectral reflectance. *Applied Optics*, Vol.16, No.5, pp.1151-1156

Tyree, M. & Sperry, J. (1988). Do woody plants operate near the point of catastrophic xylem dysfunction caused by dynamic water stress? *Plant Physiology* Vol.88, pp.574–580

Tyree, M. (2003). Hydraulic limits on tree performance: transpiration, carbon gain and growth of trees. *Trees–Structure and Function* Vol.17, pp.95–100

Van den Honert, T. (1948). Water transport in plants as a catenary process. Discussion of the Faraday Society, Vol.3, pp.146-153

Vidale, P.; Lüthi, D.; Wegmann, R. & Schär, C. (2007). European climate variability in a heterogeneous multi-model ensemble. *Climatic Change*, Vol.81, pp.209-232

Wanjura, D.; Hatfield, J. & Upchurch, D. (1990). Crop water stress index relationship with crop productivity. *Irrigation Science*, Vol.11, pp.93–99

Wanjura, D.; Upchurch, D.; Sassenrath-Cole, G.; DeTar, W. (1995). Calculating time-thresholds for irrigation scheduling. Proceedings Beltwide Cotton Conferences (Jan 4-7, San Antonio, Texas), pp. 449-452

Wanjura, D. & Upchurch, D. (1996). Time thresholds for canopy temperature-based irrigation. In: Camp CR, Sadler EJ, Yoder RE (eds) Evapotranspiration and Irrigation Scheduling. Proceedings of the International Conf. (San Antonio, Texas, 3-6 Nov), ASAE/IA/ICID, pp. 295-303

Wiegand C. & Namken, L. (1966). Influence of plant moisture stress,solar radiation and air temperature on cotton leaf temperatures, *Agronomy Journal*, Vol.58, pp.582-586

Wolfe, D. et al. (1998). Interpretation of photosynthetic acclimation to CO_2 at the whole-plant level. *Global Change Biology*, Vol.4, pp.879-893

Woolley, J. (1971). Reflectance and Transmittance of Light by Leaves. *Plant Physiology*, Vol.47, pp.656-662

Wright, G.; Smith, R. & Morgan, J. (1983). Differences between two grain sorghum genotypes in adaptation to drought stress: III. Physiological responses. *Australian Journal of Agricultural Research*, Vol.34, pp.637–651

Wullschleger, S.; Tschaplinski, T. & Norby, R. (2002). Plant water relations at elevated CO2– implications for water-limited environments. *Plant, Cell and Environment*, Vol.25,pp.319–331

Yordanov, I.; Velikova, V. & Tsonev, T. (2000). Plant responses to drought stress, and stress tolerance. *Photosynthetica*, Vol.38, No.2,pp. 171-186

Yordanov, I.; Velikova, V. & Tsonev, T. (2003). Plant reponses to drought and stress tolerance. *Bularian Journal of Plant Physiology*, Special Issue 2003, pp.187–206

Yoshimoto, M.; Oue, H. & Kobayashi, K. (2005). Energy balance and water use efficiency of rice canopies under free-air CO2 enrichment. *Agricultural and Forest Meteorology*, Vol.133, pp. 226–246

Yu, G.-R.; Miwa, T.; Nakayama, K.; Matsuoka, N. & Kon, H. (2000). A proposal for universal formulas for estimating leaf water status of herbaceous and woody plants based on spectral reflectance properties. *Plant and Soil*, Vol.227, pp.47-58

Yuan, G.; Luo, Y.; Sun, X.; Tang, D. (2004). Evaluation of a crop water stress index for detecting water stress in winter wheat in the North China Plain. *Agricultural Water Management*, Vol.64, pp.29–40

Zhang, J.; Nguyen, H. & Blum, A. (1999). Genetic analysis of osmotic adjustment in crop plants. *Journal of Experimental Botany* Vol.50, pp.291-302

Zwiggelaar, R. (1998). A review of spectral properties of plants and their potential use for crop/weed discrimination in row-crops. *Crop Protection*, Vol.17, No.3, pp.189-206

8

Effects of Salinity on Vegetable Growth and Nutrients Uptake

Ivana Maksimovic and Žarko Ilin
University of Novi Sad, Faculty of Agriculture
Serbia

1. Introduction

Irrigation of vegetables during both air and soil-born drought has very positive effect on growth, development and yield. The impact of irrigation on the metabolism of plants is very complex because the increase in soil moisture affects plant physical, chemical and biological properties. Increasing the amount of water in the soil increases its heath capacity. Irrigated (wet) soil is cooler in summer and warmer during the cold weather in comparison with dry soil. Irrigation increases heat conduction in soil and relative humidity of ground layers of the atmosphere. These changes reduce the temperature fluctuations of ground layers of the air and soil during the day and night. Therefore, irrigation suppresses harmful effects of the weak spring and autumn frosts. The water in the soil acts as a solvent and as an environment in which chemical reactions take place. With increasing soil moisture concentration of soil solution is diminishing and the power of water to dissolve different substances is increasing. The CO_2 dissolved in the water contributes significantly to this process by lowering the pH value of soil solution. In weak acidic solution mineral elements present in the soil become more soluble, which may alter plant nutrition. However, if there is any excess of water in the soil, than easily soluble salts of potassium, nitrogen and phosphorus from the upper layers of soil can be leached into deeper soil layers, even outside the zone of the root system, which is very undesirable. In general, irrigation increases soil pH, which should be borne in mind because optimum acidity (pH) for cultivation of most vegetables is between 6.0 and 6.5. Irrigation has a positive effect on microbial processes in the soil which in turn affects nutrient availability to plants. Moreover, irrigation stimulates aerobic processes that may be temporarily slowed down or replaced by anaerobic if the soil is too wet.

In agricultural practice it is not always possible to provide sufficient quantities of irrigation water of good quality. Often more mineralized water and processed waste waters are used (Kalavrouziotis et al., 2010). Therefore, by irrigation the soil can be enriched with useful or harmful salts and various other compounds, depending on water quality. In the irrigation water too high concentration of ions that make it saline is present very often (Table 1).

Salinity of arable land is a problem that is becoming more and more important in many areas where irrigation is a regular agro-technical measure, and in semi-arid and arid regions in the world where atmospheric precipitations are not sufficient to flush the salts from the root zone. When irrigation water is of inadequate quality, the occurrence of chlorosis between leaf veins is commonly observed, leaf tissue necrosis often develops and flowering may not happen at all. Very great importance is ascribed to the time during which the plants were exposed to different concentrations of salts. As explained by Munns (2002), effects of salinity on salt-tolerant plants are the same as effects of water deficiency. Within minutes and hours of exposure to salinity salt-specific effects are not visible. If the exposure to salts lasts many days, salt-induced injuries become apparent on older leaves of salt-sensitive plants, in addition to reduced rate of leaf emergence and heavier impact on leaves than on roots, which are symptoms typical for water-stress. After weeks of exposure to salts older leaves of sensitive genotypes die and if exposure lasts several months younger leaves die and the whole plant may die before seed maturation. Differences between short-term/temporary effects of salinization vs. long-term effects (either single irrigation seasons or multiple years) are often overlooked (Maggio et al., 2011).

Water designation	Total dissolved salts (ppm)	EC (dSm^{-1})	Common name	Botanical name	Threshold EC (dS m^{-1})	Rating*
Fresh water	< 500	< 0.6				
Slightly brackish	500–1000	0.6–1.5	Bean	*Phaseolus vulgaris* L.	1.0	S
			Carrot	*Daucus carota* L.	1.0	S
			Muskmelon	*Cucumis melo* L.	1.0	MS
			Eggplant	*Solanum melongena* L.	1.1	MS
			Onion	*Allium cepa* L.	1.2	S
			Radish	*Raphanus sativus* L.	1.2	MS
			Lettuce	*Lactuca sativa* L.	1.3	MS
			Pepper	*Capsicum annuum* L.	1.5	MS
Brackish	1000–2000	1.5–3.0	Garlic	*Allium sativum* L.	1.7	MS
			Potato	*Solanum tuberosum* L.	1.7	MS
			Cabbage	*Brassica oleracea* L.	1.8	MS
			Celery	*Apium graveolens* L.	1.8	MS
			Spinach	*Spinacia oleracea* L.	2.0	MS
			Cucumber	*Cucumis sativus* L.	2.5	MS
			Tomato	*Lycopersicon esculentum* L.	2.5	MS
Moderately saline	2000-5000	3.0–8.0	Pea	*Pisum sativum* L.	3.4	MS
			Red beet	*Beta vulgaris* L.	4.0	MT
			Asparagus	*Asparagus officinalis* L.	4.1	T
Saline	5000-10000	8.0–15.0				
Highly saline	10000-35000	15.0–45.0				

Table 1. Classification of waters with respect to total salt concentration and salt tolerance of selected vegetables (after Maas 1990, Francois & Maas, 1994 & Hillel, 2000). Asterix (*) marks the level of tolerance: sensitive (S), moderately sensitive (MS), moderately tolerant (MT), tolerant (T).

Stress caused by lack of water has similarities with the stress caused by excess salt in the soil solution, although there are differences (Munns, 2002). The stress caused by the presence of salts is one of the biggest problems that accompany agricultural production (Francois & Maas, 1994). Vegetables are not immune to the increased concentration of salts and therefore their presence significantly affects their quality and yield. According to FAO (1997) saline soil is the soil whose soil solution electrical conductivity (ECE) is 4 dS m^{-1} and higher, whereas soils whose ECE exceeds 15 dS m^{-1} are considered highly saline. Cations that are most commonly associated with saline soils are Na$^+$, Ca^{2+} and Mg^{2+}, accompanied by anions Cl$^-$, SO$_4^{2-}$ and HCO$_3^-$. However, major ions are considered to be Na$^+$ and Cl$^-$ since both are toxic to plants (Dudley, 1992, Hasegawa et al., 2000). Sulfates, which are usually involved in metabolic processes of plants as an integral part of proteins and enzymes, can also disrupt the metabolism of plants. They are usually more toxic than chlorides.

Stress caused by increased concentrations of salts effect the metabolism of plants and the final outcome of crop production in many ways. Excess salt has an osmotic effect, which means that the presence of salts reduces the amount of water accessible to plants. Some ions may be toxic to some processes in plants. Increased salt concentrations can lead to disturbances in mineral nutrition of plants, the plant hormone imbalances, and to the formation of reactive compounds such as different types of oxygen and other free radicals that damage cell membranes (Marschner, 1995). Plant cells, tissues, organs, individuals and entire ecosystems can have and/or can develop mechanisms by which they protect themselves against adverse effects of elevated salt concentrations (Pitman & Läuchli, 2002).

2. Effect of excess salinity on plant growth

Response of vegetables to the presence of increased amounts of salts is primarily stunted growth (Romero-Aranda et al., 2001). The ultimate impact of excess salts is of course very dependent on the other environmental factors such as humidity, temperature, light and air pollution (Shannon et al., 1994). The accumulation of salts in the leaves cause premature aging, reduces the supply of plant parts with nutrients and products of carbon assimilation of the fastest-growing plant parts and thus impair the growth of the entire plant. In the more sensitive genotypes salts accumulate more rapidly and because cells are not able to isolate the salt ions in vacuoles to the same extent as more tolerant genotypes, the leaves of more sensitive genotypes usually die faster (Munns, 2002). Neumann (1997) suggests that growth inhibition due to excessive salt concentration in the leaves reduces the volume of new leaf tissue in which excess salts can accumulate and therefore, in combination with the continuous accumulation of salts, it can lead to an increase in salt concentration in the tissue. It is often difficult to determine the relative influence of osmotic effect and the effect of the toxicity of specific ions on vegetable yield. In any case, yield losses due to osmotic stress can be very significant even before symptoms of toxicity on leaves become noticeable. Under the influence of salt stress growth of many species of vegetables is reduced, such as tomato (Romero-Aranda et al., 2001, Maggio et al., 2004), pepper (De Pascale et al., 2003b), celery (De Pascale et al., 2003a) and peas (Maksimovic et al., 2008, Maksimovic et al ., 2010). There are significant differences in salt tolerance between plant species and genotypes and similar goes for the ability to tolerate water deficiency (Munns, 2002; Lukovic et al., 2009). Table 1 shows in parallel the classification of waters with respect to the total concentration of salts and tolerance of selected vegetable species to salts. Salinity causes anatomical changes in leaves of many plant species. For example, the epidermis and mesophyll leaves of beans, cotton and *Attriplex* become thick, length of palisade mesophyll cells and diameter of spongy mesophyll cells increase and thickness of palisade and spongy layers and increasing as well (Longstreth & Nobel, 1979). In some other plant species were recorded adverse effects. In spinach leaves the presence of salt reduces the intercellular spaces (Delfine et al., 1998) and stomatal density in tomato (Romero-Aranda et al., 2001), but it increases stomatal density in pea (Maksimović et al., 2010) (Fig. 1).

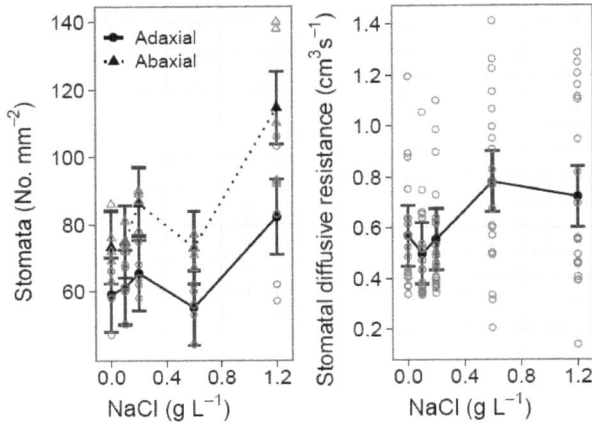

Fig. 1. Stomatal density and stomatal diffusive resistance in the presence of 0, 0.1, 0.2, 0.6, or 1.2 g NaCl L-1 in the nutrient solution (Maksimovic et al. 2010).

3. Effect of excess salinity on the water regime of plants

The main cause of reduced plant growth in the presence of salt can be impairment of water regime. Increasing the salt concentration in the soil increases the osmotic pressure of the soil solution and plants can not uptake the water as easily as in the case of relatively non-saline soils. Therefore, as the concentration of salt i.e. soil EC increases, water becomes less accessible to plants, even if the soil contains significant amounts of water and looks wet. Osmotic pressure depends on the number of particles contained in the solution and the temperature. Osmotic pressure (OP) of extracted soil solution can be expressed by the following empirical formula: OP = 0.36 x EC (dS m-1). At a pressure of about 1.44 bar, corresponding to the EC of 4 dS m-1, the plants start to show signs of physiological stress caused by water shortage. Therefore, in saline soils, despite the fact that water can be physically present, it becomes inaccessible to plants and the phenomenon is known as physiological drought (Ayers & Westcot, 1994).

The first effects of soil salinity, especially when it comes to low and moderate salt concentrations, can be attributed to the increase of osmotic value of the soil solution (Munns & Termaat, 1986). With the increasing salinity of soil solution, uptake of water through the root system becomes more difficult which leads to decreased evapotranspiration and yield. There are several reasons why evapotranspiration decreases with increase in soil salinity. Due to decreased accessibility of water to the root system root growth is reduced wich leads to a reduction in the total absorption area for water uptake. At the same time, total leaf area e.g. transpiration surface is reduced. As one of the mechanisms by which plants protect their cells from harmful effect of high concentration of salts is dilution, then increasing of water retention in the tissues of the plant further reduces transpiration. These factors reduce the efficiency of water usage and ultimately result in reduction of vegetable growth and yield. The vegetation period is shortened, water regime of plants is disrupted and the uptake and distribution of essential elements in both semi-controlled and field conditions is altered (Maksimovic et al., 2008, Maksimovic et al., 2010).

At very low soil water potential, the uptake of water and maintenance of turgor pressure in the tissues becomes very difficult. Water potential of leaves of plants well provided with water ranges from -0.2 to about -0.6 MPa, but the leaves of plants in arid regions can have significantly lower values, from -0.2 to 5 MPa even in extreme conditions (Taiz & Zeiger, 2006). Since the uptake of water is spontaneous process, the water potential of root cells must be more negative than potential of soil solution. If, due to increased salt concentration, the difference between water potential of soil solution and of root cells differs very slightly, plants may adapt osmoticaly by accumulation of so-called compatible osmolites in their cells. In that way, water potential of plant cells is kept more negative in relation to the soil solution water potential, thus permitting continuous uptake of water (Guerrier, 2006; Ghoulam et al., 2002).

Increasing the concentration of salt in a medium in which is the root leads to a reduction in the osmotic potential of leaves (Sohan et al., 1999, Romero-Aranda et al., 2001). Reduced osmotic potential of leaves is reflected in many processes in plants. Several authors have reported that water and osmotic potential of plants become more negative with increase in soil salinity, while turgor pressure concomitantly increases (Meloni et al., 2001, Romero-Aranda et al., 2001). Ashraf (2001) found that leaf water potential and evapotranspiration significantly decreased with increasing salt concentration in six species of the genus *Brassica*. At 200 mM NaCl *B. campestris* and *B. carinata* held a significantly higher water potential of leaves than other species in their experiment and therefore can be considered more tolerant to stress caused by salts. According to Sohan et al. (1999), the decrease in water potential can be explained by: 1) the influence of high concentrations of salts due to which plants accumulate more NaCl in the leaves than usual, and 2) by the reduced flow of water from root to aboveground organs due to the reduction of water conductivity, causing water stress in the tissues of leaves.

After Katerji et al. (1997), a decrease in RWC indicates loss of turgor which occurs due to disturbances in the increase in the area of individual leaves, in other words in leaf expansion. The connection between the impact of salt on gas exchange in leaves and growth is not completely understood. Many experimental results indicate that gas exchange in leaves of plants remains unchanged under the influence of soil water potential, until it reaches a certain threshold value (Ritchie, 1981). Results of Shalhevet (1994) suggest that the expansion of leaves is the most affected by osmotic stress and that there was a linear relationship between transpiration and the synthesis of organic matter in different agro-ecological conditions. The slope of this function represents the efficiency of water utilization by plants (water use efficiency, WUE). More recently, stomatal traits have been proven to critically affect WUE. In absence of stress, it has been demonstrated that low stomatal density reduces transpirational water fluxes (Zhang et al., 2001) and improves water use efficiency (Masle et al., 2005).

4. Accumulation of compatible osmolites increases vegetable tolerance to osmotic stress

One of the ways plants can adapt to conditions of osmotic stress is the accumulation of salt ions, if these salts are isolated in individual cell compartments by which their involvement in metabolism is prevented. The ability to regulate the concentration of salts through compartimentation is an important aspect of tolerance to increased salt concentrations (Romero-Aranda et al., 2001). In the presence of salts plants often accumulate low molecular weight substances which are called compatible osmolites. These substances do not interfere with normal biochemical reactions in cells (Hasegawa et al., 2000, Ashraf & Foolad, 2007). Compatible osmolites are low molecular weight molecules such as proline and glycine betaine (Ghoulam et al., 2002, Ashraf & Foolad, 2007). It is believed that under conditions of stress, proline has a role in osmotic adjustment of cells, enzymes and membrane protection and also as a source of nitrogen for a moment when conditions of stress are over (Ashraf & Foolad, 2007). The role of glycine betaine is also in maintaining pH of the cells, cell detoxification and binding of free radicals. Conditions of salt stress also lead to the accumulation of the other nitrogen compounds such as amino acids, amides, proteins and polyamines, which is often correlated with tolerance to salt (Mansour, 2000). Another group of compatible osmolytes are carbohydrates, both simple sugars (glucose, fructose, sucrose, fruktani), and starch. Their most important roles, beside in osmotic adjustment, is carbon storage and neutralization of free radicals (Parida et al., 2002). A similar role is attributed to the polyols that may accumulate under conditions of salt stress as well (Bohnert et al., 1995).

Ionic status of plants is highly correlated with tolerance to salts so that it can serve as a selection criterion in breeding to help create genotypes more tolerant to excess salt (Ashraf & Khanum, 1997).

5. Effect of excess salinity on mineral nutrition of plants

Increased salt concentration in the vicinity of the root system can interfere with mineral nutrition of plants and limit vegetable yield due to salinity or osmotic value of the soil solution. Salinity affects nutrient availability to plants in many ways. It modifies binding, retention and transformation of nutrients in the soil and affects the uptake and/or absorption of nutrients by the root system due to antagonism of ions and reduced root growth. It disrupts the metabolism of nutrients in the plant, primarily through water stress, thus reducing the efficiency of utilization of nutrients. In the presence of increasing concentrations of salts some species-specific symptoms may be present, such as necrosis and burns of leaf edges due to the accumulation of Na^+ and Cl^- ions (Wahome, 2001). The high concentration of ions can disrupt the structure and function of cell membranes. Mineral nutrition of plants depends on the activity of membrane transporters which participate in the transfer of ions from the soil into the plant and regulate their distribution within and between cells (Marschner, 1995; Tester & Davenport, 2003, Epstein & Bloom, 2005). Changes in membranes may finally lead to disturbances in chemical composition of cells and can therefore be displayed as symptoms of deficiency of some essential elements, similarly as it happens in the absence of salts (Grattan & Grieve, 1999).

High concentrations of NaCl act antagonistically to the uptake of the other nutrients, such as K^+, Ca^{2+}, N, P (Cramer et al., 1991, Grattan & Grieve, 1999). Increased concentrations of

NaCl increase concentrations of Na^+ and Cl^- and reduce concentrations of Ca^{2+}, K^+ and Mg^{2+} in many plant species (Bayuelo-Jimenez et al., 2003). In the presence of NaCl, the concentration of K^+, Ca^{2+} and P in vegetative parts decreased and in pods and grains increased (Fig. 2A). Ratios between concentrations of essential cations is changed as well (Fig. 2B). It was found that deleterious effects of salinity on tomato biomass production can be ameliorated by an enhanced supply of calcium. Similarly to the effect on the uptake of macroelements, salt stress can exert stimulatory and inhibitory influence on the uptake of some trace elements (Grattan & Grieve, 1999).

Irrigation waters sometimes contain increased concentration of boron, which can lead to reduced yields of crops (Nable et al., 1997; Kastori et al., 2008). Numerous studies have been done in order to find suitable methods for early assessment of tolerance to excess boron and finding sources of tolerance to high concentrations of this element (Nable et al., 1997, Yau & Ryan, 2008; Brdar-Jokanovic et al., 2010).

It is often considered that the use of fertilizers may aggravate problems that exist due to the presence of excessive amounts of salts in the soil. However, the lack of essential elements in accessible forms is a very common reason for poor productivity on such soils. When the saline soils are for the purpose of remediation washed with large quantities of water, some essential elements may concomitantly be washed away. Therefore, the application of fertilizers in appropriate doses is necessary to attain higher yields.

5.1 Effect of excess salinity on nitrogen, phosphorus and potassium uptake and metabolism

Under the conditions of salt stress, the uptake of nitrogen is often disrupted and numerous studies have shown that excess salts can reduce the accumulation of nitrogen in plants (Pardossi et al., 1999, Silveira et al., 2001, Wahid et al., 2004). Increase in uptake and accumulation of Cl^- is accompanied by a reduction in the concentration of NO_3^- in eggplant (Savvas & Lenz, 2000). There are authors who have attributed this reduction to the antagonism between Cl^- and NO_3^- (Bar et al., 1997) and those who explain it by reduced water uptake (Lea-Cox & Syvertsen, 1993). The rate of nitrate uptake or interactions between NO_3^- and Cl^- is associated with tolerance of examined plant species to salts. Kafkafi et al. (1992) found that tomato and melon varieties tolerant of salts have a higher flow rate of NO_3^- ions than more sensitive varieties.

Nitrogen fertilization on saline soils is often necessary because in such soils there is a lack of accessible nitrogen and also because losses of nitrogen due to leaching typical for nitrate form (Yin et al., 2007, Abdelgadir et al., 2010). In addition, rate of nitrification of ammonia is often significantly reduced due to the large direct toxic effects of Cl^- and the total amount of salt on the activity of nitrifying bacteria (Stark & Firestone, 1995).

Level of salinity does not affect necessarily the overall uptake of nitrogen by plants which may continue to accumulate nitrogen in the presence of excess salts despite a reduction in yield of dry matter. With the increase in soil salinity, total removal of nitrogen through the yield often decreases. Reduction in nitrogen fertilizer use efficiency is primarily a result of reduction of plant growth rate rather than the reduction of nitrogen uptake rate.

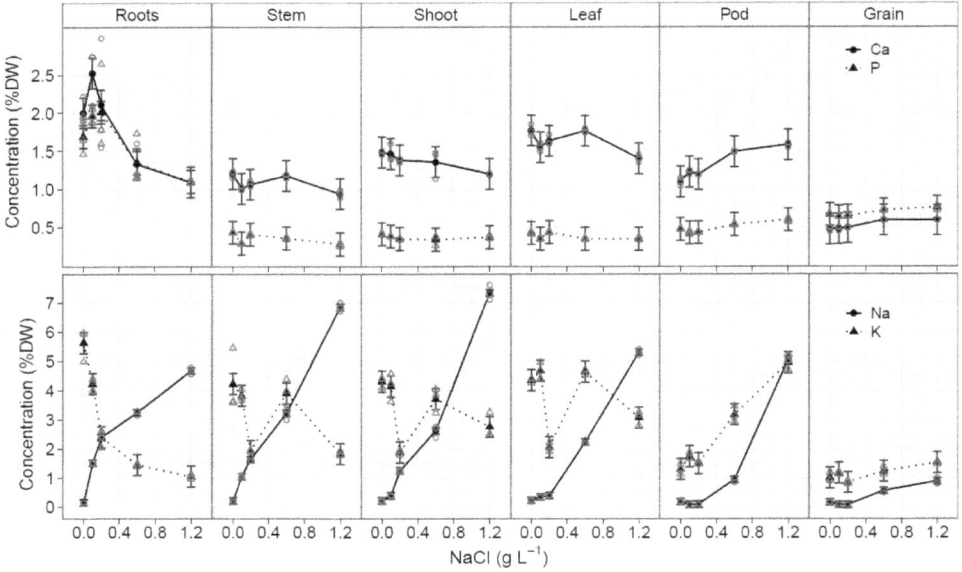

(A)

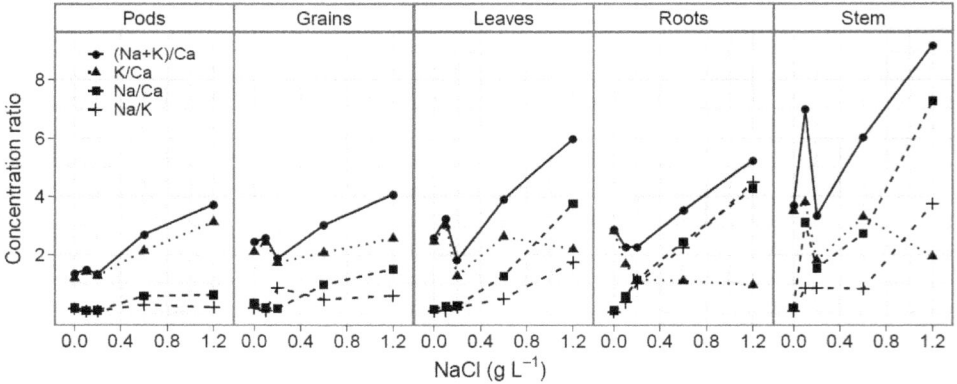

(B)

Fig. 2. Concentration of Ca^{2+}, P, Na^+ and K^+ (% in dry weight) (A) and ratios between concentrations of cations in different tissues of peas ((K^++Na^+)/Ca^{2+}, K^+/Ca^{2+}, Na^+/Ca^{2+}, and Na^+/K^+, in relative units) (B) grown in the presence of 0, 0.1, 0.2, 0.6, or 1.2 g NaCl L^{-1} in the nutrient solution (Maksimović et al., 2010).

Excess water and poor aeration that lead to anaerobic conditions can reduce the accessibility and absorption of nitrogen through the root system. In anaerobic conditions the intensity of reduction of NO_3^- to NO_2^- is higher. Woodruff et al. (1984) concluded that poor soil drainage, associated with high levels of the water table, led to a lack of nitrogen and the crop grown on such soils require more nitrogen fertilizer. Similarly, after the heavy rains that cause water retention and the appearance of anaerobic conditions, plants are often chlorotic, which is symptom characteristic of nitrogen deficiency. Such plants require the application of high doses of fertilizers in order to neutralize the effect of water lodging.

Nitrogen fixing bacteria affect nitrogen nutrition as well. Due to the toxic effects of salts on rhizobium the metabolism of nodulating bacteria can drastically alter (Hua et al., 1982) and number and activity of root nodules may be reduced (Singleton et al., 1982). Graham & Parker (1964) found that the highest EC that can tolerate *Rhizobium* strain compatible with pea is of 4.5 dS m^{-1}. On the basis of tolerance to salt concentration, Elsiddig & Elsheikh (1998) proposed the division of strains of bacteria from the genera *Rhizobium* and *Bradyrhizobium* in four groups: sensitive strains, 0-200 mM; moderately sensitive, 200-500 mM; tolerant, 500-800 mM; and highly tolerant, more than 800 mM. Such a classification should be considered with precaution, as a great influence on the overall tolerance to salts has pH value of the soil solution, temperature and carbon source that bacteria use.

Water stress, that is the result of high osmotic pressure of the soil solution, leads to the disturbance of nitrogen metabolism in plant tissues. In the presence of excess salts the synthesis of proteins is disturbed as well. Nightingale & Farnham (1936) found that with increase in osmotic pressure the amount of soluble organic nitrogen and proteins in sweet peas decreased, while the nitrate form of nitrogen accumulated. Similar results were obtained in the works of Ben-Zioni et al. (1967) and Kahane & Poljakoff-Mayber (1968) who concluded that the lack of water through a salt stress caused may result in slowing down the metabolism of plants grown on saline soils. It is considered that the fact that the because of the impact that disruption in water supply has on the total plant nitrogen status, this indicator (the total amount of nitrogen in plants) does not accurately reflect the impact of excess salts on plants. Protein nitrogen, but not total nitrogen, is therefore the appropriate indicator of plant nitrogen status in the soils in which salinity is increased. In plants tolerant to salts synthesis of proteins is mainly undisturbed. In general, on saline soils, the effects of nitrogen application were observed at lower doses than on normal and alkaline soils. Sharma (1980) found that the yield of eggplant fruits grown on saline soils in Yemen increased from the application of 40 kg N ha^{-1}, while higher doses, up to 120 kg N ha^{-1}, did not significantly affect the increase of yield.

Besides the dose, form of nitrogen, time and method of application significantly affect the efficiency of fertilization on saline soils. Foliar application of nitrogen fertilizers (e.g. 3% solution of urea, 20 kg N ha^{-1}) along with the application to the soil is very economical and effective on saline soils. This also saves irrigation, which would otherwise be necessary to carry out in order to dissolve and distribute well the fertilizer, and it is very important wherever the quality of water is not sufficiently good. Sharma (1980) found that application over the soil (75%) followed by foliar application (25%) gives better results than the application of total amount of nitrogen fertilizers in one of the two ways. Regarding the form of nitrogen in the fertilizer, urea gives better results than CAN and $(NH_4)_2SO_4$ in the soils of low to moderate salinity. However, on the more mineralized soils, it is better to use fertilizers containing NO_3^- than NH_4^+.

In some parts of the world, for example in parts of India, California, Pakistan and Egypt, saline ground water that is used for irrigation already contain significant amounts of NO_3^-. While this may be beneficial for plants, it is not always the case because these waters are sometimes toxic to plants. Long-term irrigation with such waters stimulates the development of vegetative tissues, ripening may be delayed and effect on seed filling may be adverse. Grains are often empty and of poor quality. In such regions, at least two irrigations have to be done with non-saline canal water, and production of forage crops is preferable to production of grain crops.

The final impact of salinity of soil solution on the concentration of phosphorus in plants depends heavily on plant species, phase of ontogenesis, the type and level of salinity and concentration of phospohorus that is already present in the soil (Grattan & Grieve, 1999). In most cases, excess of salts in soil solution leads to a reduction in phosphorus concentration in the tissues of plants, but the results of some studies show that salinity may increase but that does not affect the uptake and accumulation of phosphorus (Sonneveld & de Kreij, 1999, Kaya et al. 2001). Kochian (2000) suggests that the reduction of the availability of phosphorus in saline soils is the result of the activity of ions-antagonists, which can reduce the activity of phosphate and phosphate transporters of both high and low affinity, which are necessary for the uptake of phosphorus (Kochian, 2000). Reduced uptake of phosphorus can also be a consequence of the strong influence of sorption processes that control the concentration of phosphorus in the soil and low solubility of Ca-P minerals (Marschner, 1995).

In the saline soil phosphorus availability is to a greater extent dependent on the length and area of the root system (which is limited due to salinity) and antagonistic effects of excess phosphorus chloride on the uptake of phosphorus by the root system. Plant response to phosphorus fertilizers depends on the degree of soil salinity. In general, the use of phosphorus fertilizers in saline soils helps to increase vegetable yields directly by adding phosphorus and by reducing absorption of toxic elements such as chlorine Cl^- (Chabra et al., 1976) and fluorine F^- (Singh et al., 1995). Therefore, it is usually necessary to apply phosphorus fertilizers on saline soils.

In conditions of high salinity plants may show signs of potassium deficiency due to antagonistic effects of Na^+ and Ca^{2+} on K^+ absorption and/or abnormal Na^+/K^+ or Ca^{2+}/K^+ ratio (an example is given in Fig. 2). In such circumstances, the application of potassium fertilizers can increase the yield of plants. There is evidence that sodium in the aboveground parts of plants influences the transport of potassium from the roots to aboveground parts, or at least the concentration of potassium in leaves (Song & Fujiyama, 1998). It has been shown that increased concentrations of Na^+ block channel protein used for the uptake of K^+, AKT1, and in this way reduce the uptake of K^+. Inhibitory effect of Na^+ on transport of K^+ through channels in the membranes is probably more important in the phase of uptake of K^+ from the soil solution than in the phase of K^+ transport to the xylem (Qi & Spadling, 2004). The degree of tolerance of plants to the salinity is higher if they have a more efficient system for the selective uptake of K^+ instead of N^+ (Ashraf, 2004, Carden et al., 2003).

6. Effect of excess salinity on photosynthesis

Since plant growth directly depends on photosynthesis, stress factors that affect plant growth, affect the photosynthesis as well (Taiz & Zeiger, 2006). The effect of irrigation on production of organic matter and yield of vegetables is irreplaceable, as illustrated in Table 2. The capacity of the photosynthetic apparatus is reduced in the presence of excess salts (Ashraf, 2001, Romero-Aranda et al., 2001). However, the intensity of photosynthesis and yield are not correlated in the same way in different plant species. In asparagus and cotton a positive correlation was found (Pettigrew & Meredith, 1994; Faville et al., 1999), while in wheat there was practically no correlation (Hawkins & Lewis, 1993). In any case, the ultimate effect of salts on photosynthesis depends on the concentration of salts and plant species. It is possible that low salinity increases photosynthesis while high salinity reduces it. Connection between photosynthesis and stomatal conductivity is not obvious because stomatal conductivity may remain unchanged both in the case of low and high salinity (Parida et al., 2004).

Biological yield (t ha^{-1})		Yield of tubers (t ha^{-1})				Harvest index (%)		Dry matter in tubers (%)			
		Young		Mature				Young		Mature	
NI	I	NI	I	NI	I	NI	I	NI	I	NI	I
29.30	40.16	13.14	17.09	21.13	28.27	42.73	46.88	17.74	19.01	22.65	20.66
Starch content (%)				NO$_3$ concentration (mg kg^{-1})				NO$_2$ concentration (mg kg^{-1})			
Young		Mature		Young		Mature		Young		Mature	
NI	I	NI	I	NI	I	NI	I	NI	I	NI	I
15.10	15.09	19.41	19.35	338.2	281.2	213.7	235.8	0.81	0.78	0.65	0.67

Table 2. Effect of irrigation in the three year trial with potato. Sprinkling irrigation was employed at the technical minimum of 70% FWC and irrigation norm of 40 mm. Mean values are given for 11 fertilization treatments: unfertilized, manure, manure+NPK(80:80:80), manure+NPK(120:80:80), NPK(80:80:80), NPK(120:80:80), NPK(160:80:80), NPK(200:80:80), NPK(120:80:120), NPK(120:80:160), and NPK(120:80:200). Young and Mature stand for young and physiologically mature tubers, NI stands for not irrigated and I for irrigated potato. Adapted from Ilin (1993).

According to Iyengar & Reddy (1996) reduction of the intensity of photosynthesis in the presence of salts may be a result of dehydration of cell membranes which reduces their permeability to CO_2. Osmotic stress, caused by reduction of water potential, ultimately inactivates photosynthetic electron transport by reducing intercellular spaces. Salt toxicity is primarily caused by the Na^+ and Cl^- ions themselves. Ion Cl^- inhibits photosynthesis by inhibiting the uptake of NO_3^- form of nitrogen by the root system (Fisarakis et al., 2001). Reduced uptake of nitrate in combination with osmotic stress may explain the inhibitory effect of excess salts on photosynthesis. The presence of salt leads to the closure of stomata (Fig. 1). Reduced stomatal conductance results in a decrease in uptake of CO_2 that can be used in carboxylation reactions (Brugnoli & Bjorkman, 1992). Closure of stomata in the presence of increasing concentrations of salts is a way for plants to reduce water losses in the process of transpiration. This, however, affects the antenna system of chloroplasts, the biochemical reactions that take place in them and the entire system of energy transformation in chloroplasts (Iyengar & Reddy, 1996). Higher stomatal conductance in plants increases the diffusion of CO_2 in the leaves and thus increases the intensity of photosynthesis. However, the results which for net assimilation rate and stomatal conductance reported Ashraf (2001) for six different *Brassica* species have shown that between these two phenomena there is no connection. There are also experimental data on inhibition of photosynthesis in the presence of excess salts that are not accompanied by changes in the functioning of stomata. Iyengar & Reddy (1996) found that the cause of the non-stomatal inhibition of photosynthesis is actually increased resistance to diffusion of CO_2 through the liquid phase of mesophyll cell walls to the point of reduction of CO_2 in the chloroplast, and reduced efficiency of the enzyme RuBPC/O. In addition to these reasons, reduction of the intensity of photosynthesis in the presence of excess salts may be explained by the rapid aging of leaves, and changes in activities and actions of other enzymes, beside RuBPC/O, that are involved in photosynthesis, which leads to changes in the structure of the entire cell cytoplasm of photosynthetic tissues. At the same time, slower transport of the products of photosynthesis from the source to the sink also leads to a slowing down of photosynthesis (Iyengar & Reddy, 1996). Although salt stress reduces the intensity of photosynthesis, the results of several studies show that this is not a cause of reduced expansion rate of leaf cells. According to Alarcon et al. (1994) growth of the leaf is reduced more rapidly than photosynthesis at lower concentration of sodium. This means that in plants may happen a certain loss of intensity of photosynthesis without any effects on the cell growth itself.

In the presence of large amounts of salts in the soil, in plants occur many changes that lead to increased tolerance to salts and preservation of photosynthetic activity. Such changes are necessary for plants to maintain a balance between photosynthesis and growth. These mechanisms are not fully understood and may be somewhat specific to the species/genotype. Some plants can adapt to higher salinity by biochemical changes in the photosynthetic pathway. For example, facultative halophyte *Mesembryanthemum crystallinum* instead of the usual C3 uses CAM pathway (Cushman et al., 2008), while another species is tolerant of salts due to property that photosynthetic path runs along the C4 instead of C3 biochemical pathway (Zhu & Meinzer, 1999). Understanding the mechanisms by which salinity affects photosynthesis would help to improve conditions for growing vegetables and increase their yield, and would provide a useful tool for future genetic engineering.

7. Effect of excess salinity on amino acid composition, hormonal balance, antioxidant system and quality of vegetables

Changes in electrical conductivity of water, the sodium adsorption ratio (SAR) and the concentration of boron in water can affect the amino acid composition of plants. Totawat & Saxena (1974) in the greenhouse experiment found that with increasing SAR and/or boron, regardless of the electrical conductivity of water, significantly decrease the synthesis of amino acids for the species *Vigna catjang*. Synthesis of arginine, histidine, aspartate, glutamine, methionine and phenylalanine decreased, and the synthesis of lysine and valine increased. In addition to the synthesis of amino acids, which was reduced two to three times, in this experiment excess salts reduced the total amount of nitrogen in plants. Both phenomena can be explained by inhibition of the synthesis of RNA and DNA.

Salts affect the level of hormones in plants. The responses of tomato to salt stress conditions are largely determined by the concentration of endogenous ABA (Chen et al. 2003; Mulholland et al. 2003). However ABA is only one component of composite hormonal control (Ross & O'Neill, 2001; Nemhauser et al., 2006). It was found that the concentration of ABA and cytokinins increase with increasing in salt concentration (Aldeuquy & Ibrahim, 2001, Vaidyanathan et al., 1999). Abscisic acid may modify the expression of genes affected by salts, and these genes likely play a significant role in the mechanisms of tobacco tolerance to excess salts (Boussiba & Richmond, 1976). Itai et al. (1968) have noted a significant reduction in the concentration of cytokinins in the exudates of plants exposed to elevated concentrations of salt. In tissues exposed to stress, the incorporation of L-leucine into proteins decreased. Also, kinetin pretreatment can partially neutralize the effect of salts on leucine incorporation into proteins, suggesting that in tissues exposed to salt stress concentrations of endogenous cytokinins, essential for normal metabolism of above-ground plant parts, are lower. Treatment by cytokinins can lighten the effect of stress caused by insufficient nitrogen supply (Stoparić & Maksimović, 2008). Inhibitory effect of NaCl on photosynthesis, growth and transport of the products of photosynthesis may be reduced under the influence of ABA. In tomato, different levels of endogenous GA3 and ABA on water fluxes may reduce or enhance plant salt tolerance (Maggio et al., 2010). Although the nature of the ABA receptor is unknown, there is evidence that ABA is involved in reversible protein phosphorylation and modification of pH and concentration of Ca^{2+} in the cytosol (Leung & Giraudat, 1998). Increase in the concentration of ABA results in the uptake and increased concentration of Ca^{2+} which contributes to preservation of membrane integrity during longer periods of time and allows the plant to regulate uptake and transport of ions. Genotypes that are tolerant of salt in the presence of salts also accumulate jasmonates (Pedranzani et al., 2003) which mainly affect the transmission of signals. Similar to other types of stresses, salt stress leads to the formation of free radicals which activate antioxidant system of plants. The ability to survive under conditions of salt stress is largely dependent on the efficiency of this antioxidant system (Spychalla & Desborough, 1990, Gossett et al. 1994a,b). A mild salt stress may also improve both lypophilic and hydrophilic antioxidant activities. Exposure to moderate salinity (4.4 dS m^{-1}) can increase up to 40% the concentration of carotenoids in tomato fruits (De Pascale et al., 2001). After Garratt et al. (2002), the most important antioxidant enzymes are catalase (CAT), glutathione reductase (GR), superoxide dismutase (SOD) and glutathione S-transferase (GST). The mechanism through which excess salts affect antioxidant response of plants is not completely understood. Meneguzzo et al. (1999) found that it the most likely occurs over the toxic effects of Cl^{-} on photosystem II or through changes that occur in the membranes under the influence of higher Na^{+}/Ca^{2+} ratio.

8. Irrigation with treated municipal wastewaters in vegetable production

Treated municipal wastewaters, in addition to being a source of water, are a source of nutrients for plants (Jimenez-Cisneros, 1995) and heavy metals (Alloway, 1995). The use of such water often enriches the soil with heavy metals and thereby stimulates their accumulation in the soil-plant system (Kalavrouziotis & Arslan-Alaton, 2008). In principle, there is a positive correlation between the concentration of metals in the soil and in plant tissues (Kabata-Pendias & Pendias, 1995). The concentration of metals in plants is affected by many factors and one of the most important is the genetic basis of plants, i.e. genotype (Woolhouse, 1983). Heavy metals can affect the anatomy and uptake and distribution of essential elements in plants (Maksimovic et al., 2007). The impact of treated wastewaters on growth, yield and metabolism of vegetables intensively studied Kalavrouziotis. Particular attention is devoted to the study of accumulation of essential elements and heavy metals in soil and in plant tissues. This question is important because it is directly related to food safety and health of consumers. Also, the results of his research should also be used to determine limits for the concentration of certain elements in treated municipal wastewaters.

It was found that irrigation of broccoli and brussel sprouts with treated municipal wastewater contributed differently to nutrition of different plant parts (roots, stems and sprouts) and that interaction of elements is quite complicated. However, the authors finally concluded that these differences between the control (deionized water irrigated plants) and plants irrigated with treated wastewater were not statistically significant and that such waters can be used for irrigation of vegetables which is very important because of the limited availability of high quality fresh water for irrigation and the importance of irrigation for vegetable production. A particular risk is the presence of heavy metals such as Co, Ni and Pb in these waters. Irrigation of broccoli with treated wastewater increased synergism between Ca and Fe, K, P and Fe and Ni in the roots and because of that plants accumulated large amounts of Fe, P and Ni (Kalavrouziotis et al., 2009). At the same time, in the roots, leaves and flowers an interaction between Ni and Zn and P and Zn were found, in roots and leaves between Ca and Ni, K and B and P and K, and in the bud and blossom between Fe and Ni, Mn and B, and Fe and Ca. In general, in the roots of broccoli higher concentration of nutrients in relation to the leaves and flowers have accumulated, and especially higher concentration of Fe (1200 mg g^{-1}) was found, which indicates that this vegetable species is accumulator of Fe. The lowest accumulation of nutrients under the influence of treated municipal waste water was found in the blossom, the plant part used in diet.

9. Conclusion

Irrigation is often conducted by using to certain extent mineralized water and treated municipal waste waters. In this way, various beneficial or harmful salts and various other compounds can enter into the soil. The final outcome of the influence of impurities in the water for irrigation on the metabolism of vegetables depends on the nature of the substance, the time of exposure and the ability of plants to adapt. This ability is genetically determined but also depends on the combination of agro-ecological conditions in which the plant develops.

The stress provoked by excess salts in the soil solution has similarities with the stress caused by lack of water, although there are differences. Excess salt has an osmotic effect, which means that the amount of water accessible for plants is reduced. Yield losses due to osmotic stress can be very significant before toxicity symptoms on plants become apparent. Increased salt concentrations can lead to a reduction in evapotranspiration, disturbances in mineral nutrition of plants, the plant hormone imbalances, and to the formation of free radicals that damage cell membranes. Expansion of leaves may be impaired and their anatomical properties altered. The high concentration of salts in the soil solution may reduce the removal of nitrogen, phosphorus and potassium so it is necessary to add these elements in the form of fertilizers.

Understanding the mechanisms by which salinity affects photosynthesis and other physiological processes would help to improve conditions for growing vegetables and increase their yield and quality, and would provide a useful tool for future genetic engineering.

10. Acknowledgement

This work was supported by the Ministry of Education and Science of the Republic of Serbia, Project No. TR31036 (2011-2014) and Provincial Secretariat for Science and

Technological Development of the Autonomous Province of Vojvodina, Project No. 114-451-2218/2011 (2011-2014).

11. References

Abdelgadir, E.M., Fadul, E.M., Fageer, E.A. & Ali, E.A. (2010). Response of wheat to nitrogen fertilizer at reclaimed high terrace salt-affected soils in Sudan. *Journal of Agriculture & Social Sciences*, 6, 43-47, ISSN 1813 2235.

Alarcon, J.J., Sanchez-Blanco, M.J., Bolarin, M.C. & Torrecillas, A. (1994). Growth and osmotic adjustment of two tomato cultivars during and after saline stress. *Plant and Soil* 166, 75-82, ISSN 0032 079X.

Aldeuquy, H.S. & Ibrahim, A.H. (2001). Water relations, abscisic acid and yield of wheat plants in relation to the interactive effect of seawater and growth bio-regulators. *Journal of Agronomy and Crop Science*, 187, 97-104, ISSN 0931 2250.

Alloway, B.J. (1995). *Heavy metals in soils*, Blackie Academic and Professional, pp. 284–305, ISBN 07514 01986.

Ashraf, M. & Foolad M.R. (2007). Roles of glycine betaine and proline in improving plant abiotic stress resistance. *Environmental and Experimental Botany*, 59, 206–216, ISSN 0098 8472.

Ashraf, M. & Khanum, A. (1997). Relationship between ion accumulation and growth in two spring wheat lines differing in salt tolerance at different growth stages. *Journal of Agronomy and Crop Science*, 178, 39-51, ISSN 0931 2250.

Ashraf, M. (2001). Relationships between growth and gas exchange characteristics in some salt-tolerant amphidiploid *Brassica* species in relation to their diploid parents. *Environmental and Experimental Botany*, 45, 155-163, ISSN 0098 8472.

Ashraf, M. (2004). Some important physiological criteria for salt tolerance in plants. *Flora* 199, 361–376, ISSN 0367 2530.

Ayers, R.S. & Westcot, D.W. (1994). Water quality for agriculture. Irrigation and drainage paper 29, FAO, Rome, 174 p, ISBN 92510 22631.

Bar, Y., Apelbaum, A., Kafkafi, U. & Goren, R. (1997). Relationship between chloride and nitrate and its effect on growth and mineral composition of avocado and citrus plants. *Journal of Plant Nutrition*, 20, 715-731, ISSN 0190 4167.

Bayuelo-Jimenez, J.S., Debouck, D.G. & Lynch, J.P. (2003). Growth, gas exchange, water relations, and ion composition of *Phaseolus* species grown under saline conditions. *Field Crops Research*, 80, 207-22, ISSN 0378 4290.

Ben-Zioni, A., Itai, C. & Vaadia, Y. (1967). Water and salt stress, kinetin and protein synthesis in tobacco leaves. *Plant Physiology*, 42, 361-365, ISSN 0032 0889.

Bohnert, H.J., Nelson, D.E. & Jensen, R.G. (1995). Adaptation to environmental stresses. *Plant Cell*, 7, 1099-1111, ISSN 1040 4651.

Boussiba, S. & Richmond, A.E. (1976). Abscisic acid and the after-effect of stress in tobacco plants. *Planta*, 129, 217-219, ISSN 0032 0935.

Brdar-Jokanović, M., Maksimović, I., Nikolić-Đorić, E., Kraljević-Balalić, M. & Kobiljski, B. (2010). Selection criterion to assess wheat boron tolerance at seedling stage, primary vs. total root length. *Pakistan Journal of Botany*, 42, 3939-3947, ISSN 0556 3321.

Brugnoli, E. & Björkman, O. (1992). Chloroplast movements in leaves, Influence on chlorophyll fluorescence and measurements of light-induced absorbance changes related to ΔpH and zeaxanthin formation. *Photosynthesis Research, 32*, 23–35, ISSN 0166 8595.

Carden, D.E., Walker, D.J., Flowers, T.J. & Miller, A.J. (2003). Single-cell measurements of the contributions of cytosolic Na$^+$ and K$^+$ to salt tolerance. *Plant Physiology,* 131, 676–683, ISSN 0032 0889.

Chabra, R., Ringoet, A. & Lamberts, D. (1976). Kinetic and interaction of chloride and phosphate ahsorption hy intact tomato plants from a dilute nutrient solution. *Zeitschrift fur Pflanzenphysiologie 78,* 253-261, ISSN 0044 328X.

Chen, G., Fu, X., Herman Lips, S. & Sagi, M. (2003). Control of plant growth resides in the shoot, and not in the root, in reciprocal grafts of flacca and wild-type tomato (*Lysopersicon esculentum*), in the presence and absence of salinity stress. *Plant and Soil* 256, 205-215, ISSN 0032 079X.

Cramer, G.R., Epstein, E., & Läuchli, A. (1991). Effects of sodium, potassium and calcium on salt-stressed barley. 2. Elemental analysis. *Physiologia Plantarum,* 81, 197–202, ISSN 0031 9317.

Cushman, J.C., Agarie, S., Albion, R.L., Elliot, S.M., Taybi, T. & Borland, A.M. (2008). Isolation and characterization of mutants of common ice plant deficient in Crassulacean Acid Metabolism. *Plant Physiology,* 147, 228–238, ISSN 0032 0889.

De Pascale S., Maggio A., Fogliano V., Ambrosino P. & Ritieni, A. (2001). Irrigation with saline water improves carotenoids content and antioxidant activity of tomato. *Journal of Horticultural Science & Biotechnology,* 76, 447-453, ISSN 1462 0316.

De Pascale S., Maggio A., Ruggiero C., Barbieri G. (2003a) Growth, water relations, and ion content of field grown celery under saline irrigation (*Apium graveolens* L. var. dulce [Mill.] pers.). *Journal of the American Society for Horticultural Science,* 128, 136-143, ISSN 0003 1062.

De Pascale S., Ruggiero C., Barbieri G., Maggio A. (2003b) Physiological response of pepper (*Capsicum annuum* L.) to salinity and drought. *Journal of the American Society for Horticultural Science,* 128, 48-54, ISSN 0003 1062.

Delfine, S., Alvino, A., Zacchini M., & Loreto, F. (1998). Consequences of salt stress on conductance to CO_2 diffusion, Rubisco characteristics and anatomy of spinach leaves. *Australian Journal of Plant Physiology,* 25, 395–402, ISSN 0310 7841.

Dudley L.M. (1992). Salinity in the soil environment. In: *Handbook of Plant and Crop Stress,* M. Pessarakli (Ed.), Marcel Dekker, New York, 13–30, ISBN, 97808 24719 487.

Elsiddig, A.E. & Elsheikh, E.A.E. (1998). Effects of salt on rhizobia and bradyrhizobia, a review. *Annals of Applied Biology,* 132, 507–524, ISSN 0003 4746.

Epstein, E. & Bloom, A.J. (2005). Mineral Nutrition of Plants, Principles and Perspectives. 2nd Edn. Sunderland, MA. Sinauer Associates, ISBN 97808 78931 729.

FAO (1997). Small-scale irrigation for arid zones. Principles and options. FAO, Development Series 2. Food and Agriculture Organisation of the United Nations, FAO, Rome, Italy, 51 pp., ISBN 97892 51038 963.

Faville, M.J., Silvester, W.B. & Green, T.G.A. (1999). Partitioning of ^{13}C-label in mature asparagus (*Asparagus officinalis* L.) plants. *New Zealand Journal of Crop and Horticultural Science,* 27, 53-61, ISSN 0114 0671.

Fisarakis, I., Chartzoulakis, K. & Stabrakas, D. (2001). Response of Sultana vines (*V. vinifera* L.) on six rootstocks to NaCl salinity exposure and recovery. *Agricultural Water Management,* 51, 13–27, ISSN 0378 3774.

Francois, L.E. & Maas, E.V. (1994). Crop response and management on salt-affected soils. p. 149-181. In: *Handbook of Crop Stress,* M. Pessarakli (Ed.), Marcel Dekker, N.Y. [reprinted in Second Edition, p. 169-201. 1999], ISBN 08247 89873.

Garratt, L.C., Power J.B. & Davey, M.R. (2002). Improving the shelf-life of vegetables by genetic manipulation, p. 267-287. In: *Fruit and Vegetable Processing, Improving Quality,* W. Jongen (Ed.), Woodhead Publishing Ltd, Abingdon, Cambridge, ISBN 18557 35482.

Ghoulam, C., Foursy, A. & Fares, K. (2002). Effect of salt stress on growth, inorganic ions and proline accumulation in relation of osmotic adjustment in five sugar beet cultivarts. *Environmental and Experimental Botany,* 47, 39-50, ISSN 0098 8472.

Gossett, D.R., Millhollon, E.P. & Lucas, M.C. (1994a). Antioxidant response to NaCl stress in salt-tolerant and salt-sensitive cultivars of cotton. *Crop Science,* 34, 706-714, ISSN 0011 183X.

Gossett, D.R., Millhollon, E.P., Lucas, M.C., Banks, S.W. & Marney, M.M. (1994b). The effects of NaCl on antioxidant enzyme activities in callus tissue of salt-tolerant and salt-sensitive cultivars of cotton. *Plant Cell Reports,* 13, 498-503, ISSN 0721 7714.

Graham, P.H. & Parker, C.A. (1964). Diagonistic features in the characterization of root nodule bacteria of legumes. *Plant and Soil,* 20, 383-396, ISSN 0032 079X.

Grattan, S.R. & Grieve, C.M. (1999). Mineral nutrient acquisition and response of plants grown in saline environments. In: *Handbook of Plant and Crop Stress,* M. Pessarakli, (Ed.), Marcel Dekker Press Inc., New York, pp. 203-229, ISBN 08247 89873.

Guerrier, G. (2006). Fluxes of Na^+, K^+ and Cl^-, and osmotic adjustment in *Lycopersicon pimpinellifolium* and *L. esculentum* during short- and long-term exposures to NaCl *Physiologia Plantarum,* 97, 583–591, ISSN 0031 9317.

Hasegawa, P.M., Bressan, R.A., Zhu, J.-K. & Bohnert, H.J. (2000). Plant cellular and molecular responses to high salinity. *Annual Review of Plant Physiology and Plant Molecular Biology,* 51, 463-499, ISSN 0066 4294.

Hawkins, H.J. & Lewis, O.A.M. (1993). Combination effect of NaCl salinity, nitrogen form and calcium concentration on the growth, ionic content and gaseous exchange properties of *Triticum aestivum* L. cv. Gamtoos. *New Phytologist,* 124, 161-170, ISSN 0028 646X.

Hillel, D. (2000). *Salinity Management for Sustainable Irrigation.* The World Bank, Washington, D.C., ISBN 08213 4773X.

Hua, S.S.T., Tsai, V.Y., Lichens, G.M. & Noma, A.T. (1982). Accumulation of amino acids in *Rhizobium* sp. strain WR1001 in response to NaCl salinity. *Applied and Environmental Microbiology,* 44, 135-140, ISSN 0099 2240.

Ilin Ž. (1993) *Effect of fertilization and irrigation on potato yield and quality.* PhD. Thesis, Faculty of Agriculture, University of Novi Sad, Serbia (in Serbian, abstract in English).

Itai, C., Richmond, A. & Vaadia, Y. (1968). The role of root cytokinins during water and salinity stress. *Israel Journal of Botany,* 17, 187-195, ISSN 0021 213X.

Iyengar, E.R.R. & Reddy, M.P. (1996). Photosynthesis in high salt-tolerant plants. In: *Hand Book of Photosynthesis,* M. Pesserkali (Ed.), Marshal Dekar, Baten Rose, USA, pp. 56–65, ISBN 08247 97086.

Jiménez-Cisneros, B. (1995). Wastewater reuse to increase soil productivity. *Water Science and Technology,* 32, 173–180, ISSN 0273 1223.

Kabata-Pendias, A. & Pendias, H. (1995). Trace Elements in Soils and Plants. CRC Press. Boca Raton, USA. 2nd Edition, ISBN 08493 66437.

Kafkafi, U., Siddiqi, M.Y., Ritchie, R.J., Glass, A.D.M. & Ruth, T.J. (1992). Reduction of nitrate ($^{13}NO_3$) influx and nitrogen (^{13}N) translocation by tomato and melon varieties after short exposure to calcium and potassium chloride salts. *Journal of Plant Nutrition*, 15, 959-975, ISSN 0190 4167.

Kahane, I. & Poljakoff-Mayber, A. (1968). Effect of substrate salinity on the ability of protein synthesis in pea roots. *Plant Physiology*, 43, 1115–1119, ISSN 0032 0889.

Kalavrouziotis, I.K. & Arslan-Alaton, I. (2008). Reuse of urban wastewater and sewage sludge in the Mediterranean countries, case studies from Greece and Turkey. *Fresenius Environmental Bulletin*, 17, 625-639, ISSN 1018 4619.

Kalavrouziotis, I.K., Koukoulakis, P.H. & Mehra, A. (2010). Quantification of elemental interaction effects on Brussels sprouts under treated municipal wastewater. *Desalination*, 254, 6–11, ISSN 0011 9164.

Kalavrouziotis, I.K., Koukoulakis, P.H., Sakellarkou-Makrantonaki, M. & Papanikolaou, C., (2009). Effect of treated municipal wasrewater on the essentaia nutrient interactions in the plant of *Brassica oleracea* var. Italica. *Desalination*, 242, 297-312, ISSN 0011 9164.

Kastori, R., Maksimović, I., Kraljević-Balalić, M. & Kobiljski, B. (2008). Physiological and genetic basis of plant tolerance of excess boron. *Matica Srpska Proceedings for Natural Sciences*, 114, 41-51, ISSN 0352 4906.

Katerji, N., van Hoorn, J.W., Hamdy, A., Mastrorilli, M. & Mou Karzel, E. (1997). Osmotic adjustment of sugar beets in response to soil salinity and its influence on stomatal conductance, growth and yield. *Agricultural Water Management*, 34, 57-69, ISSN 0378 3774.

Kaya, C., Kirnak, H. & Higgs, D. (2001). Enhancement of growth and normal growth parameters by foliar application of potassium and phosphorus in tomato cultivars grown at high (NaCl) salinity. *Journal of Plant Nutrition*, 24, 357-367, ISSN 0190 4167.

Kochian, L.V. (2000). Molecular physiology of mineral nutrient acquisition, transport and utilization. In: *Biochemistry and Molecular Biology of Plants*. B.B. Buchan, W. Gruissen, R.L. Jones (Eds.) American Society of Plant Physiology, Rockville, EUA, pp. 1204-1249, ISBN 09430 88399.

Lea-Cox, J.D. & Syvertsen J.P. (1993). Salinity reduces water-use and nitrate-N-use efficiency of Citrus. *Annals of Botany*, 72, 47-54, ISSN 0305 7364.

Leung, J. & Giraudat, J. (1998). Abscisic acid signal transduction *Annual Review of Plant Physiology and Plant Molecular Biology*, 49, 199-122, ISSN 0066 4294.

Longstreth, D.J. & Nobel, P.S. (1979). Salinity effects on leaf anatomy. *Plant Physiology*, 63, 700–703, ISSN 0032 0889.

Luković, J., Maksimović, I., Zorić, L., Nagl, N., Perčić, M., Polić, D. & Putnik-Delić, M. (2009). Histological characteristics of sugar beet leaves potentially linked to drought tolerance. *Industrial Crops and Products*, 30, 281-286, ISSN 0926 6690.

Maas, E.V. (1990). Crop salt tolerance, pp. 262-304. In: *Agricultural Salinity Assessment and Management*, K.K. Tanji (Ed.), Amer. Soc. Civil Engrs., New York, ISBN 08726 27624.

Maggio A., De Pascale S., Angelino G., Ruggiero C. & Barbieri G. (2004). Physiological response of tomato to saline irrigation in long-term salinized soils. *European Journal of Agronomy* 21, 149-159, ISSN 1161 0301.

Maggio, A., Barbieri, G., Raimondi, G. & De Pascale, S. (2010). Contrasting effects of GA3 reatments on tomato plants exposed to increasing salinity. *Journal of Plant Growth Regulation*, 29, 63-72, ISSN 0721 7595.

Maggio, A., De Pascale, S., Fagnano, M. & Barbieri, G. (2011). Saline agriculture in Mediterranean environments. *Italian journal of Agronomy*, 6, 36-43, ISSN 1125 4718.

Maksimović, I., Belić, S., Putnik-Delić, M. & Gani, I. (2008). The effect of sodium concentration in the irrigation water on pea yield and composition. *Proceedings of ECO Conference 2008*, Novi Sad, pp. 231-235, ISBN 97886 83117 356.

Maksimović, I., Kastori, R., Krstić, L. & Luković, J. (2007). Steady presence of Cd and Ni affects root anatomy, accumulation and distribution of essential ions in maize seedlings. *Biologia Plantarum*, 51, 589-592, ISSN 0006 3134.

Maksimović, I., Putnik-Delić, M., Gani, I., Marić, J. & Ilin, Ž. (2010). Growth, ion composition, and stomatal conductance of peas exposed to salinity. *Central European Journal of Biology*, 5, 682-691, ISSN 1895 104X.

Mansour, M.M.F. (2000). Nitrogen containing compounds and adaptation of plants to salinity stress. *Biologia Plantarum*, 43, 491–500, ISSN 0006 3134.

Marschner, H. (1995). *Mineral nutrition of higher plants*, 2nd Ed., Academic Press, London. ISBN 01247 35436.

Masle, J., Gilmore, S.R. & Farquhar, G.D., (2005). The ERECTA gene regulates plant transpiration efficiency in *Arabidopsis*. *Nature*, 436, 866-870, ISSN 0028 0836.

Meloni, D.A., Oliva, M.A., Ruiz, H.A. & Martinez, C.A. (2001). Contribution of proline and inorganic solutes to osmotic adjustment in cotton under salt stress. *Journal of Plant Nutrition*, 24, 599-612, ISSN 0190 4167.

Meneguzzo, S., Navari-Izzo, F. & Izzo, R. (1999). Antioxidative responses of shoots and roots of wheat to increasing NaCl concentrations. *Journal of Plant Physiology*, 155, 274–280, ISSN 0176 1617.

Mulholland, B.J., Taylor, I.B., Jackson, A.C. & Thompson, A.J. (2003). Can ABA mediate responses of salinity stressed tomato? *Environmental and Experimental Botany* 50, 17-28, ISSN 0098 8472.

Munns, R. & Termaat, A. (1986). Whole-plant responses to salinity. *Australian Journal of Plant Physiology*, 13, 143-160, ISSN 0310 7841.

Munns, R. (2002) Comparative physiology of salt and water stress *Plant, Cell and Environment* 25, 239–250 ISSN 0140 7791.

Nable, R.O., Bañuelos G.S. & Paull, J.G. (1997). Boron toxicity. *Plant and Soil*, 193, 181-198, ISSN 0032 079X.

Nemhauser, J.L., Hong, F.X. & Chory, J. (2006). Different plant hormones regulate similar processes through largely nonoverlapping transcriptional responses. *The Cell*, 126, 467-475, ISSN 0092 8674.

Neumann, P. (1997). Salinity resistance and plant growth revisited. *Plant, Cell and Environment*, 20, 1193-1198, ISSN 0140 7791.

Nightingale, G.T. & Farnham, R.B. (1936). Effects of nutrient concentration on anatomy, metabolism and bud abscision of sweet pea. *Botanical Gazette*, 97, 477-517, ISSN 0006 8071.

Pardossi, A., Bagnoli, G. Malorgio, F. Campiotti, C.A. & Tognoni, F. (1999). NaCl effects on celery (*Apium graveolens* L.) grown in NFT. *Sciencia Horticulturae*, 81, 229–42, ISSN 0304 4238.

Parida, A.K., Das, A.B. & Das, P. (2002). NaCl stress causes changes in photosynthetic pigments, proteins and other metabolic components in the leaves of a true mangrove, *Bruguiera parviflora*, in hydroponic cultures. *Journal of Plant Biology*, 45, 28–36, ISSN 1226 9239.

Parida, A.K., Das, A.B. & Mohanty, P. (2004). Defense potentials to NaCl in a mangrove, *Bruguiera parviflora*, differential changes of isoforms of some antioxidative enzymes. *Journal of Plant Physiology*, 161, 531–42, ISSN 0176 1617.

Pedranzani, H., Racagni, G., Alemano, S., Miersch, O., Ramírez, I, Peña-Cortés, H., Taleisnik, E., Machado-Domenech E. & Abdala, G. (2003). Salt tolerant tomato plants show increased levels of jasmonic acid. *Plant Growth Regulation*, 41, 149-158, ISSN 0167 6903.

Pettigrew, W.T. & Meredith, W.R. Jr. (1994). Leaf gas exchange parameters vary among cotton genotypes. *Crop Science*, 34, 700–705. ISSN 0011 183X.

Pitman, M.G. & Läuchli, A. (2002). Global impact of salinity and agricultural ecosystems In: *Salinity, Environment - Plants – Molecules*, A. Läuchli, U. Lüttge (Eds.), 3–20. Kluwer Academic Publishers, Netherlands, ISBN 14020 04923.

Qi, Z. & Spalding, E.P. (2004). Protection of plasma membrane K^+ transport by the salt overly sensitive Na^+-H^+ antiporter during salinity stress. *Plant Physiology*, 136, 2548-2555, ISSN 0032 0889.

Ritchie, J.T. (1981). Water dynamics in the soil-plant-atmosphere system. *Plant and Soil*, 58, 81–96, ISSN 0032 079X.

Romero-Aranda, R., Soria, T. & Cuartero, J. (2001). Tomato plant water uptake and plant-water relationships under saline growth conditions. *Plant Science*, 160, 265–72, ISSN 0168 9452.

Ross, J. & O'Neill, D. (2001). New interactions between classical plant hormones. *Trends in Plant Science*, 6, 2-4, ISSN 1360 1385.

Savvas, D. & Lenz, F. (2000). Effects of NaCl or nutrient-induced salinity on growth, yield, and composition of eggplants grown in rockwool. *Scientia Horticulturae*, 84, 37-47, ISSN 0304 4238.

Shalhevet, J. (1994). Using water of marginal quality for crop production, major issues. *Agricultural Water Management*, 25, 233-269, ISSN 0378 3774.

Shannon, M.C., Grieve, C.M. & Francois, L.E. (1994). Whole-plant response to salinity. In: *Plant–Environment Interactions*, R.E. Wilkinson (Ed.), Marcel Dekker, New York, pp. 199–244. ISBN 08247 03774.

Sharma, S.K. (1980). Effect of different rates and methods of nitrogen application on yield of eggplant under saline conditions of Yemen Arab Republic. *Indian Journal of Agronomy*, 25, 557-558, ISSN 0537 197X.

Silveira, J.A.G., Melo, A.R.B., Viégas, R.A. & Oliveira, J.T.A. (2001). Salinity-induced effects on nitrogen assimilation related to growth in cowpea plants. *Environmental and Experimental Botany*, 46, 171-179, ISSN 0098 8472.

Singh, V., Gupta, M.K., Rajwanshi, P., Mishra, S., Srivastava, S., Srivastava, R., Srivastava, M.M., Prakash, S. & Dass, S. (1995). Plant uptake of fluoride in irrigation water by ladyfinger (*Abelmorchus esculentus*). *Food and Chemical Toxicology*, 33, 399-402, ISSN 0278 6915.

Singleton, P.W., Elswaify, S.A. & Bohlool, B.B. (1982). Effect of salinity on *Rhizobium* growth and survival. *Applied and Environmental Microbiology*, 44, 884-890, ISSN 0099 2240.

Sohan, D., Jasoni, R. & Zajicek, J. (1999). Plant-water relations of NaCl and calcium-treated sunflower plants. *Environmental and Experimental Botany*, 42, 105-111, ISSN 0098 8472.

Song, J. & Fujiyama, H. (1998). Importance of Na content and water status for growth in Na-salinized rice and tomato plants. *Soil Science and Plant Nutrition*, 44, 197–208, ISSN 0038 0768.

Sonneveld, C. & de Kreij, C. (1999). Response of cucumber (*Cucumis sativus* L.) to an unequal distribution of salt in the root environment. *Plant and Soil*, 209, 47–56, ISSN 0032 079X.

Spychalla, J.P. & Desborough, S.L. (1990). Superoxide dismutase, catalase, and alpha-tocopherol content of stored potato tubers. *Plant Physiology*, 94, 1214-1218, ISSN 0032 0889.

Stark, J.M. & Firestone, M.K. (1995). Mechanisms for soil moisture effects on activity of nitrifying bacteria. *Applied and Environmental Microbiology*, 61, 218–221, ISSN 0099 2240.

Stoparić, G. & Maksimović, I. (2008). The effect of cytokinins on the concentration of hydroxyl radicals and the intensity of lipid peroxidation in nitrogen deficient wheat. *Cereal Research Communications*, 36, 601-609, ISSN 0133 3720.

Taiz, L. & Zeiger, E. (2006). *Plant Physiology*, 4th Edition, Sinauer Associates, Inc., ISBN 08789 38567.

Tester, M. & Davenport, R. (2003). Na+ tolerance and Na+ transport in higher plants. *Annals of Botany*, 91, 503–527, ISSN 0305 7364.

Totawat, K.L. & Saxena, S.N. (1974). Effect of the Quality of applied irrigation water on the amino-acid makeup of *Vigna*-Catjang. *Botanical Gazette*, 135, 1-4, ISSN 0006 8071.

Vaidyanathan, R., Kuruvilla, S. & Thomas, G. (1999). Characterization and expression pattern of an abscisic acid and osmotic stress responsive gene from rice. *Plant Science*, 140, 25–36, ISSN 0168 9452.

Wahid, A., Hameed, M. & Rasul, E. (2004). Salt-induced injury symptom, changes in nutrient and pigment composition and yield characteristics of mungbean. *International Journalof Agricultural Biology*, 6, 1143–52, ISSN 1560 8530.

Wahome, P.K. (2001). Mechanisms of salt stress tolerance in two rose rootstocks, *Rosa chinensis* 'Major' and *R. rubiginosa*. *Sciencia Horticulturae*, 87, 207-216, ISSN 0304 4238.

Woodruff, J.R., Ligon, J.T. & Smith, B.R. (1984). Water table interaction with nitrogen rates in subirrigated corn. *Agronomy Journal*, 76, 280-283, ISSN 0002 1962.

Woolhouse, H.W. (1983). *Toxicity and tolerance in the responses of plants to metals*. In: O.L. Lange, P.S. Nobel, C.B. Osmond, H. Ziegler (Eds.), Physiological Plant Ecology III. Responses to the Chemical and Biological Environment, Vol 12C. Springer-Verlag, Berlin. pp 799, ISBN 03871 09072.

Yau, S.K. & Ryan, J. (2008). Boron toxicity tolerance in crops, a viable alternative to soil amelioration. *Crop Science*, 48, 854-865. ISSN 0011 183X.

Yin, F., Fu, B. & Mao, R. (2007). Effects of Nitrogen Fertilizer Application Rates on Nitrate Nitrogen Distribution in Saline Soil in the Hai River Basin, China. *Journal of Soils and Sediments*, 7, 136–142, ISSN 1439 0108.

Zhang, H.X., Hodson, J.N., Williams J.P. & Blumwald, E. (2001). Engineering salt-tolerant *Brassica* plants: characterization of yield and seed oil quality in transgenic plants with increased vacuolar sodium accumulation. Proceedings of the National Academy of Sciences of the United States of America, 98, 12832-12836, ISSN 0027 8424.

Zhu, J. & Meinzer, F.C. (1999). Efficiency of C4 photosynthesis in *Atriplex lentiformis* under salinity stress. *Australian Journal of Plant Physiology*, 26, 79–86, ISSN 0310 7841.

9

Sustainable Rice Yield in Water-Short Drought-Prone Environments: Conventional and Molecular Approaches

B. P. Mallikarjuna Swamy and Arvind Kumar

Plant Breeding, Genetics, and Biotechnology Division,
International Rice Research Institute (IRRI), Metro Manila,
Philippines

1. Introduction

The growing human population is putting enormous pressure on food security. Agriculture has to respond to this increased food demand by producing more from shrinking land and water resources (Pimentel et al., 1997). The climate change process has caused unpredictable and uneven rainfall patterns, resulting in more competition for water resources for crop cultivation and other socioeconomic uses (Hossain, 1995; Alcamo et al., 1999). Rice is one of the major staple food crops for more than a third of the world's population (David, 1991). It is mostly grown in well-puddled and irrigated conditions, and requires two to three times more water than other food crops such as wheat or maize. In total, rice production in the world uses about 1,600 km^3 of water, which accounts for 30% of the fresh water used worldwide (Gleick, 1993). There is an urgent need to increase rice production to meet global demand. It is estimated that the world needs to produce 40% more rice to feed the population by 2025 (FAO , 2002). With the grain yield in irrigated areas reaching stagnation, a large portion of the predicted increase has to come from the water-short drought-prone rainfed lowland and upland rice areas. Rainfed lowland and upland areas occupy about 38% of the total cropped rice area but contribute only 21% to total rice production (Khush, 1997). The lower contribution of the water-short drought-prone rainfed areas to total rice production is due to the lower per unit productivity primarily caused by the frequent occurrence of drought due to a failure of rain or a long spell between two rains. Drought has been identified as the key factor for low productivity in the rainfed ecosystem (Zeigler & Puckridge, 1995). Drought is particularly more frequent in South and Southeast Asia and sub-Saharan Africa, and the ongoing climatic change process is likely to further worsen the scenario in these rice-growing areas. In Asia alone, about 34 million ha of rainfed lowland and 8 million ha of rainfed upland rice experience drought stress of varying intensities at different stages of the crop almost every year (Wopereis et al., 1996; Huke & Huke, 1997). Most of the rice varieties presently cultivated in rainfed areas were developed for irrigated conditions. These varieties are highly popular among farmers because of their high yield potential and good grain quality but are highly susceptible to drought, causing substantial yield losses during years of drought. Breeding rice varieties with increased yield under drought is an important research area to achieve sustainable rice production under water-short situations. The drought breeding program at the International Rice Research Institute (IRRI) aims to develop high-yielding, drought-tolerant cultivars with good grain quality for rainfed lowland and upland ecosystems by following two-pronged modified conventional and molecular breeding approaches. In this chapter, we discuss in detail the progress achieved in the application of conventional and molecular approaches to develop drought-tolerant rice cultivars suitable for cultivation in water-short rainfed areas.

2. The major drought-prone rice ecosystems

Rice ecosystems are mainly classified into four types: irrigated, rainfed upland, rainfed lowland, and deepwater. Rainfed upland and lowland ecosystems are highly drought prone because of their uneven topography and heavy dependence on rainfall for the water source. Rainfed upland rice constitutes 13% of global rice production area and is generally the lowest yielding ecosystem. The main features of upland area are non-bunded fields, no/poor accumulation of water, low-input production practices, and cultivation of rice by direct seeding. Most of the rice varieties grown in upland areas are traditional landraces/varieties, low-yielding, prone to lodging, but adapted to aerobic soils. Upland rice is highly prone to drought because of poor accumulation of water in the field due to uneven upper toposequence, absence of bunds, and lower water-holding capacity of the soil (Serraj et al., 2009). Grain yield in the upland ecosystem is the lowest because of poor soil fertility, frequent occurrence of drought, high weed infestation, and low input use. Upland rice is mostly grown by small or subsistence farmers in the poorest region of Asia, Central and West Africa, and Latin America. India and China are the two countries with the largest area under upland. Indonesia, Thailand, the Philippines, Vietnam, Laos, and Myanmar (Burma) are the other countries practicing rice cultivation in the upland ecosystem.

Rainfed lowland rice is the second most important rice ecosystem after the irrigated ecosystem and it represents about 25% of total rice production area. The main physical features of the rainfed lowland rice areas are bunded fields, complete reliance on rainfall or drainage from higher lands for a water source, and variable soil water-table depth depending on the position of the field in the toposequence. Most of the rainfed lowland area is located in South and Southeast Asia, with the largest area of approximately 14.4 million hectares in India. Thailand, Bangladesh, Indonesia, and the Philippines are other countries with large rainfed lowland rice cultivation area. More than 90% of the area planted to rainfed lowland rice is in Asia. The yield of rainfed lowland rice is low and varies from 1.5 t ha^{-1} in Cambodia to 3.0 t ha^{-1} in Indonesia. In East Africa, around 70% of the cultivated rice area is rainfed lowland. Depending upon the toposequence, rainfall pattern, and amount of total rainfall received, rainfed lowland has been further classified as drought-prone shallow lowland, favorable shallow lowland, drought-prone mid-lowland, favorable mid-lowland, and flood-prone mid-lowland. Favorable rainfed areas account for about 20% of the total rainfed lowland area (Mackill et al., 1996). The frequency and severity of drought decreases from shallow to mid-lowland. Within the rainfed lowland ecosystem, certain regions in eastern India and Bangladesh are prone to both flood and drought at different periods in the same season or in different years depending upon the rainfall pattern. In the rainfed lowland ecosystem, a high association between percentage of rice area under irrigation and rice yield and poverty has been reported in India (Hossain, 1995).

3. Conventional breeding approaches

Drought is the primary limiting factor for rice production in rainfed ecosystems and drought at the reproductive stage particularly causes a severe reduction in grain yield (Cruz & O'Toole, 1984). Due to the inherently high variable nature of rainfed environments, breeding for drought-prone rainfed ecosystems requires strong research commitments for a longer period to see changes in farmers' fields. There are not many examples of concerted and product-oriented long-term breeding efforts that led to a significant yield increase at the farmers' level in rainfed water-short drought-prone environments. Even now, most of the varieties grown in rainfed environments are either high-yielding but drought-susceptible varieties or traditional local varieties/landraces that are drought tolerant but low yielding and poor in grain quality. The varietal characteristics that need to be considered during the development of genotypes for rainfed environments are farmers' preferred grain quality traits, high yielding ability with good drought tolerance, and resistance to/tolerance of prevalent biotic stresses (Atlin, 2003; Kumar et al., 2008).

3.1 Precise phenotyping under drought stress

Standardized phenotyping under well-managed water stress conditions is a prerequisite for the success of a drought breeding program. Drought screening (DS), which clearly distinguishes high-yielding but drought-susceptible lines from drought-tolerant lines, is necessary for effective selection for higher grain yield under drought. A breeding program has to develop a suitable screening methodology depending upon the crop growth stage (seedling/vegetative/reproductive) at which drought occurrence is more frequent and severe in the region. Most breeding programs carry out large-scale screening of breeding/mapping populations in natural field conditions and make an adjustment in date of sowing and planting to synchronize the crop growth stage with the period most likely to have the least rainfall. In recent years, some advanced programs have developed a rainout shelter facility for uninterrupted drought screening. In some cases, a line-source sprinkler irrigation system that allows the development of different severity of stress, along the line away from the source, is also followed (Cruz & O'Toole, 1984; Fukai et al., 1996; Lanceras et al., 2004). The crop growth stage when drought should be imposed in the field depends upon the water-holding capacity of the soil. In general, in upland, drought appears in severe form in 7-10 days after stopping irrigation whereas in lowland it takes 18-21 days for severe drought occurrence. The severity of drought to be applied in a drought screen should be very close to the drought situation prevalent in the region. In a region with moderate drought stress every year, screening under severe drought stress that leads to a significant yield reduction will not be appropriate.

In lowland, in screening experiments, a yield reduction of 30% for regions with mild drought occurrence, a yield reduction of 31-65% for regions with moderate drought occurrence, and a yield reduction of 65-85% for regions with severe drought occurrence have been reported to be appropriate (Kumar et al., 2007). In upland, due to prevalent conditions, drought mostly occurs in severe form and screening under severe drought conditions is recommended. However, for places with favorable rainfall, screening under conditions that bring a yield reduction of 40% or less for regions with mild drought occurrence, a yield reduction of 41-75% for regions with moderate drought occurrence, and a yield reduction of 76-90% for regions with severe drought occurrence will be appropriate.

In drought screens, adequate precautions need to be taken to irrigate the experiments at an appropriate time before drought is too severe to cause a loss of all genetic variability and affect an efficient selection strategy.

Severe drought screening at the reproductive stage is targeted in drought screens at IRRI. In upland screens, to initiate stress, experiments are drained at 35-45 days after seeding depending upon the duration of varieties. Stress plots are irrigated only when the soil water tension fall below -50 kPa at 30-cm soil depth. At this soil water potential, most lines wilt and exhibit leaf drying. This type of cyclic stress is considered to be efficient in screening for drought tolerance in populations consisting of genotypes with a broad range of growth duration (Lafitte, 2003) and ensures that all lines receive adequate stress during reproductive development. In lowland, to initiate stress, the field is drained 25-30 days after transplanting and experiments are irrigated only when the soil water tension falls below -60 kPa at 30-cm depth or alternatively when drought-susceptible checks along with around 70% of the lines show severe leaf rolling at 10 a.m. At IRRI, every season, several mapping and breeding populations are screened under severe reproductive-stage drought stress and the desired yield reductions as compared to yield in yield in irrigated control situation are achieved (Table 1).

S. no.	Population	Sesaon/year	% yield reduction
1	N22/Swarna	2010DS	78
2	N22/IR64	2010DS	82
3	IR77298-5-6-18/IR64	2009DS	80
4	IR77298-14-1-2-10/IR64	2010DS	63
5	Dhagad desi/Swarna	2010DS	65
6	Dhagad desi/IR64	2010DS	66

Table 1. Yield reduction in different populations under managed drought stress

3.2 Use of traditional and improved donors

The selection of parents is an important critical step in breeding for drought tolerance. A large variation exists for drought tolerance in the available rice germplasm (Mackill et al., 1996). Traditionally grown landraces and wild species accessions have high adaptability to drought, and the hidden genetic potential available in these accessions provides a better opportunity to improve the drought tolerance of popular rice varieties. However, most of the drought-tolerant donors used in the breeding programs so far are either from the cultivated rice gene pool or traditional landraces. Even though wild species can offer better genes for drought tolerance, no population with wild species has been developed for drought studies. Wild species such as *Oryza rufipogon, O. australiensis, O. glaberrima*, and several accessions of rice germplasm consisting of traditional varieties, landraces, and breeding lines were screened at IRRI under drought over the seasons/years to identify new donors. Efforts are also being made to develop wild species-derived populations for selection and qunatitaive trait loci (QTL) mapping. Several important drought-tolerant donors such as Apo, Aday Sel, N22, Dular, and Dhagad Desi were identified (Table 2) and used effectively in developing populations to select high-yielding lines under drought following a pedigree breeding approach.

Traditional donors have largely been low yielding and they carry undesirable linkage drag. Improved donors through pre-breeding between traditional donors and improved cultivars have been developed. At IRRI, traditional donors are used in developing mapping populations for the identification of major QTLs for grain yield under drought and better segregants combining appropriate plant type, high yield, and drought tolerance identified from such mapping populations are used in breeding programs. These improved pre-breeding lines are crossed with high-yielding drought-susceptible popular varieties and appropriate segregants are selected. With popular high-yielding varieties, such poor-combining donors can also be used in a backcross breeding program and appropriate segregants can be selected from BC_1F_2 or BC_2F_2 generation onwards. Some of the traditional and improved donors that showed consistent performance in drought screening at IRRI are listed in Table 2.

S. no.	Designation	Plant height (cm), non-stress	Days to 50% flowering, non-stress	Grain yield (kg ha^{-1}), drought
1	IR86931-B-414	95	76	2827
2	IR77298-14-1-2	102	82	2725
3	Aus Bak Tulsi	128	76	2472
4	IR86918-B-382	95	77	2405
5	Aus 299	125	76	2353
6	IR77298-5-6-18	96	85	1171
7	IR77298-14-1-2-10	85	85	1183
8	Kali Aus	103	76	2126
9	IR55419-04	117	69	2128
10	IR80461-B-7-1	120	80	1574
11	IR74371-70-1-1	95	77	1971
13	IR74371-54-4-1	104	78	2131
14	IR81896-B-B-74	120	84	1984
15	IR80461-B-7-1	106	78	1684
16	IR84878-B-60-4-1	78	85	2418
17	IR83380-B-B124-1	76	87	1482
18	IR83373-B-B-24-4	84	82	2271
19	IR71525-19-1-1	127	85	1036
20	Apo	110	84	1125
21	N22	120	75	2992
22	Dhagad Desi	93	84	1482
23	Dular	136	76	1539
24	Swarna (check)	85	115	178
25	IR64 (check)	81	115	342
26	MTU1010 (check)	97	85	542

Table 2. Traditional and improved donors for drought tolerance

3.3 Development of populations

While developing a breeding or mapping population, one of the parents should be a drought-resistant donor and the other parent should be a well-adapted local cultivar. The selected parents should have good combining ability and complement each other to overcome their specific drawbacks in terms of pest and disease resistance or grain quality. Different breeding populations such as recombinant inbred lines (RILs), backcross inbred lines (BILs), doubled haploids (DH), chromosomal segment substitution lines (CSSLs), and multiparent advanced generation inter-cross (MAGIC)-derived lines are used for breeding drought tolerance in rice. It is always desirable to have a large breeding population as it provides a better chance for breaking linkage drag and to obtain segregants with the right combinations of traits to meet local conditions and farmers' preferences. The population size runs up to 2500 to 5000 F_2 plants in breeding populations and 400 to 500 plants in mapping populations. Backcross-derived populations are more preferred for both breeding and mapping, as they help in the recovery of all the good traits of recipient parents and up to a certain extent break undesirable linkages. MAGIC populations are also being used in drought breeding but the development of populations is laborious. However, this offers the advantages of testing multiple drought-tolerant alleles in a single population and broadened genetic base of the recipient parent. Populations developed by marker-aided recurrent selection (MARS) and genome-wide selection are also being used in drought breeding.

3.4 Grain yield as a selection parameter under drought

In the past, secondary yield component traits and physiological traits such as harvest index, spikelet fertility, root length, root dry weight, root volume, osmotic adjustment, stomatal conductance, and relative water content were used as selection criteria for improving grain yield under drought in rice (Jongdee et al., 2002; Pantuwan et al., 2002). Through selection for secondary traits, desired improvement in grain yield under drought could not be achieved. Several experiments conducted at IRRI and elsewhere have clearly demonstrated that grain yield has moderate to high broad-sense heritability under drought. Further, the heritability of grain yield was comparable with the heritability of secondary yield traits or physiological traits under drought stress (Kumar et al., 2007). An increase in heritability of grain yield with an increase in severity of stress has also been reported (Kumar et al., 2007). Recently, grain yield has been suggested as a selection parameter in drought breeding programs under both upland and lowland situations (Bernier et al., 2008; Kumar et al., 2007; Kumar et al., 2008). By using this approach, improved breeding lines are developed at IRRI and in partner countries such as the Philippines, India, Nepal, and Bangladesh (Verulkar et al., 2010; Mandal et al., 2010). In Brazil and Thailand also, direct selection for yield has been successfully applied in breeding for drought tolerance in upland and rainfed lowland conditions (Ouk et al., 2006).

4. Combining high yield potential with good yield under drought

Farmers in the water-short rainfed drought-prone environment need rice varieties that provide high yield in seasons with good rainfall and good yield in seasons with moderate to severe drought. Moreover, these varieties need to have resistance against prevalent insects and diseases and farmers' preferred quality traits. This requires that segregating populations be exposed to selection under normal irrigated situations, drought screens, diseases, insects, and grain quality traits. Drought conditions favor rice infestation by blast (caused by *Pyricularia oryzae*) and brown spot (*Helminthosporium oryzae*) (Shrivastava & Verulkar, 2009). One of the parents used in developing segregating generations should possess tolerance of blast and brown spot and selection for blast and brown spot is practiced in segregating and advanced generations. A description of the detailed selection protocol followed is provided in Appendix 1.

Rainfed upland and lowland drought-prone environments are highly variable. Large genotype × environment interactions lead to differential performance of breeding lines in different environments. This requires that advanced breeding lines be tested at several locations and for several seasons to identify lines suitable to different regions. Further, breeding lines need to be tested for several seasons to generate reliable data on their performance under different severity of drought and normal irrigated situations. From such multilocation-multiseason evaluation, identified lines providing high yield under normal irrigated situations, good yield under drought, acceptable grain quality traits, and resistance to insects and diseases should be tested in on-farm participatory varietal selection (PVS) experiments to identify breeding lines preferred by farmers. Farmers' preferred breeding lines should be promoted for release as varieties through the national system.

Drought-tolerant breeding lines developed at IRRI are tested in multilocation trials under a drought breeding network that runs across India, Bangladesh, and Nepal in South Asia; Laos and Cambodia in Southeast Asia; and Tanzania and Mozambique in East Africa. High-yielding, drought-tolerant, farmers' preferred lines are promoted to national systems for release as varieties. The developed breeding lines are shared with other national systems in other countries and are also tested under the International Network for Genetic Evaluation of Rice (INGER) in different countries.

5. Molecular breeding approaches

Crop genetic improvement for environmental stresses such as drought is challenging due to its complex genetic nature and poor understanding of the physiological and molecular mechanisms associated with drought (Blum, 1988; Sinclair, 2011). It is a challenging task for any breeder to manipulate grain yield under drought with precision by traditional breeding approaches. The availability of complete rice genomic sequence information, rice linkage maps, and molecular marker technology has made it possible to dissect complex traits into individual quantitative trait loci (QTLs) (Temnykh et al., 2001; McCouch et al., 2002; Tuberosa, 2004; Tuberosa & Salvi, 2004, 2006). Linkage mapping, association mapping, nested association mapping, marker-aided recurrent selection (MARS), and genome-wide selection (GWS) are different approaches presently followed for the mapping and introgression of QTLs for drought tolerance (Tsonev et al., 2009). Major-effect QTLs can be exploited in maker-assisted selection more precisely to breed for drought tolerance. The drought molecular breeding program at IRRI aims at identifying and introgressing major-effect QTLs for grain yield under drought in rice mega-varieties.

5.1 Mapping of major-effect QTLs for grain yield under drought

Molecular breeding approaches are a fast-track approach to improve grain yield under drought of popular high-yielding varieties. The complex nature of drought, multigenic inheritance of grain yield, and the complexity to maintain quality and other desirable traits of well-established varieties necessitate the use of markers to track QTLs/genes controlling these traits. Molecular approaches were suggested to be appropriate for effective drought-tolerance improvement in rice (Price et al., 2002b, 200b; Tuberosa 2004; Tuberosa & Salvi, 2004, 2006). Earlier, several efforts made to improve grain yield under drought in rice through the introgression of the identified QTLs for yield-related secondary traits did not yield the desired results. The identification and introgression of major-effect QTLs for grain yield under reproductive-stage drought stress in drought-susceptible rice varieties could be a suitable marker-assisted breeding (MAB) strategy. For marker-assisted introgression to be successful, the identification of QTLs with a high and stable effect in an improved genetic background is one of the essential requirements. Understanding all the sources of genetic variations such as QTL main effects, QTL × QTL, and QTL × environment interactions is very important before embarking on marker-assisted selection (MAS) of drought QTLs (Xing et al., 2002). To date, several QTLs with large and consistent effects on grain yield under drought stress have been identified at IRRI. Bernier et al. (2007) reported $qDTY_{12.1}$, a QTL on chromosome 12 in the Vandana/Way Rarem population, explaining about 51% of the genetic variance for yield under severe upland drought stress over two years. A consistent major-effect QTL for grain yield, $qDTY_{1.1}$, was identified to show an effect in several populations. Vikram et al. (2009) detected $qDTY_{1.1}$ in three RIL populations derived from the donor N22 crossed to drought-susceptible varieties Swarna, IR64, and MTU1010 consistently over two years Ghimire et al. (2011) also detected $qDTY_{1.1}$ in two RIL populations derived from donor Dhagad deshi crossed to Swarna and IR64 consistently over two seasons. Two major-effect QTLs, $qDTY_{3.1}$ and $qDTY_{2.1}$, were identified in an Apo/2*Swarna population explaining 30% and 15% of the phenotypic variance, respectively (Venuprasad et al., 2009). Four major-effect QTLs, $qDTY_{2.2}$, $qDTY_{4.1}$, $qDTY_{9.1}$, and $qDTY_{10.1}$, were identified in Adaysel/IR64-derived populations (Swamy et al., 2011b, unpublished). It is important to mention that most of the major-effect QTLs were consistent across water regimes and are suitable for improving varieties for different severities of water stress (Zou et al., 2005; Kamoshita et al., 2008). The major-effect QTLs and their additive effect on grain yield are provided in Table 3. Epistatic QTLs were also detected for grain yield under drought; they can be simultaneously introgressed together with the major-effect QTLs to

Genetic background	QTL	Ecosystem	Additive effect (%) over trial mean
Vandana	$qDTY_{12.1}$	Upland	47
IR64	$qDTY_{1.1}$	Lowland	32
IR64	$qDTY_{9.1}$	Lowland	27
IR64	$qDTY_{10.1}$	Lowland	22
IR64	$qDTY_{2.2}$	Lowland	13
IR64	$qDTY_{4.1}$	Lowland	14
Swarna	$qDTY_{1.1}$	Lowland	25
Swarna	$qDTY_{2.1}$	Lowland	19
Swarna	$qDTY_{3.1}$	Lowland	25
Swarna	$qDTY_{8.1}$	Lowland	16
MTU1010	$qDTY_{1.1}$	Lowland	17

Table 3. Major-effect QTLs for grain yield under drought identified at IRRI

improve the yield of the recipient parent (Xing et al., 2002; Lanceras et al., 2004; Zou et al., 2005; Dixit et al., 2011). Many more such QTLs are likely to exist and need to be extracted from novel drought-tolerant donors. Wild progenitor species are an appropriate source for exploiting naturally occurring variation to harness QTL alleles for drought tolerance in rice. The AB-QTL analysis approach has been successful in exploiting favorable alleles for various traits of agronomic importance (Xiao et al., 1998; Zhang et al., 2006; Mc Couch et al., 2007).

5.2 Selective genotyping and bulk segregant analysis

The genotyping of large populations to identify QTLs involves high costs, and is laborious. To reduce the cost and time associated with whole genotyping, alternative approaches such as selective genotyping and bulk segregant analysis (BSA) were proposed (Darvasi & Soller, 1992; Bernier et al., 2007; Venuprasad et al., 2009; Vikram et al., 2009). In selective genotyping, only tail lines selected based on phenotype are genotyped and used for QTL detection. Major-effect QTL $qDTY_{12.1}$, which was initially identified in a population of 436 lines, was detected by selective genotyping of only 169 lines (Navabi et al., 2009). Vikram et al. (2009) detected two major-effect QTLs, $qDTY_{1.1}$ and $qDTY_{8.1}$, explaining up to 11.9% and 15.6% of the phenotypic variance by genotyping only 36.5% of the lines from the populations. BSA is a cost-effective and efficient genotyping method to detect major-effect QTLs (Michelmore et al., 1991). It has been used successfully in identifying QTLs for grain yield under drought in rice by Shashidhar et al. (2005). Venuprasad et al. (2009) used BSA to identify two large-effect QTLs, $qDTY_{2.1}$ and $qDTY_{3.1}$. Vikram et al. (2009) detected two major-effect QTLs, $qDTY_{1.1}$ and $qDTY_{8.1}$. Ghimire et al. (2011) detected a major-effect QTL, $qDTY_{1.1}$, for grain yield under drought in more than one population. These cost-saving genotyping methods are more efficient and accurate in identifying major QTLs for grain yield under drought. With the availability of genome-wide polymorphic SNP markers, Single nucleotide polymorphism (SNP) genotypic arrays, and cheaper SNP assays, marker-assisted breeding can become a common breeding practice (McNally et al., 2009; Fukuoka et al., 2010).

5.3 Fine mapping of major-effect grain-yield QTLs

Fine mapping is necessary to further narrow down the confidence interval of major-effect QTLs for precise linkage-drag-free introgression of the QTLs in marker-assisted selection. Fine mapping is carried out either in the same QTL mapping population or in a large advanced backcross population segregating only for the QTL region and genotyping the population with additional markers in the QTL region. There are several reports of fine mapping of QTLs for different traits under drought. A QTL for leaf rolling and leaf drying on chromosome 1 was fine-mapped to 3.8 cM region (Salunkhe et al., 2011). Ding et al. (2011) fine-mapped a major-effect QTL ($qFSR4$) for root volume per tiller to a 38-kb region using NILs. Nguyen et al. (2004) fine-mapped a QTL ($oa3.1$) for osmotic adjustment under drought and identified the candidate genes. A major-effect QTL for drought recovery score (DRS) was fine-mapped on chromosome 9 (Lang & Buu, 2008). Efforts were also made at IRRI to fine-map four large-effect grain-yield QTLs under drought such as $qDTY_{2.1}$, $qDTY_{2.2}$, $qDTY_{9.1}$, and $qDTY_{12.1}$ in backcross-derived advanced populations (Bernier et al., 2007;

Venuprasad et al., 2009, Dixit et al., 2011). Uga et al. (2011) fine-mapped a major-effect QTL (*Dro1*) on chromosome 9 to a 608.4-kb region.

5.4 Validation of major-effect QTLs

The effect of major-effect grain-yield QTLs indentified at IRRI is validated across different backgrounds and environments. All these QTLs showed a consistent effect across different environments. The first major-effect QTL identified at IRRI, $qDTY_{12.1}$, was evaluated in 21 field trials in the Philippines and eastern India. The effect of this QTL was consistent across the environments (Bernier et al., 2009). It was also interesting to note that the effect of $qDTY_{12.1}$ increased with an increase in severity of stress (Bernier et al., 2009). The major-effect QTLs $qDTY_{3.1}$, $qDTY_{2.2}$, $qDTY_{4.1}$, q $DTY_{9.1}$, and $qDTY_{10.1}$ were evaluated at IRRI during 2010DS and 2011DS and the effects were consistent. The markers linked to the major-effect QTLs were validated on a panel of drought-tolerant lines to confirm their presence in a larger set of lines. It is notable that major-effect QTL $qDTY_{12.1}$ was present in 85% of the lines. $qDTY_{3.2}$, $qDTY_{2.1}$, $qDTY_{3.1}$, $qDTY_{1.1}$, $qDTY_{8.1}$, and $qDTY_{1.2}$ were present in more than 50% of the lines. The study also indicated the presence of at least one major-effect grain-yield QTL in every drought panel line.

5.5 Meta analysis and comparative genomics of grain-yield QTLs

The identification of the most accurate and precise major-effect QTLs across genetic backgrounds and environments through Meta analysis is a prerequisite for the successful use of QTLs in MAS across different genetic backgrounds. Khowaja et al. (2009) carried out Meta analysis of QTLs for various phenotypic traits identified under drought stress. Courtois et al. (2009) identified the meta-QTLs for root traits under drought. The number of QTLs was considerably reduced to a few consistent meta-QTLs and confidence intervals of QTL regions also decreased. MQTL regions with small genetic and physical intervals are important regions for MAS, fine mapping, candidate gene identification, and functional analysis. Meta analysis of 53 grain-yield QTLs identified from 15 reports resulted in 14 meta-QTLs (Swamy et al., 2011a). There were seven meta-QTLs of around 1.3 Mb and they corresponded to a reasonably small genetic distance of 6 cM and they are suitable for use in marker-assisted selection. These QTLs can be introgressed in popular rice mega-varieties to develop drought-tolerant and high-yielding lines. Meta-QTL regions were compared for synteny in other cereal crops. The major-effect $MQTL_{1.4}$ was also found in maize on chromosome 3 near marker *msu2*, in wheat on chromosome 4B near marker *Rht-b1*, and in barley on chromosome 6H near marker *Bmac0316*, while major-effect $MQTL_{3.2}$ was also found in maize on chromosome 1 near marker *Umc107a* (Swamy et al., 2011a).

5.6 Marker-assisted backcross to improve drought tolerance

Even though many QTLs have been identified for various drought-related traits in rice, there are few efforts to introgress them to develop improved breeding lines (Price et al., 2002b; Courtois et al., 2003). The lack of success may be due to the use of small-effect QTLs for secondary traits/physiological traits in introgression or the unavailability of tightly linked markers (Shen et al., 2001). Now, with the detection of several major-effect, consistent

QTLs for grain yield under drought, the introgression of yield QTLs may produce desired results. The identified major-effect QTLs for grain yield under drought $qDTY_{12.1}$, $qDTY_{3.1}$, $qDTY_{2.2}$, $qDTY_{9.1}$, $qDTY_{10.1}$, $qDTY_{4.1}$, and $qDTY_{1.1}$ increased grain yield under stress conditions and did not have any adverse effect on grain yield under non-stress conditions. The introgression of these QTLs by marker-assisted backcross breeding provides an opportunity to improve the drought tolerance of well-adapted high-yielding but drought-susceptible popular rice mega-varieties of South and Southeast Asia, and Africa. The main objectives of MAS are to introgress and pyramid these QTLs and enhance yield by 1.0-1.2 t ha^{-1} under drought. $qDTY_{12.1}$ has been successfully introgressed in the background of Vandana. The Vandana-introgressed lines with $qDTY_{12.1}$ showed a yield advantage of 0.5 t ha^{-1} over drought-tolerant cultivar Vandana under drought and yielded similar to Vandana under normal irrigated situations. $qDTY_{2.2}$, $qDTY_{4.1}$, $qDTY_{9.1}$, and $qDTY_{10.1}$ have been introgressed in an IR64 background.

The major-effect QTLs $qDTY_{1.1}$, $qDTY_{2.1}$, $qDTY_{2.2}$, $qDTY_{3.1}$, $qDTY_{4.1}$, $qDTY_{9.1}$, $qDTY_{10.1}$, and $qDTY_{12.1}$ are being introgressed in several popular rice mega-varieties. In general, the major-effect QTLs identified for grain yield under drought have a genetic gain of 10% to 30%, with a yield advantage of 150 to 500 kg ha^{-1} over recipient parents. However, considering significant economical benefit to farmers, the development of drought-tolerant rice varieties with a yield advantage of at least 1.0 t ha^{-1} could be the desired target for rice breeders. Marker-aided QTL pyramiding of major-effect QTLs through backcross breeding can be considered as an option for achieving this target. We have successfully pyramided four QTLs ($qDTY_{2.2}$, $qDTY_{4.1}$, $qDTY_{9.1}$, and $qDTY_{10.1}$) and introgressed lines with four, three, and two QTLs in an IR64 background developed. Introgressed lines with three and two QTLs in an IR64 background showed a yield advantage of 1.2-2.0 t ha^{-1} under drought, yielded similar to IR64 under normal irrigated situations, and possessed quality traits similar to those of IR64. Efforts are also under way to combine drought and submergence tolerance in Swarna-Sub1 and IR64-Sub1 backgrounds. The improved drought-tolerant NILs of IR64 and Vandana are ready for field testing by farmers (Kumar, 2011). The protocol procedure for pyramiding three QTLs is described in Appendix 2.

6. Marker-aided recurrent selection for grain yield under drought

Marker-aided recurrent selection helps in the introgression of all the significant QTLs detected for a set of traits in a population to develop an improved cultivar. The MARS scheme involves several cycles of selection and crossing based on the genotype. The gains achieved by this approach are suggested to be more than with phenotype-based breeding and other marker-assisted breeding approaches (Charmet et al., 1999; Ribaut & Ragot, 2007). MARS is highly useful in breeding combinations of traits, including grain yield under drought, yield potential under non-stress conditions, disease-insect resistance/tolerance, and grain quality characteristics. This approach has been successfully used in maize (Johnson, 2004; Crosbie et al., 2006). We have initiated programs to apply MARS for improving drought tolerance in rice at IRRI. Two populations, IR55419-04/Sambha Mahsuri and CT9993-5-10-1-M/Sambha Mahsuri, are currently being used to apply MARS.

7. Development of QTL-NILs for physiological and molecular studies

Understanding the physiological and molecular mechanisms associated with drought tolerance of major-effect QTLs is one of the major components of drought breeding. For this purpose, near-isogenic lines for QTLs (QTL-NILs) are being developed through marker-assisted introgression and background recovery of the recipient genome. The NILs differing only at QTLs and similar in the background genotype are used to understand the physiological and molecular mechanisms of increased yield under drought provided by QTLs.

8. Success stories of drought breeding at IRRI

The drought breeding program at IRRI standardized phenotyping under managed drought stress in field conditions. Direct selection for grain yield under drought has been applied as a criterion for developing improved cultivars through conventional as well as molecular approaches. Grain yield under drought has also been used in mapping and introgressing QTLs. Several major-effect, consistent QTLs have been identified in the background of popular rice mega-varieties. Following conventional breeding approaches, many varieties have been developed and released for cultivation in partner countries. During 2009 and 2010, three breeding lines, IR74371-70-1-1 as Sahbhagi dhan in India, IR74371-54-1-1 as Sahod ulan in the Philippines, and IR80411-B-49-1 as Tarharra 1 in Nepal, were released as drought-tolerant varieties. In 2011, three varieties, IR74371-3-1-1, IR74371-46-1-1, and IR74371-70-1-1, were released in Nepal as Sookha dhan 1, Sookha dhan 2, and Sookha dhan 3, respectively. IR74371-70-1-1 has been release as BRRI dhan56 in Bangladesh. IR77080-B-34-3 and IR81047-B-3-4 are breeding lines identified as promising in Mozambique. Through molecular approaches, major-effect QTLs have been introgressed and pyramided in the background of Vandana and IR64. Many improved lines developed through conventional and molecular approaches are being evaluated through participatory varietal selection in different countries.

9. Conclusions and future perspectives

Grain yield under drought is a very complex trait. Long-term and consistent efforts are required to make a significant improvement in the drought tolerance of rice varieties. A prior knowledge of the target area with respect to drought pattern, farmers' preferences, and prevailing pests and diseases is essential. The selection of parents and breeding strategies plays an important role in a successful breeding program. The conventional breeding approach with grain yield as a selection parameter has been successful in developing improved drought-tolerant varieties. Through molecular breeding approaches, major QTLs for grain yield under drought have been identified and introgressed in popular rice varieties. Meta analysis and QTL validation of major-effect grain-yield QTLs indicate their consistent effect across environments and variable but significant effect against different genetic backgrounds. The drought breeding program at IRRI has been successful in developing and releasing drought-tolerant varieties in partner countries that showed a yield advantage of 0.8-1.2 t ha^{-1} under drought over presently cultivated varieties. The challenges ahead are the effective use of the drought QTLs and their combinations in breeding for drought tolerance, fine mapping of QTLs to facilitate precise introgression without undesirable linkages, and understanding the physiological and molecular mechanisms associated with major-effect grain-yield QTLs under drought.

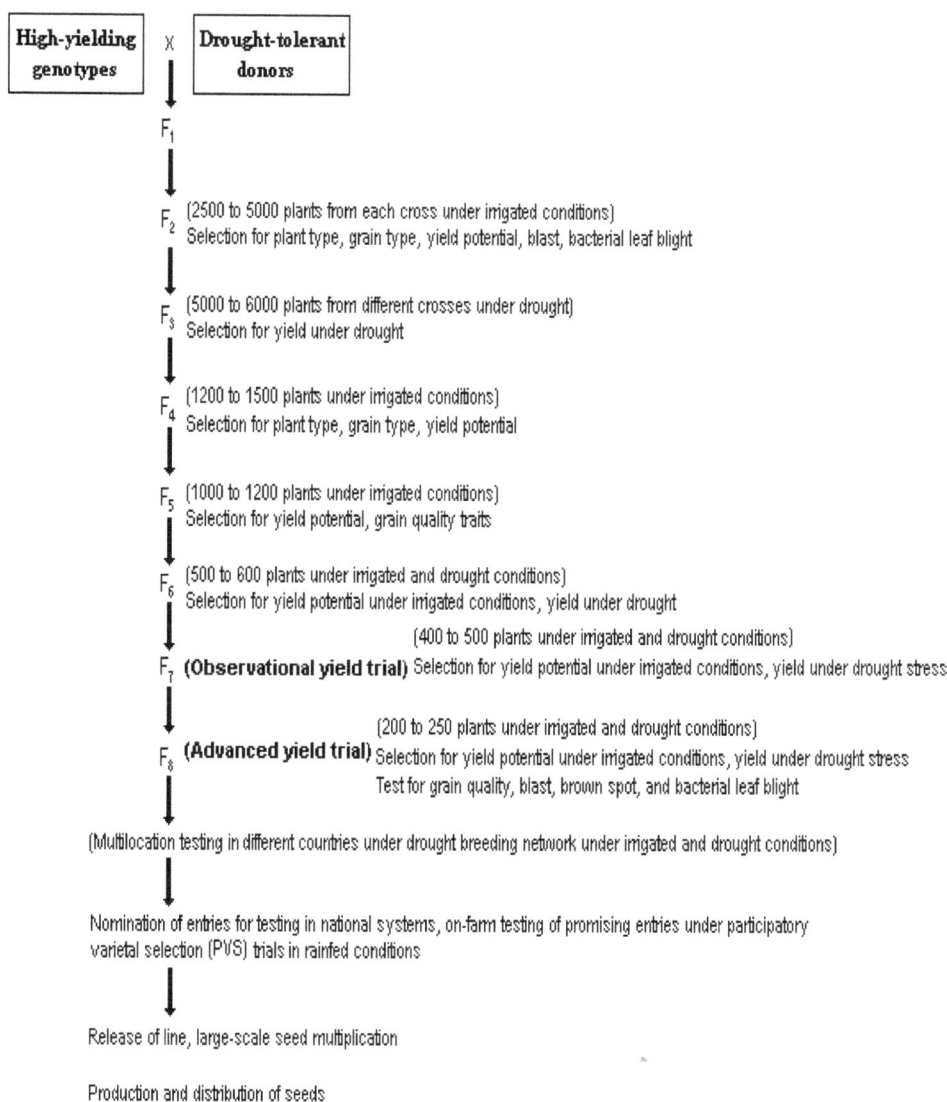

Appendix 1. Conventional breeding approach for developing drought-tolerant rice varieties

Generation	No. of plants	Genotyping	Cross
Selection of parents	-	Foreground selection $QTL_{1.1}$ – M1, M2, M3 (DA1) $QTL_{1.2}$ – M4, M5, M6, M7, M8, M9 (DA2)	Make crosses between plants with $QTL_{1.1}$ and $QTL_{1.2}$
F_1 (two QTLs)	50	Confirm F_1 $QTL_{1.1}$ –M1, M2, M3 (DA1) $QTL_{1.2}$ – M4, M5, M6, M7, M8, M9 (DA2) Select plants with QTL	Make crosses between plants having $QTL_{1.1}$ and $QTL_{1.2}$ with plants having $QTL_{1.3}$
F_1 (three QTLs)	100	Confirm F_1 $QTL_{1.1}$ – M1, M2, M3 (DA1) $QTL_{1.2}$ – M4, M5, M6, M7, M8, M9 (DA2) $QTL_{1.3}$ – M10, M11, M12, M13, M14 (DA3) Select plants with all the QTLs	Cross F_1 plants with three QTLs to recipient parent (RP)
BC_1F_1	1000	Confirm BC_1F_1 $QTL_{1.1}$ – M1, M2, M3 (DA1) $QTL_{1.2}$ – M4, M5, M6, M7, M8, M9 (DA2) $QTL_{1.3}$ – M10, M11, M12, M13, M14 (DA3) Select plants with all three QTLs Background selection using 100 SSR markers uniformly distributed on all the chromosomes in selected plants	Cross BC_1F_1 with three QTLs to RP
BC_2F_1	2000	Confirm BC_2F_1 $QTL_{1.1}$ – M1, M2, M3 (DA1) $QTL_{1.2}$ – M4, M5, M6, M7, M8, M9 (DA2) $QTL_{1.3}$ – M10, M11, M12, M13, M14 (DA3) Select plants with all three QTLs Background selection for segregating SSR markers in selected plants	Self the selected plants

| BC$_2$F$_2$ | 4000 | Confirm F$_2$ QTL$_{1.1}$ – M1, M2, M3 (DA1) QTL$_{1.2}$ – M4, M5, M6, M7, M8, M9 (DA2) QTL$_{1.3}$ – M10, M11, M12, M13, M14 (DA3) Select plants with all three QTLs in homozygous condition Background selection for segregating SSR markers in selected plants Check for QTL flanking markers (RA) Select plants with QTLs in homozygous condition, flanks with recipient allele, and having more recipient genome background | Select the plants with all three QTLs, **RP** flanking, and more recipient background Advance the lines |
| BC$_2$F$_3$ | 60 | Screen QTL+ and QTL– lines under drought | |

Note* M = marker, DA = Donor allele, RA = Recipient allele

Appendix 2. Protocol for pyramiding of three major-effect grain yield QTLs under drought

10. References

Alcamo, J., Leemans, R., & Kreileman, E. (1999). Global Change Scenarios of the 21th Century: Results from the Image 2.1 Model. *Elsevier Science*, The Netherlands

Atlin, G. (2003). Improving drought tolerance by selecting for yield. In: Fischer, K.S., Lafitte, R., Fukai, S., Atlin, G., Hardy, B. (eds.). Breeding rice for drought-prone environments. Los Baños (Philippines): International Rice Research Institute. pp. 14-22

Bernier, J., Kumar, A., Venuprasad, R., Spaner, D., & Atlin, G.N. (2007). A large-effect QTL for grain yield under reproductive-stage drought stress in upland rice. *Crop Sci.*, Vol. 47, 507-516

Bernier, J., Atlin, G.N., Serraj, R., Kumar, A., & Spaner, D. (2008). Breeding upland rice for drought resistance (a review). *J. Sci. Food Agric.*, Vol. 88, 927-939

Bernier, J., Kumar, A., Venuprasad, R., Spaner, D., Verulkar, S., Mandal, N.P., Sinha, P.K., Peeraju, P., Dongre, P.R., Mahto, R.N., & Atlin, G. (2009). Characterization of the effect of a QTL for drought resistance in rice, *qtl12.1*, over a range of environments in the Philippines and eastern India. *Euphytica*, Vol. 166, 207-217

Blum, A. (1988). Plant breeding for stress environments. Boca Raton, Florida (USA): CRC Press. 222 p

Charmet, G., Robert, N., Perrtant, M.R., Gay, G., Sourdille, P., Groos, C., Bernard, S., & Bernard, M. (1999). Marker-assisted recurrent selection for cumulating additive and interactive QTLs in recombinant inbred lines. *Theor. Appl. Genet.*, Vol. 99, 1143-1148

Courtois, B., Ahmadi, N., Khowaja, F., Price, A., Rami, J.F., Frouin, J., Hamelin, C., & Ruiz, M. (2009). Rice root genetic architecture: meta-analysis from a QTL database improves resolution to a few candidate genes. *Rice*, Vol. 2, 115-128

Courtois, B., Shen, L., Petalcorin, W., Carandang, S., Mauleon, R., & Li, Z. (2003). Locating QTLs controlling constitutive root traits in the rice population IAC 165_Co39. *Euphytica*, Vol. 134, 335-345

Crosbie, T.M., Eathington, S.R. & Johnson, G.R. (2006). Plant breeding: past, present, and future. In: Lamkey, K.R., Lee, M. (eds.). Plant breeding: the Arnel R. Hallauer international symposium. Ames, IA: Blackwell Publishing.

Cruz, R.T. & O'Toole, J.C. (1984). Dry land rice response to an irrigation gradient at flowering stage. *Agron. J.*, Vol. 76, 178-183

Darvasi, A. & Soller, M. (1992). Selective genotyping for determination of linkage between a marker locus and a quantitative trait locus. *Theor. Appl. Genet.*, Vol. 85, 353-359

David, C.C. (1991). The world rice economy: challenges ahead. In: Khush, G.S. and Toenniessen, G.H. (eds.). *Rice Biotechnology*, CAB International, UK, pp 19-54

Ding, X., Li, X. & Xiong, L. (2011). Evaluation of near-isogenic lines for drought resistance QTL and fine mapping of a locus affecting flag leaf width, spikelet number, and root volume in rice. *Theor. Appl. Genet.*, DOI: 10.1007/s00122-011-1629-1O

Dixit, S., Swamy, B.P.M., Vikram, P., Ahmed, H.U., Sta Cruz, M.T., Amante, M., Atri, D., Leung, H., & Kumar, A. (2011). Fine mapping four large-effect QTLs for rice grain yield under drought reveals sub-QTLs conferring response to variable drought severities. *Theor. Appl. Genet.*, (Submitted)

FAO (2002). Crops and drops – making the best use of water for agriculture. Food and Agriculture Organization of the United Nations. Rome

Fukuoka, S., Ebana, K., Yamamoto, T., & Yano, M. (2010). Integration of genomics into rice breeding. *Rice*, Vol. 3, 131-137

Fukai, S., Cooper, M., & Salisbury, J. (eds.) (1996). Breeding Strategies for Rainfed Lowland Rice in Drought-prone Environments. Proceedings of an International Workshop held at Ubon Ratchathani, Thailand, 5-8 November 1996. ACIAR Proceedings No. 77, 260 p

Ghimire, K.H., Quiatchon, L.A., Vikram, P., Swamy, B.P.M., Hernandez, J.E., Borromeo, T.H., & Kumar, A. (2011). A large-effect drought grain yield QTL from Indian landrace in background of rice mega-varieties. (Unpublished).

Gleick, P.H. (1993). Water and conflict: fresh water resources and international security. *Int. Security*, Vol. 18(1), 79-112

Hossain, M. (1995). Sustaining food security for fragile environments in Asia: achievements, challenges, and implications for rice research. In: Fragile lives in fragile ecosystems. Proceedings of the International Rice Research Conference; 13-17 Feb 1995; Los Baños, Philippines. Manila (Philippines): International Rice Research Institute. p 3-23

Huke, R.E., & Huke, E.H. (1997). Rice area by type of culture: South, Southeast, and East Asia. IRRI, Los Baños, Philippines

Johnson, G.R. (2004). Marker assisted selection. *Plant Breed. Rev.*, Vol. 24, 293-310

Jongdee, B., Fukai, S., & Cooper, M. (2002). Leaf water potential and osmotic adjustment as physiological traits to improve drought tolerance in rice. *Field Crops Res.*, Vol. 76, 153-163

Kamoshita, A., Babu R.C., Boopathi, N.M., & Fukai, S. (2008). Phenotypic and genotypic analysis of drought-resistance traits for development of rice cultivars adapted to rain fed environments. *Field Crops Res.*, Vol. 109, 1-23

Khowaja, F.S., Norton, G.J., Courtois, B., & Price, A.H. (2009). Improved resolution in the position of drought-related QTLs in a single mapping population of rice by meta-analysis. *BMC Genomics*, Vol. 10, 276

Khush, G.S. (1997). Origin, dispersal, cultivation and variation of rice. *Plant Mol. Biol.*, Vol. 35, 25-34

Kumar, A., Bernier, J., Verulkar, S., Lafitte, H.R., & Atlin, G.N. (2008). Breeding for drought tolerance: direct selection for yield, response to selection and use of drought tolerant donors in upland and lowland-adapted populations. *Field Crops Res.*, Vol. 107, 221-231

Kumar, A. (2011). Making rice less thirsty: progress at IRRI. Annual Seminar series, March 3, 2011

Kumar, R., Venuprasad, R., & Atlin, G.N. (2007). Genetic analysis of rainfed lowland rice drought tolerance under naturally-occurring stress in eastern India: heritability and QTL effects. *Field Crops Res.*, Vol. 103, 42-52

Lafitte, R. (2003). Managing water for controlled drought in breeding plots. In: Fischer, K.S., Lafitte, R., Fukai, S., Atlin, G., Hardy, B. (eds.). Breeding rice for drought-prone environments. Los Baños (Philippines): International Rice Research Institute. pp. 23–26.

Lanceras, J.C., Pantuwan, G.P., Jongdee, B., & Toojinda, T. (2004). Quantitative trait loci associated with drought tolerance at reproductive stage in rice. *Plant Physiol.*, Vol. 135, 384-399

Lang, N.T. & Buu, B.C. (2008). Fine mapping for drought tolerance in rice (*Oryza sativa* L.). *Omonrice*, Vol. 16, 9-15

Mackill, D.J., Coffman, W.R., & Garrity, D.P. (1996). Rain-fed Lowland Rice Improvement International Rice Research Institute, Los Baños, Philippines, 242 p

Mandal, N.P., Sinha, P.K., Variar, M., Shukla, V.D., Perraju, P., Mehta, A., Pathak, A.R., Dwivedi, J.L., Rathi, S.P.S., Bhandarkar, S., Singh, B.N., Singh, D.N., Panda, S., Mishra, N.C., Singh, Y.V., Pandya, R., Singh, M.K., Sanger, R.B.S., Bhatt, J.C., Sharma, R.K., Raman, A., Kumar, A., & Atlin, G. (2010). Implications of genotype × input interactions in breeding superior genotypes for favorable and unfavorable rainfed upland environments. *Field Crops Res.*, Vol. 118, 135-144

McCouch, S.R., Teytelman, L., Xu, Y., Lobos, K.B., Clare, K., Walton, M., Fu, B., Maghirang, R., Li, Z., Xing, Y., Zhang, Q., Kono, I., Yano, M., Fjellstrom, R., Declerck, G., Schneider, D., Cartinhour, S., Ware, D., & Stein, L. (2002). Development and mapping of 2240 new SSR markers for rice (*Oryza sativa* L.). *DNA Res.*, Vol. 9, 199-207

McCouch, S.R., Sweeney, M., Li, J., Jiang, H., Thomson, M., Septiningsih, E., Edwards, J., Moncada, P., Xiao, J., Garris, A., Tai, T., Martinez, C., Tohme, J., Sugiono, M., McClung, A., Yuan, L.P., & Ahn, S.N. (2007). Through the genetic bottleneck: *O.*

rufipogon as a source of trait-enhancing alleles for *O. sativa*. *Euphytica*, Vol. 154, 317-339

McNally, K.L., Childs, K.L., Bohnert, R., Davidson, R.M., Zhao, K., Ulat, V.J., Zeller, G., Clark, R.M., Hoen, D.R., Bureau, T.E., Stokowski, R., & Ballinger, D.G. (2009). Genome-wide SNP variation reveals relationships among landraces and modern varieties of rice. *Proc. Natl. Acad. Sci.* USA, Vol. 106, 12273-12278

Michelmoore, R.W., Paran, I., & Kesseli, R.V. (1991). Identification of markers linked to disease resistance genes by bulked segregant analysis: a rapid method to detect markers in specific genomic regions by using segregating populations. *Proc. Natl. Acad. Sci.* USA, Vol. 88, 9828-9832

Navabi, A., Mather, D.E, Bernier, J., Spaner, D.M., & Atlin, G.N. (2009). QTL detection with bidirectional and unidirectional selective genotyping: marker-based and trait-based analyses. *Theor. Appl. Genet.*, Vol. 118, 347-358

Nguyen, T.T.T., Klueva, N., Chamareck, V., Aarti, A., Magpantay, G., Millena, A.C.M., Pathan, M.S., & Nguyen, H.T. (2004). Saturation mapping of QTL regions and identification of putative candidate genes for drought tolerance in rice. *Mol. Gen. Genomics*, Vol. 272, 35-46

Ouk, M., Basnayake, J., Tsubo, M., Fukai, S., Fischer, K.S., Cooper, M., & Nesbitt, H. (2006). Use of drought response index for identification of drought tolerant genotypes in rain fed lowland rice. *Field Crops Res.*, Vol. 99, 48-58

Pantuwan, G., Fukai, S., Cooper, M., Rajatasereekul, S., & O'Toole, J.C. (2002). Yield response of rice (*Oryza sativa* L.) genotypes to different types of drought under rainfed lowlands. Part 3. Plant factors contributing to drought resistance. *Field Crops Res.*, Vol. 73, 181-200

Pimentel, D., Houser, J., Preiss, E., White, O., Fang, H., Mesnick, L., Barsky, T., Tariche, S., Schreck, J., & Alpert, S. (1997). Water resources: agriculture, the environment, and society. *BioScience*, Vol. 47, 97-106

Price, A.H., Cairns, J.E., Horton, P., Jones, H.G., & Griffiths, H. (2002a). Linking drought-resistance mechanisms in upland rice using a QTL approach: progress and new opportunities to integrate stomatal and mesophyll responses. *J. Exp. Bot.*, Vol. 53, 989-1004

Price, A.H., Townend, J., Jones, M.P., Audebert, A., & Courtois, B. (2002b). Mapping QTLs associated with drought avoidance in upland rice grown in the Philippines and West Africa. *Plant Mol. Biol.*, Vol. 48, 683-695

Ribaut, J.M. & Ragot, M. (2007). Marker-assisted selection to improve drought adaptation in maize: the backcross approcah, perspectives, limitations and alternatives. *J. Exp. Bot.*, Vol. 58, 351-360

Salunkhe, A.S., Poornima, R., Silvas, K., Prince, J., Kanagaraj, P., Sheeba, A., Amudha, K., Suji K.K., Senthil, A., & Chandra Babu, R. (2011). Fine mapping QTL for drought resistance traits in rice (*Oryza sativa* L.) using bulk segregant analysis. *Mol. Biotechnol.*, Vol. 49, 90-95

Serraj, R., Kumar A., McNally, K.L., Slamet-Loedin, I., Bruskiewich, R., Mauleon, R. Cairns, J., & Hijmans, R.J. (2009). Improvement of drought resistance in rice. *Adv. Agron.*, Vol. 103, 41-99

Shashidhar, H.E., Vinod, M.S., Sudhir, N., Sharma, G.V., & Krishna-Murthy, K. (2005). Markers linked to grain yield using bulked segregant analysis approach in rice (*Oryza sativa* L.). *Rice Genet. Newsl.*, Vol. 22, 69-71

Shen, L., Courtois, B., McNally, K.L., Robin, S., & Li, Z. (2001). Evaluation of near-isogenic lines of rice introgressed with QTLs for root depth through marker-aided selection. *Theor. Appl. Genet.*, Vol. 103, 75-83

Shrivastava, M.N. & Verulkar, S.B. (2009). Progress in crop improvement research. Limited Proceedings. Edited by Hossain, M., Bennett, J., Mackill, D., Hardy, B. Los Baños, Laguna (Philippines): International Rice Research Institute (IRRI), Vol. 14, 3-12

Sinclair, T.R. (2011). Challenges in breeding for yield increase for drought. *Trends Plant Sci.*, Vol. 16, 289-293

Swamy, B.P.M, Vikram, P., Dixit, S., Ahmed, H.U., & Kumar, A. (2011a). Meta-analysis of grain yield QTL identified during agricultural drought in grasses showed consensus. *BMC Genomics*, Vol. 12, 319

Swamy, B.P.M., Ahmed, H.U., Henry, A., Dixit, S., Vikram, P., Mauleon, R., Vera Cruz, C., Kouji, S., Ali, M., Ramiah, V., Shoshi, K., Leung H., & Kumar, A. (2011b) Convergence of mapping, physiological, and gene expression analyses reveal multiple QTL that enhance transpiration and yield under drought in rice (Unpublished)

Temnykh, S., Declerck, G., Lukashova, A., Lipovich, L., Cartinhour, S., & McCouch, S. (2001). Computational and experimental analysis of microsatellites in rice (*Oryza sativa* L.): frequency, length variation, transposon associations and genetic marker potential. *Genome Res.*, Vol. 11, 1441-1452

Tsonev, S., Todorovska, E., Avramova, V., Kolev, S., Abu-Mhadi, N., & Christov, N.K. (2009). Genomics assisted improvement of drought tolerance in maize: QTL approaches. *Biotechnol. & Biotechnol.*, Vol. 23, 1410-1413

Tuberosa, R. (2004). Molecular approaches to unravel the genetic basis of water use efficiency. *In*: M.A. Bacon (Ed.), Water use efficiency in plant biology. Blackwell, Oxford, UK. pp. 228-301

Tuberosa, R. & Salvi, S. (2004). QTLs and genes for tolerance to abiotic stress in cereals. *In*: Gupta, P.K., Varshney, R. (Eds.). Cereal genomics. Kluwer, The Netherlands. pp. 253-315

Tuberosa, R. & Salvi, S. (2006). Genomics-based approaches to improve drought tolerance of crops. *Trends Plant Sci.*, Vol. 11, 405-412

Uga, Y., Okuno, K., & Yano, M. (2011). *Dro1*, a major QTL involved in deep rooting of rice under upland field conditions. *J. Exp. Bot.* doi:10.1093/jxb/erq429

Venuprasad, R., Dalid, C.O., Del Valle, M., Zhao, D., Espiritu, M., Sta Cruz, M.T., Amante, M., Kumar, A., & Atlin, G.N. (2009). Identification and characterization of large-effect quantitative trait loci for grain yield under lowland drought stress in rice using bulk-segregant analysis. *Theor. Appl. Genet.*, Vol. 120, 177-190

Verulkar, S.B., Mandal, N.P., Dwivedi, J.L., Singh, B.N., Sinha, P.K., Mahato, R.N., Dongre, P., Singh, O.N., Bosee, L.K. Swaine, P., Robin, S., Chandrababu, R., Senthil, S., Jain, A., Shashidhar, H.E., Hittalmani, S., Vera Cruzi, C., Paris, T., Ramani, A., Haefelei, S., Serraji, R., Atlin, G., & Kumar, A. (2010). Breeding resilient and productive genotypes adapted to drought-prone rainfed ecosystem of India. *Field Crops Res.*, Vol. 117, 169-266

Vikram, P., Sta Cruz, T., Espiritu, M., Valle, M.D., Singh, A.K., & Kumar, A. (2009). Major effect QTLs for grain yield under lowland drought stress. Poster presented at International Rice Genetics Symposium, 16-19 Nov. 2009, Manila, Philippines

Wopereis, M.C.S., Kropff, M.J., Maligaya, A.R., & Tuong, T.P. (1996). Drought-stress responses of two lowland rice cultivars to soil water status. *Field Crops Res.*, Vol. 46, 21-39

Xiao, J., Li, J., Grandillo, S., Ahn, S.N., Yuan, L., Tanksley, S.D., & McCouch, S.R. (1998). Identification of trait-improving quantitative trait loci alleles from a wild rice relative, *Oryza rufipogon*. *Genetics*, Vol. 150, 899-909

Xing, Y.Z., Tan, Y.F., Hua, J.P., Sun, X.L., Xu, C.G., & Zhang, Q. (2002). Characterization of the main effects, epistatic effects and their environmental interactions of QTLs on the genetic basis of yield traits in rice. *Theor. Appl. Genet.*, Vol. 105, 248-257

Zeigler, R.S. & Puckridge, D.W. (1995). Improving sustainable productivity in rice based rainfed lowland systems of south and Southeast Asia. Feeding four billion people. The challenge for rice research in the 21st century. *GeoJournal*, Vol. 35, 307-324

Zhang, X., Zhou, S., Fu, Y., Su, Z., Wang, X., & Sun, C. (2006) Identification of a drought tolerant introgression line derived from Dongxiang common wild rice (*O. rufipogon* Griff.). *Plant Mol. Biol.*, Vol. 62, 247-259

Zou, G.H., Mei, H.W., Liu, Y., Liu, G.L., Hu, S.P., Yu, X.Q., Li, M.S., Wu, J.H., & Luo, L.J. (2005). Grain yield responses to moisture regimes in a rice population: association among traits and genetic markers. *Theor. Appl. Genet.*, Vol. 112, 106-113

Part 2

Irrigation Systems and Water Regime Management

10

Effects of Irrigation-Fertilization and Irrigation-Mycorrhization on the Alimentary and Nutraceutical Properties of Tomatoes

Luigi Francesco Di Cesare[1], Carmela Migliori[1], Valentino Ferrari[2], Mario Parisi[3], Gabriele Campanelli[2], Vincenzo Candido[4] and Domenico Perrone[3]

[1]CRA-IAA, Milan
[2]CRA-ORA, Monsampolo del Tronto (AP)
[3]CRA-ORT, Pontecagnano (SA)
[4]Dipartimento di Scienze dei Sistemi Colturali,
Forestali e dell'Ambiente, Università della Basilicata, Potenza
Italy

1. Introduction

Tomato, a key vegetable in the Italian Mediterranean diet, has recently gaining been attention in relation to the prevention of some human diseases. This interest is due to the presence of carotenoids and particularly lycopene, which is an unsaturated alkylic compound, that appears to be an active compound in the prevention of cancer, cardiovascular risk and in slowing down cellular aging, owing to its high antioxidant and antiradical power (Gerster 1997; Giovannucci et al. 1995). Lycopene is found in fresh, red-ripe tomatoes as all-trans (79-91 %) and cis- (9-21%) isomers (Boileau et al., 2002; Shi et al., 1999; Stahh & Sies, 1992).

In this paper, lycopene was used as a measure of the nutraceutical quality, while flavour volatiles, soluble sugars, organic acids, dry matter and pH as expression of the nutritional quality.

In red-ripe tomato fruit, many volatile compounds have been identified [2(E)-hexenal, hexanal, 3(Z)-hexen-1-ol, β-ionone, 2(E)4(Z)-decadienal, 2-isobutylthiazole, 3(Z)-hexenol, linalool, methylsalicylate, 2-methoxyphenol, 6-methyl-5-hepten-2-one, 6-methyl-5-hepten-2-ol, 2,3-epoxygeranial, neral, geranial, nerylacetone, β-damascenone, α-terpineol etc.] (Di Cesare et al., 2003).

Hexanal, 2(E)-hexenal, 2-isobutylthiazole, are considered to be responsible for the fresh tomato flavour (Dirinck et al., 1976). Besides, according to other authors (Buttery et al., 1989; Buttery & Ling, 1993) 3(Z)-hexenal, β-ionone, β-damescenone, 1-penten-3-one, 2 and 3-methylbutanols, 2-isobutylcyanide, 2(E)-heptenal, phenylacetaldehyde, 6-methyl-5-hepten-2-one, 3(Z)-hexenol, 2-phenylethanol, methylsalycilate are considered to be important contributors to tomato flavour.

Soluble sugars and organic acids play an important role in the characterization of the tomato taste. Sugars (glucose, fructose and traces of sucrose) and organic acids (citric, malic and oxalic) represent half of the total dry matter of tomato fruit (Silvestri & Siviero, 1991).

The irrigation and the potassium (K) fertilization are two among the agronomical factors that mostly influence the alimentary and nutraceutical quality of tomato. Full irrigation regimes enhance the crop yield, however, cause a reduction of total and soluble solids and less colourful and firmness of fruits (Branthôme et al., 1994; Colla et al., 1999; Colla et al., 2001; Dumas et al., 1994; Favati et al., 2009). Among the several studies, Veit Kohler et al. (1999) reported that the higher levels of sugars, titratable acidity, aroma volatiles, and vitamin C are responsible for the higher tomato fruit quality under conditions of limited water supply. Taking into account the effects of irrigation on agronomic performances, a negative trend in response to increasing soil water deficit was observed for fruit yield and size (Candido et al., 2000; Patanè & Cosentino, 2010). Pernice et al. (2010) have also shown that the effects of irrigation on yield and unitary weight of berries are genotype-dependent. Finally, some studies demonstrated that irrigation generally reduces green fruit yield and blossom-end rot (Candido et al., 2000; Warner et al. (2007).

The nutritional quality of tomatoes may be affected by the amount of water applied, regardless of fertilizer management, and their irrigation system. For example, heavy rainfall may reduce the oxygen concentration in the soil, and indirectly affect the nutritional value of fruit (Dorais et al., 2008). Considering the effect of irrigation on ascorbic acid content of fruit, depending on cultivar, low soil water tension generally could decrease the content of this antioxidant compound (Rudich et al., 1977). Zushi & Matsuzoe (1998) also showed that the effects of soil water deficits on the vitamin C content (fm basis) may be positive or null, depending on the cultivar.

The effects of water availability on the synthesis of carotenoids have been studied, but results are sometimes contradictory, and the data often incomplete.

Dumas et al. (2003) reviewed several studies that looked at agronomic and environmental factors that influenced lycopene concentrations in tomatoes and reported that moisture stress, for example, reduced lycopene content in some tomato varieties but increased it as well as β-carotene content in others. On the other hand, it is reported that the increase of abscisic acid, induced by limited water supply, may affect the ethylene production and then the carotenoids synthesis (Dorais et al., 2008). Serio et al. (2006) confirmed the increase of lycopene (and ascorbate) content at moderate levels of water stress. In recent studies performed on two tomato ecotypes, Pernice et al. (2010) found an higher amount of total carotenoids content of fruits under no irrigation conditions, nevertheless no variation for antioxidant activity of carotenoids extracts were detected. Furthermore, this behaviour was also ascertained after canning process.

In soilless culture, however, higher plant water availability provided by a capillary system did not reduce the lycopene content or the antioxidant activity, compared to tomato plants grown either on rockwool or sawdust and irrigated according to solar radiation (Dorais, 2007). However, Dorais (2007) also found that lycopene content of fruit from plants grown in sawdust andpeat at low levels of moisture stress was less than that of fruit grown on rockwool and irrigated according to solar radiation. No effect of irrigation treatment on antioxidant activity detected.

In general, increasing the water supply increased tomato fruit yield but reduced fruit quality attributes, because of high fruit water content (Dorais et al. 2001). Consequently, on a fresh weight basis, vitamin, mineral and carotenoid contents were generally lower under higher water supply although not all nutraceutical compounds responded to soil moisture variation (Dorais et al., 2008).

Mineral nutrients (nitrogen, phosphorous, potassium and calcium) and their supply through fertilization can affect yield and quality of tomato crops.

K is needed in stomatal movement for water regulation in the plant. It activates enzymes and is required for carbohydrate metabolism and translocation, nitrogen metabolism and protein synthesis, and regulation of cell sap concentration. This essential nutrient helps in vigorous growth of tomato and stimulates in early flowering and setting of fruits, thereby increasing the number and production of tomato berries per plant (Bergmann, 1992; Ruiz & Romero, 2002).

In tomato, as well as in other fruit and vegetables, K improves fruit quality: its levels are positively correlated with a good fruit shape, a reduction in ripening disorders (puffiness, gray-wall, blotchy ripening, green back and yellow shoulder), while it simultaneously plays an important role as a counter-ion to organic acids, maintaing electroneutrality in fruit (Dorais et al., 2001). Several studies have shown that K-fertilization increased several quality parameters in tomato fruit such as marketable yield, total and soluble solids contents (°Brix), sugars, acidity, red colour and leads to a reduction of the sugar/acid ratio (Ghebbi Si-smail et al., 2007; 2005; Oded & Uzi, 2003; Wuzhong, 2002). Wright and Harris (1985) pointed out that flavour scores indicated that increasing N and K fertilization had a detrimental effects on tomato flavour. An increase in titratable acidity and soluble solids was found with increasing fertilization. Concentrations of hexanal, 2-hexanone, benzaldehyde, phenylacetaldehyde, β-ionone and 6-methyl-5-hepten-2-one increased with increasing N+K levels.

Trudel and Ozbun (1970; 1971) reported that K enhanced the colour of the fruit, increasing the lycopene content and reducing that of β-carotene. The positive effect of K fertilization on lycopene content, has been confirmed more recently by other authors (Oded & Uzi, 2003; Serio et al., 2007), as it is shown that K deficiency can lead to a reduction in the synthesis of carotenoids, particularly lycopene (Dumas et al., 2003). With regard to accumulation of this antioxidant compound in tomato fruits, Taber et al. (2008) have found that the response to high fertilizer dose of K is cultivar-dependent.

The vesicular-arbuscular mychorrhizae (VAM) are most common in nature but also more interesting for application in agriculture, as they can colonize most cultivated plants.

In addition to improving plant mineral nutrition (mainly phosphate) (Elia et al., 2006; Conversa et al., 2007), VAM fungi can promote photosynthetic activity, improve resistance to root pathogens and drought stress-saline and soil properties too (Turk et al., 2006). Besides, VAM are largely beneficial to crops, since they increase absorption of micro elements (van der Heijden et al., 2006) and water use efficiency (Davies et al., 2002) in the plant.

The importance of mycorrhizal symbiosis for plant growth and health is raising a growing interest in the use of these fungi as bio-fertilizers, bio-regulators and bio-protectors (Akhtar et al., 2008), thus reducing the input of fertilizers and pesticides (Gianinazzi et al., 2003).

Many studies conducted to evaluate the use of VAM fungi in vegetable crops have provided for the distribution fungal inoculum at transplantation time near plant roots; the need to produce the fungal inoculation of host plants through breeding, is still an obstacle for wider application in the field.

The most promising method of VAM application in vegetable crops is represented by inoculation of seedlings (pre-inoculation) during their nursery growth in containers (Azcona-Aguilar & Barea, 1997). The pre-inoculation of VAM fungi reduced the mortality of seedlings, resulted in greater uniformity of crop growth (Waterer & Coltman, 1988) and increased yields too (Sorensen et al., 2008).

Although most vegetables (except *Brassicaeae* and *Chenopodiaceae*) are colonized by VAM fungi, it is known that the effectiveness of mycorrhization much depends on the specificity of fungus-host plant (species and/or cultivars) (Sensoy et al., 2007); sometimes VAM colonization does not improve growth and nutritional status of the plant but also a negative effect of the symbiosis may occur (Gosling et al., 2006).

Some studies have also found that VAM influence positively some quality yield parameters, such as essential oils in aromatic plants (Copetta et al., 2006).

VAM fungi colonize the same area of root tissue and therefore often occur together in the rhizosphere and roots of plants. This contemporary occurrence raises a number of reciprocal interactions between these two groups of organisms, among which also a VAM suppressiveness against phytoparasitic nematodes populations (Gera & Cook, 2005). Competition for nutrients or induction of morphological changes or systemic resistance in root tissues were suggested as mechanisms for the nematicidal effect of VAM (De la Peña et al., 2006; Elsen et al., 2008).

Identification of VAM species is difficult because they do not possess a sexual stage and do not grow in host absence. The identification is generally based on morphological characters, in particular, those of spores (Morton & Bentivenga, 1994) which are, however, very similar and result not useful for reliable identification to species level (Jacquot et al., 2000). Molecular techniques (PCR, Nested-PCR analysis of nucleotide sequences, RFLP, RAPD) were used to identify and groped distinguish different species of mycorrhizal fungi in planta and in soil (van Tuinen et al., 1998). These methods have proved much more rapid and sensitive in identifying the species with respect to classical techniques.

2. Materials and methods

2.1 Irrigation

K-fertilization trials

The experimental trial was carried out using a split-plot scheme with 3 replicates, in the plain of Battipaglia (Salerno, South Italy, 40° 58 ' 45" 61''' Lat. N, 14° 98' 32" 27''' Long. E, 12 m a.s.l.), during spring-summer of 2009, aimed to compare two irrigation regimes in combination with two levels of K-fertilization on two tomato ecotypes (Corbarino and Vesuviano).

The soil of the experimental plots was clay loam (43.8% sand, 27.8% silt, 28.3% clay), with poor organic matter content (1.4%) and very low salt concentration. Cationic exchange capacity was high (C.E.C. of 20,5 meq/100 g of soil). With regard to K, exchangeable and available fractions were very high (345 ppm and 0.72 meq/100 g of soil) too.

Effects of Irrigation-Fertilization and Irrigation-Mycorrhization on the Alimentary
and Nutraceutical Properties of Tomatoes

197

The two tomatoes studied, Corbarino (C) and Vesuviano (V), were ecotypes originating from Campania region (Italy), respectively from plain of Corbara (Agro Sarnese-Nocerino area) and slopes of Vesuvius Volcano. They were different in many morphological: the first one showed oval fruits tending to an extended shape (ratio between two diameters of 1.52) with unitary weight of approximately 13-18 g.; the latter, instead, was characterised by elongated, pear-shaped fruits (ratio between two diameters of 1.85) with a clearly pronounced mucro (pointed shape at blossom end) and a weight of about 20-25 g (Parisi et al., 2006).

Transplant was performed, on May 16th, in single rows with a density of 4.0 plants/m². Microirrigation was started from fruit-setting (40 days after transplanting) adopting two different water regimes: "reduced" irrigation (I_1) (total irrigation volume 300 m³/ha) and "normal irrigation" (I_2) (total irrigation volume 1500 m³/ha). These established water regimes were applied in quantities of 50 m³/ha for I_1 level or 250 m³/ha for I_2 level, once a week for six weeks.

Regard to K-fertilization, to the natural endowment of land (K_0) was compared K_1 level (200 kg/ha of K_2O, added to soil before transplant).

Each parcel was made up of 5 rows having an extension of 7 m. The main morpho-physiological surveys were conducted for the plants of all the parcels; while, at harvesting, the main productive and qualitative aspects of the berries were estimated on the plants of the central row (Giordano et al., 2000).

Fruit harvest was performed at full ripening on August 20th and September 24th; productive data were submitted to analysis of the variance with the program Mstat-C using the scheme "Randomized Complete Block Design for Factor A (watering regime), with Factor B (ecotype) as a Split Plot on A and Factor C (K-fertilization) as a Split Plot on B". Means were evaluated with Duncan's test.

2.2 Irrigation-Mycorrhization trials

The experimental trial was carried out using a split-plot scheme with 3 replicates, in the plain of Metaponto (Matera, South Italy 40° 24′N; 16° 48′E; 10 m a.s.l.), during spring-summer of 2009, aimed to compare 3 watering regimes in combination with two different mycorrhization systems on Faino (Syngenta) F1 hybrid.

The soil of the experimental plots had a mixed composition, containing 51.3% of silt, 29.0% of sand and 19.7% of clay, with a slight alkaline reaction (pH 7.68), poor in total N (0.8 g /kg), with average organic matter (9.2 g / kg) and well equipped in exchangeable P (21.2 mg / kg) and K (215 mg / kg), in open field conditions.

Transplant was performed on May 28th 2009 in double rows with a density of 4.94 plants/m².

"Oval fruit-like" tomatoes were irrigated and mycorrhized as follows:

- V_{100}= full restoration (100%) of Maximum Crop EvapoTranspiration (ETc);
- V_{50}= half restoration (50 %) of ETc;
- V_0= non irrigated control (irrigated only at transplantation);
- M_1 (MICOSAT F)= endomycorrhizal fungi of the genus Glomus, rhizosphere bacteria and saprophytic fungi;
- M_2 (VAM)= only endomycorrhizal fungi of the genus Glomus.

3. Results and discussion

3.1 Effects of irrigation - K-fertilization interactions

3.1.1 Agronomical and morphological data

The main agronomical and morphological results are reported in Table 1. The lack of natural rainfall during the period from flowering to harvest, together with the adoption of irrigation schemes of medium and small-scale to the needs of the tomato crop have been the main cause of low yields for both ecotypes (26.5 t/ha for Corbarino and 27.3 t/ha for Vesuviano tomato).

	Yield		Ripe fruits with defects	Fruit weight	Size homogeneity	Fruit firmness
FACTORS	Total (t/ha)	Waste (%)	(%)	(g)	(1)	(1)
Watering regime (I)						
"normal" (I_1)	32,9 b	9,8 b	15,8 a	12,8 b	3,1 b	3,1 a
"reduced" (I_2)	20,9 a	5,8 a	14,1 a	10,4 a	2,8 a	3,8 b
Signif.	**	**	n.s.	*	*	**
Genotype (G)						
Corbarino	26,5 a	8,2 a	16,8 a	10,5 a	2,9 a	3,1 a
Vesuviano	27,3 a	7,4 a	13,1 b	12,6 b	3,0 a	3,8 b
Signif.	*n.s.*	*n.s.*	**	**	*n.s.*	**
K fertilization (K)						
(K_0)	27,1 a	8,3 a	17,6 a	12,1 a	3,0 a	3,4 a
(K_1)	26,3 a	7,0 a	12,6 b	11,3 a	3,1 a	3,5 a
Signif.	*n.s.*	*n.s.*	*	*n.s.*	*n.s.*	*n.s.*
Interactions						
(I) x (G)	*n.s.*	*n.s.*	**	*n.s.*	*n.s.*	**
(I) x (K)	*n.s.*	**	*	*n.s.*	*n.s.*	*n.s.*
(G) x (K)	*n.s.*	*n.s.*	n.s.	*n.s.*	*n.s.*	*n.s.*
(I) x (G) x (K)	*	*n.s.*	n.s.	*n.s.*	*n.s.*	*n.s.*

(1) value from 1 (worst) to 5 (best). ** P=0.01; *P=0.05; n.s.= not significant

Table 1. Effects of irrigation/K-fertilization on agromical and qualitative traits of Corbarino and Vesuviano tomato.

Indeed a previous research in the same area indicated better agronomic performance for these local varieties (Parisi et al., 2006).

Effects of Irrigation-Fertilization and Irrigation-Mycorrhization on the Alimentary
and Nutraceutical Properties of Tomatoes

199

In this research, yields were highly influenced by the water regime; in particular moving from no irrigation to normal irrigation condition, increases in total production of 41.5% for Corbarino and 64.4% for Vesuviano tomato were found. For the first local variety the results were in agreement with those of Pernice et al. (2010).

No significant effects of ecotype and K-fertilization on total yield were recorded; instead, the interactions genotype x irrigation x K-fertilization and that between water regime x K-fertilization were significant for total yield and waste percent, respectively. The incidence of ripe fruit with defects (such as cracking, virus infections and damage by other biotic agents) seemed to have decreased with K-fertilization (from 17.6% to 12.6%), moreover the interactions between irrigation ×genotype and between irrigation × K fertilization were significant.

Increasing in water supply resulted in better values of size homogeneity and, according to Pernice et al., 2010, in worsening of fruit firmness.

3.1.2 Alimentary quality

The most of volatiles substances were found in both genotypes, while their concentrations, expressed as µg/100 g d.m., seemed to be influenced by both agronomical treatments (Table 2).

	(µg/100g d.m.)							
	I_1K_0C	I_2K_0C	I_1K_1C	I_2K_1C	I_1K_0V	I_2K_0V	I_1K_1V	I_2K_1V
ALCOHOLS								
1-penten-3-ol	92,93 a A	89,09 a A	102,81 a A	194,58 b B	118,02 b B	37,37 a A	46,49 a A	83,75 b B
2-pentanol	0 a A	0 a A	115,48 b B	0 a A	253,92 b B	74,80 a B	0 a A	0 a A
3-methyl-1-butanol	1187,99 b A	799,73 a A	1245,44 b A	1130,44 a B	1011,95 b B	321,54 a A	289,99 a A	370,37 a A
2-methyl-1-butanol	134,06 a A	139,25 a A	183,98 a A	214,75 b B	148,76 b B	74,17 a A	73,13 a A	78,25 a A
1-pentanol	183,08 a A	172,42 a A	226,92 a A	227,43 a B	206,04 b B	70,16 a A	69,73 a A	81,43 b A
3(Z)-hexen-1-ol	138,895 b B	4,685 a A	16,155 a A	21,955 a A	40,26 b B	9,685 a A	1,045 a A	11,59 b A
1-hexanol	18,125 a B	10,83 a A	0 a A	32,395 b B	20,455 b B	6,51 a A	0,915 a A	7,2 b A
6-methy-5-hepten-2-ol	1,075 a A	9,015 b A	16,355 a B	11,285 a A	26,56 b A	9,53 a A	0,815 a A	9,995 b A
benzyl alcohol	2506,37 b A	1349,205 a B	1329,405 b A	347,055 a A	1114,895 a A	1100,62 a B	2931,885 b B	496,715 a A
phenethyl alcohol	214,73 b A	130,99 a A	369,92 a B	387,255 a B	210,475 a B	208,025 a B	0 a A	62,205 b A
Total Amount	4477,26	2705,22	3604,47	2567,15	3151,34	1912,41	3414,00	1201,51

CARBONYL COMPOUNDS								
3-methyl-3-buten-2-one	7,16 b A	2,01 a A	9,97 b B	4,51 a B	16,04 a B	14,74 a A	7,96 a A	15,93 b A
1-penten-3-one	62,49 a B	89,2 a A	48,935 a A	62,90 b A	62,44 b B	13,60 a B	16,78 b A	7,19 a A
pentanal	68,78 a A	63,56 a B	55,64 b A	1,68 a A	89,66 b B	20,73 a B	31,87 b A	0 a A
2(E)-pentenal	73,28 a A	67,83 a A	80,43 a A	92,02 b A	84,69 b B	0 a A	26,42 a A	67,51 b B
hexanal	246,61 a A	272,42 a A	249,08 a A	282,59 a A	362,33 b B	154,54 a A	146,915 a A	129,555 b A
2(E)-hexenal	44,5 a A	62,60 b B	49,51 a A	77,78 b B	176,245 b B	69,3 a A	79,32 b A	50,245 b A
n-heptanal	17,18 a A	17,58 a A	17,815 a A	22,74 a A	35,605 b B	12,11 a A	14,56 a A	12,425 a A
benzaldehyde	40,635 a A	30,58 a A	69,695 a B	49,6 a A	47,49 b A	23,13 a A	38,46 a A	20,535 a A
2(E)-heptenal	60,495 a A	62,3 a A	70,55 a A	52,975 a A	79,2 b B	35,94 a A	27,945 a A	41,17 b A
1-octen-3-one	0 a A	1,65 b A	9,64 a B	9,145 a B	10,775 a B	7,89 a A	0 a A	4,32 b A
6-methyl-5-epten-2-one	462,965 a A	671,315 b A	600,45 a B	667,465 a A	1060,27 b B	590,135 a A	468,485 a A	568,795 b A
2(E),4(E)-eptadienal	11,29 a A	13,585 a A	14,335 a A	16,9 a A	17,96 b B	9,45 a B	1,03 a A	4,15 b A
2(E),4(Z)-eptadienal	0,695 a A	16,69 b A	8,275 a B	19,865 b A	21,075 b A	5,895 a A	1,465 a A	12,065 b B
2(E)-octenal	177,875 a A	226,565 b A	249,415 a B	203,04 a A	309,04 b B	103,905 a A	80,355 a A	140,865 b A
2,6-dimethyl-5-eptenal	2,015 a B	2,775 a A	0 a A	15,395 b B	1,995 a A	29,45 b B	28,26 b B	4,62 a A
2(E)-nonenal	46,9 a A	62,665 a A	62,485 a A	65,795 a A	90,15 a B	61,38 a A	40,22 a A	47,08 a A
6-methyl-3,5-eptadien-2-one	31,66 a A	49,775 a A	49,11 a A	65,81 a A	102,355 b B	53,435 a A	40,295 a A	42,365 a A
2(E),4(E)-decadienal	127,38 a A	121,955 a A	166,6 b A	109,84 a A	158,17 b B	90,825 a A	76,05 a A	62,655 a A
2(E),4(Z)-decadienal	115,605 a A	126,765 a A	176,355 a A	136,97 a A	136,255 a B	83,57 a A	66,605 a A	80,585 a A
Total Amount	1597,52	1961,82	1988,29	1957,02	2861,75	1380,03	1193,00	1312,06
HETEROCYCLIC DERIVATIVE								
2-isobutylthiazole	10,94 aA	265,52 b A	190,13 aB	791,57 b B	828,095 b B	323,195 a B	409,855 b A	37,2 a A
TERPENS								
2,3-epoxigeranial	50,55 a A	100,59 b A	77,7 a A	77,845 a A	163,03 b B	90,71 a A	70,99 a A	71,145 a A

Effects of Irrigation-Fertilization and Irrigation-Mycorrhization on the Alimentary
and Nutraceutical Properties of Tomatoes

201

neral	62,075 a A	89,89 a A	87,45 a A	88,755 a A	148,96 b B	80,985 a A	61,04 a A	73,7 a A
geranial	109,105 a A	208,765 b B	177,42 a A	151,89 a A	340,515 b B	194,81 a A	152,62 a A	140,59 a A
β-damascenone	61,66 a A	57,555 a A	80,175 a A	78,315 a A	111,885 a B	73,29 a A	43,3 a A	58,795 a A
neryl acetone	77,93 a A	85,335 a A	100,84 a A	90,43 a A	108,265 a B	82,385 a B	50,6 a A	49,06 a A
β(Z)-ionone	2,015 a A	7,895 b A	14,44 a B	16,485 a B	22,52 b B	9,475 a A	1,605 a A	10,03 b A
β(E)-ionone	36,83 a A	39,605 a A	38,93 a A	38,475 a A	44,79 a B	30,46 a A	21,635 a A	20,905 a A
(E,E)-pseudoionone	34,76 a A	51,73 a A	47,13 a A	55,225 a A	86,845 a A	55,575 a A	207,655 b B	35,395 a A
Total Amount	434,93	641,37	624,09	597,42	1026,81	617,69	548,41	459,62
PHENOLIC DERIVATIVES								
2- methoxyphenol	60,085 a A	113,435 b A	97,48 a A	103,905 a A	121,06 a B	89,41 a A	0,605 a A	60,72 b A
methyl salicylate	202,96 a A	319,375 b A	204,495 a A	248,93 a B	161,73 a A	170,735 a B	207,925 b A	85,78 a A
eugenol	49,32 b A	0 a A	47,96 a A	67,625 a B	54,56 b A	29,15 a B	29,58 b A	0 a A
Total Amount	312,37	432,81	349,94	420,94	337,35	289,30	238,11	146,50

Table 2. Quali-quantitative composition of identified volatile components.

Different letters indicate significant difference ($p \leq 0.05$). Small letters concern the statistical analysis within the same genotype, keeping as fixed variable the K-fertilization levels, while the capital letters concern the statistical analysis within the same genotype, keeping as fixed variable the watering regimes.

In order to better estimate the effects of two agronomical methods on the tomato flavour, all the identified volatile compounds were distinguished in the volatile characteristic compounds (hexanal, 2(E)-hexenal and 2-isobutylthiazole), responsible of the tomato impact aroma, and the contributor volatile flavour such as alcohols, carbonyl compounds (aldehydes and chetones), phenolic derivatives and terpens. Fig. 1 showed that this effect was more evident for the volatile characteristic compounds and particularly for 2-isobutylthiazole in the sample treated with the highest watering volumes and K-fertilization. On the contrary, in Vesuviano genotype, the volatile characteristic compounds had the highest concentration in the control. In fact the increasing watering volumes and K-levels caused a decrease of the values of the three compounds respect of the control, especially for 2-isobutylthiazole. The negative effects of the two cultural methods on the characteristic volatile compounds found in Vesuviano tomato were in accord to Veit-Khöler et al. (1999), who noted that an increase of watering regimes caused a decrease of six carbon atoms aldehydes as hexanal, 3(Z)-hexenal and 2(E)-hexenal in tomato samples.

Fig. 2 showed the influence of agronomical treatments on the contributor volatile compounds. In the Corbarino tomato, the increase of watering levels, with or without K-fertilization, caused a decrease of alcohols, an enhancement of carbonyl compounds and terpens; while the content of phenolic derivatives were quite similar in all the samples. In Vesuviano genotype, the unirrigated sample had the highest values of all the classes of contributor substances, especially of the alcohols. Besides, the K-fertilization caused a decrease of contributor compounds and this behaviour was more evident in the sample irrigated with the highest watering regime, except of carbonyl compounds.

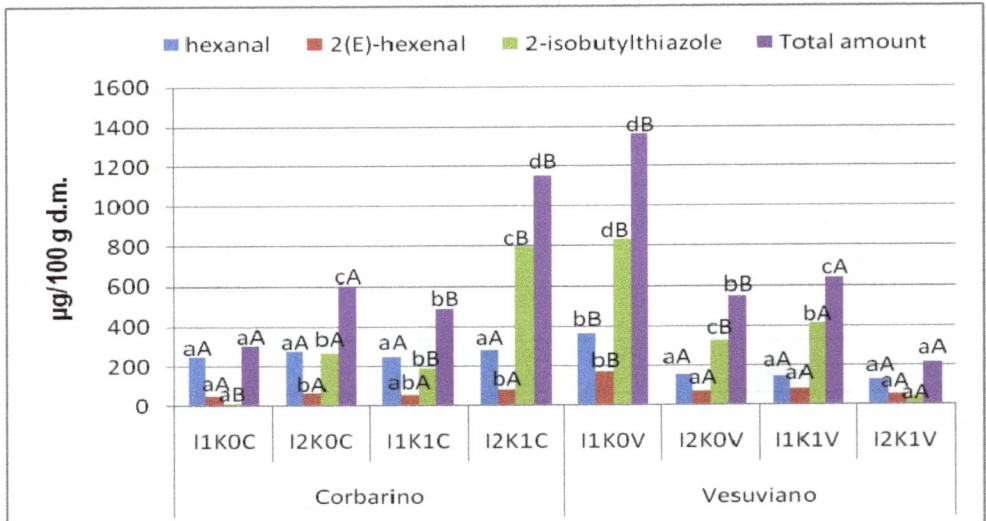

Small letters concern the statistical analysis within the same genotype, keeping as fixed variable the K-fertilization level, while the capital letters concern the statistical analysis within the same genotype, keeping as fixed variable the watering regimes.

Fig. 1. Characteristic volatile compounds in irrigated and K-fertilized "Corbarino" and "Vesuviano" tomato genotypes. Different letters indicate significant difference ($p \leq 0.05$).

In Table 3 the other parameters of alimentary quality of two genotypes were reported. Dry matter was always similar in all the samples of Corbarino genotype, but it decreased in the Vesuviano samples irrigated with the highest water levels, according to Fontes et al (2000) and Helyes et al. (2009), who observed a positive effect of K-fertilization at different watering regimes. The pH of both genotypes was not modified by different watering regimes and K-fertilization, according to Ghebbi Si-smail et al. (2007), even if Fontes et al. (2000) reported a decrease of pH with increasing K-levels. As concern the soluble sugars, glucose decreased in the samples irrigated with the highest levels only in Vesuviano genotype, according to Parisi et al. (2006) and Mitchell et al. (1991). On the contrary fructose was not affected by the two cultural methods in both genotypes.

As regards the organic acids, oxalic acid was not influenced by the cultural techniques in both genotypes, citric acid did not show any variation in all the Corbarino samples, while it was positively influenced by fertilization in Vesuviano genotype, according to Mitchell et al. (1991). Malic acid was negatively affected only by watering irrigation.

Effects of Irrigation-Fertilization and Irrigation-Mycorrhization on the Alimentary
and Nutraceutical Properties of Tomatoes

203

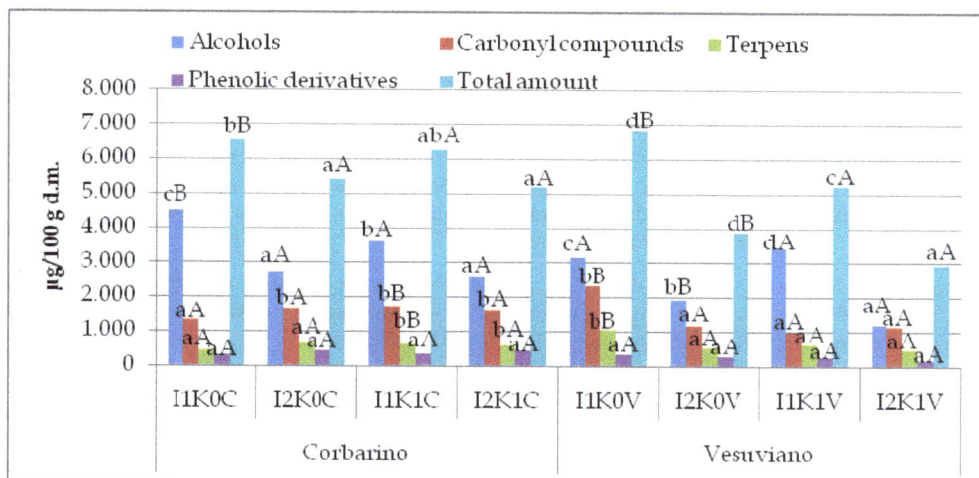

Small letters concern the statistical analysis within the same genotype, keeping as fixed variable the K-fertilization level, while the capital letters concern the statistical analysis within the same genotype, keeping as fixed variable the watering regimes.

Fig. 2. Contributor volatile compounds in irrigated and K-fertilized Corbarino and Vesuviano tomato genotypes. Different letters indicate significant difference ($\rho \leq 0.05$).

	d.m.%	pH	SOLUBLE SUGARS (g/100g p.f.)		ORGANIC ACIDS (mg/100g p.f.)		
			Glucose	Fructose	Oxalic	Citric	Malic
I_1K_0C	10.92 aA	3.92 aA	3.17 aA	2.67 aA	2.06 aB	477.95 aA	154.66 bB
I_2K_0C	10.12 aA	4.10 aA	3.15 aA	2.73 aA	2.02 aB	474.22 aA	123.84 aA
I_1K_1C	10.88 aA	4.24 aA	2.92 aA	2.66 aA	2.08 aB	487.13 aA	138.76 bB
I_2K_1C	10.49 aA	4.31 aA	3.18 aA	2.88 aA	2.01 aB	495.12 aA	118.17 aA
I_1K_0V	11.05 bB	3.93 aA	3.23 bB	2.87 aA	1.71 aA	457.92 aA	162.46 bB
I_2K_0V	10.00 aA	3.97 aA	2.79 aA	2.65 aA	1.73 aA	467.83 aA	136.45 aB
I_1K_1V	11.38 bB	3.82 aA	3.08 bB	2.77 aA	1.74 aA	582.81 bB	178.07 bB
I_2K_1V	10.56 aA	4.11 aA	2.78 aA	2.54 aA	1.74 aA	553.84 aB	155.70 aB

Different letters indicate significant difference ($\rho \leq 0.05$). Small letters concern the statistical analysis within the same genotype, keeping as fixed variable the K-fertilization level, while the capital letters concern the statistical analysis within the same genotype, keeping as fixed variable the watering regimes.

Table 3. Effects of irrigation and K- fertilization on chemical-physical parameters of two tomato genotypes.

3.1.3 Nutraceutical quality

Fig. 3 showed the content of total lycopene in two treated genotypes. In both Corbarino and Vesuviano tomato, the lycopene content increased in samples treated with the highest watering regimes, with or without K-fertilization. So the lycopene content in both genotypes seemed to be mainly influenced by watering regimes than fertilization. In literature, contrasting results about total lycopene in relation with irrigation regimes and K-fertilization are reported.

Achilea & Kafkafi (2003) evaluated, in a series of experiments performed in different countries, the specific contribution of potassium nitrate to yields and quality parameters of processing tomatoes. They noted that potassium nitrate was found to be the best form for maximum lycopene concentration in the fresh fruit.

In another experiment, Fontes et al. (2000) studied the effect of K-fertilizers rates on the fruit size, mineral composition and quality of the trickle-irrigated tomatoes. Lycopene and other analytical parameters in the fruits were not affected by K-rates.

On the contrary, Serio et al. (2007) determined the influence of K-levels in the nutrient solution on lycopene content of tomato plants grown in a soil-less system using rockwool slabs as substrate. Two growing seasons were studied with the aim of comparing three K-levels: low, medium and high (corresponding to 150, 300 and 450 mg K t^{-1} in the nutrient solution). The lycopene content increased linearly with increasing K-level in the nutrient solution.

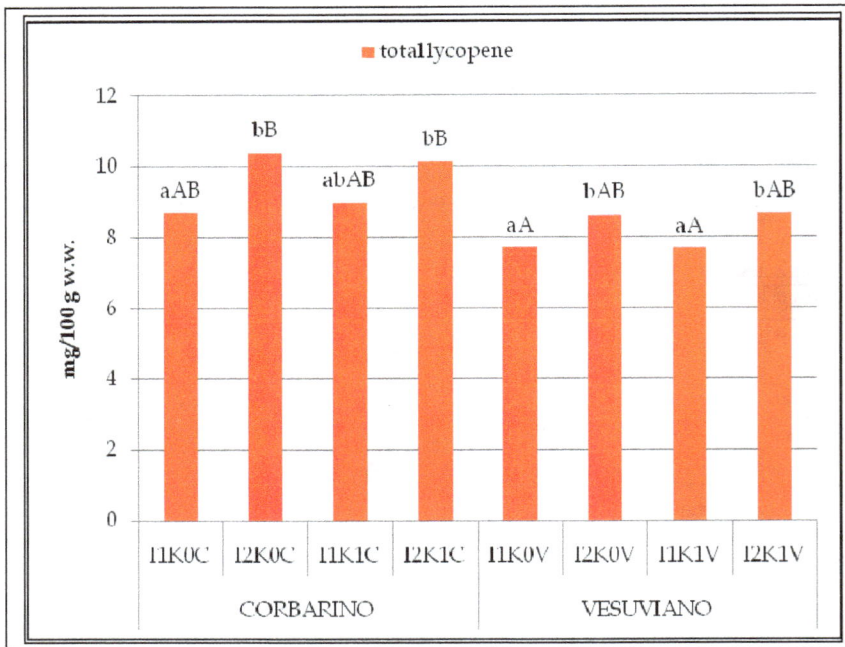

Fig. 3. Total lycopene in irrigated and K-fertilized Corbarino and Vesuviano tomato genotypes. Different letters indicate significant difference ($p \leq 0.05$). Small letters concern the statistical analysis within the same genotype, keeping as fixed variable the K-fertilization level, while the capital letters concern the statistical analysis within the same genotype, keeping as fixed variable the watering regimes.

Effects of Irrigation-Fertilization and Irrigation-Mycorrhization on the Alimentary
and Nutraceutical Properties of Tomatoes

205

Ghebbi Si-smail. et al. (2007) studied the effect of three levels of K-supply on the total lycopene content of two tomato cultivars, during two seasons in open field. The plants were grown in loamy soil poor in organic matter (1.18%) and in K (75 ppm K_2O). Results about lycopene concentration seemed to be negatively affected by K-fertilization.

Zdravković et al. (2007), in order to research whether lycopene production depended upon mineral macro-nutrient, used industrial tomato cv. Narvik SPF as a material. The content of lycopene regarding the fertilizer formulation was from 33.69 mg/kg in control to 56.92 mg/kg in plants treated with increased content of potassium.

Taber et al. (2008), studying the response of tomato cultivars with fruit of average and high lycopene to increased K-fertilization, demonstrated that K-fertilization could affect carotenoids biosynthesis, and the response of tomato to an high K-rate was genotype dependent.

Finally, Helyes et al. (2009) in an open field experiment, studied the effects of irrigation (regularly irrigated, irrigation cut-off 30 days before harvest, and unirrigated) and two K-supplementation (454 and 555 kg ha[-1] respectively) on lycopene and other analytical parameters of tomato fruit. The high K-rate increased the lycopene content of tomato in the irrigation cut-off and unirrigated treatments.

In our research, the scarce influence of K-fertilization on the lycopene biosynthesis, in contrast with other authors (Achilea & Kafkafi, 2003; Helyes et al., 2009; Serio et al., 2007; Zdravkovic et al., 2007;), could be explained with the high natural content of K in the soil, where the tested tomatoes were grown, as reported in materials and methods. On the other hand, this behaviour is in accordance with other authors (Dumas et al., 2003; Panagiotopoulos & Fordham, 1995;), who found a positive response of the lycopene content to fertilization levels only in tomato grown in soil treated with excessively high dose of fertilizers, in contrast to the modern cultural management, whose target is the reduction of the crops chemical input.

Besides, our researches showed that the lycopene content seemed not to be genotype-dependent because Corbarino and Vesuviano showed the same trend in relation with K-fertilization, in contrast with Taber et al. (2008).

The effect of watering supply on lycopene content of tomato was investigated by other authors. Brandt et al. (2003) found that the lycopene content of tomato harvested in greenhouse, supplied with 50% of optimal water intake, was higher than that of fruits supplied with 100% of optimal intake. Riggi et al. (2008) studied the influence of two watering regimes (a fully irrigated treatment receiving 100% of evapotranspiration for the whole growing season and an unirrigated control watered up to plant establishment only) on lycopene and β-carotene accumulation during fruit ripening in a field-grown processing tomato. Higher amounts of lycopene were measured in the well watered treatment. On the contrary, Favati et al. (2009) in a two-years research found an higher fruit content of carotenoids (lycopene and β-carotene) under water deficit irrigation respect to well irrigated processing tomatoes. In other studies, on the effects of four irrigation regimes (40, 50, 60 and 70% depletion of the available soil moisture) on three tomato cvs, Naphade (1993) ascertained that the lycopene content decreased in response to moisture stress, whereas Matsuzoe et al. (1998) in red and pink cherry tomato cvs found that lycopene increased when there were soil water deficits. In red and pink large-fruited tomatoes, soil water deficits also tended to increase the amount of lycopene in the region of the outer pericarp (Zushi and Matsuzoe, 1998). However, the effects of water availability seemed to need further studies, particularly in relation with the fruit environmental conditions (Dumas et al., 2003).

3.2 Effects of Irrigation – Mycorrhization Interactions

3.2.1 Agronomical and morphological data

Table 4 showed that the watering regimes had positively influenced the commercial yields. In fact, the total yield had highlighted an increase of 36.1 t/ha and 59.6 t/ha respectively for V_{50} and V_{100}. The higher productivity levels of the watered thesis were also accompanied by a significant increase in weight of the berries. Besides, the increase of water volumes caused an increase of waste berries, from 12.6% of the control (V_0) to 15.1% and 16.7% respectively in V_{50} and V_{100}.

	Yield		Fruit weight	Fruit firmness
FACTORS	Total (t/ha)	Waste (%)	(g)	(Kg/cm^2)
Watering regime (IR)				
V_0	47,6 a	12,6 a	65,6 a	1,0 a
V_{50}	83,7 b	15,1 b	93,2 b	1,0 a
V_{100}	107,2 c	16,7 c	104,6 c	0,9 a
Signif.	**	**	**	*n.s.*
Mycorrhization (M)				
M_1	82,5 a	14,3 a	90,4 a	1,01 a
M_2	81,7 a	14,5 a	88,5 c	0,95 a
Signif.	*n.s.*	*n.s.*	*n.s.*	*n.s.*
Interactions				
(IR) x (M)	*n.s.*	*n.s.*	*n.s.*	*n.s.*

** P=0.01; *P=0.05; n.s.= not significant.

Table 4. Effects of irrigation/Mycorrhization on agronomical traits of Faino F_1 cultivar.

The yields of the mycorrhized thesis with Micosat F and VAM did not show significant differences and the yields of the two mycorrhizae had similar values in comparison with unmycorrhized and irrigated samples (V_{50}).

Take into account the interaction between irrigation × mycorrhization, the agronomical parameters had not statistically significant differences. As concern firmness, the reported values were similar among all the studied theses, irrigated, mycorrhized and irrigated/mycorrhized samples.

Effects of Irrigation-Fertilization and Irrigation-Mycorrhization on the Alimentary
and Nutraceutical Properties of Tomatoes

207

3.2.2 Alimentary quality

Table 5 showed the composition and the values ($\mu g/100$ g d.m.) of tomato aromatic profiles treated with different watering regimes and mycorrhizae. The single volatile compounds were identified in all the analyzed thesis and the influence of the agronomical treatments were only ascertained for their content. Even in this section, as for the irrigation/K fertilization one, the volatile compounds were subdivided in characteristic and contributor volatile compounds of the tomato aroma.

	(μg/100g d.m.)								
	V0	V50	V100	M1V0	M1V50	M1V100	M2V0	M2V50	M2V100
ALCOHOLS									
1-penten-3-ol	11,99 b	3,08 a	12,54 b	3,06 a	3,14 a	4,05 a	3,14 a	4,87 a	5,09 a
3-methyl-1-butanol	400,27	450,16 bc	592,53 c	248,48 b	363,73 bc	551,36	138,20 a	214,98 b	246,33 b
2-methyl-1-butanol	70,12 ab	150,12 c	28,32 a	84,66 ab	111,66 b	100,24 ab	22,37 a	28,84 a	28,02 a
1-pentanol	24,00 bc	10,65 b	42,96 d	5,58 a	18,96 bc	23,03 bc	4,88 a	7,99 a	10,21 b
3(Z)-hexen-1-ol	n.d.	n.d. a	n.d. a	n.d. a	2,15 a	2,12 b	n.d. a	3,15 bc	5,12 c
1-hexanol	n.d.	n.d.	n.d.	n.d.	3,08 b	n.d.	n.d.	5,03 c	4,15 c
6-methy-5-hepten-2-ol	28,85 c	47,34	34,38 bc	5,12 a	15,14 b	32,23 bc	2,19 a	3,93 a	4,81 a
benzyl alcohol	116,21 b	162,87 c	142,27 c	71,19 a	127,74 d	103,74 d	52,80 a	50,70 a	45,46 a
phenethyl alcohol	105,19 d	173,45 de	217,29 de	127,17 b	135,05 d	148,22 d	76,86 c	43,02 b	13,99 a
Total amount	**756,63 b**	**997,67 c**	**1070,29 c**	**545,26 ab**	**780,65 b**	**964,99 c**	**300,44 a**	**362,51 a**	**363,18 a**
CARBONYL COMPOUNDS									
3-methylbutanal	62,76 bc	71,51 c	83,76 d	71,58 c	64,44 bc	57,74 b	28,13 a	41,62 b	45,15 b
3-methyl-3-buten-2-one	18,95 b	3,15 a	31,25 c	4,38 a	1,13 a	35,54 c	1,15 a	24,40 bc	22,38 b
1-penten-3-one	2,15 a	2,05 a	5,22 c	2,15 a	2,88 ab	3,12 b	2,16 a	3,58 b	3,12 b
pentanal	3,12 a	5,74 b	15,62 c	4,12 ab	2,97 a	5,20 b	3,59 a	3,84 ab	4,87 b
benzaldehyde	13,20 b	21,92 c	12,65 b	8,87 ab	24,49 c	14,88 b	3,16 a	7,59 ab	5,18 a
2(E)-heptenal	40,40 c	35,77 d	33,60 c	19,77 b	38,20 c	25,86 bc	5,12 a	4,37 b	6,15 a
6-methyl-5-epten-2-one	308,56 c	438,03 d	310,46 c	182,92 b	261,43 c	310,69 c	99,84 a	150,42 a	200,59 b
2(E)-octenal	41,29 b	8,62 a	46,81 b	2,46 a	32,20 b	38,00 b	3,88 a	5,26 a	2,88 a
2,6-dimethyl-5-eptenal	22,73 cd	31,60 b	25,51 cd	4,55 b	19,90 cd	5,03 b	2,16 a	2,37 a	4,35 b
6-methyl-3,5-eptadien-2-one	17,14 c	22,48 c	17,32 c	3,75 a	3,71 a	2,75 a	2,09 a	7,14 b	5,04 ab
2(E)-nonenal	78,15 c	131,36 c	97,38 bc	31,74 ab	79,90 b	69,81 b	24,22 a	21,32 a	20,84 a
2(E),4(E)-decadienal	34,25 b	38,15 b	54,94 c	4,12 a	2,87 a	32,75 b	3,12 a	4,72 a	2,18 a

2(E),4(Z)-decadienal	45,51 cd	50,91 cd	64,01 d	21,29 c	35,42 c	38,51 c	6,14 b	5,00 b	1,15 a
Total amount	688,21	861,29 d	798,53 d	361,70 bc	569,54 c	639,88 cd	184,76 a	281,63 b	323,88 bc
TERPENS									
linalool	113,94	161,51 c	136,51 c	n.d. a	132,72 c	96,89 b	n.d. a	n.d. a	n.d. a
a-terpineol	n.d.	n.d.	n.d.	n.d.	n.d.	n.d.	n.d.	n.d.	n.d.
2,3-epoxigeranial	33,39	54,40 c	8,04 a	22,73 b	27,15 b	29,80 b	8,15 a	7,38 a	5,12 a
neral	40,51	63,45 c	37,60 b	25,88 ab	39,82 b	37,87 b	11,18 a	8,99a	6,12 a
geranial	75,97 b	110,66 c	95,83 c	37,60 ab	64,64 b	73,50 b	22,90 a	24,09 a	34,59 ab
β-damascenone	52,42	70,96 d	67,10 d	29,75 b	42,60 c	41,90 c	15,94 ab	9,54 a	4,15 a
neryl acetone	29,34 b	57,08 c	68,11 c	17,65 b	40,39	46,89	5,12 a	20,71 ab	33,05 b
β(Z)-ionone	7,08 a	34,86 c	149,65 d	3,88 a	35,18 c	26,00 b	28,35 b	24,24 b	25,43 b
β(E)-ionone	20,29 bc	35,49 d	40,53 d	2,16 a	20,75 bc	21,47 bc	15,16 b	17,32 b	10,16 b
(E,E)-pseudoionone	22,95 b	27,00 bc	33,06 c	4,14 a	15,44 b	18,83 b	5,04 a	20,08 b	4,74 a
Total amount	395,89 b	615,41 c	636,43 c	143,79 ab	418,69 b	393,15 b	111,84 a	132,35 ab	123,36 a
PHENOLIC DERIVATIVES									
2- methoxyphenol	4,45b	9,15 c	26,23 d	2,88 a	2,01 a	4,84 b	1,08 a	28,60 d	1,18 a
eugenol	119,19 c	69,18 b	132,02 c	60,51 b	58,73 b	36,64 ab	20,12 a	66,91 b	80,81 bc
methyl salicylate	42,82 b	103,99 c	216,55 d	5,14 a	164,21 cd	113,79 c	5,14 a	196,54 d	44,48 b
Total amount	166,46 b	182,32 b	374,80 d	68,53 b	224,95 c	155,27 b	26,34 a	292,05	126,47 b

Table 5. Quali-quantitative composition of identified volatile components in Faino F_1 tomatoes. n.d.= not detected.

As concern the characteristic volatile compounds, Fig. 4 showed that hexanal had the highest value in unmycorrhized samples V_{50} and V_{100} with respect to the untreated sample (V_0). A constant content of 2(E)-hexenal was noted for unmycorrhized and irrigated samples. 2-isobutylthiazole, among the same samples, had the highest concentration in V_{50} and the lowest in V_{100}. In the irrigated and treated with two different kinds of mycorrhization samples, a decrease of all the characteristic volatile compounds was noted, especially for M_2 treatment. If the total content of characteristic volatile compounds was considered, the sample with the highest content was V_{50}, while M_1V_0 and M_2V_0 showed the lowest levels.

Fig. 5, concerning the contributor volatile compounds, showed that in the unmycorrhized irrigated samples alcohols, carbonyl compounds and terpenes had the highest content in V_{50} and V_{100}, except for phenolic derivatives, only in V_{50}. Besides, in the irrigated and mycorrhized samples, contributor volatile compounds were always lower than unmycorrhized and irrigated samples. This trend was more evident when the total content of contributor volatile compounds was considered.

Effects of Irrigation-Fertilization and Irrigation-Mycorrhization on the Alimentary
and Nutraceutical Properties of Tomatoes

209

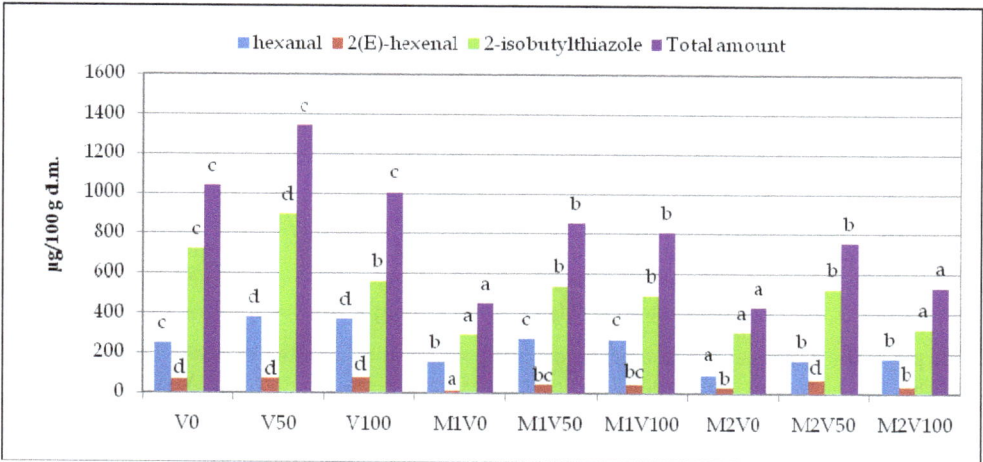

Different letters indicate significant difference (ρ≤ 0.05).

Fig. 4. Effects of irrigation and mycorrhization on characteristic volatile compounds of Faino
F_1 tomatoes.

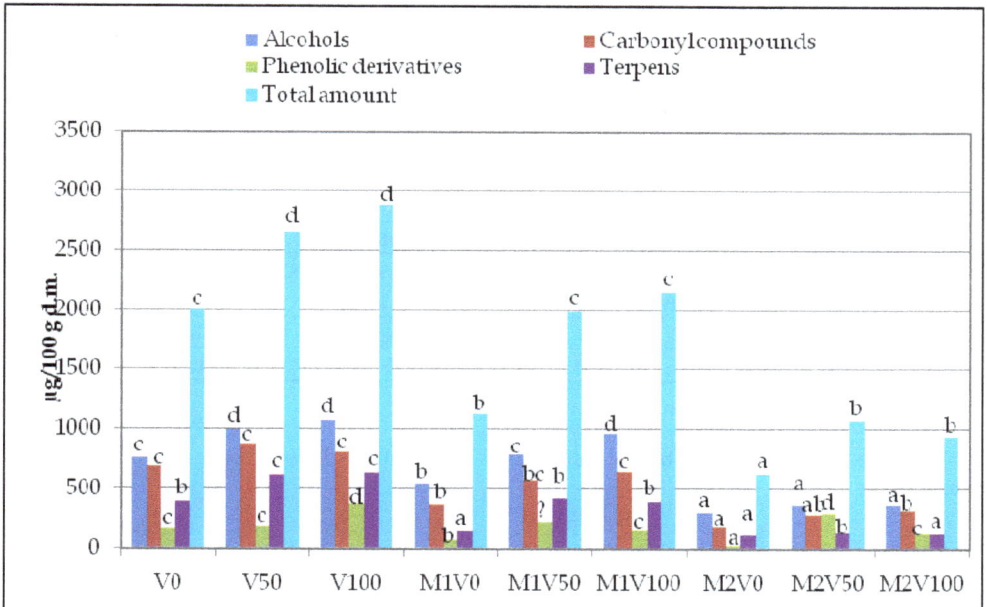

Different letters indicate significant difference (ρ≤ 0.05).

Fig. 5. Effects of irrigation and mycorrhization on contributor volatile compounds of Faino
F_1 tomatoes.

The other alimentary quality parameters were reported in Table 6.

The dry matter (d.m.) showed similar values in the irrigated and unmycorrhized samples. Among the samples treated with M_1, the value of M_1V_0 was slightly higher than the other samples (M_1V_{50} and M_1V_{100}), but with little statistically significant differences. Among all examined thesis, M_2V_0 had the highest d.m. value, while M_2V_{100} the lowest one. So, a light positive influence of mycorrhization system, and particularly of M_2, was noted in the unirrigated samples.

Among the unmycorrhized samples, V_0 had the lowest value of pH, while V_{100} the highest one. A similar trend was observed for irrigated and mycorrhized samples. Among all the samples, the lowest pH was noted for M_2V_0, while the highest values in V_{100} and M_1V_{100}.

Since pH is the expression of the organic acids content, the same table showed that to higher organic acids content corresponded lower pH, as demonstrated for M_2V_0. These two parameters are very important in processed tomatoes, because a pH=4.3 ensures a microbiological stability during the conservation in the sterilized products (Silvestri & Siviero, 1991).

As concern soluble sugars, both glucose and fructose showed the highest content in the unirrigated sample (V_0) and mycorrhized samples (M_1V_0 and M_2V_0). Furthermore, their content was always higher in mycorrhized sample than untreated one.

| | d.m. (%) | pH | SOLUBLE SUGARS (g/100 g w.w.) | | | ORGANIC ACIDS (mg/100 g w.w.) | | | |
			Glucose	Fructose	Total amount	Oxalic	Citric	Malic	Total amount
V_0	7,05 a	4,38 a	1,22 b	1,42 b	2,64 b	3,72 a	501,94 bc	30,58 b	536,23 bc
V_{50}	6,88 a	4,44 a	1,10 ab	1,47 b	2,58 b	4,86 b	460,63 b	18,41 a	483,89 b
V_{100}	7,15 b	5,66 b	1,00 a	1,31 ab	2,31 ab	3,94 ab	454,88 b	17,65 a	476,47 b
M_1V_0	7,21 b	4,35 a	1,44 b	1,56 b	3,00 b	3,79 a	548,36 c	32,24 b	584,39 bc
M_1V_{50}	6,81 a	4,86 a	0,91 a	1,19 a	2,10 a	3,86 a	454,36 b	24,91 ab	483,12 b
M_1V_{100}	6,98 a	5,33 b	0,91 a	1,21 a	2,12 a	3,69 a	403,73 a	17,08 a	424,49 a
M_2V_0	7,63 b	4,02 a	1,35 b	1,53 b	2,88 b	4,11 b	586,38 c	27,89 b	618,37 c
M_2V_{50}	7,21 b	4,42 a	0,89 a	1,14 a	2,03 a	3,94 ab	424,04 b	16,93 a	444,90 ab
M_2V_{100}	6,32 a	4,34 a	0,91 a	1,20 a	2,11 a	3,95 ab	457,33 b	21,37 ab	482,64 b

Different letters indicate significant difference ($\rho \leq 0.05$).

Table 6. Effects of mycorrhization and irrigation on chemical-physical parameters of Faino F_1 tomatoes.

3.2.3 Nutraceutical quality

From Fig. 6 it could be deduced that in unmycorrhized and irrigated samples (V_{50} and V_{100}) the total lycopene content significantly increased, in comparison with the control, respectively of 6,58% and 18,54% and this increase was directly proportional to the watering volume. Among the unirrigated samples, M_1V_0 presented the same content respect of V_0, while M_2V_0 showed a deep decrease (-31%). Utilizing the mycorrhizae with a watering regimes corresponding to 50%, the lycopene content decreased both in M_1V_{50} (-18%) and M_2V_{50} (-45%) with respect to V_{50}. In the full irrigated samples (V_{100}), the lycopene content in M_1V_{100} and M_2V_{100} showed a decrease of about -53% in comparison with V_{100}.

From reported data, it could be deduced that the highest concentration of lycopene was obtained in fully irrigated samples without mycorrhization. This trend was more evident for M_2 mycorrhized samples.

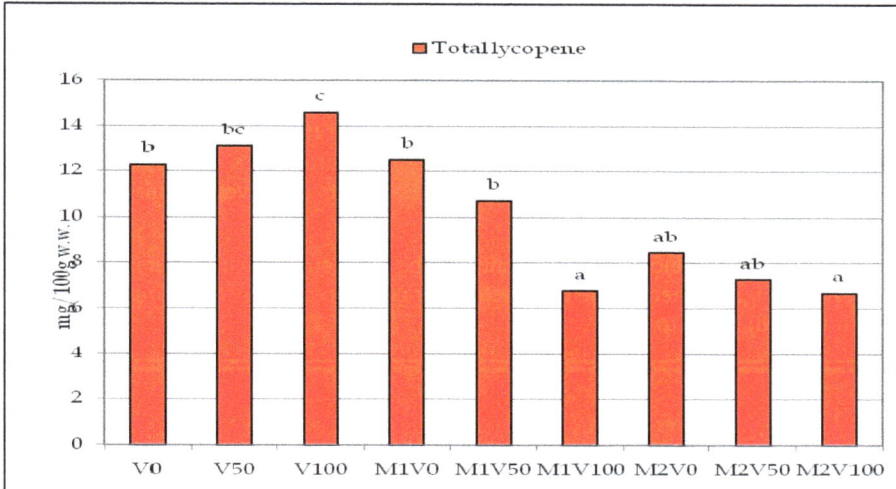

Different letters indicate significant difference ($\rho \le 0.05$).

Fig. 6. Effects of irrigation and mycorrhization on total lycopene content of Faino F_1 tomatoes.

Recent studies pointed out that lycopene, as well as other carotenoids, are synthesized from isoprenoids that are products of mevalonic acid (MVA) and methylerythritol-4-phosphate (MEP) pathway (Botella-Pavia & Rodriguez-Concepcion, 2006). The use of M_1 and M_2 mycorrhizae could most likely decrease the action of enzymes involved in these pathways such as deoxyxylulose-5-phosphate synthase. This hypothesis seems to be the most plausible to explain the negative interaction of mycorrhization vs. lycopene synthesis, because of the scarce available literature about this behaviour.

However our results about the negative effect of mycorrhization on lycopene content were in contrast with other authors. Ordookhani & Zare (2011) investigated the effects of inoculating tomato (*Lycopersicon esculentum* F1 Hybrid, Delba) roots with plant growth-promoting rhizobacteria (PGPR) and Arbuscular Mycorrhiza Fungi (AMF) on lycopene and other parameters of tomato fruit. PGPR treatments were inoculated with *Pseudomonas putida, Azotobacter chroococcum* and the AMF treatment was inoculated with *Glomus mossea*. In comparison to the untreated sample, lycopene of fruit was increased by PGPR and AMF treatments. In the PGPR × AMF treatment maximum lycopene were found in plants of the *Pseudomonas + Azotobacter + AMF* treatment. In the PGPR treatment maximum lycopene were found in plants of the *Pseudomonas + Azotobacter* treatment. Data showed that lycopene increased when AMF added to PGPR treatments. Probably, the different behaviour of lycopene vs. Mycorrhization could depend on different composition of the mycorrhizae.

On the other hand, a positive effect of high watering regimes on lycopene content was in accordance with the other authors, as reported in the section relative to the Irrigation/ K-fertilization.

4. Conclusion

It can be deduced that agronomic parameters (yields, waste %, fruit weight etc.) are influenced by agronomic treatments when are separately analysed. Irrigation gives a better commercial yield and higher weight of berries in all examined genotypes. A different behaviour is showed by the other experimental treatments (K-fertilization and mycorrhization). Besides, poor effects on the same parameters were observed for the interactions between the different agronomical treatments and even in this case they seem not to depend on the genotype.

The influence of the agronomical treatments on the biochemical parameters can be detected from Fig. 7 and Fig. 8, where these parameters are compared against the control (I_1K_0 for fig. 7 and V_0 for Fig. 8) equal to 100.

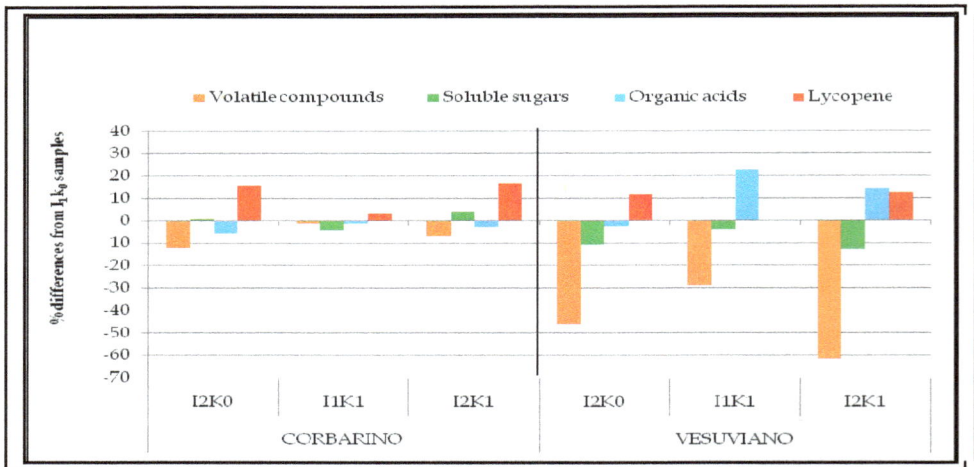

Fig. 7. Comparison of alimentary and nutraceutical parameters of "Corbarino" and "Vesuviano" tomatoes cultivated with different watering regimes and different K-fertilization levels. The percent changes with respect to the control (I_1K_0), assumed equal to 100%, are reported.

In Corbarino and Vesuviano tomato, the irrigation and irrigation/K-fertilization can reduce the flavour synthesis; this effect is more evident in Vesuviano tomato.

In Faino (F1) tomato hybrid, the irrigation enhances the volatile compounds content, while, in combination with mycorrhization, it cause a decrease of the volatile substances, especially with M2 (VAM).

Effects of Irrigation-Fertilization and Irrigation-Mycorrhization on the Alimentary
and Nutraceutical Properties of Tomatoes

213

Autochthon tomato varieties, as Corbarino and Vesuviano, are already adapted to the arid climate of the Vesuvius volcano slopes, characterized by low rainfall and difficulties in water supply. In this case, organoleptic characteristic are enhanced when they are cultivated in soil where the lonely water support is represented by rainfall. Besides, in new genotypes, as well as Faino hybrid tomato, the same characteristic are enhanced by high watering regimes. The interaction, as irrigation/K-fertilization and irrigation/mycorrhization, seems to have poor effects on aromatic profiles. A correlation among agronomical treatments and their effects on sugars and organic acids is very difficult. However, the lonely positive effect, on sugar and organic acids content, due to mycorrhization, is evident only in rainfall watered fields.

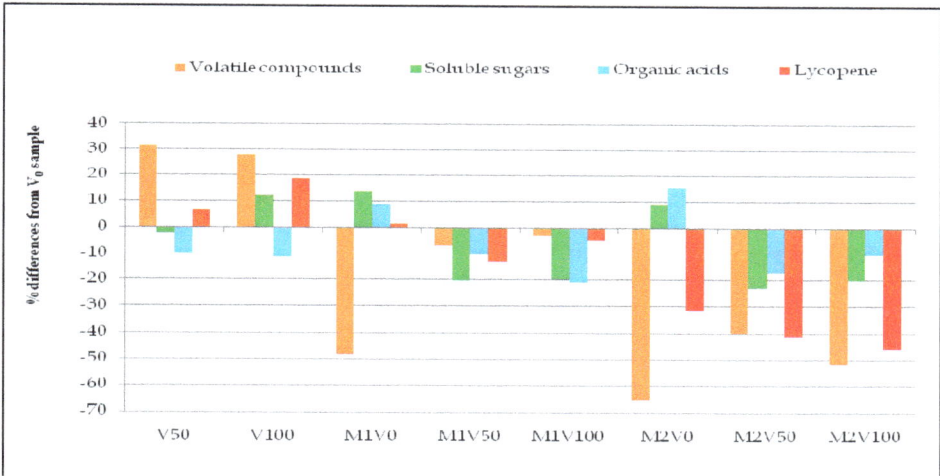

Fig. 8. Comparison of alimentary and nutraceutical parameters of " Faino F_1 tomatoes. The percent changes with respect to the control (V_0) ,assumed equal to 100%, are reported.

As concern the nutraceutical quality, high watering regimes can cause an increase of lycopene in three genotypes. Moreover, in Corbarino and Vesuviano tomato, irrigation x K-fertilization considerably enhances the lycopene content in the theses with the highest level of both irrigation and fertilization. On the contrary, a considerable decrease of the lycopene content is noted in the Faino tomato in the theses with irrigation x mycorrhization.

In conclusion, the irrigation seems to be the best agronomical treatment to enhance the organoleptic, nutraceutical and agronomical characteristics of tomatoes, regardless of genotype and origin countries. Besides, the interactions of irrigation with K-fertilization and mycorrhization cause poor effects on biochemical and agronomical characteristics.

5. References

Achilea, O. & Kafkafi, U. (2003). Enhanced performance of processing tomatoes by potassium nitrate-based nutrition. *Acta horticulturae A.*, No. 616, pp. 81-87, ISSN 0567-7572.

Akhtar, M.S. & Siddiqui, Z.A. (2008). Arbuscular mycorrhizal fungi as potential bioprotectants against plant pathogens. In: *Mycorrhizae: Sustainable Agriculture and Forestry* (Siddiqui Z.A., Akhtar M.S. and Futai K. Eds), Springer Netherlands. pp. 61-97, ISBN 978-1-4020-8769-1.

Azcon-Aguilar, C. & Barea, J.M. (1997). Applying mycorrhiza biotechnology to horticulture: significance and potentials. *Scientia Horticulturae*, Vol. 68, No. 1-4, (March 1997), pp. 1-24, ISSN 0304-4238.

Bergmann, W. (1992). Causes, development and diagnosis of symptoms resulting from mineral. Macronutrients: Potassium. In: *Nutritional Disorders of Plants: Development, Visual and Analytical Diagnosis.* (Bergmann, W. and Fischer, G., Eds.). Gustav Fisher Verlag, Jena, Germany.

Boileau, T.W.M.; Boileau, A.C. & Erdman, J.W. jr. (2002). Bioavailability of all-trans and cis-isomers of lycopene. *Experimental Biology and Medicine*, Vol. 227, No. 10, (November 2002), pp. 914-919, ISSN: 1535-3702.

Botella-Pavía, P. & Rodríguez-Concepción, M. (2006). Carotenoid biotechnology in plants for nutritionally improved foods. *Physiologia Plantarum*, Vol. 126, No. 3, (March 2006), pp. 369–381, ISSN 0031-9317.

Brandt, S.; Lugasi, A.; Barna, E.; Hovari, J.; Pek, Z. & Helyes, L. (2003). Effects of the growing methods and conditions on the lycopene content of tomato fruits. *Acta Alimentaria*, Vol. 32, No. 3, (August 2003), pp. 269-278, ISSN 0139-3006.

Branthôme, X.; Plè, Y. & Machado, J.R. (1994). Influence of drip-irrigation on the technological characterisation of processing tomatoes. *Acta Horticulturae*, Vol. 376, pp. 285-290, ISSN 0567-7572.

Buttery, R. G.; Teranishi, R.; Flath, R. A. & Ling, L. C. (February 1989). Fresh tomato volatiles. Composition and sensory studies. In *Flavor Chemistry, Trends and Developments*; Teranishi, R., Buttery, R. G., Shahidi, F., Eds.; ACS Symposium Series 388; ISBN 9780841215702, American Chemical Society: Washington, DC, pp. 213-222.

Buttery, R. & G.; Ling, L. C. (1993). Volatile components of tomato fruit and plant parts. Relationship and biogenesis. In *Bioactive Volatile Compounds from Plants*; Teranishi, R., Buttery, R. G., Sugisawa, H., Eds.; ACS Symposium Series 525; ISBN 9780841226395 American Chemical Society: Washington, DC, pp. 23-34.

Candido, V.; Miccolis, V. & Perniola, M. (2000). Effects of irrigation regime on yield and quality of processing tomato (*Lycopersicon esculentum* Mill.) cultivars. *Acta Horticulturae*, Vol. 537, No. 2, pp. 779-788. ISSN: 0567-7572

Colla, G.; Battistelli, A.; Moscatello, S.; Proietti, S.; Casa, R.; Lo Casco, B. & Leoni, C. (2001). Effect of reduced irrigation and nitrogen fertirigation rate on yield, carbohydrate accumulation, and quality of processing tomatoes. *Acta Horticulturae*, Vol. 542, pp. 187-196, ISSN 0567-7572.

Colla, G.; Casa, R.; Lo Cascio, B.; Saccardo, F.; Temperini, O. & Leoni C. (1999). Responses of processing tomato to water regime and fertilization in Central Italy. *Acta Horticulturae*, Vol. 487, pp. 531-536, ISSN 0567-7572.

Conversa, G.; Elia, A. & La Rotonda, P.; (2007). Mycorrhizal Inoculation and Phosphorus Fertilization Effect on Growth and Yield of Processing Tomato. *Acta Horticulturae*, Vol. 758, pp. 333-338, ISSN 0567-7572.

Effects of Irrigation-Fertilization and Irrigation-Mycorrhization on the Alimentary
and Nutraceutical Properties of Tomatoes

215

Copetta, A.; Lingua, G.; Berta, G.; Bardi, L. & Masoero, G. (2006). Three arbuscular mycorrhizal fungi differently affect growth, distribution of glandular trichomes and essential oil composition, appreciated by NIR spectroscopy and electronic nose, in *Ocimum basilicum* var. Genovese. *Acta Horticulturae,* Vol. 723, pp. 151-156, ISSN 0567-7572.

Davies, Jr. F.T.; Olalde-Portugal, V.; Aguilera-Gomez, L.; Alvarado, M.J.; Ferrera-Cerrato, R.C. & Boutton, T.W. (2002). Alleviation of drought stress of Chile ancho pepper (Capsicum annuum L. cv. San Luis) with arbuscolar mycorrhiza indigenous to Mexico. *Scientia Horticulturae,* Vol. 92, No. 3, (February 2002), pp. 347-359, ISSN 0304-4238.

De la Peña, E.; Rodríguez Echeverría, S.; van der Putten, W.H.; Freitas, H. & Moens M. (2006). Mechanism of control of root-feeding nematodes by mycorrhizal fungi in the dune grass *Ammophila arenaria. New Phytologist,* Vol. 169, pp. 829-840, ISSN 0028-646X.

Di Cesare, L.F.; Forni, E.; Viscardi, D. & Ferrari, V. (2003). Valutazione della composizione aromatica e chimico-fisica di alcune cultivar di pomodoro da impiegare per il consumo diretto o per la trasformazione. *Industria Conserve,* Vol. 78, No. 2, pp. 195-206, ISSN 0019 7483.

Dirinck, P.; Schreyen, L.; van Wassenhove, F. & Schamp, N. (1976). Flavour quality of tomatoes. *J Sci Food Agric.,* Vol. 27, No 6, (June 1976), pp. 499-508, ISSN 0022-5142.

Dorais, M.; Papadopoulos, A.P. & Gosselin A. (2001). Greenhouse tomato fruit quality: the influence of environmental and cultural factors. *Hortic Rev,* Vol. 26, pp.239–319, ISSN 0163-7851.

Dorais, M. (2007). Effect of cultural management on tomato fruit health qualities. FAV Health (2005). QC, Canada. *Acta Horticulturae,* Vol. 744, pp. 279–293, ISSN 0567-7572.

Dorais, M.; Ehret, D. L. & Papadopoulos, A. P. (2008). Tomato (*Solanum lycopersicum*) health components: from the seed to the consumer. *Phytochem Rev,* Vol. 7, pp. 231–250, ISSN 1568-7767.

Dumas, Y.; Leoni, C.; Portas, C. A. M. & Bièche, B. (1994). Influence of water and nitrogen availability on yield and quality of processing tomato in the European Union Countries. *Acta Horticulturae,* Vol. 376, pp. 185-192, ISSN 0567-7572.

Dumas, Y.; Dadomo, M.; Di Lucca, G. & Grolier P. (2003). Effects of environmental factors and agricultural techniques on antioxidant content of tomatoes. *J. Sci. Food Agric.,* Vol. 83, No. 5, (April 2003), pp. 369-382, ISSN 0022-5142.

Elia, A.; Conversa, G. & La Rotonda, P. (2006). Effect of mycorrhizal inoculation and P rates on yield and quality of processing tomato. *Abstracts V International Conference on Mycorrhyza "Mycorrhiza for Science and Society",* Granada (Spagna), 23-27 July 2006, pp. 217.

Elsen, A.; Gervacio, D.; Swennen, R. & De Waele, D. (2008). AMF-induced biocontrol against plant parasitic nematodes in *Musa* sp.: a systemic effect. *Mycorrhiza,* Vol. 18, pp. 251-256, ISSN 0940-6360.

Favati, F.; Lovelli, S.; Galgano, F.; Miccolis, V.; Di Tommaso, T. & Candido, V. (2009) Processing tomato quality as affected by irrigation scheduling. *Scientia Horticulturae,* Vol. 122, No. 4, (November 2009), pp. 562–571, ISSN: 0304-4238.

Fontes, P.C.R.; Sampalo, R.A. & Finger, F.L. (2000). Fruit size, mineral composition and quality of trickle-irrigated tomatoes as affected by potassium rates. *Pesq. Agropec. Bras. Brasilia*, Vol. 35, No. 1, (January 2000), pp. 21-25, ISSN 0100-204X.

Gera Hol, W.H. & Cook, R. (2005). An overview of arbuscular mycorrhizal fungi-nematode interactions. *Basic and Applied Ecology*, Vol. 6, pp. 489-503, ISSN 1439-1791.

Gerster, H. (1997). The potential role of lycopene for human health. *Journal of American college of Nutrition*, Vol.16, pp. 109-126, ISSN 0731-5724.

Ghebbi Si-smail, K.; Bellal, M. & Halladj, F. (2007). Effect of potassium supply on the behaviour of two processing tomato cultivars and on the changes of fruit technological characteristics. *Acta Horticulturae*, Vol. 758, pp. 269-274, ISSN 0567-7572.

Gianinazzi, S.; Oubaha, L.; Chahbandar, M.; Blal, B. & Lemoine, M.C. (2003). Biotization of microplants for improbe performance. *Acta Horticulturae*, Vol. 625, pp. 165-172, ISSN 0567-7572.

Giordano, I.; Pentangelo, A.; Villari, G.; Fasanaro, G. & Castaldo D. (2000). Caratteristiche bio-agronomiche e idoneità alla trasformazione di pomodori dell'ecotipo "Corbarino". *Industria Conserve*, Vol. 75, pp. 317-329, ISSN 0019 7483.

Giovannucci, E.; Ascherio, A.; Rimm, E.B.; Stampfer, M.J.; Colditz, G.A. & Willett, W.C. (1995). *Journal of National Cancer Institute*, Vol. 87 (special issue), pp. 1767-1776, ISSN 0027-8874.

Gosling, P.; Hodge, A.; Goodlass, G. & Bending, G.D. (2006). Arbuscular mycorrizal fungi and organic farming. *Agriculture, Ecosistem and Environment*, Vol. 113, pp. 17-35, ISSN 0167-8809.

Helyes, L.; Dimény, J.; Bőcs, A.; Schober, G. & Pék, Z. (2009). The effect of water and potassium supplement on yield and lycopene content of processing tomato. *Proceedings of the XIth International Symposium on the Processing Tomato: Toronto, Canada, June 9-11 2008.* Leuven, Belgium: International Society for Horticultural Science. In *Acta Horticulturae*, No.823, pp, ISSN 0567-7572.

Jacquot, E.; van Tuinen, D.; Gianinazzi, S. & Gianinazzi-Pearson, V. (2000). Monitoring species of arbuscular mycorrhizal fungi in planta and in soil by nested PCR: application to the study of the impact of sewage slude. *Plan. and Soi.*, Vol. 226, pp. 179-188, ISSN 0032-079X.

Matsuzoe, N.; Zushi, K. & Johjima, T. (1998). Effect of soil water deficit on coloring and carotene formation in fruits of red, pink and yellow type cherry tomatoes. *J Jpn Soc Hortic Sci*, Vol. 67, No. 4, pp. 600–606, ISSN: 0013-7626.

Mitchell, J.P.; Shennan, C.; Grattan, S.R. & May, D.M. (1991). Tomato fruit yields and quality under water deficit and salinity. *Journal of the American Society for Horticultural Science*, Vol. 116, No. 2, pp. 215-221, (Mar 1991), ISSN 0003-1062.

Morton, J. N. & Bentivenga, S.P. (1994). Levels of diversity in endomycorrhyzal fungi (*Glomales, Zygomycetes*) and their role in defining taxonomic and non-taxonomic groups. *Plan. and Soi.*, Vol. 159, pp. 47-59, ISSN 0032-079X

Naphade, A.S. (1993).Effect of water regime on the quality of tomato. *Maharashtra J Hortic* Vol. 7, No.2, pp. 55–60, ISSN 0970-2873

Oded, A. & Uzi, K. (2003). Enhanced performance of processing tomatoes by potassium nitrate based nutrition. *Acta Horticulturae*, Vol. 613, pp. 81–87, ISSN 0567-7572.

Effects of Irrigation-Fertilization and Irrigation-Mycorrhization on the Alimentary
and Nutraceutical Properties of Tomatoes
217

Ordookhani, K. & Zare M. (2011). Effect of Pseudomonas, Azotobacter and Arbuscular Mycorrhiza Fungi on Lycopene, Antioxidant Activity and Total Soluble Solid in Tomato (*Lycopersicon Esculentum* F1 Hybrid, Delba). *Advances in Environmental Biology*, Vol. 5, No. 6. pp. 1290-1294. ISSN 1995-0756.

Panagiotopoulos, L.J. & Fordham, R. (1995). Effects of water stress and potassium fertilisation on yield and quality (flavour) of table tomatoes (*lycopersicon esculentum mill.*). *Acta Hort. (ISHS)*, No. 379, (June 1995), pp. 113-120, ISSN 0567-7572.

Parisi, M; Pentangelo, A.; D'Onofrio, B.; Villari, G. & Giordano, I. (2006). Studi su ecotipi campani di pomodorino "Corbarino" e "Vesuviano" in due ambienti. *Italus Hortus*, Vol. 13, No. 2, pp. 775-778, ISSN 1127-3496.

Patanè, C. & Cosentino, S.L. (2010). Effects of soil water deficit on yield and quality of processing tomato under a Mediterranean climate. *Agricultural water management*, Vol. 97, No. 1, pp.131-138, ISSN 0378-3774.

Pernice, R.; Parisi, M.; Giordano, I.; Pentangelo, A.; Graziani, G.; Gallo, M.; Fogliano, V. & Ritieni, A. (2010). Antioxidant profile of small tomato fruits: effect of water regime and industrial process. *Scientia Horticulturae*, Vol. 126, No. 2, pp. 156-163, ISSN 0304-4238.

Riggi, E.; Patané, C. & Ruberto, G. (2008). Content of carotenoids at different ripening stages in processing tomato in relation to soil water availability. *Australian Journal of Agricultural Research*, Vol. 59, No. 4 (April 2008), pp. 348–353, ISSN: 00049409.

Rudich, J.; Kalmar, C.; Geizenberg, C. & Harel, S. (1977) Low water tensions in defined growth stages of processing tomato plants and their effects on yield and quality. *J Hort Sci*, Vol. 52, pp. 391–399, ISSN 1611-4426

Ruiz, J. M. & Romero, L. (2002). Relationship between potassium fertilisation and nitrate assimilation in leaves and fruits of cucumber (*Cucumis sativus*) plants. *Annals of Applied Biology*, Vol. 140, pp. 241–245, 0003-4746

Sensoy, S.; Demir, S.; Turkmen, O.; Erdinc, C. & Savur, O.B. (2007). Responses of some different pepper (*Capsicum annuum* L.) genotypes to inoculation with two different arbuscular mycorrhizal fungi. *Scientia Horticulturae*, Vol. 113, pp. 92-95, ISSN 0304-4238.

Serio, F.; Ayala, O.; Bonasia, A. & Santamaria, P. (2006). Antioxidant properties and health benefits of tomato. In: *Recent Progress in Medicinal Plants. Volume 13. Search for Natural Drugs*. (Govil, J. N., Sing, V. K. and Arunachalam, C., Eds.). Studium Press, Houston, TX, USA. 163–183.

Serio, F.; Leo, L.; Parente, A. & Santamaria, P. (2007). Potassium nutrition increases the lycopene content of tomato fruit. *Journal of Horticultural Science & Biotechnology*, Vol. 82, No. 6, pp. 941–945, ISSN 14620316.

Shi, J.; Le Maguer, M. & Niekamp F. (1999). Lycopene degradation and isomerization in tomato dehydration. *Food Research*, Vol. 32, pp. 15-21, 0963-9969

Silvestri, G. & Siviero, P. (1991). *La coltivazione del pomodoro da industria*. Edizioni l'informatore Agrario, ISBN 887220030X, Verona (Italy).

Sorensen, J.N.; Larsen, J.M. & Jakobsen, I. (2008). Pre-inoculation with arbuscular mycorrhizal fungi increases early nutrient concentration and growth of field-grown leeks under high productivity conditions. *Plant Soil*, Vol. 307, pp. 135-147, ISSN 0032-079X

Stahh, W. & Sies, H. (1992). Uptake of lycopene and its geometrical isomers is greater from heat-processed than from unprocessed tomato juice in humans. *J. Nutr.,* Vol. 122, pp. 2161-2166, ISSN 0022-3166.

Taber, H.; Perkins-Veazie, P.; Shanshan, L.; White, W.; Rodermel, S. & Yang X. (2008). Enhancement of tomato fruit lycopene by potassium is cultivar dependent. *HortScience,* Vol. 43, No. 1, (February 2008), pp. 159-165, ISSN 0018-5345

Trudel, M. J. & Ozbun, J. L. (1970). Relationship between chlorophyll and carotenoids of ripening tomato fruits as influenced by K nutrition. *Journal of Experimental Botany,* Vol. 21, pp. 881–886, ISSN 0022-0957.

Trudel, M. J. & Ozbun, J. L. (1971). Influence of potassium on carotenoid content of tomato fruit. *Journal of the American Society for Horticultural Science,* Vol. 96, pp. 763–765, ISSN 003-1062.

Turk, M.A.; Assaf, T.A.; Hameed, K.M. & Al-Tawaha, A.M. (2006). Significance of Mycorrhizae. *World journal of Agricultural Sciences,* Vol. 2, No. 1, pp. 16-20, ISSN 1817-3047.

van der Heijden, M.G.A.; Streitwolf-Engel, R.; Siegrist, S.; Neudecker, A.; Ineichen, K.; Boller, T.; Wiemken, A. & Sanders I.R. (2006). The micorrhizal contribution to plant productiviy, plant nutrition and soil structure in experimental grassland. *New Phytol.,* Vol. 172, pp. 739-752, ISSN 0028-646X.

van Tuinen, D.; Jacquot, E.; Zhao, B.; Gollotte, A. & Giannizzi-Pearson, V. (1998). Characterization of root colonization profiles by a microcosm community of arbuscular mycorrhizal fungi using 25S rDNA-targeted nested PCR. *Mol. Ecol.,* Vol. 7, pp. 879-887, ISSN 0962-1083.

Veit-Köhler, U.; Krumbein, A. & Kosegarten, H. (1999). Effect of different water supply on plant growth and fruit quality of *Lycopersicon esculentum. Journal of Plant Nutrition and Soil Science,* Vol. 162, No.6, pp. 583–588, ISSN 0718-9508.

Warner, J.; Tan, C. S. & Zhang, T. Q. (2007). Water management strategies to enhance fruit solids and yield of drip irrigated processing tomato. *Canadian Journal of Plant Science,* Vol. 87, pp. 345-353, ISSN 0008-4220.

Waterer, D.R. & Coltman, R.R. (1988). Phosphorus concentration and application interval influence growth and mycorrhizal infection of tomato and onion transplants. *J. Am. Soc. Hortic. Sci.,* Vol. 113, pp. 704–798, ISSN 0003-1062.

Wright, D.H. & Harris, N.D. (1985). Effect of nitrogen and potassium fertilization on tomato flavour. *J. Agric. Food Chem.,* Vol. 33, pp. 355-358, ISSN 0021-8561.

Wuzhong, N. (2002). Yield and quality of fruits of solanaceous crops as affected by potassium fertilization. *Better Crops International,* Vol. 16, No. 1, pp. 6-8, ISSN 0006-0089.

Zdravković, J.; Marković, Z.; Zdravković, M.; Damjanović, M. & Pavlović, N. (2007). Proceeding of "III Balkan Symposium on Vegetables and Potatoes", *Acta Horticulturae,* No. 729, pp. 177-182, ISSN 0567-7572.

Zushi, K & Matsuzoe, N. (1998). Effect of soil water deficit on vitamin C, sugar, organic acid, amino acid and carotene contents of large-fruited tomatoes. *J Jpn Soc Hort Sci,* Vol. 67, pp. 927–933, ISSN 0013-7626.

11

Experiments on Alleviating Arsenic Accumulation in Rice Through Irrigation Management

Shayeb Shahariar[1] and S. M. Imamul Huq[2]
[1]Soil, Agronomy and Environment Section, Biological Research Division, Bangladesh Council of Scientific and Industrial Research (BCSIR), Dhanmondi
[2]Department of Soil, Water and Environment, University of Dhaka
Bangladesh

1. Introduction

Arsenic (As) in groundwater is a major health concern in Bangladesh and the risks of As ingestion using shallow tubewells (STWs) for drinking-water was identified in the deltaic region, particularly in the Gangetic alluvium of Bengal including Bangladesh and West Bengal of India during the early nineties. It has been termed the world's biggest natural calamity in known human history. Ground water is the primary source of drinking water for approximately 90% of the total 147 million people (WHO 2001). More than 35 million people of Bangladesh are exposed to an As contamination in drinking water exceeding the national standard of 50 µg L⁻¹ while an estimated 57 million people are at the risk of exposure to As contamination exceeding the WHO guideline of 10 µg L⁻¹ (BGS/DPHE 2001). Extensive contamination in Bangladesh was confirmed in 1995, when additional survey showed contamination of mostly shallow tube-wells (STWs) across much of southern and central Bangladesh (Imamul Huq et al. 2006a). However, a few instances of deep tube-wells (DTWs) contamination are also in report. Approximately 27% of STWs and 1% of DTWs in 270 upazillas (sub-districts) of the country are contaminated with As at Bangladesh standard whereas about 46% of STWs are contaminated at WHO standard. So far 38,000 persons have been diagnosed with an additional of 30 million people at risk of As exposure (APSU 2005). Concentrations of arsenic exceeding 1,000 µg L⁻¹ in shallow tube-wells were reported from 17 districts in Bangladesh (Ahmed et al. 2006). Efforts are being directed towards ensuring safe drinking water either through mitigation technique or through finding alternative sources. Even if an As-safe drinking water supply could be ensured, the same groundwater will continue to be used for irrigation purpose, leaving a risk of soil accumulation of this toxic element and eventual exposure to the food-chain through plant uptake and animal consumption (Imamul Huq 2008). Given the studies on As uptake by crops (Imamul Huq et al. 2001; Abedin et al. 2002; Ali et al. 2003; Islam et al. 2005) there is much potential for the transfer of As present in groundwater to crops. The use of groundwater for irrigation has increased abruptly over the last couple of decades. About 86% of the total groundwater withdrawn is utilized in the agricultural sector

(Imamul Huq et al. 2005). There has been a gradual increase in the use of ground water for irrigation over the last two decades. The increase from 1999 – 2000 to 2006-2007 has been more than 22 per cent. In the *Boro* (dry) season of 2004, 75% of the irrigation water was from ground water (BADC 2005), which was 41% of the total in 1982-83. About 40% of total arable land of Bangladesh is now under irrigation facilities and more than 60% of this irrigation need are met from groundwater extracted by deep tube-well (DTW), shallow tube-well (STW) or hand tube-well. Of the total area of 4 million ha under irrigation, 2.4 million ha is covered via STWs and 0.6 million ha is covered via 23,000 DTWs. In the dry season, 3.5 million ha is used for *Boro* rice (FAO 2006).

It has been estimated that water extraction from the shallow aquifer for irrigation adds 1 million kg of As per year to the arable soil in Bangladesh, mainly in the paddy fields. The background level of As in soils is 4 to 8 mg kg^{-1}. In areas irrigated with As contaminated water, the soil level can reach up to 58 mg kg^{-1} (Imamul Huq and Naidu 2003).

It is of concern that a number of studies from Bangladesh have reported increased As concentrations in soils and crops because of irrigation with As-contaminated groundwater. The increase in soil concentrations may finally result in a reduction of soil quality and crop yields. Recent data on total and inorganic As in rice and vegetables from Bangladesh (Williams et al. 2006; Williams et al. 2005) indicate that rice contributes significantly to the daily intake. A positive correlation between As in groundwater resources, soil and rice has been reported, indicating that food chain contamination takes place because of prolonged irrigation with contaminated water (Correll et al. 2006). The risks of land degradation are likely to increase with the accumulation of As in the soil.

Management options should, therefore, focus on preventing and minimizing As input to soils. Farmers often use more irrigation water than needed. Optimizing water input would be a sound option to reduce As input while saving water. Furthermore, aerobic growth conditions in paddy fields may reduce bioavailability and uptake of As in rice. Other possible options include breeding crops tolerant to As and/or low accumulation of As in grains, and shifting from rice in the dry season to crops that demand less water, where feasible.

This paper aims at devising remedial measures through water management to minimize As toxicity in rice by making more oxidized rice rhizosphere (oxidizing arsenite) as well as to reduce the entry of arsenic into food chain.

2. Materials and methods

Sampling site

Soil where groundwater contamination by Arsenic has not been reported was sought. As such, a field from Block no-9, research station of Bangladesh Jute Research Institution (BJRI), Jagir, Manikganj was selected for sampling. The whole sampling site was divided into two sampling spots. The soil thus selected belongs to the Sonatola soil series under the AEZ-8 (Young Brahmaputra and Jamuna Floodplains) and land type was Medium Highland. The geolocation of the sampling spot is 23° 53.034' North and 90° 02.265' East.

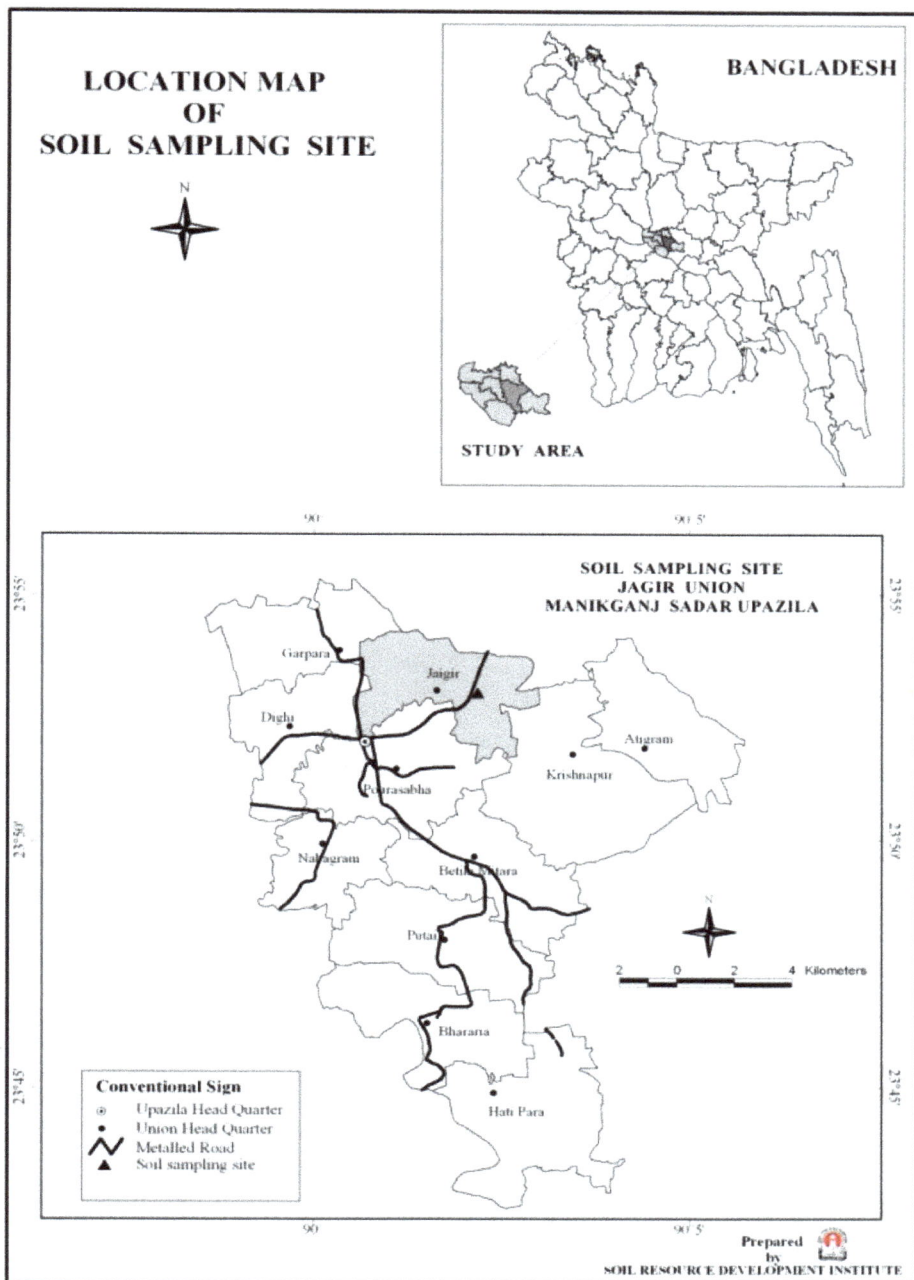

Fig. 1. Location map showing the geographic position of the sampling site from where soils were collected for the experiment

Collection and preparation of soil samples

The bulk of soil samples representing 0-15 cm depth from the surface were collected by composite soil sampling method as suggested by soil survey staff of the United States Department of Agriculture (USDA 1951). For laboratory analyses, the samples were collected from top to bottom with the help of an auger and mixed thoroughly. Samples were collected from the two sampling sites and put in polythene bags, tagged with rubber band and labeled. For the bulk portion, samples were collected with spade into jute made large bags and carried to the net house. The collected soil samples were dried in air for 3 days (40° C) by spreading in a thin layer on a clean piece of paper. Visible roots and debris were removed from the soil sample and discarded. For hastening the drying process, the soil samples were exposed to sunlight. After air drying, a portion of the larger and massive aggregates were broken by gently crushing them with a wooden hammer. Ground samples were screened to pass through a 2 mm stainless steel sieve. The sieved samples were then mixed thoroughly for making the composite sample. Soil samples were preserved in plastic containers and labeled properly showing the soil number, sample number, date of collection. These soil samples were used for various physical analyses. Another portion of soil samples (2 mm sieved) was further ground and screened to pass through a 0.5 mm sieve. The sieved sample were mixed thoroughly to make composite samples and persevered in the same way as above. These soils were used for chemical and physicochemical analyses. The bulk soil sample collected for pot experiments were air dried, cleared off the debris and crushed to make the bigger clods smaller. The crushed soil samples were screened through a 5 mm sieve.

Experimental set-up

A pot experiment was carried out in the net house of the Department of Soil, Water and Environment, University of Dhaka, Bangladesh. In the experiment, three water regimes and two sources of As salts were chosen. The water regimes maintained were 100%, 75% and 50% of field capacity. The FC value has been predetermined in the fields and the value was used in the pot experiments. For maintaining the required water regime, the pots were measured every alternate days and the required amount of water was added to the pots. The two sources of arsenic used were As^{III} and As^V. The salts used were sodium meta arsenite ($NaAsO_2$) as a source of As^{III} and hydrated sodium meta arsenate ($Na_2HAsO_4.7H_2O$) as a source of As^V. The application rates were 0 (control) and 30 mg As/kg soil from both the sources. All experiments were done in triplicates. The pots were arranged in the net house in three blocks with randomization in each block. The three blocks belonged to the three moisture regimes. Two varieties of rice seeds *viz.*, BRRI dhan-28 and BRRI dhan-29 were collected from the Bangladesh Rice Research Institute (BRRI). The background level of As in soil was 1.62 mg/kg which was taken as control. There were a total of 36 pots for each variety 3 (2×2×3) i.e. 3 replication, 2 sources of arsenic salt, 2 doses of arsenic and 3 moisture regimes. Thus, a total of 72 pots were used for the experiment. Earthen pots of 5 L volume were taken having no holes on the bottom and marked in accordance with water regime, arsenic dose, sources of salt and variety of rice plants. Each pot was filled with 4.5 kg soil. Fertilizer requirement for the rice plants were calculated on the basis of soil test values as described in Fertilizer Recommendation Guide 2005 (BARC 2005). The whole amount of required P (TSP) and K (MP) and 1/3 of the N (Urea) doses as estimated for the soil of each pot were mixed with the soil thoroughly before the transplantation of seedlings into the pots. Arsenic was applied to the soil as solution before transplantation of the seedlings.

Method of rice cultivation

Seeds of BRRI dhan-28 and BRRI dhan-29 were at first taken into two separate pots, dipped in water and kept overnight. Then seeds of the two varieties were spread on two separate seedbeds made of the same soil used for the pot experiment. It took 2 to 3 days for germination while the seedbeds were kept under dark by covering with wet jute bag. After the germination of the seeds, the seedbeds were taken into the net house and kept under natural condition. After one month, the seedlings were transplanted into the pots. The pots were thinned to 5(five) plants after seedling were established. Second 1/3 dose of N (Urea) was applied 35 days after transplantation and the final 1/3 was applied during panicle initiation stage of rice plants. Weeds were removed manually. Agronomic characters like plant height, tiller numbers, panicle numbers, leaf colors etc. were observed during the growth period.

Collection of plant samples

Plants of BRRI Dhan-28 and BRRI Dhan-29 were harvested at the age of 130 and 140 days of the transplants respectively. The plants were harvested by uprooting. The unfilled grain samples of the two varieties were collected before two days of harvesting. The harvested roots were washed with deionized water several times to remove ion from the ion free space as well as to dislodge any adhering particles on the root surface. The upper parts of the plants were also washed. The height of the plant samples were measured from the top leaf blade to bottom of the plant from where root starts. Then the plant samples are wrapped with tissue paper to remove the extra water and dried in air for half an hour. Then fresh weight of whole plants was taken. The plant samples were separated replaced into two parts, root and straw. The plant samples were then air dried before putting to oven drying at 75±5°C for 48 hours and the dry weight of plant samples were noted. The dried plant samples were then ground and were sifted through a 0.2 mm sieve. The ground plant samples of two varieties were preserved in small plastic bottles separately. After harvesting, soil samples in each pot were collected from the rhizosphere. The samples were air dried and homogenized and was screened to pass through a 0.5 mm sieve for chemical analysis. The soil samples were preserved in plastic bags and labeled.

Laboratory analysis

Various physical, chemical and physiochemical properties of the soils were determined following procedures described in Imamul Huq and Alam (2005). Both plants and soil were analyzed for total arsenic by hydride generation atomic absorption spectrometry (HG-AAS) with the help of 5% potassium iodide (KI) and 10% urea in acid medium. The hydride was generated using 6N HCl and 1.2% $NaBH_4$ and 1% NaOH in deionized water. The arsenic from the plant samples was extracted with HNO_3, and from the soil with aqua regia solution (Portman and Riley 1964). 0.5 g of plant sample was weighed separately into 100 ml Pyrex glass beaker. 15 ml of nitric acid (HNO_3) was added and the beaker left for half an hour. Then the sample beakers were placed on the hotplate. At first the beakers were heated at low temperature (50-75°C) before increasing the temperature to 140°C for the final dissolution of organic material. After dissolution was complete; samples were diluted to 25 ml into volumetric flask, shaken and filtered into plastic bottles. This extract was also used for the determination of total As, Fe, P and K content of plant samples. Certified reference materials were used throughout the digestion and analyzed as part of the quality assurance/quality control protocol. Reagent blanks and internal standards were used where appropriate, to ensure accuracy and precision in the analysis of arsenic. Each batch of ten samples was accompanied by reference standard samples to ensure strict QA/QC procedures. The amount of As uptake (mg/100 plants) by different plant parts and the plant as a whole were calculated. The uptake was calculated using the As concentration in the dry matter and the dry weight of plant parts, and the result was expressed as mg/100 plants.

Uptake (As) = Concentration (As) in dry matter × dry weight of plant part for 100 plants

The Transfer Co-efficient in root, straw and husk of plants was determined using the following formula:

$$T.F = \frac{(mg\ of\ the\ elements\ /\ kg\ dry\ weight\ of\ plants)}{(mg\ of\ the\ elements\ /\ kg\ dry\ weight\ of\ soil)}$$

3. Results and discussion

The collected soil sample was analyzed in the laboratory before setup of the experiment to see the nutrient status of the soil. Background level of As (arsenic) was also determined in the soil sample. Some important physical, physicochemical and chemical properties of the experimental soil sample are listed in the table 1.

Soil properties	Value
Particle size analysis	
Sand	17.456 %
Silt	62.246 %
Clay	20.298 %
Textural class	Silt loam
pH	7.2
Organic carbon	0.65 %
Organic matter	1.11 %
Field capacity	35.89 %
Moisture content	8.25 %
Total nitrogen	0.0567 %
Total phosphorous	768.50 mg/kg
Available phosphorous	30 mg/kg
Total potassium	2121.93 mg/kg
Available potassium	63.33 mg/kg
Arsenic	1.62 mg/kg
Iron	21289.06 mg/kg

Table 1. Some physical, physicochemical and chemical properties of the soil sample used for the experiment

Agronomic parameters

Symptoms of any abnormality in the rice plants were noted during the experiment in order to assess the phytotoxicity of As. Both varieties showed some symptoms of toxic effects at As treated soil (30 mg/kg As concentrations) and the symptoms became more pronounced with time of exposure of the plants to arsenic. The symptoms were: reduced plant growth; yellowing and wilting of leaves. Brown necrotic spots were also observed on old leaves of the plants of both the varieties growing on As treated soil. Red brown necrotic spots on old leaves, tips and margins of rice, due to arsenic toxicity, have also been reported (Aller et al. 1990; Marin et al. 1992). Plant heights were measured from time to time and finally at maturity. At the initial stage of growth, plant height did not differ from the control plants but at maturity, plant heights decreased in arsenic treated pots with decreasing moisture level. Plants of both the rice varieties grown on 30 mg As/kg-treated soil under 50% moisture regimes showed the shortest plant height. The decreasing trend of plant height with decreasing moisture regime was not statistically significant except for BRRI dhan-29 growing under arsenate (As^V) treated soil. It was clear that arsenic did not readily cause plant height reduction, but the progressive accumulation of arsenic in plants with time of exposure might have caused the plant height reduction. A reduction in plant height with increasing As concentration has also been reported in rice plants (Yamare 1989; Barrachina et al. 1995; Islam 1999). Fresh as well as dry matter production of the two varieties did not show any appreciable difference for arsenite or arsenate treatments under the three moisture regimes. Maximum weights were noted for the control plants of 100% moisture level, whereas the minimum values were for plants growing at 75% moisture level. At 50% moisture regime higher value in both fresh and dry weights was found than that of 75% moisture regime in both arsenite and arsenate treated soils. Significant difference in fresh and dry weight was observed between the moisture regimes only in case of BRRI dhan-28 under 30 mg As/kg arsenate (As^V) treated soil. The value showed that, both fresh and dry matter production was higher in 50% FC in some cases than 75% FC but lower than 100% FC for BRRI dhan-28 grown under 30 mg As/kg arsenate (As^V) treated soil.

Arsenic accumulation

Arsenic concentrations in different parts (root, straw and husk) of the rice plants at different moisture levels (100%, 75% and 50% FC) of the two varieties are presented in Figure 2 (a & b) and Figure 3 (a & b). It is important to note that in the control plants of both varieties there were some As accumulation which could be due to the presence of the background As (1.62 mg/kg) in soil.

BRRI dhan-28

Root

Arsenic concentration in roots of both arsenite (As^{III}) and arsenate (As^V) treated soil increased with increasing moisture level. Arsenite (As^{III}) salt contributed more towards As accumulation at 100%, 75% and 50% FC moisture level than arsenate (As^V). This possibly could be due to high mobility of As^{III} than the As^V into the soil through pore spaces. For 30mgAs/kg arsenic treated soil, the concentration of As was 10.04 mg/kg d.w. for As^{III} treatment, while the value was 9.51 mg/kg d.w. for As^V treatment in the roots of BRRI dhan-28 at 100% FC moisture level but at 50% FC As concentrations were 4.78 mg/kg d.w. 3.02 mg/kg d.w. for arsenite and arsenate treatment respectively (Figure 2 a & b). However, the maximum values were found to be 10.04 mg/kg d.w. and 9.51 mg/kg d.w. in roots of BRRI dhan-28 from As^{III} and As^V sources respectively (Figure 2 a & b). Both at 75% and 50% FC As concentration in the roots of BRRI dhan-28 was lower than in those of 100% FC for both As sources. These values clearly show that with the reduction of moisture level As accumulation in roots reduced. However, arsenic concentration in the roots of BRRI dhan-28 at different moisture level, either from As^{III} or As^V, was not statistically significant. In general, the roots accumulated higher As than the other plant parts. Indeed, the As concentration in the roots of BRRI dhan-28 was higher at 100% FC than 75% and 50% FC for both As sources.

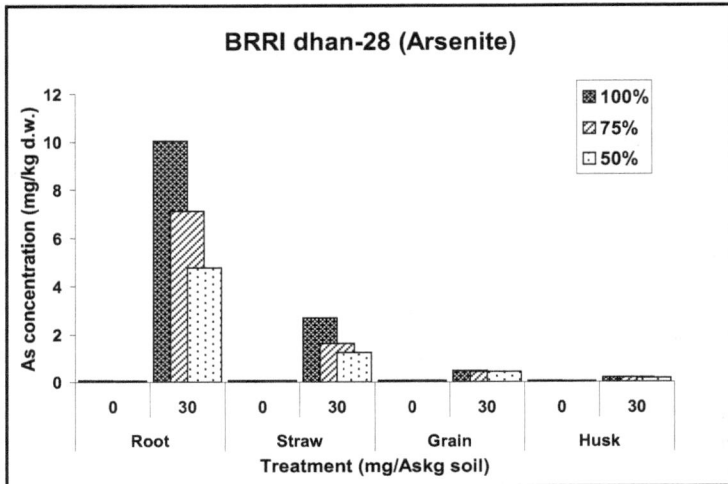

(a)

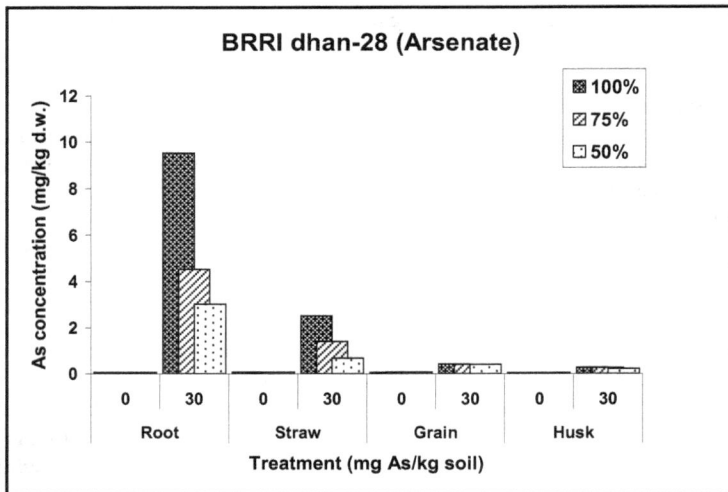

(b)

Fig. 2. Arsenic concentration in different parts of BRRI dhan-28 under the three moisture regimes as affected by (a) arsenite (As^{III}); (b) arsenate (As^V)

Straw

Arsenic concentration in straw of BRRI dhan-28 followed the same pattern as for roots in both arsenite and arsenate treated soil. The magnitude of the increasing trend varied considerably between the moisture levels. The maximum As concentration (2.70 mg/kg d.w.) was found in straw of BRRI dhan-28 treated with 30 mg As/kg soil from As^{III} at 100% FC, while for 50% FC it was 1.25 mg/kg d.w. (Figure 2a). The arsenic concentration for As^V treated soil in straw of BRRI dhan-28 was 2.52 mg/kg d.w. at 100% FC and 0.66 mg/kg d.w. at 50% FC (Figure 2b). The accumulation did not differ significantly for the two sources in this variety. As accumulation in straw was higher for As^{III} treated soil at all the moisture regimes.

Grain

The arsenic concentrations in grain were almost similar, irrespective of the moisture regime and source of arsenic. The maximum value was found 0.48 mg/kg d.w. for arsenite treated soil while for arsenate treated soil it was 0.42 mg/kg d.w. at 100% FC. However, from none of the arsenic sources did the grain As concentration exceed the maximum permissible limit of 1.0 mg of As/kg in the grain of rice (National Food Authority 1993). This Australian standard is taken as a reference, as no standard has yet been adopted in Bangladesh. Arsenic accumulation in grain was not significant for any of the variables. In a previous work with two moisture regimes, it has been observed that arsenic accumulates more under 100% FC than under 75% of FC (Imamul Huq et al. 2006b).

Husk

Arsenic accumulations in rice husk were almost similar for the two sources of arsenic, though there was a variation between the moisture regimes. The maximum value was found 0.27 mg/kg d.w. in husk of BRRI dhan-28 treated with 30 mg As/kg soil from As^V at 100% FC, and the concentration decreased very slightly with decreasing moisture level. The husk As concentration did not exceed the maximum permissible limit of 1.0 mg of As/kg– similar to what was observed for the moisture regime at 75% and 50% FC.

BRRI dhan-29

Root

In this variety too, arsenic concentration in the roots from both the sources (As^{III} and As^V) increased with increasing moisture level. However, the values were relatively lower than what was observed for BRRI dhan-28. The maximum As concentration in roots of BRRI dhan-29 was 8.55 mg/kg d.w. for As^{III}, and 8.08 mg/kg d.w. for As^V treated soil at 100% FC (Figure 3 a & b) whereas at 50% FC the values were 2.99 mg/kg d.w. and 1.57 mg/kg d.w. for arsenite and arsenate treatment respectively. Of the three moisture regimes, at 100% FC arsenic accumulation was more in roots from both As^{III} and As^V sources than 75% and 50% FC. These values clearly show that with the reduction of moisture level As accumulation in roots reduced. However, arsenic concentration in the roots of BRRI dhan-29 at different moisture level, either from As^{III} or As^V, was not statistically different. Arsenic concentration in roots of both arsenite (As^{III}) and arsenate (As^V) treated soil increased with increasing moisture level. In general, with the 100%, 75% and 50% moisture regime, the roots of BRRI dhan-29 from As^V treated soil accumulated lesser amounts of As than As^{III} treated soil.

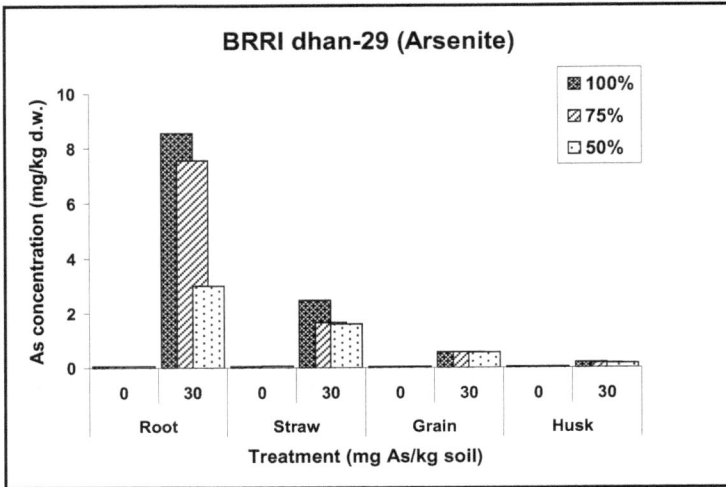

(a)

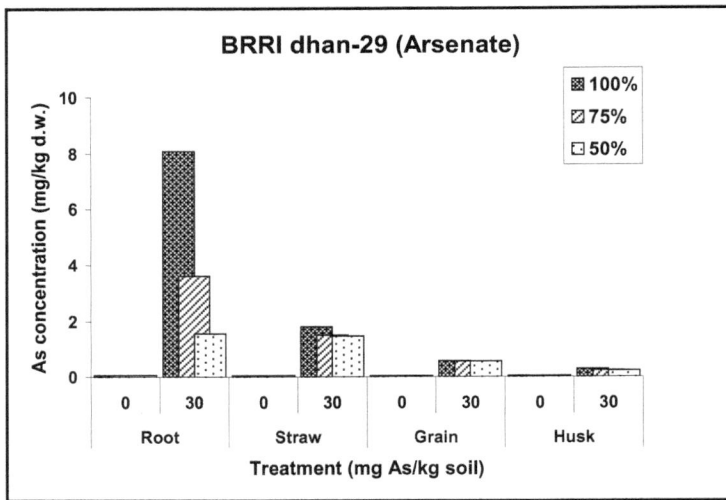

(b)

Fig. 3. Arsenic concentration in different parts of BRRI dhan-29 under the three moisture regimes as affected by (a) arsenite (As^{III}); (b) arsenate (As^V)

Straw

The As concentration in straw of BRRI dhan-29 increased with increasing moisture regime in the growth medium. The magnitude of the increasing trend varied considerably between the sources of arsenic. The maximum As concentration (2.47 mg/kg d.w.) was found in BRRI dhan-29 treated with As^{III} at 100% FC, while the value was 1.81 mg/kg d.w. for As^V at 100% FC (Figure 3 a & b). A similar trend was also observed for BRRI dhan-28 (Figure 2 a &

b for BRRI dhan-28). Under 50% FC the As^{III} treated plants accumulated 1.60 mg As/kg d.w., while for As^V treated plants it was 1.48 mg/kg d.w. Of the three moisture regimes, straw of BRRI dhan-29 accumulated more As from As^{III}, than from As^V.

Grain

The arsenic concentrations in grain were almost similar, irrespective of the moisture regime and source of arsenic. In BRRI dhan-29 the maximum value was found to be 0.58 mg/kg d.w. for arsenite treated soil while for arsenate treated soil it was 0.57 mg/kg d.w. at 100% FC. The concentration decreased very slightly with decreasing moisture level (Figure 3 a & b). However, from none of the arsenic sources did the grain As concentration exceed the maximum permissible limit of 1.0 mg of As/kg in the grain of rice (National Food Authority 1993). Arsenic accumulation in grain was not significant for any of the variables.

Husk

Arsenic accumulations in rice husk were almost similar for the two sources of arsenic, though there was a variation between the moisture regimes. The maximum value was found to be 0.28 mg/kg d.w. in the husk treated with 30 mg As/kg soil from As^V at 100% FC, and the concentration decreased very slightly with decreasing moisture level. The husk As concentration did not exceed the maximum permissible limit of 1.0 mg of As/kg– similar to what was observed for the moisture regime at 75% and 50% FC.

Comparison among the three moisture regimes

In both varieties, As accumulation was found to be reduced at 75% of field capacity and more reduced at 50% of field capacity than the 100% FC. In our previous experiment with two water regimes, similar observations have been made (Imamul Huq et al, 2006b). Reducing moisture up to 50% did not cause any significant yield difference in terms of biomass production while it reduced the uptake and accumulation of arsenic in all the plant parts of rice.

Comparison among the plant parts

Maximum As accumulation was observed in roots, followed by straw and husk. Similar observations have been reported earlier (Imamul Huq and Naidu, 2005, Imamul Huq et al. 2006a, Imamul Huq et al. 2007, Marin et al. 1992, Xie and Huang 1998). Abedin et al. (2002) showed the rice tissue As concentration in the order: root > straw > husk > grain. Roots of BRRI dhan-28 accumulated more As than BRRI dhan-29 under the three moisture regimes, and more from As^{III} than As^V. In both BRRI dhan-28 and BRRI dhan-29 As concentrated more in the roots and was transferred to the upper parts of the plant. Although As in straw was comparatively less than the roots, there are high possibilities to accumulate As in grains in both the varieties. Transfer factor values (Farrago and Mehra 1992) greater than 0.1 indicated that the rice plant has a strong affinity to As accumulation for all treatments. Transfer factor values also showed greater affinity for As^{III} than for As^V, irrespective of the variety.

Comparison between two arsenic sources

Arsenic accumulation in all parts of both the varieties was little bit higher for As^{III} than for As^V under three moisture regime (100%, 75% and 50% of field capacity). Similar observations have also been noted earlier (Imamul Huq et al. 2006b). However, there was no significant difference in arsenic accumulation between the two arsenic sources.

Comparison between the two varieties

Arsenic accumulation varied between the different plant parts of the two varieties, but there was no significant difference in As accumulation between the two varieties. Roots of the rice variety BRRI dhan-28 concentrated more As than BRRI dhan-29, resulting in higher accumulation in the straw of BRRI dhan-28 under three moisture regimes, irrespective of the source of the arsenic.

Arsenic uptake

It was observed that As uptake decreased with decreasing moisture regime in both the As (As^{III} and As^V) treated plants irrespective of variety. However, As uptake was higher in the As^{III}-treated plants than in As^V. It was clear that in all parts of both varieties, As uptake was higher at 100% field capacity. It has been found that BRRI dhan-29 is more susceptible to As accumulation than BRRI dhan-28.

4. Conclusions

From this experiment it was observed that growing rice at reduced moisture level could alleviate As toxicity without significant biomass yield reduction. Reduction in moisture level or irrigation water could reduce As uptake in rice plant by reducing it's phytoavailability, thereby helping to reduce its entry into the food chain to some extent.

5. Acknowledgements

The authors would like to acknowledge the Bangladesh Australia Centre for Environmental Research (BACER-DU), University of Dhaka, for providing laboratory facilities and financial assistance, and M. N. Goni, Principal Scientific Officer and A. K. M. Maksudul Alam, Principal Scientific Officer of Soil Science Division, Bangladesh Jute Research Institute (BJRI) for their cooperation in collecting soil samples for the pot experiments.

6. References

Abedin, M.J., Cressner, M.S., Meharg, A.A., Feldmann, J. and Cotter-Howells, J., 2002. Arsenic accumulation and metabolism in rice (*Oryza sativa* L.). Environ. Sci. Tech., 36: 962-968.

Ahmed, K.M., Imamul Huq, S.M. and Naidu, R., 2006. Extent and severity of arsenic poisoning in Bangladesh. In: Naidu, R., Smith. E., Owens, G., Bhattacharya, P. and Nadebaum, P. (Eds.), Managing Arsenic in the Environment: From Soil to Human Health. CSIRO Publishing, Melbourne, Australia, ISBN 0-643-06868-6. pp. 525-540.

Ali, M.A., Badruzzaman, A.B.M., Jalil, M.A., Hossain, M.D., Ahmed, M.F., Masud, A.A., Kamruzzaman, M. and Rahman, M.A., 2003. Arsenic in plant and soil environment of Bangladesh. In: Ahmed, M.F., Ashraf, A.M, and Adeel, Z. (Eds.), Fate of Arsenic in the Environment. Bangladesh University of Engineering and Technology, Dhaka, Bangladesh and United Nations University, Tokyo, Japan. ISBN 984-32-0507-3. pp. 85-112.

Aller, A.J., Bernal, J.L., Nazal, Del. M.J. and Deban, L., 1990. Effect of selected tree elements on plant growth. J. Sci. Food and Agriculture, 51: 447-472.

APSU, 2005. The response to arsenic contamination in Bangladesh: A position paper, published by DPHE, Dhaka, Bangladesh. p. 55.

BADC, 2005. Survey report on irrigation equipment and irrigated area in Boro/2004 season, Bangladesh Agricultural Development Corporation, Dhaka.

BARC, 2005. Fertilizer Recommendation Guide 2005, Miah, M.U., Farid, A.T.M., Miah, M.A.M., Jahiruddin, M., Rahman, S.M.K., Quayyum, M.A., Sattar, M.A., Motalib, M.A., Islam, M.F., Ahsan, M. and Sultana, R. (Eds.), BARC Soil Publication no.-45, ISBN 984-32-3166-X. pp. 1-260.

Barrachina, A.C., Carbonell, F.B. and Beneyto, J.M., 1995. Arsenic uptake and distribution and accumulation in tomato plants: effects of arsenic on plant growth and yield. Journal of Plant Nutrition, 18: 1237-1250.

BGS/DPHE, 2001. Arsenic Contamination of Groundwater in Bangladesh. In: Kinniburg, D.G. and Smedley, P.L. (Eds.), Final Report. British Geological Survey Report, WC/00/19, Vol. 2.

Correll, R., Imamul Huq, S.M., Smith, E., Owens, G. and Naidu, R., 2006. Dietary intake of arsenic from crops. In: Managing Arsenic in the Environment: From Soil to Human Health. Naidu, R., Smith, E., Owens, G., Bhattacharya, P. and Nadebaum, P., CSIRO Publishing, Melbourne, Australia. pp. 255-271.

FAO, 2006. Arsenic contamination of irrigation water, soil and crops in Bangladesh: Risk implications for sustainable agriculture and food safety in Asia. RAP Publication 2006/20. FAO regional Office for Asia and the Pacific, Bangkok. pp. viii+1-38.

Farrago, M.E. and Mehra, A., 1992. Uptake of elements by the copper tolerant plant *Armeria maritima*. Metal Compounds in Environment and Life, 4 (Interrelation between Chemistry and Biology). Science and Technology Letters, Northwood.

Imamul Huq, S.M., 2008. Fate of arsenic in irrigation water and its potential impact on food chain. In: Arsenic Contamination of Groundwater: Mechanism, Analysis, and Remediation (Eds.). Satinder Ahuja, John Wiley & Sons, Inc.

Imamul Huq, S.M., Haque, H.A., Joardar, J.C. and Hossain, M.S.A., 2007. Arsenic accumulation in rice grown in *Aman* and *Boro* seasons. Dhaka Univ. J. Biol. Sci. 16(2): 91-97.

Imamul Huq, S.M., Correll, R. and Naidu, R., 2006a. Arsenic Accumulation in Food Sources in Bangladesh: Variability with Soil Type. In: Naidu, R., Smith, E., Owens, G., Bhattacharya, P. and Nadebaum, P. (Eds.), Managing Arsenic in the Environment: From Soil to Human Health. CSIRO Publishing, Melbourne, Australia. ISBN 0-643-06868-6. pp. 283-293.

Imamul Huq, S.M., Shila, U.K. and Joardar, J.C., 2006b. Arsenic mitigation strategy for rice using water regime management. Land Cont. & Reclam. 14(4): 805-813.

Imamul Huq, S.M. and Alam, M.D. (Eds.), 2005. A Handbook on Analysis of Soil, Plant and Water. BACER-DU, University of Dhaka, Bangladesh. pp. xxii+1-246.

Imamul Huq, S.M., Bulbul, A., Choudhury, M.S., Alam, S. and Kawai, S., 2005. Arsenic bioaccumulation in a green algae and its subsequent recycling in soils of Bangladesh. In: Bhattacharya, B. and Chandrashkharam (Eds.), Natural Arsenic in Ground Water: Occurrence, Remediation and Management. Taylor and Francis Group, London. ISBN 04-1536-700 X. pp. 119-124.

Imamul Huq, S.M. and Naidu, R., 2005. Arsenic in ground water and contamination of the food chain: Bangladesh scenario. In: Natural Arsenic in Ground water: Occurrence, Remediation and Management (Eds.), Bundschuh, J., Bhattacharya, P. and Chandrasekharam, D., Balkema, Leiden, The Netherlands. pp. 95-101.

Imamul Huq, S.M. and Naidu, R., 2003. Arsenic in ground water of Bangladesh: Contamination in the food chain. In: Ahmed, M.F. (Eds.), Arsenic contamination: Bangladesh perspective. ITN-Bangladesh, Centre for water supply and waste management, BUET, Dhaka, Bangladesh.

Imamul Huq, S.M., Smith, E., Correll, R., Smith, L., Smith, J., Ahmed, M., Roy, S., Barnes, M. and Naidu, R., 2001. Arsenic transfer in water soil crop environments in Bangladesh I: Assessing potential arsenic exposure pathways in Bangladesh. In: Arsenic in the Asia Pacific region workshop, Adelaide, Australia. pp. 50-52.

Islam, M.R., Jahiruddin, M., Rahman, G.K.M.M., Miah, M.A.M., Farid, A.T.M., Panaullah, G.M., Loeppert, R.H., Duxbury, J.M. and Meisner, C.A., 2005. Arsenic in paddy soils of Bangladesh: levels, distribution and contribution of irrigation and sediments. In: Behavior of arsenic in aquifers, soils and plants: Implications for Management. International Symposium held in Dhaka, Bangladesh during January 16-18, 2005 and organized by CIMMYT, CU, TAMU, USGS and GSB.

Islam, M.S., 1999. Arsenic Toxicity Remediation in Rice Plants. M.Sc thesis, Dept. of Soil Science, University of Dhaka, Bangladesh. pp. 3-120.

Marin, A.R., Masscheleya, P.H. and Parrick, Jr, W.H., 1992. The influence of chemicals from and concentration of arsenic on rice growth and tissue arsenic concentration. Plant and Soil, 139: 175–183.

National Food Authority, 1993. Australian Food Standard Code. Australian Government Publication Service, Canberra.

Portman, J.E. and Riley, J.P., 1964. Determination of Arsenic in seawater, marine plants & silicate and carbonate sediments. Anal. Chem. Acta., 31: 509-519.

USDA (United States Department of Agriculture), 1951. Soil Survey Manual by Soil Survey Staff, Bureau of Plant Industry. Soil and Agricultural Engineering, Handbook No.-18, pp. 205.

WHO, 2001. Environmental Health Criteria 224: Arsenic and Arsenic Compounds. WHO, Geneva.

Williams, P.N., Islam, M.R., Adomako, E.E., Raab, A., Hossain, S.A., Zhu, Y.G. and Meharg, A.A., 2006. Increase in rice grain arsenic for regions of Bangladesh irrigating paddies with elevated arsenic in groundwater. Environ. Sci. Tech., 40: 4903-4908.

Williams, P.N., Price, A.H., Raab, A., Hossain, S.A., Feldmann, J. and Meharg, A.A., 2005. Variation in arsenic speciation and concentration in paddy rice related to dietary exposure. Environ. Sci. Tech., 39: 5531-5540.

Xie, Z.M. and Huang, C.Y., 1998. Control of arsenic toxicity in rice plants grown on an arsenic-polluted paddy soil. Soil Sci. Plant Anal., 4: 2471-2477.

Yamare, T., 1989. Mechanism and counter measures of arsenic toxicity to rice plant. Bulletin of the Shimane Agricultural Experiment Station.

12

Experimentation on Cultivation of Rice Irrigated with a Center Pivot System

Gene Stevens, Earl Vories, Jim Heiser and Matthew Rhine
*University of Missouri and United States Dept. Agric.-Agricultural Research Service**
USA

1. Introduction

1.1 Feeding the world

Rice is the staple food for one half of the world's population. Consumed mainly by humans rather than fed to livestock, rice is an efficient food for supplying carbohydrates, vitamins, and nutrients in diets. Demographers predict the Earth's population will increase to nine billion people by 2045 (United Nations, 2004). To keep pace with increased food demand, rice farmers will need higher yields, increased hectares of rice production, and more efficient use of water resources. Unfortunately, traditional rice production uses large amounts of water. Irrigation practices are needed to grow rice with less water and on well-drained soils that are not currently used for traditional flooded rice culture.

Most rice cultivars do not tolerate extended periods of water stress. For optimum yields, rice is produced with irrigation supplied from rivers, lakes, or groundwater aquifers. In the dry-seeded, delayed-flood culture, rice is flooded at approximately the V-4 growth stage (Counce et al., 2000); and the flood is maintained continously until after heading. Insufficient pumping capacitiy sometimes results in dry portions of the fields, leading to nitrogen losses and low yields. Excessive pumping wastes water and energy and increases pressure on levees; furthermore, soil, fertilizers, and pesticides may be carried in the runoff from fields.

1.2 Sprinkler irrigation

This chapter is focused on center pivot sprinkler irrigation systems (Figure 1) for rice production. However, the principles are relavent to other sprinkler equipment such as linear move systems or floppy sprinkler systems with emitters attached to lateral cables stretched between tall poles. Sprinkler irrigation for rice production can be a water-saving alternative to conventional flood irrigation. Although center pivot systems may intially be more costly

---

* Gene Stevens, Jim Heiser, and Matt Rhine are plant scientists at the University of Missouri. Earl Vories is an agricultural engineer with the United States Dept. of Agric.-Agricultural Research Service. Mention of trade names or commercial products in this publication is solely for the purpose of providing specific information and does not imply recommendation or endorsement by the University of Missouri or the U.S. Department of Agriculture.

to install, annual labor expenses for rice are reduced because maintaining leeves and gates to control flood water depth is no longer needed.

Fig. 1. Center pivot irrigation system equipped with a backflow prevention valve used to apply irrigation water, and an injection pump for applying liquid nitrogen fertilizer and fungicides for controlling diseases.

Several farmers in the southeast United States conducted field-scale evaluations of rice production with center pivot irrigation in the 1980's. However, most rice growers became discouraged with the system because of management problems. The three main obstacles were (1) diseases, (2) weeds, and (3) ruts from wheels on the outside spans. Since that time, improvements have been made in rice genetics, herbicides, fungicides, and irrigation equipment (Figure 2), which now make it a feasible option for rice production.

Although several scientific studies were conducted with center pivot rice production in the 1980's, there is little recent information available (McCauley, 1990; Westcott and Vines, 1986). Rice research under center pivot systems was started at the University of Missouri-Delta Research Center at Portageville, Missouri in 2008 (Vories et al., 2010). In addition, center pivot manufacturers are working with producers in several countries to demonstrate center pivot rice production and the Missouri team is conducting water use research in South Africa in Limpopo Province. We recently reported on a nitrogen fertigation rice experiment with center pivot irrigation (Rhine et al., 2011) and we are helping several farmers in Missouri grow rice with center pivot irrigation through extension education programs.

In this chapter, we will share our experiences growing under center pivots with regard to rice cultivars and hybrids , herbicide programs and problem weeds, nitrogen fertilization methods, rates, and timing, and chemigation with fungicide to control brown spot and blast diseases. Information will be included for using local weather station data, reference evapotranspiration (ET), and suggest a crop adjustment coefficient for rice to manage irrigation for center pivot rice.

| standard two wheels | rubber tracks | four-wheel drive |

Fig. 2. Traction alternatives include (a) standard two-wheel drives, which can create deep ruts and cause the system to stall; the problem is most common on clay soils. For better flotation, (b) tracks can be added or (c) more tires used where high water flow rates occur.

2. Rice cultivars and hybrids

2.1 Asian and African rice

The most widespread rice species grown by farmers are *Oryza sativa* L. and *O. glaberrima*. Cultivation of *O. sativa* began in southern China, Laos, and Thailand (Jones, 2003). Also called "Asian" rice, *O. sativa* is now grown in over 100 countries. Selection and isolation has resulted in indica and japonica subgroups. Generally, indica cultivars have longer grain size and are better adapted to hot climates than japonica cultivars. In Asian countries, India, and the Philippines, *O. sativa* is usually cultivated by transplanting seedlings into flooded paddies. In North, Central, and South America, Europe, Australia, and the Middle East, farmers direct seed rice with a grain drill planter or water seed pre-germinated seed broadcast by airplane.

In countries with rain-fed production, paddy rice can become completely submerged by flash flooding. A few weeks later, rice in the same fields may be subjected to drought stress from lack of rainfall. This is a challenging environment for rice farmers and is not conducive for producing high yields. However, selection has produced rice lines with more drought tolerance than rice from areas with uniform flood culture. Developing rice more tolerant to dry soil conditions would be helpful for reducing short-term water stress. Accessions, such as FR13A, were collected from rain-fed regions that can survive both flooded and dry extremes (Bailey-Serres et al., 2010; Reddy et al., 2010). On the molecular level, the rice lines have an extra ethylene responsive factor called SUB1A which helps them tolerate stress. Marker assisted-breeding was used to develop a "near isogenic SUB1 introgression line" of a high-yielding japonica inbred rice cultivar, M202 (Fukao et al., 2011). Survival rates of M202 (SUB1) plants were higher, compared to the parent line without SUB1. Plants from both groups were exposed to both submergence and low soil moisture. Currently, it is not known whether this is just a survival mechanism or if SUB1 will help rice plants produce higher yields by tolerating short-term water stress. Marker-assisted breeding technologies will play an important role in helping scientists improve rice tolerance to dry soil conditions. Marker assisted breeding can be done without "genetic engineering" since in most cases, a natural marker can be found without using transgenes. The process saves breeders time and increases accuracy by providing an effective screening test for desired genes on progeny after a cross is made.

Cultivars with *O. glaberrima* (sometimes called "African rice") in their pedigree are generally more tolerant of low soil moisture conditions than *O. sativa*. Deepwater and upland ecotypes of *O. glaberrima* are grown in West Africa (Mclean et al., 2002). Deepwater African rice cultivars are produced in flooded river basins and deltas. Upland African rice is usually grown in rainfed culture. When adequate irrigation is available, Asian rice cultivars generally have greater yield potential than African rice, but, in non-irrigated rice fields in West Africa, Asian rice is more susceptible to drought stress, soil nematodes, insects, and viruses. Monty Jones, co-recipient of the World Food Prize of 2004, cross-bred *O. sativa* and *O. glaberrima* to produce interspecific hybrids. Several challenges were overcome including "sexual incompatibility and hybrid sterility" (Sarla and Swamy, 2005). One of the goals was to combine the yield potential and grain quality of Asian rice with the pest and drought resistance of African rice. The Africa Rice Center (formerly WARDA) releases interspecific hybrids called "Nerica" (Figure 4), which were produced by crossing the two species.

Fig. 3. Nerica rice cultivar from the Africa Rice Center grown with center pivot irrigation in Limpopo Province, South Africa.

2.2 Rice cultivars and diseases

Rice plants are stronger and less susceptible to diseases when adequate water and nutrients are available. An effective disease control program for sprinkler irrigated rice begins with selecting cultivars or hybrids with the greatest disease resistance. The program should include regular field scouting for diseases and the option to apply fungicides, preferably by chemigation.

The most devastating disease to rice in North America is blast [*Pyricularia grisea* (Cooke) Sacc.]. Blast is spread by airborne spores. The severity of blast in a region varies greatly from year to year due to weather conditions. Crop rotation is an important control practice because the fungus can survive on straw residue from the previous year. When highly susceptible cultivars are grown with flood irrigation, plant pathologists recommend maintaining a deep flood (>10 cm, *Rice Production Handbook,* 2001). Since sprinkler irrigation is not flooded, of course, this is not possible. Blast disease was common in many Mid-South United States rice fields in 2009 (Figure 4, Table 1).

Fig. 4. In 2009, blast infection on 'Francis' cultivar with no fungicide applied resulted in complete crop loss under center pivot irrigation at Portageville, Missouri.

Center pivot irrigated rice planted with moderately resistant cultivars such as Templeton (Lee, 2009) combined with fungicide chemigation was effective in controlling the disease. Currently, there are no commercial cultivars in the United States that are 100% resistant to the disease. Crop rotation and straw management between growing seasons helps control the disease, but blast can mutate to a different race and cause serious injury to a previously resistant rice cultivar. Therefore, it is important that fields be scouted closely and fungicide applications made before the disease becomes severe.

Some debate exists among farmers on how to best manage sprinkler irrigation on rice in the presence of a fungal disease. Should more or less water be applied? Keeping rice leaves wet with frequent irrigations could intensify the problem. However, brown spot disease, (*Bipolaris oryzae*, Breda de Haan Shoemaker), in particular, tends to flare up when rice is stressed from lack of soil water. Reducing irrigation could actually increase the disease. Water-stress from lack of water makes rice plants more susceptible to diseases than plants with adequate soil moisture. In Missouri in 2008 center pivot trials, brown spot disease occurred on susceptible rice cultivars (e.g., 'Wells') when we failed to maintain soil moisture tensions below 50 centibars. Wells and Francis were the only cultivars out of six tested where we observed brown spot disease. After applying fungicide by chemigation and increasing the irrigation amount, the brown spots were not found on new growth upper leaves.

Rice cultivar	No fungicide	Fungicide applied
	--------------------------kg rice grain ha-1----------------------	
Templeton	7,950	8,335
Cocodrie	5,550	8,000
Taggart	5,400	7,880
Francis	0	7,345
Wells	850	7,220
Catahoula	7,350	5,990

Table 1. Rough rice grain yields in 2009 (high blast environment) from cultivars with and without azoxystobin fungicide applied by chemigation through the center pivot system.

Shealth blight (*Rhizoctonia solani*, Kuhn) is less likely to be a problem with center pivot than flood irrigation (A. Wrather, personal communication). Although brown spot and blast may increase with center pivot irrigated rice, the incidence of sheath blight diminishes in this cropping system. In flood irrigated rice, the soil borne fungus moves in the water and attacks the rice stems. Without a flood, we have not found shealth blight in rice.

2.3 Hybrid rice

In flood irrigated trials across the southeastern United States, rice hybrids averaged 17 to 20% higher yields than inbred cultivars (Walker et al., 2008). Hybrid rice lines usually produce more tillers and have greater blast resistance than inbred rice cultivars under center pivot irrigation. In a high blast disease environment, hybrids can also respond to fungicide applications. Planted in the same field as inbreds in 2009 (see Table 1), RiceTec CLXL729 yielded 8,300 kg ha-1 without fungicide and 9,950 kg ha-1 with fungicide.

Hybrids are better suited for sprinkler irrigation than inbreds because of their vigorous root growth. If a sprinkler irrigation application is delayed, rice hybrids are less likely to show leaf curling. However, hybrid seeds tend to cost more per bag than inbreds. Some of this cost is offset by the lower seeding rate. Because rice hybrids produce more tillers per seed than inbreds, fewer seeds per hectare are needed. Typically, hybrid seed is drill planted at 28 to 34 kg ha-1 versus inbred seed planted at 100 kg ha-1.

Making two 45° angle drill passes at one-half rates helps avoid gaps in the plant canopy, although most growers may not want to spend the extra planting time. The low seeding rate for hybrids makes plant population uniformity especially critical. Scout the field shortly after emergence to assess the stand. If a field has a history of bird damage, applying a bird repellent to the seeds before planting is a good prevention measure. Early in the season, a thin plant population makes weed control more difficult. Fortunately, by mid-season, the robust tillering of hybrids helps them fill in the open areas. If there any large skips in the field, replant those areas as soon as possible for better weed control. Uneven stands make weed control difficult later, especially in fields irrigated by center pivot systems. In conventional rice culture, flooding prevents water stress and suppresses many weeds. A complete crop canopy helps control weeds since any place in a field there is not a rice plant growing to shade the soil, weeds will emerge and grow.

Fig. 5. Seeding rate for inbred rice cultivars is usually 60 to 70% higher than hybrid rice.

3. Irrigation management

3.1 Flood versus sprinkler

Most types of rice do not tolerate low soil moisture even for short periods of time. For rice grown under center pivot irrigation, this is a special concern since there is not a paddy full of water to act as a buffer. Depending on the pumping capacity and irrigation rate, large pivot systems can require more than two days to complete a revolution. If the soils have low water holding capacities, such as low organic matter sandy soils, rice plants are more likely to become stressed between water applications. Also, irrigation systems in humid regions are often under-designed assuming some rainfall will occur. With these systems in a dry growing season, farmers cannot keep up with evapotranspiration (ET, the combined water loss from soil and plants) demand of the crop.

No reports were found in the literature for side-by-side comparisons of water use between flood irrigated and center pivot irrigated rice. Burt et al. (2000) found the potential irrigation application efficiency for continuous flood irrigation is 80% under practical conditions, which is within the range they reported for center pivot systems (75 - 90%). However, they added that surface irrigation systems "require the most 'art' of all the irrigation methods, both to obtain a high distribution uniformity and a high application efficiency. In general, people have not learned the art." In practice, large quantities of water can be lost from fields using surface irrigation, including continuous flood.

Rice farmers experience less-than-optimal flood irrigation application efficiencies for many reasons. Leaky levees and sandy areas that do not retain flood water are common problems. In the United States, farming operations are typically spread over large areas, requiring farmers to simultaneously manage numerous irrigation systems at different locations. One worker is often responsible for managing pumps in several fields, requiring him/her to move from field to field to determine when to begin water application and return to the field to determine if the irrigation is complete and shut off the water supply. Each field waters differently and often differences are observed within fields due to factors like highly variable soils. Farmers often produce multiple crops, so they may be harvesting wheat and planting double-crop soybeans at the same time they are applying the initial flood to their rice.

With flood irrigation in the US Mid-South, rice requires considerably more water than other crops (Hogan et al., 2007). In Arkansas, the Rice Production Handbook reported typical values for the amount of irrigation water applied to rice on Arkansas soils ranged from 610 to 1220 mm (Tacker et al., 2001). Similarly, Vories et al. (2006) reported a range of 460 to 1435 mm observed for 33 Arkansas rice fields during the 2003 through 2005 growing seasons and Smith et al. (2006) reported values from 382 to 1034 mm in Mississippi in 2003 and 2004. Even at the low end of the Handbook range (610 mm), rice production in Arkansas over the ten years from 2000 through 2009, based on harvested cropland hectares from USDA-NASS (2010), required an average of 3.6 billion m^3 of irrigation water per year.

The large amount of water applied to flooded rice has resulted in two problems: the energy costs associated with pumping make up a significant portion of the rice production budget, and the cost is greatly influenced by fluctuations in energy prices; and water shortages are being observed in some rice-producing areas. The US Army Corps of Engineers (2000) reported that by 1915, only about a decade after commercial rice farming began in parts of Arkansas, the main alluvial aquifer was already being tapped at a rate that exceeded its ability to recharge in some areas. The aquifer serves as the principal water source for agriculture in eastern Arkansas and surrounding areas, and similar problems have been encountered with some surface water sources in the region.

Reducing the water requirements for rice has been a goal of farmers and researchers for many years. Vories et al. (2005) reported that a multiple inlet approach required 24% less irrigation water than conventional flooding and the method has been widely adopted. Producing rice in a row-crop culture with furrow irrigation rather than with continuous flood was also investigated. Vories et al. (2002) compared furrow irrigation of rice with conventional flooding and reported consistently lower yields. Studies during the 1980s addressed sprinkler irrigation of rice in Louisiana (Westcott and Vines, 1986) and Texas (McCauley, 1990) and reported large yield reductions compared with flooded production. Producers will not readily abandon flooded production for an alternative system that produces lower yields. However, more recently, Vories et al. (2010) reported comparable yields between center-pivot irrigated and flooded rice on a producer's field in Arkansas.

3.2 Center pivot irrigation field tests

In response to renewed interest in center pivot rice, the University of Missouri and USDA-ARS began to investigate center pivot irrigation of rice in Portageville in 2008. The project was in conjunction with the Missouri Department of Natural Resources and Valmont Irrigation, Valley, Nebraska.

Much of Southeast Missouri farmland has abundant groundwater near the soil surface. For rice farmers in this region, the advantages of center pivot rice production are less about water savings than being able to grow rice in fields that do not hold flood water. Producers are looking for additional options in crop choice on fields not suited to flood irrigation and ways to cut energy costs for pumping. In 1811-1812 A.D., four earthquakes occurred in the Mid-South United States exceeding 8 on the Richter scale. The events shattered the alluvial plains between the St. Francis River and the Mississippi River, extending south of New Madrid, Missouri, to Marked Tree, Arkansas. During the quakes, sand was extruded from the water-saturated subsoil and flowed upward in geysers to the surface forming sand blows. The impacted counties in Northeast Arkansas and Southeast Missouri are now part of a major rice producing region of the United States, but farmers must be cautious when selecting fields to produce rice because some fields in do not hold flood water very well. Freeland et al. (2008) discussed sand blows and fissures, which are similar to sand blows but linear in nature, and despite land grading efforts by farmers, these features still persist in fields and affect the abilities of soils to hold flood irrigation water for rice production.

Rainfall in the Mid-South United States is sufficient for rainfed crop production, but periods of drought during the growing season make irrigation essential for optimum yields of all widely produced summer crops. Furthermore, climate change is expected to increase the frequency and severity of drought in the region. However, irrigation scheduling, the correct timing of irrigation during the growing season, is more difficult in sub-humid regions like the Mid-South than in arid locations. Factors such as cloudy weather, rainfall, and temperature swings caused by the movement of weather fronts all complicate irrigation scheduling. Weather conditions in sub-humid regions vary greatly from year to year and even within a year and the variability must be accounted for in the scheduling system. Most commonly used methods either measure or estimate soil water content. Although many types of instruments have been developed to measure soil water content, using many different kinds of technology, all have drawbacks. In addition, the highly variable soils in the region have limited the use of soil water measurements for irrigation scheduling.

In our first series of experiments, we focused on nitrogen management and the typical irrigation schedule was 12 mm of irrigation water applied every other day unless rainfall occurred. WaterMark ® (Irrometer, Co., Inc., Riverside, CA) soil moisture sensors were used to help manage irrigation water applications (Figure 6). The sensors were installed at depths of 15 and 30 cm below the soil surface and buried wires from each sensor were connected to a central datalogger that transmitted the data by radio to a server and the internet. Dr. John Travlos and Greg Rotert at the University of Missouri developed an alert system which notified us by email and cell phone when average soil water potential dropped below -50 centibars (Figure 7). The system functioned satisfactorily, but sensor to sensor readings were variable, partly due to differences in the soil profile among the sensor locations.

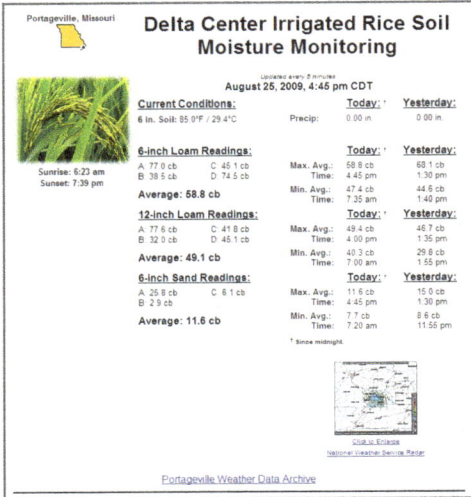

(c) internet website for soil sensor readings

datalogger and transmitter

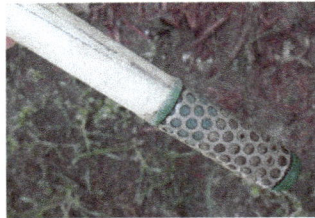

soil moisture sensor glued to PVC pipe

Fig. 6. WaterMark soil sensor (a), datalogger and transmitter (b), and webpage (c) displaying soil moisture updated every 5 minute intervals.

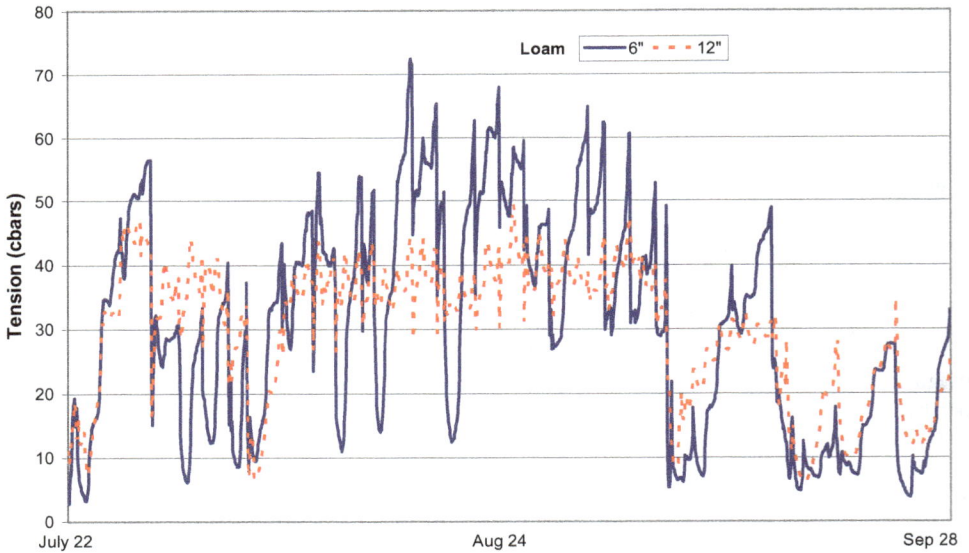

Fig. 7. Soil water tension on silt loam soil under center pivot irrigation in 2009 at Portageville, Missouri.

3.3 Water balance models

If crop water use estimates are available, daily soil water in the rooting zone can be tracked from the beginning of the growing season until harvest. *Soil water deficit* is the difference between the amount of water that is currently in a soil and the amount needed to fill the root zone back up to capacity (called *field capacity* or *well drained upper limit*). It pays to keep a balance sheet for soil water deficits, similar to a checkbook registry where rainfall and irrigation amounts are added and crop water use is subtracted, to determine when and how much to irrigate. Some farmers adjust water rate each time they irrigate and have a fixed schedule (e.g. twice a week). Others keep the same irrigation rate and vary the length of time between irrigations. The amount of water roots take up from soil between rainfalls or irrigations is affected by soil texture and rooting depth. As soils become drier, roots are not able to pull as much of the remaining water from the films surrounding soil particles. Generally, fine soil particles hold water tighter than coarse particles and the allowable soil water deficit to attain high yield varies between clays, loams, and sandy soils. Roots cannot grow as deep on soils with compacted layers called "pans" and allowable deficits must also account for this. Growers should record rainfall near their fields. Weather station measurements of solar radiation (sunlight), air temperature, humidity, and wind speed can be used to estimate a reference ET. ET from weather station data is typically referenced to a short grass (ET_o) or tall grass (ET_r) and is converted to ET for the specific crop (ET_c) by multiplying by an adjustment factor called a *crop coefficient*. Most methods that estimate soil water content rely on a crop coefficient to relate ET_c to a reference evapotranspiration at different growth stages. Allen et al., (1998) presented procedures for estimating crop coefficients.

The University of Missouri recently completed a web based ET advisory system for maize, soybean, rice and cotton (Fig 8) . A link to the site can be found at http://agebb.missouri.edu/ weather/ realtime/portageville.asp. Local farmers can access the information on portable computers or cell phones with internet access, allowing farmers to make decisions in the field about irrigation timing and rates by comparing weekly rainfall versus crop water use.

Crop coefficient values change during the season based on crop growth stage (Figure 9). The short-grass-reference coefficient curve shown in Figure 9 is experimental and is being evaluated in field trials in Missouri and South Africa. Heat units (Degree Day 50's) are used to predict rice growth stages for the x-axis. During early vegetative growth stages, the crop adjustment factor increases linearly, which increases the calculated daily ET_c. After the crop reaches about 80% canopy coverage, the adjustment curve flattens, but daily ET_c still varies based on temperature, cloudiness, humidity level and wind speed fluctuations. After a rain or irrigation when rice plants are small, reference ET_o may be a better estimate of total ET in the field due to increased evaporation from the wet soil and plants. Daily ET_o values are available on each University of Missouri real-time weather station website.

To schedule high frequency irrigation with systems such as center pivots, Allen et al. (1998) recommended using seperate coefficients for estimating crop transpiration and soil evaporation. In this system, a basal crop coefficient (K_{cb}) describes plant transpiration and a soil water evaporation coefficient (K_e) describes evaporation from the soil surface. The Arkansas Irrigation Scheduler (AIS; Cahoon et al, 1990) uses a dual crop coefficient approach to calculate a water balance to use in scheduling irrigation. Rooting depth is not used explicitly in the program, but is implicit in the choice of a maximum allowable SWD or management allowed depletion (MAD). Cahoon et al. (1990) provided a detailed description of the program and Vories et al. (2009) provided information about changes to the program since the earlier publication, including allowing the user to input a locally determined ET_o. However, since US rice is almost always produced with flood irrigation, much less work has been devoted to irrigation scheduling for rice than for other crops.

Crop Water Use Calculator

Station:	Portageville
Crop:	Rice
Weather Date:	8/4/2011
Planting Date:	5/5/2011
Output Format:	Millimeters

Rainfall totals:
Last 3 days: **0.00 mm**
Last 7 days: **8.64 mm**

Rice ET totals:
Last 3 days: **18.93 mm**
Last 7 days: **38.98 mm**

Date	Rainfall
8-04	0.00 mm
8-03	0.00 mm
8-02	0.00 mm
8-01	0.00 mm
7-31	4.32 mm
7-30	0.00 mm
7-29	4.32 mm

Date	ETcrop
8-04	4.81 mm
8-03	7.05 mm
8-02	7.07 mm
8-01	6.21 mm
7-31	4.72 mm
7-30	5.92 mm
7-29	3.22 mm

This report was created 8/5/2011 5:24 p.m. using Portageville-Delta Center data for Rice planted on 5/5/2011.

This information is provided by the University of Missouri Commercial Agriculture Automated Weather Station Network. The Commercial Agriculture Automated Weather Station Network and the University of Missouri give no warranty as to the accuracy, reliability, utility, or completeness of this information. While we use care to provide accurate weather/climatic information, errors may occur because of equipment or other failure. Users of this weather/climate data do so at their own risk, and are advised to use independent judgement as to whether to verify the data presented.

The University of Missouri expressly disclaims all warranties, express and implied, including, but not limited to, the implied warranty of merchantability and fitness for a particular purpose. Under no circumstances including negligence, shall the University of Missouri, or its Board of Curators, officers, employees or agents, be liable for any incidental, indirect, special or consequential damages (including damages for loss of business profits, business interruption, and the like) arising out of the use, misuse or inability to use the historical weather database or related documentation.

[AgEBB Home Page] - [Missouri Weather Stations] - [Comments]

Site maintained by people at AgEBB
agebb@missouri.edu

Fig. 8. Output from Missouri Crop Water Use program accessed by smart phone.

When tracking soil water deficits, use common sense to decide how many inches or millimeters to credit rainfall in the balance sheet. "Effective" rainfall amounts can be less than the total recorded by the weather station. Farmers may have a good idea of how much rain is lost to runoff water from a particular field, but runoff is affected by duration and intensity of the rain storm, soil properties such as texture, structure, initial water content, and slope, and vegetative and residue cover. Runoff is most likely when a large amount of rainfall occurs in a short time period or when the soil was already saturated before the rain started. Reduce the rainfall entered as a credit based on your estimate of water staying in the field.

Rice ET coefficient, K

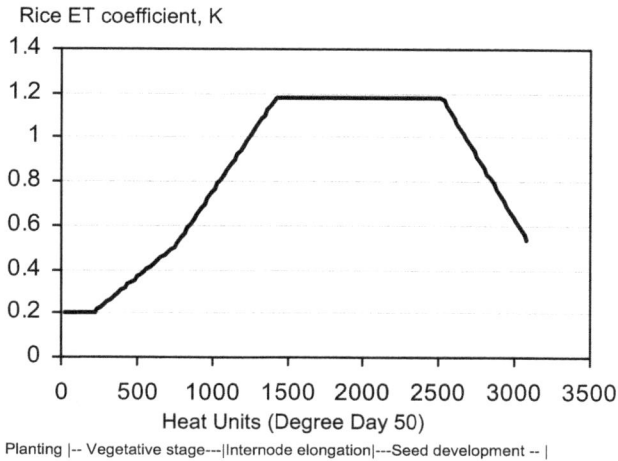

Planting |-- Vegetative stage---|Internode elongation|---Seed development -- |

Fig. 9. Experimental rice evapotranspiration adjustment coefficient (K).

4. Weed control

The most difficult part of growing rice without flooding for most farmers is weed control. In 2008, blackbirds fed on seeds after planting. The reduced plant stand made weed control more difficult later. As the surviving seedling grew, tillers helped fill in the plant skips, but in places where there were gaps in the canopy, new weeds continued to emerge all season and as a result we had to hand weed to keep the field clean. In 2009 and 2010, we treated seed with a bird repellant and planted the rice seed 1 cm deeper to inhibit bird feeding.

4.1 Herbicides

Herbicides are used in traditional flood irrigated drill seeded rice, but the suppression effect provided by flooding is a major component of rice weed control. The goal of farmers with center pivot irrigated rice should be to start with a clean field, spray weeds while they are still small (Figure 10), and manage the crop to acheive a solid rice canopy as soon as possible, similar to producing other non-flooded crops. The biggest challenge in 2008, our first year growing rice under center pivot irrigation, was controlling palmer amaranth (pigweed; *Amaranthus palmeri*, S. Wats.). We conducted weed control studies using conventional herbicides and Clearfield technology. Clearfield technology uses imazethapyr and imazamox herbicide resistant rice cultivars (not genetically engineered). When Newpath® (imazethapyr) herbicide was sprayed early in the season, we found that pigweed in the field were not killed because they were resistant to ALS (acetlactate synthase) herbicides. The pigweed apparently became ALS resistant when imazaquin (Scepter®) was repeatedly sprayed on soybean in the field in the 1990's.

Clearfield technology works well in fields without ALS resistance. Fortunately, we were able to develop an alternative program using clomazone, propanil, quinclorac, halosulfuron, acifluorfen, and bentazon. Spray timings are very critical for weed control and a successful program was clomazone pre-emergence followed by applications of propanil and quinclorac when the pigweeds were in the 2 to 4 leaf stage (Table 2). Our last herbicide treatment of the season was acifluorfen/bentazon applied to control any weeds not killed just prior to crop canopy closure.

Fig. 10. The key to maintaining weed control in rice is starting with a uniform plant population and spraying weeds while they are still small.

Weeds will be different at every location. In South Africa, our major weeds were wandering jew (*Tradescantia fluminensis*, Vell.) and wild watermelon (*Citrullus lanatus*, Thunb).

Application timing Weed stage	Herbicide‡	USD cost/hectare
Preemergence	clomazone	$28.75
2-4 leaf †	propanil + quinclorac + halosulfuron	$127.00
4-5 leaf	propanil + quinclorac	$84.57
4-5 leaf	acifluorfen + bentazon	$24.85
Total		$265.17

† Leaf growth stages were from palmer amaranth pigweed.
‡ Trade names for herbicides are clomazone (Command), propanil (Stam), quinclorac (Facet), halosulfuron (Permit), and acifluorfen + bentazon (Storm). Mention of trade names should not be considered an endorsement of these products by the University of Missouri.

Table 2. Herbicides and costs per hectare for chemicals applied in non-weed control test areas under the center pivot irrigated rice at Portageville, MO in 2009.

Of course, weed control in center pivot rice depends on having good herbicides and rice herbicide availability is an issue in some African countries. Center pivot irrigation may give farmers the opportunity to plant rice in countries where it has never been produced on a large scale. While this is great, it can create short-term logistical problems for farmers purchasing rice herbicides. It is a "chicken and egg" scenario in that until chemical companies see a potential for future returns from large rice plantings, they will not apply for herbicide labels, but farmers cannot start growing rice without the herbicides. If governments want to promote rice production in their countries, procedures should be made to allow easier experimental use testing of standard rice herbicide such as propanil and quinclorac which have been safely used in other countries for decades.

5. Nitrogen fertilization and timing

Nitrogen in flood irrigated rice systems is usually applied as granular urea (46 % N) by ground spreader equipment early in the season and supplemented by airplane after fields are flooded. The average cost of applying N by airplane in Missouri is about $7 per 45 kg of fertilizer. In center pivot irrigated rice, fertigation using an injection pump is an option for splitting total N into several applications (Figure 11). Injection pump equipment can also be used for fungicide chemigation for blast to save additional aerial application costs. In flood irrigated fields, dividing total N from urea into several split applications usually does not increase rice yields unless there is a problem maintaining flooded conditions. In Missouri experiments with furrow- irrigated rice on sandy loam soils, splitting total N into three applications produced significantly higher yields than a single application (Hefner and Tracy, 1991). However, in Arkansas, applying additional N on furrow-irrigated rice with clay soil after R0 growth stage did not increase yields (Vories et al., 2002).

Most of the research reports for rice nitrogen management are from flooded culture. Extension recommendations for total N rates in flood irrigated rice fields are based on empirical N field tests and adjustments made for specific cultivars, crop rotation, and soil texture. Rice farmers using delayed flood irrigation sometimes split total nitrogen (N) between two or three N applications. The first urea N application is made at V4 growth stage on dry soils followed immediately with a permanent flood. The goal is to push the urea below the soil surface with irrigation water and maintain a consistent 2 to 4 inch flood depth until the field is drained before harvest. Without oxygen, soil bacteria do not convert ammonium from urea to nitrate. Denitrification occurs mainly in rice fields when inconsistent flooding causes alternating aerobic and anaerobic soil conditions. Midseason N at R0 growth stage applied via airplane is beneficial for increasing rice yield when there is a problem maintaining floodwater from levee leaks or sandy areas in fields.

Two rice cultivars and one hybrid were grown from 2008 to 2010 at Portageville, Missouri under center pivot irrigation. Plots were fertilized with urea (46 % N) broadcasted at V4 growth stage, followed by five weekly applications of liquid urea – ammonium nitrate (UAN, 32 % N) applied to simulate fertigation. Total N rates were 0, 50, 101, 151, 202 and 252 kg N ha^{-1} with 50% of total N from urea/50% from UAN or 25% of N from urea/75% from UAN. An interaction was found for yield between total N and rice genetics (cultivars and hybrids). Economic optimum N rates for most cultivars and hybrids were from 124 to 168 kg N ha-1. When 151 kg total N ha-1 was applied, hybrid rice produced 5,167 to 10,156 kg ha-1 while cultivars produced 3,648 to 9,110 kg ha-1 (Table 3). In 2009 and 2010, 151 kg total N kg-1 yielded significantly more grain when 75% of N was applied as UAN fertigations compared to 50% as UAN. However, for most total N rates, application programs did not make a significant difference in yield. For percent milled head rice, an interaction was found between year and rice genetics. However, head rice was not affected by nitrogen management.

Cultivar/hybrid	Total N	2009	2010
	kg N ha-1	----------kg rice ha-1-----	
Templeton	0	6,527	4,876
	50	8,045	5,393
	101	7,031	3,648
	151	9,110	5,827
	202	9,242	6,760
	252	8,020	6,514
RT CLXL729	0	7,812	4,385
	50	9,249	6,759
	101	9,387	7,465
	151	10,156	5,167
	202	10,092	7,056
	252	10,263	4,448

Table 3. Rice yields from center pivot irrigated research plots with different total nitrogen rates averaged across fertigation timing treatments at Portageville, Missouri.

For Missouri fields rotated with soybean, a recommended program for center pivot rice is 35 kg N ha-1 broadcast at first tiller growth stage followed by five weekly fertigation applications at 22 kg N ha-1. For the first tiller application, 17.5 kg N as urea and 17.5 kg N as ammonium sulfate are blended and broadcast dry. In tests on flood irrigated rice, the S in ammonium sulfate promotes early tillering and more leaves to shade weeds. Boron at (0.8 kg ha-1) is usually added to the blend to prevent B deficiency. The most common liquid fertilizer for fertigation is UAN (32%N). If the field was planted in a non-legume grass crop such as corn or rice, an additional sixth fertigation application of 22 kg N ha-1 is recommended.

6. Fertigation and chemigation

Acheiving good mixing of fertilizer or fungicide with irrigation water is critical to making uniform fertigation and chemigation applications. In South Africa, we injected ASN liquid fertilizer (ammonium sulfate nitrate) at the base of the main vertical pivot pipe without an atomizer (e.g., Mister Mist'r®) inserted in the injection port. The atomizer helps to spread the fertilizer in the irrigation stream, which is especially important when there is a short mixing length. Based on the size and green color of the rice, most of the nitrogen was applied in the nearest span to the pivot point. Liquid fertilizer is heavier than water and tends to flow on the bottom of the main pipe and may not have gone past the first span. To prevent this from happening, use of an atomizer and injecting at a port on the water supply pipe that will get the most turbulence before flowing up the pivot point (Figure 11 a) is highly recommended.

Other equipment required for fertigation are a backflow prevention valve, hoses, a chemical storage tank and injection pumps. A backflow prevention valve is necessary to prevent accidental chemical contamination of the irrigation water source (lake, river, well, etc.). Some atomizers come with an internal mechanical spring for a two way check valve to prevent water from the irrigation system from flowing into the chemical supply tank. If not part of the atomizer, a separate check valve is needed to prevent a fertilizer or chemical spill in case of a system malfunction.

Each year in the N experiment, azoxystrobin fungicide was applied through the center pivot irrigation system using an injection pump at R2 and R4 growth stages (Figure 12).

(a) Backflow prevention valve, chemical diffuser, and hose from injection pump

(b) Liquid supply tank for fertilizer or fungicide used for chemigation

Fig. 11. Essential equipment required for chemigation or fertigation for center pivot irrigated rice.

Two types of injection pumps, piston and diaphragm, are available for chemigation and fertigation (Figure 13). Piston type injection pumps are often more difficult to change rate settings but after they are calibrated, they provide very accurate delivery. Be sure the injection pump can be calibrated to match the rate per hectare and size of the field. If an endgun will be used be sure to include that area in the calculations and realize that the system flowrate will vary depending on the endgun status. If you have a pump that is not accurate at low rates for fungicide, it may be possible to dilute the chemical in water in the chemical supply tank to acheive the desire rate of active ingredient per hectare.

7. Insect control

Most of the insects that can attack flood irrigated rice will have the same or perhaps more severe impact in center pivot irrigated rice. The Missouri Rice Degree Day 50 program (http://agebb.missouri.edu/rice/ricemodel.htm) shows when to scout for specific insects and pesticide control options in rice.

Insect pressure was light in Missouri center pivot rice from 2008 to 2011. We sprayed one time for fall armyworm (*Spodoptera frugiperda*, J.E. Smith) control but rice farmers were also spraying flooded fields in the region for the pest at the same time. We are concerned that billbugs (*Sphenophorus* spp.) could be a problem in the future. Insecticide seed treatment is used to prevent billbugs from multiplying on young rice. We found several billbug larvae, tiny white worms, feeding on lower stems and roots late in the 2011 season in a field in continous rice since 2009 (Figure 13a). We have not found any grape colaspis (*Colaspis brunnea*, Fabricius) but based on their behavior in flood irrigated rice, we suspect they might be a potential problem in center pivot irrigated rice.

Fig. 12. Two types of injection pumps commonly used for chemigation. The pump on the left is a piston type and the pump on the right is a diaphragm type.

In South Africa, spotted maize beetle (*Astylus atromaculatus*, Blanchard) migrated in large numbers to the center pivot rice field from maize fields that had finished pollination (Figure 13b). The beetle mainly feeds on plant pollen. It generally is not an economic concern in maize because the abundance of maize pollen in the air. In maize, the tassel (male flower) is seperate from the ear and silks (female flower). But in rice the male and female are together and the majority of plants are self-pollinated. The cutting and chewing of the rice flowers by the spotted beetles to get the pollen appeared to cause sterility in some of the rice plants in our field. More research will need to be done to determine the economic impact of these beetles on center pivot rice.

8. Soil compaction

In flooded rice, a soil "traffic pan" is not a problem because water is readily available to the plant. But, in center pivot irrigated rice, compaction which prevents root growth to lower soil layers makes rice plants more prone to water stress between irrigations. Wet soils are especially vulnerable to soil compaction from heavy equipment. We are currently measuring soil compaction in a center pivot irrigation timing test in Missouri to determine the relationship between soil compaction and rice yields in different areas of the field.

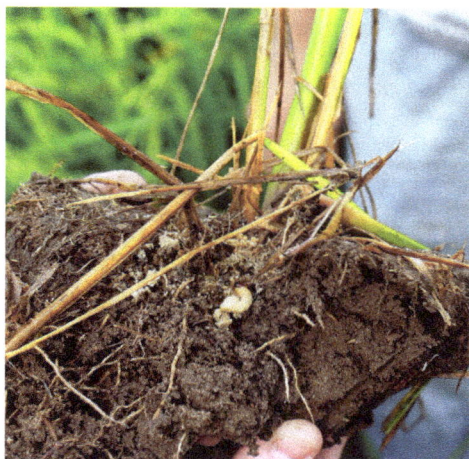

(a) billbug larvae on rice roots   (b) spotted beetle on maize tassel

Fig. 13. (a) Billbug larve feeding on roots and lower stem of center pivot rice in Missouri and (b) spotted beetles migrated from neighboring maize field to center pivot irrigation rice in South Africa.

9. Conclusions

As the world population increases more rice for food will be needed. Center pivot sprinkler irrigation is a viable method for producing rice, particularly in fields that are not suited for flood irrigation. Irrigation scheduling based on ET or soil moisture sensors is recommended to avoid excessive or inadequate applications of water and produce optimum yields. Selecting rice cultivars and hybrids with disease resistance and making timely application of chemicals for pest control are critical for producing a successful crop.

10. Acknowledgment

The authors wish to thank the Missouri Department of Natural Resources and Howard G. Buffett Foundation for funding the center pivot rice research in Missouri and South Africa. A special acknowledgement goes to Fred Ferrell. Without his vision and MidValley Irrigation this research would not have been done. We appreciate Valmont Industries and Jake LaRue for providing irrigation equipment, guidance, and recommendations. We appreciate the help from Andy Murdock and Blake Onken and personnel at Lindsay Corporation in South Africa. Finally a special thanks to Dr. John Travlos, Dr. Pat Guinan, and Greg Rotert at the University of Missouri for help with soil moisture sensor and weather station installation and maintenance.

11. References

Allen, R. G, L. S. Pereira, D. Raes, and M. Smith. 1998. Crop evapotranspiration: Guidelines for computing crop water requirements. *Irrigation and Drainage Paper No. 56*. Rome, Italy: U.N. Food and Agriculture Organization.

ASCE-EWRI, 2004. The ASCE standardized reference evapotranspiration equation. *Technical Committee report to the Environmental and Water Resources Institute of the American Society of Civil Engineers from the Task Committee on Standardization of Reference Evapotranspiration*. 173 p.

Bailey-Serres, J., T. Fukao, P. Ronald, A. Ismail, S. Heuer, and D. Mackill. 2010. Submergence tolerant rice : SUB1's journey from landrace to modern cultivar. Rice 3 :138-147.

Burt, C. M., A. J. Clemmens, R. Bliesner, J. L. Merriam, and L. Hardy. 2000. Selection of irrigation methods for agriculture. Reston, Va.: ASCE.

Cahoon, J., J. Ferguson, D. Edwards, and P. Tacker. 1990. A microcomputer-based irrigation scheduler for the humid mid-South region. *Applied Eng. in Agric.* 6(3):289-295.

Counce, P. A., T. C. Keisling, and A. J. Mitchell. 2000. A uniform, objective and adaptive system for expressing rice development. *Crop Sci.* 40(2):436-443.

Freeland, R. S., J. T. Ammons, and C. L. Wirwa. 2008. Ground penetrating radar mapping of agricultural landforms within the New Madrid seismic zone of the Mississippi embayment. *Applied Eng. in Agric.* 24(1):115-122.

Fukao T, Yeung E, Bailey-Serres J, 2011. The submergence tolerance regulator SUB1A mediates crosstalk between submergence and drought tolerance in rice. Plant Cell. 23(1): 412-27.

Hefner, S. and P. Tracy. 1991. The effect of nitrogen quantity and application timing on furrow-irrigated rice. J. Prod. Agric. 4: 541-546.

Hogan, R., S. Stiles, P. Tacker, E. Vories, and K. J. Bryant. 2007. Estimating irrigation costs. Little Rock, Ark.: Ark. Coop. Ext. Serv. FSA28-PD-6-07RV. Available at: www.uaex.edu/Other_Areas/publications/pdf/FSA-28.pdf.   Accessed   17 February, 2010.

Jones, J.B., 2003. Agronomic handbook: management of crops, soils and fertility. CRC Press LLC. New York.

Lee, F. 2009. New varieties have superior blast resistance. Rice Research News. 1(3): 5.

Luo, L. 2010. Breeding for water-savings and drought resistance (WDR) in China. J. Exp. Botany doi 10.1093/jxb/erq185

McCauley, G. N. 1990. Sprinkler vs. flood irrigation in traditional rice production regions of southeast Texas. *Agron. J.* 82(4):677-683.

Mclean, J.L., D.C. Dawe, B. Hardy, and G.P. Hettel. 2002. Rice Almanac: source book for the most important economic activity on earth. International Rice Research Institute. CABI Publishing, Wallingford, UK.

Reddy, C.S., A. Babn, B.P. Swamy, K. Kaladhar, and N. Sarla. 2010. ISSR markers based on GA and AG repeats reveal genetic relationships among rice varieties tolerant to drought, flood, or salinity. J. Zhejiang Univ.-Science B 10:133-141.

Rhine, M. Stevens, G., J. W. Heiser, and E. Vories, 2011. Nitrogen management for center pivot sprinkler irrigated rice. Crop Management. *In press.*

Rice Production Handbook. 2001. Cooperative Extension Service, University of Arkansas MP192-10M- 1-01RV.

Sarla, N. and B.P.M. Swamy. 2005. *Oryza glaberrima* : A source for the improvement of *Oryza sativa*. Current Science 89 : 955-962.

Smith, M. C., J. H. Massey, J. Branson, J. W. Epting, D. Pennington, P. L. Tacker, J. Thomas, E. D. Vories, and C. Wilson. 2006. Water use estimates for various rice production systems in Mississippi and Arkansas. *Irrig. Sci.* 25(2):141-147.

Snyder, R. L. 2001. PMday. Davis, Calif.: Regents of the University of California. Available at: http://biomet.ucdavis.edu/. Accessed 7 May 2006.

Tacker, P., E. Vories, C. Wilson, Jr., and N. Slaton. 2001. 9 - Water management. In *Rice Production Handbook*, 75-86. ed. N. A. Slaton. Little Rock, Ark.: Univ. of Ark. Coop. Ext. Serv. MP192-10M- 1-01RV.

United States Army Corps of Engineers. 2000. Grand Prairie Area Demonstration Project. Available at:
www.mvm.usace.army.mil/grandprairie/overview/default.asp.

United Nations. 2004. World population to 2300. Department of Economic and Social Affairs. Population Division. New York.

USDA-NASS. 2010. *Quick Stats.* Washington, DC: USDA National Agricultural Statistics Service. Available at: http://www.nass.usda.gov/.

Vories, E. D., P. A. Counce, and T. C. Keisling. 2002. Comparison of flooded and furrow-irrigated rice on clay. *Irrig. Sci.* 21(3):139-144.

Vories, E. D., M. Mccarty, G. Stevens, P. Tacker, and S. Haidar. 2010. Comparison of flooded and sprinkler irrigated rice production. ASABE Paper No. IRR10-9851.

Vories, E. D., P. Tacker, and S. Hall. 2009. The Arkansas irrigation scheduler. *Proceedings of World Enviromental and Water Resources Congress.* p. 3998-4007.

Vories, E. D., P. L. Tacker, and R. Hogan. 2005. Multiple inlet approach to reduce water requirements for rice production. *Applied Eng. in Agric.* 21(4):611-616.

Vories, E. D., P. L. Tacker, C. Wilson, S. Runsick, and J. Branson. 2006. Water use measurements from the Arkansas rice research verification program. In *Proc. 31st Rice Tech. Working Group*, 136. Baton Rouge, La.: LSU Ag. Center.

Walker, T., J. Bond, B. Ottis, P. Gerard, and D. Harrell. 2008. Hyrid rice response to nitrogen fertilization for Midsouthern United States rice production. Agron J. 100:381-386.

Westcott, M. P., and K. W. Vines. 1986. A comparison of sprinkler and flood irrigation for rice. *Agron. J.* 78(4): 637-640.

13

Sustainable Irrigation Practices in India

Rajapure V. A.[1] and Kothari R. M.[2]
[1]Dept. Of Microbiology, Sikkim University, Sikkim
[2]Rajiv Gandhi Institute of IT and BT, Bharati Vidyapeeth Deemed University, Pune
India

1. Introduction

During 1940-50's, water was amply available, its table was hardly 10-15 ft below the surface and needs of water were indeed meager due to simple life-style for India's population of 350 million. Therefore, with limited education, scanty knowledge of agronomic practices, insufficient financial resources and poor market information, nobody applied mind as to what should be done to make agriculture a sustainable profession, especially when water for irrigation from local rivulets, reservoirs and wells was available in the desired quantity. A few resourceful farmers used galvanized iron (GI) pipes for creating a network on the farm for carrying water from the source to different places. At that time, many considered this achievement a significant technological revolution, as it permitted to use the same quantity of water to take 2-3 crops per year by minimizing evaporational losses during the conveyance. This arrangement appeared to give a ray of hope to reduce total dependence on rain god *Varun*, who has not been kind enough in subsequent years for providing desired quantum of rain, at desired frequency and for the desired crop. Since then, GI pipes are totally replaced by economical PVC pipes to irrigate small or large farms economically and most practicably.

Table 1 has summarized the percentage of area brought under irrigation as a function of years. The kinetics of growth in irrigation may appear impressive at the first glance. However, a closer look reveals that it took 40 years (see data of 1990-91) to double the acreage of land under irrigation. The momentum picked up for irrigation since then is impressive and projected area under irrigation by 2010-11 is tentatively 47.5%.

Year	Area under irrigation (%)
1950-51	18.1
1960-61	19.1
1970-71	24.1
1980-81	29.7
1990-91	35.1
2000-01	43.4
2010-11	47.5

*Source: Dept. of Agriculture and Cooperation, Agricultural Census Division, 2010.

Table 1. Post-independence percent area under irrigation*

A closer look from a different angle summarized in Table 2 clearly points out that marginal land holders have mustered the courage, meager resources and efforts to bring maximum acreage under irrigation, no matter which was the source of water. This was probably for their sustainable livelihood without which food/financial security was almost impossible. This rationale appears holding true from the acreage brought under irrigation by large size land holders too, who had an alternative source of livelihood and agriculture was just another avenue of supplementary income.

Size Class	Canals	Tanks	Wells	Tube-wells	Others	Total
Marginal	3405	855	1296	5419	1409	12384
Small	2929	587	1971	4335	1069	10891
Semi-medium	3219	463	2500	4740	1010	11932
Medium	3447	276	2511	4502	811	11547
Large	1578	77	952	1715	550	4872
Total	14578	2258	9231	20711	4849	51627

All figures in `000 ha.
*Source: Dept. of Agriculture and Cooperation, Agricultural Census Division, 2000-2001.

Table 2. Distribution of irrigation sources for various classes of land holdings*

At this juncture, it was considered worthwhile to take a stock of different water resources.

1.1 Monsoon – A sustainable source of water

Presently, India cultivates annually 1-3 crops on its 125 million hectares of agricultural land, solely depending upon the availability of water for irrigation. Majority of the farmers undertake a single crop under rain-fed conditions using monsoon rain, generated by vast aerial circulations over the Bay of Bengal and Arabian Sea, facing east and west coast of India. These monsoon-generated rains, precipitated over 52-72 days from the last week of June until the last week of October, not only irrigate the crops, but they also replenish in a large measure sustainable source of surface and sub-surface water.

1.2 Surface sources of water

The surface source comprises thousands of small rivulets (locally known as *nullahs*), which merge in locally flowing minor rivers, turning into major rivers, which provide throughout yearly source of water for drinking, irrigation and industry. These major rivers are Sutlej, Ganga, Yamuna in north India, Teesta and Brahmaputra in N-E India, Narmada and Tapti in central India and Krishna, Godavari and Kaveri in peninsular India. The availability of water from these major rivers is guaranteed for the whole year for drinking, industry and hydroelectricity; for irrigation, it is available only to a section of farmers who have abundant resources to generate capital-intensive infrastructure for pumping water over 1-10 km distance.

Surplus water flowing through the above mentioned major rivers is diverted and collected in major dams like Bhakra-Nangal, Saradar Sarovar, Hirakud, Nagarjun Sagar, Koyna,

Damodar Valley and several locally constructed minor dams for supply to urban settlements, local industry and canal irrigation for agriculture.

1.3 Harvested sources of water

Rain water harvesting is not a new concept in India as historical excavations confirm the existence of village tanks, *bandharas*, bench terraces etc. to retard the flow rate of water and channelize it for storage. Water harvesting, which has been re-discovered and popularized, is borne out of sheer necessity.

It is on record that millions of Rajasthani families migrated to different parts of India due to chronic hardships experienced by them as a result of continued water scarcity. The grandma used to tell us the story as to why there was no alternative to migration, leaving farms, homes and immovable hereditary property behind. Water was so scarce that a child used to get 2-3 liters water for bath, while the elders were getting about 5 lit. Turn-by-turn, all family members used to take bath in a shallow stone tub with a small outlet at the bottom for the collection of used water in an underlying drum for its subsequent use in washing the clothes. The effluent after clothe washing was once again used for wiping the floor in the home and after that it was finally used in the evening for either spraying on the terrace to render it cool for over-night sound sleep in the absence of electricity or surplus effluent used for deficit irrigation. Thus, each drop of water was recycled 4 times and when availability of rains and harvested little water was in question, a momentous decision was taken to migrate (Jain, 2003).

The success of watershed and irrigation depends largely on 2 factors: (i) to harvest water and store it by constructing economical earthen percolation reservoirs or dams and (ii) to use it effectively through micro-irrigation system (MIS) displayed on demonstration farms for the cost and benefits to the farmers, who have a faith in 'seeing is believing', rather than a faith in formal agriculture education in universities.

Water harvesting has been made successful at the foot of hillocks in a totally degraded land by creating a reliable, captive and sustainable water storage and recharge mechanism through open larger reservoirs, which came into existence on the basis of topography of land. Water transiently collected in them was gradually transferred in the dug-out open wells for recharge by virtue of slope and seepage. This system has enabled to harvest and store about 1200 million liters of water under the scheme Watershed Development.

In simple terms, watershed development comprises of (i) allowing sufficient percolation of rain water in the catchment areas (recharge zone), (ii) permitting its further percolation in the command area (transition zone) and (iii) collecting flowing water at safe flow rate for judicious use in the delta area (discharge zone) (Jain 2003). This has been made possible by creating a network of terraces and trenches in the hilly region so that soil erosion is controlled, water percolation enhanced on the terraces for useful plantation to stabilize the soil strata and extra water harnessed through the network for the year-round use. Its pictorial presentation is depicted in Fig. 1-4.

This strategy has enabled Jain Irrigation Systems to convert barren hilly wastelands into irrigated, lush green, high yielding and sustainable farms through recharge and recovery of the ground water through 25 dug-out wells and 33 bore wells (Fig. 4).

Fig. 1. Hill terracing undertaken at the Jain watershed

Fig. 2. Jain Sagar reflecting the efforts in watershed development

Contour trenches are useful for increasing soil moisture and impounding the run-off surplus water over the slopes (Fig. 1). During bench terracing, small streams generated in the direction of the slope are diverted by a technique known as gully plugging, by using rubble structure of locally available stones or earthen embankment. The rain water allowed to percolate on bench terraces increases its agro-potential, yields more crop and surplus water is collected in the vast open reservoirs and percolation tanks (Fig. 2) to help natural recharge of under-ground aquifers (wells) through pumping or overflowing to fall in the well. There is a criticism that watershed development is problem-centric and location-specific. However, this has little merit, if its transplantation is done with modification(s) of design and structures for water harvesting, storage and use for benefit under altered conditions.

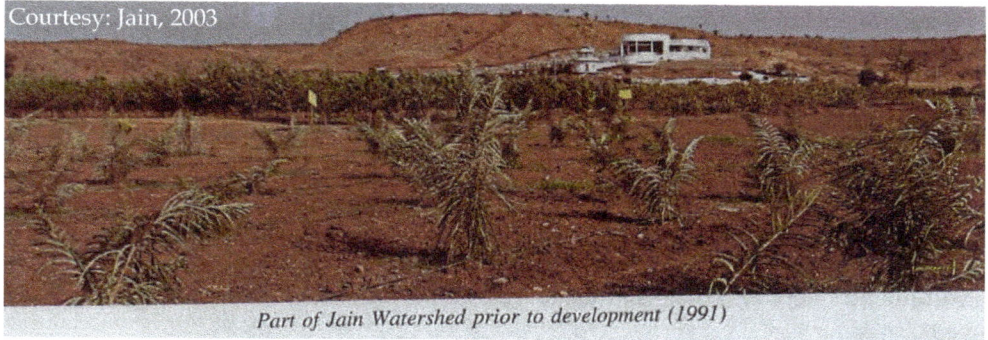

Part of Jain Watershed prior to development (1991)

Fig. 3.

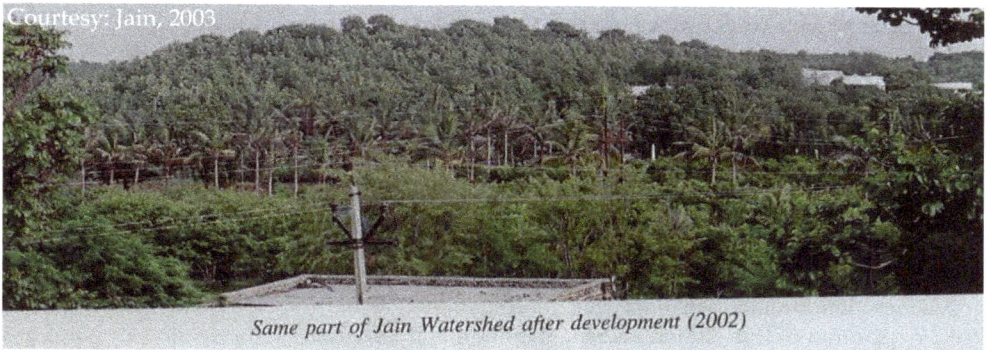

Same part of Jain Watershed after development (2002)

Fig. 4.

Watershed management in practice involved:

- Controlling, channelizing and collecting the surface run-off water
- Reducing the impact of more rainfall on soil erosion
- Decreasing the speed of flowing water by increasing its local infiltration
- Enhancing the moisture-holding capacity of soil
- Improving the soil texture and fertility
- Minimizing the chance of over-irrigation
- Arresting the ecological degradation and
- Increasing the productivity through crop intensity per unit area, per unit time and per unit of water utilized

This success of watershed development has been put to use and practiced on a large-scale by banana growers in Jalgaon region of Maharashtra who initially thought that there is no alternative to flood irrigation. However, watershed development and its imaginative conveyance for drip irrigation and ultimate use for tissue-cultured banana has doubled the acreage as irrigation by Micro-irrigation systems (MIS) requires almost 50% less water and secondly, maturity (harvesting) period of banana crop reduced from 18 to 12 months.

Presently, 90 million hectares of land is totally dependent upon annual rainfall to grow any crop, which provides livelihood to about 65% of our rural population. Rain water harvesting has the potential to add 30 million hectares to the irrigated area which ranges in 38-42% of the total land under irrigation.

1.4 Sub-surface sources of water

In the process of water percolation, seepage in short term might appear as wastage of water, but due to deeper penetration it is a safe storage of water, recharging the ground aquifers for long term usage. Therefore, another major source of water for irrigation has been underground reservoirs (wells and tube-wells) which are largely created by self-financing by the farmers and have their huge regional significance for enabling the farmers to take as many as 3-4 crops, the fourth crop being short duration vegetables, spices and condiments. Such artificial recharge by first obstructing the flow of water and then injecting in the wells/tube-wells merely by gravity has the following advantages:

- Creates reliable source of water
- Minimizes depth of water to pumping level
- Increases discharge rates
- Prevents decline of water table in the region
- Maintains safe ground water balance for future use

1.5 Recyclable sources of water

Vast stretches of land are irrigated to produce substantial amount of vegetables throughout the year in urban India, using unprocessed municipal waters, leaving the quality of vegetables doubtful in nutrition, due to contamination of pesticides and heavy metals. Precise quantum of this effluent, acreage under it and quantum of vegetables produced are not known as this work is being undertaken by the unorganized sector.

Thus, through these four resources, Indian farmers undertake about 39-43% of cultivable area into irrigation to generate about 225 million tons of cereals, 15 million tons of pulses, 337 million tons of sugar, almost equal quantity of oilseeds, besides vegetables, spices, fruits and medicinal plants. Among the commercial water guzzling crops are rice, sugarcane, banana, jute, bamboo etc.

2. Objectives of irrigation systems

The type of irrigation undertaken has largely depended on the physical texture of soil, climatic conditions (high/low temperature, rainfall, relative humidity, nearby vegetation, photo-period), suitability of soil for cultivating a particular crop and objectives for which irrigation is sought. To illustrate these factors, following examples are given.

2.1 Rice nursery

While monsoon rains start in the last week of June, major requirement of irrigation arises for preparing rice plantlets in the middle of scorching heat of May, to be ready for large scale cultivation almost throughout India, especially in Punjab, Haryana, Western Uttar Pradesh (Rice bowl of India), coastal regions and other high yielding provinces. Irrigation at this juncture is highly critical when sun is scorching and yet, production of rice plantlets is utmost necessary. Any failure in irrigation during this period or subsequent late arrival of rains generates not only hardships for farmers, but also radically alters the prosperity of rural India and their purchasing power, promoting higher growth rate of Gross Domestic Product (GDP), due to partial loss of rice yields.

2.2 Nursery for horti-/floricultural crops

The cultivation of nurseries is not only imperative for rice, but it is equally imperative for sugarcane, banana, horticultural crops and floriculture, both for local consumption as well as value-added exports. The crops raised under nurseries include green bell pepper, black pepper, cardamom, several vegetables such as ochre, brinjal, capsicum, garlic, onion, potatoes, variety of high yielding grafts of several fruit species (mango, sapota, pomegranate, lemon, pineapple, oranges, etc.), flowers such as marigold, variety of cut roses, tuberose, gerbera and ornamental plants etc. (Singh *et al.*, 2002). Production of seeds for vegetable crops and raising tissue culture derived plantlets of fruit crops like banana need huge amount of water for irrigation (Fig.5).

Fig. 5. Greenhouse depicting nursery operations

2.3 Wheat farm irrigation

After rice, the second major consumed cereal is wheat, which too requires at least 4-6 irrigations during 120-135 days crop cycle for quality and quantity of yield. Since wheat cultivation is undertaken during November-April period, contribution of rain-fed irrigation is almost nil to minimal (except for the residual moisture in the soil). Therefore, vast stretches of wheat irrigation largely depend upon canal network, wells/tube-wells and percolation tanks, frequency of irrigation being dependent upon temperature, photoperiods, evapo-transpiration rate and quality of soil as well as variety of seeds sown. Being the second largest cultivated cereal crop after rice, irrigation requirement is huge.

2.4 Value-added crop cycle

A highly value-added crop such as saffron is cultivated throughout its crop duration either in green houses or under extremely cold climate where irrigation is minimally required.

However, farmers do not take any chance due to unpredictable frequency of raining and at times extreme precipitations, thereby creating a necessity of foggers, sprinklers or drip irrigation.

2.5 Fruit crops

While India has legitimate claim to be world's number one or two producer and processor of several fruit crops, it is not without huge investment in drip irrigation, where the frequency, quantum and duration of irrigation is largely determined by the climatic conditions and sustainability of the crop without compromising with the quality and quantity of output. For example, different types of berries are 4-5 months duration crop and yet they require regular irrigation for optimal yields. Similarly, banana crop yield suffers due to either scorching heat or extremely windy climate inducing high rate of evapo-transpiration (Allen *et al.*, 1998) prior to onset of rainy season, thereby requiring protection through not only irrigation or mulching (Fukuoka, 1978), but also through shade imparted by rapidly growing leguminous bushy plants (wind breakers). This is applicable in equal measure for commercially grown crops like grapes, oranges, guava, pomegranate etc.

From the above narration, it is abundantly clear that no matter which crop, soil texture and unpredictability in the climate, irrigation should be the mainstay of supplementing necessary moisture for keeping the plants turgid, healthy and photosynthetically active for partitioning nutrients from the soil, its micro-flora, soil conditioners and soil moisture into higher yields of cereals, legumes, tubers, roots, fibers etc. Therefore, suitability of a particular type of irrigation as a function of crops needs consideration.

3. Types of irrigation

Irrigation already accounts for about 55-70% use of water in India. Due to unawareness of the cost: benefit ratio in using modern irrigation practices, out of available water, almost 60% used in irrigation is wasted. According to WHO estimate, in India agriculture uses as much as 70-90% of all renewable water resources that are now gradually diverted to human use. Following types of irrigation are used in India.

3.1 Rain-fed irrigation

Rain-fed irrigation accounts for 75-90 million hectares of land in India. Limitations of rain-fed irrigation are highlighted through the following illustration of different leguminous crops of which India is the largest producer, processor, consumer and importer.

While some of the leguminous crops like Mung (*Vigna radiata*), Urad (*V. mungo*) are totally depending on rainfed irrigation, soybean needs intense irrigation for at least one month, cowpea (*Vigna unguiculate*) requiring for 2-3 months and arhar (*Cajanus cajan*) requiring for 4-5 months. A myth that semi-arid crops do not need much irrigation has always given sub-optimal yields and at times poor quality. It is on record that India produced about 12 million tons of pulses during 1950-60 when its population was between 350-450 million and due to myths associated with irrigation, it has never crossed the mark of 150 million tons production when its population has crossed 1210 million (2011 census). The reason behind such poor performance in yield of pulses was primarily due to poor (70%) germination rate, thereby losing the productivity by 30% from the beginning of cultivation. Another reason has been lack of irrigation during flowering and seed-bearing period due to which flowers either withered or seed formation was sub-optimal, substantially reducing the yield. The notable example is soybean which is produced at the rate of 1.8 tons per acre in USA, whereas it ranges between 0.6-0.9 tons in India (Patil *et al.*, 2003).

3.2 Surface/flood irrigation

This is the most common method used worldwide with certain modifications like furrow irrigation or terrace irrigation (Fig. 6a). In both cases, water control is manual and therefore it induces large variations in volumes of water applied per unit area. Under such situation, moisture distribution is uneven and deep percolation unavoidable (Jain, 2003). This is most undesirable during the period of water shortage.

Until 1970s, rains were abundant, timely and surface as well as underground reservoirs remained replenished giving the false hope that while flood irrigation is sufficient at little cost, why to resort to modern methods of irrigation which are capital-intensive. However, this myth has slowly vanished when farmers realized that flood irrigation, though economical in short term, has transformed productive soils into saline, alkaline, acidic and non-productive soils due to huge accumulation of salty, acidic and alkaline ingredients in the soil (Patil *et al.*, 2001; 2002a; 2002b).

A closer look at the soil texture revealed that it is not only the salt content, but organic carbon content in the soil and useful micro-flora totally dependent on carbon content was almost extinct from the soil, thereby substantially eroding its self-fertilization capacity (Ramamurthy *et al.*, 1996). This problem was compounded further by indiscriminate and prolonged use of pesticides of increasing toxicity to control pests of increasing virulence, which has diminished useful micro-flora (N_2-fixers, phosphate solubilizers, sulphur oxidizers, carbon composters, ecto-/endo-mycorrhizae etc.) (Chaudhari *et al.*, 2008). Therefore, to enhance the fertility of soil and commensurate productivity, farmers resorted to increasing use of chemical fertilizers, which were not only cost-intensive, but also energy- and pollution-intensive, perforce dictating an increase in the quantum and frequency of irrigation for sustenance of crops.

Over-irrigation, has suffocated ramified root system, microbial flora and earthworms with the resulting hypoxia affecting a cascade of growth-specific physiological processes and ultimately causing fungal infection not easy to overcome, thereby adversely affecting the fertility of soil, microbial flora in rhizosphere, root system and productivity. Typical examples are green bean (*Vigna radiata),* black gram (*V. mungo*), banana etc.

When surface irrigated areas are supplied water from public network of canal system (Santhi and Pundarikanthan, 2000), irrigation scheduling depends upon water delivery, its rate of discharge, duration and frequency. Since the farmer is not sure as to when he will have the next opportunity to irrigate his crop, he often indulges in over-irrigation, thereby defeating the purpose of conserving the water (Unger and Howell, 1999). This objective could be achieved with the guarantee of frequency so that irrigation is received in time and over-irrigation is curbed.

Flood irrigation, practiced on large scale, is almost abandoned, except for water guzzling monocot crops like rice (*Oryza sativa*), sugarcane (*Saccharum officinarum*), bamboo (*Bambusa arundinacea*), banana (*Musa paradisiaca*) etc . However, in view of colossal loss of water, problems created by flood irrigation and scarcity of water resources, drip irrigation (DI) has become the mainstay of moisturizing root system of plants. Its underutilization, merely to save crops, has desiccated the plants without alleviating heat stress and providing substandard productivity in quality and quantity.

Fig. 6. Types of irrigation; (a) Flood/furrow irrigation; (b) Drip irrigation; (c) Sprinkler irrigation; (d) Mist irrigation

3.3 Micro-Irrigation Systems (MIS)

This type of irrigation guarantees higher water use efficiency (WUE) at a greater benefit: cost ratio to the user (Zhang *et al.*, 1998). In this system, water is directly applied at the root zone of the plants or plantlets and in controlled quantities through a low pressure network and at desired intervals as per the requirements of the crops (Vasane and Kothari, 2006). In case, the land is at different heights, irrigation could be beneficially provided by the construction of distribution tanks at strategic locations (Jain, 2003). This permits minimizing evaporational losses, seepage and therefore provides most efficient use of water. Computation of the cost of MIS per hectare is about USD 2000 per hectare (Jain, 2003). It includes the cost of construction of water distribution tanks, cost of installing MIS and cost of replacing old material from time to time. This cost may appear a bit on higher side for the poor farmers of India. However, upon implementation in a step-wise manner, neither the capital investment is beyond the reach, nor repentance occurs for the loss of crop. In fact, from the total cost and total income, the pay-back period is about 7.5 years, in the total lifespan of MIS for 12-15 years. During this period, the farmer is assured of adequate food grains, vegetables, fruits, spices for quality life, while preserving fertility of the land and hope for a prosperous future as it permits re-growth of local micro-flora and fauna changing the ecology for better, useful to the farmer, village and nearby region (Vasane and Kothari, 2008).

Depending upon the texture as well as fertility of the soil and duration of the crop, Drip Irrigation (DI) systems (Fig. 6b) have become more sophisticated in terms of irrigating the desired volume of water required for the crop, without increasing the consumption of either electricity or manpower (Ayars et al., 1999). Basically, the principle underlying DI has been (i) to generate about 50% moisture in the rhizosphere for keeping the root system efficient for partitioning moisture along with locally available macro- and micronutrients into the shoot system, (ii) to help soil microbial flora colonize the rhizosphere so that dependence of soil solely on chemical fertilizers is reduced to the tune of 10-40% for a comparable crop yield, (iii) to permit the availability of adequate oxygen for the respiration of the root system, local micro-flora and earthworms, (iv) to allow maximum root ramification by softening the soil to increase surface area of the root system for the absorption of adequate moisture and nutrients and (v) to sustain the growth of ecto- and endo-mycorrhizae, which are inherently well-equipped to overcome the adverse effects of heavy metals, xenobiotics and agro-chemicals which are inhibitory to microbial and plant growth.

The sophistication of DI has substantially enhanced in the last five years by reducing the evaporation of moisture so that one could undertake at least 25-40% increased acreage under irrigation in the quantity of water saved and without compromising with the soil texture, microflora and sustainability in yields (Brats et al., 1987).

Indeed this system has been a harbinger of another revolution in irrigation as it brings several advantages to the farming viz., saving of water, reducing soil erosion, directed use of fertilizers and saving of energy, labor and finance with an added advantage of early harvesting period and 10-25% higher yield (Vasane and Kothari, 2008). This system has the potential to arrest dwindling levels of underground water table. A major step in this direction is taken by many state governments by subsidizing MIS to undertake 3-4 crops per annum, subject to availability of adequate water and electricity for pumping it to surface level.

From the above narration, it is abundantly clear that no matter which crop, soil texture and unpredictability in the climate, micro-irrigation should be the mainstay of supplementing necessary moisture for keeping the plants turgid, healthy and photosynthetically active for partitioning a variety of nutrients, plant protection agents and moisture into higher yields of cereals, legumes, tubers, roots and fibers.

Optimal utilization of DI as a function of texture of soil, quality of seeds and duration of crops has provided plant systems at its optimal growth, physiological functions and ultimate productivity. Vegetables (Solanum melongena), cereals (Zea mays), pulses (Pisum sativum, V. mungo), oilseeds (Arachys hypogaea), cotton (Gossypium spp.) and fruit crops (Ficus carica, Vitis vinifera, Psidium spp.) are typical examples.

3.4 Sprinkler irrigation system

Sprinkler irrigation system includes a set of traveling rain guns and continuous transfer of lateral system to the places where irrigation is required (Seginer, 1987). This type of irrigation has the merit of high application rate in shortest period, which is desired to save desiccating crop. However, it requires higher water pressure to improve the reach of water over larger surface area and therefore not appropriate for smaller fields, a reality in Indian scenario where an average farmer holds 0.5-2.5 acre of land. Secondly, sprinkler irrigation is not suitable for slopping farms, degraded land, heavy soils and windy conditions. Thirdly, during water scarcity, high water pressure is not feasible at many places in India. It is only an ideal system when application rates are high and farms are large. From field evaluation studies, a major disadvantage of sprinkler irrigation has come to our knowledge that farmers really do not control the depth of irrigation; therefore, problems of scarcity of water aggravate (Capra and Scicolone, 1998). On the contrary, since water is scarce, hence expensive, if farmers adapt to under-irrigation uniformly, they may accept potentially low yield instead of having no yield at all.

To overcome the problems of sprinkler irrigation using rain guns, micro-sprinkling has been designed for smaller farms with an irrigation frequency depending on the availability and quality of water (Fig. 6c). With this system, farmers attain improved irrigation efficiency merely by adapting irrigation schedules at desired frequency depending on the turgidity of the foliage of the crop (Pereira, 1999).

3.5 Mist irrigation

This type of irrigation has been prerequisite for tissue culture derived crops at primary hardening (Vasane and Kothari, 2008; Vasane et al., 2010) when the plantlets were indeed tender, required higher humidity level to overcome 'heating' which was equally necessary for rapid growth of plantlets and bio-acclimatization (Vasane et al., 2010) (Fig. 6d). Misting in secluded areas is also accorded to saffron crop; it is necessary for delicate ornamental plants and cut flowers prior or during the transit so that their flower/foliar conformation is retained without loss of freshness. Misting has helped fruit crops to delay maturation, to arrest over-sweetening of grapes, to overcome heat stress to banana under scorching period and to retain freshness of vegetables, fruits, sprouts etc. prior to auctioning or for enhancing the shelf-life.

3.6 Deficit or supplementary irrigation

Acute water scarcity enforces either deficit irrigation or supplementary irrigation (SI), not to the liking of the farmers, but by perforce of circumstances, they have to consider water scarcity in totality in terms of drinking for human beings and cattle, household purposes for hygienic living, economic purposes for optimizing agricultural/industrial output and environmental benefit to render life of farmers livable by improved irrigation management (English and Raja, 1996).

Deficit irrigation mode is designed (Shangguan et al., 2001) when water is scarce and yet a purposeful strategy is adapted to sustain the crop under increasing degree of water deficit; at the cost of yield, irrigation is adopted by many farmers in the case of *Jatropha curcus,* a source of non-edible oil for biodiesel production. These authors have proposed a model for regional optimal allocation of water under deficit irrigation and have illustrated the model by suitable application.

3.7 Simulation-based irrigation: A concept

In all the modifications suggested to overcome water deficit and yet sustain the crop, several computerized simulation models are available (Endate and Fipps, 2001). However, they are of little use since the farmers are neither aware of simulation concept, nor they practice such models of varying irrigation frequency. Farmers' orientation to understand this concept is a prerequisite before their application could be considered. At best, they could cultivate crops robust to sustain water scarcity by adopting minimal irrigation and other cultural practices (viz. mulching) so that they have good chance to maximize the crop yield per unit area of land per unit time (Fukuoka, 1978).

4. Problems of water scarcity during the farm management

Semi-aridity, aridity, drought and desertification are both, natural and man-made contributions arising out of increasing water scarcity over increasing duration. Under these conditions water conservation and water use efficiency needs to be considered on a priority basis, since water scarcity implies that locally available water of inferior quality (viz., municipal effluent) has to be relied upon for irrigation in an increasing manner. In addition, water distribution uniformity (WDU) would be a fundamental measure to reduce water demand for under-irrigation, scarcity irrigation, deficit irrigation, contingency irrigation etc. Therefore, regardless of the quality of water or its quantum, strategies for sustainable water management will have to be put into practice so that water scarcity in future could be readied presently and meaningfully.

4.1 Contingency irrigation

Water shortage may appear due to natural imbalance in the quantum of rainfall and its use; however, in a large measure, it is due to short-sighted human practices which indulge into its over-exploitation and reduced land use, propagating water shortage years after years. While water tables are falling, while water demand is dramatically growing at an unsustainable rate for additional food production. This situation of increasing water demand will continue to increase in the coming decades as style of living and income enhances their consumption of food. This situation invites consideration for moderate, immediate and permanent contingency irrigation models or practices.

To address this problem, large scale development of surface and underground resources is increasingly getting unacceptable due to environmental concerns. Compounding this problem, water delivery infrastructure built in recent decades is becoming obsolete due to silting of reservoirs and lack of repair in the irrigation network. The net result is falling levels of the water table by every passing day by providing water on demand to millions of farmers who tapped it using tube-wells to grow the crops. These developments clearly point towards increasing scarcity of water.

During irrigation, no matter it is deficit or contingency, the following factors need utmost consideration:

- Water harvesting and creation of water reservoirs on farm.
- An increased storage capacity on farm for local absorption of precipitated water.
- Minimization of evaporation in water reservoirs/canal networks through the use of eco-friendly chemicals (Unger and Howell, 1999).

- Farmers' orientation in thinking to take cognizance of agro-meteorological/hydro-meteorological predictions so that water scarcity does not aggravate its wasteful use, detrimental to local population.
- Agro-meteorological predictions should, enable planning for drought-tolerant crops to reduce the impact of water scarcity through mere irrigation scheduling.
- Drip irrigation is preferred to sprinkler irrigation to minima moisture evaporation.
- Uniformity of water distribution as a function of farm topography for minimal wastage and productivity variation per unit land area.
- Deep percolation of water beyond rhizosphere avoided by controlling the greed of farmers regarding uncertainty of quantity and frequency of next irrigation.
- Due to inter-dependence between supply and demand, wherever feasible, water of inferior quality used.
- Shortage of electricity and its availability at odd hours for irrigation circumvented to activate pumping from wells/tube wells by remote control, so that farmers are spared from snake/scorpion bite while trading long distances merely to start the pumps for timely irrigation.

In nutshell, outcome of irrigation is always determined by the availability of quality and quantity of water, quantum and frequency of irrigation, respiration of soil life and compositing of organic matter for plant growth promotion and protection, ultimately providing quality and quantity of food, fodder, fiber and pharma ingredients output.

4.2 Sustainable water management

It considers conservation of all water resources using appropriate technologies and their use with social acceptability, economic viability, and eco-friendliness. Under the head of social acceptability, cross subsidization of available water needs to be made into legitimate inter-regional (rural versus urban) and inter-sectoral (agricultural versus industrial) needs. Such considerations ahead of the scarcity will provide flexible practices in irrigation management. Otherwise, politically oriented and ill-considered decisions tend to aggravate human sufferings, cattle perishing, agricultural stagnation or decline and reduced industrial output, cumulatively affecting GDP adversely.

In semi-arid zones, tillage practices to control the run off of water, vegetation management to limit evaporation from plants and use of mulches for retaining moisture in the soil has the potential to play a central role. The preferred mulches are biodegradable (viz., bagasse) so that in the first season they serve as mulch and in the next season, they serve as compost. Synthetic mulch is the next alternative.

Under aridity, consideration of low moisture carrying capacity of the ecosystem in right perspective needs irrigation in small dosages and at higher frequency so that immediate hardships out of aridity are minimized (Sarwar and Bastiaanssen, 2001). This is a case of moderate contingency arising out of water scarcity.

Drought being a natural, but temporary imbalance in the availability of moisture caused by lower than the average rainfall over the years, its uncertain frequency, limited duration and severity of sunlight aggravate the rate of evapo-transpiration, resulting into diminished moisture availability for plants to sustain (Pereira, 1989). Severity of such situation needs application of soil conditioner (Chaudhari and Kothari, 2009) and DI at night time so that due to water-holding capacity of the soil conditioner, crop is sustained at a minimal loss due to evapo-transpiration, providing the hope of livelihood, especially for rural and economically weaker population. This is a case of immediate contingency due to acute water scarcity (Pereira, 1999).

Desertification in a large measure is man-made problem over longer duration, carried forward from the past in the availability of water. While drought aggravates desertification, recycling of water for human and cattle consumption and recycling for irrigation provides a workable strategy to arrest the rate of desertification and thereby human hardships. This appears an extremely difficult contingency, being permanent in nature and scope, requiring irrigation scheduling (Teixiera et al., 1995). This is a case of chronic contingency.

4.3 Demand management of water

This aspect needs holistic approach during the period of water abundance. This statement apparently appears as paradox. However, it provides time to conserve water, so that scope, duration and intensity of scarcity in future is minimized and more awareness created in the local population for water scarcity, that the available water has to be used for minimizing human/cattle hardship and economic use in agriculture/industry. This has been illustrated by Goyal et al. (2001) through an integrated approach for sustainable improvement in agro-forestry systems where demand of water is initially high for high success rate.

In fact, the forgotten merits of practicing crop rotation (Fukuoka, 1978, Patil et al., 2003), vermin-compost application (Chaudhari and Kothari, 2009), organic manuring (Ramamurthy et al., 1998) and green farming (Vasane and Kothari, 2009) have the potential to reduce the impact of water scarcity by virtue of their moisture holding capacity of the soil at no compromise with the potential of the standing crop.

4.4 Use of municipal effluent for irrigation during water scarcity

The municipal effluents generally contain carbohydrates, proteins, fats, lignin, soaps, synthetic detergents and miscellaneous products of common household usage, which are susceptible to biodegradability. Effluent with such composition can be ensured if an effluent containing chemicals from the process industry is released through the separate discharge lines so that domestic municipal wastewater could be readily utilized with either minimal treatment or without costly or time-consuming pre-treatment to minimize phytotoxic levels and health risks associated with it. Typical industrial effluents rich in pesticides, heavy metals, carcinogens, xenobiotics and extreme acidity/alkalinity have relatively less volume and industries are encouraged to treat it at the point of origin so that larger volume of industrial effluent has minimal contaminants, with enhanced suitability and feasibility of use only as a last resort when plant life is at a stake due to persistent droughts (Teixiera et al., 1995). In fact, waters contaminated with worms like ascaris, enteric bacteria and enteric viruses could be either decontaminated in shallow water ponds exposed to scorching sunlight or subjected to methanogenesis for biogas production (Suryavanshi et al., 2009) so that useful product is obtained, attainment of temperature beyond 40°C during biogas generation period has the chance of water disinfection and its use for irrigation. Thus, water treated through natural system of solarization or after biogas production could spare the field workers from direct contact with microbial infections and allow the soil to retain its useful micro-flora so that salad crops which are eaten uncooked and green fodder crops for cattle upon solarization are within the safety zones. Any other treatment prescribed in the literature does not sound logistically and economically feasible. However, monitoring of (a) microbial contamination in water to be used for irrigation upon solarization and (b) surface contamination of food and fodder is essential to minimize health and ecological risks, as prescribed by the WHO guidelines.

While the above irrigation practices provide a short term succor to make the life of a farmer livable, use of such sub-standard quality water needs to be monitored for long term effects by monitoring (a) dispersion of soil particles, (b) stability of soil aggregates, (c) permeability or infiltration rate of water, (d) exchangeable ions and (e) total dissolved solids. These measures have the potential to maintain osmotic pressure exerted by water within the limits on the growth and yield of plants, failure to which has potential of progressive decline in the yield (Sarwar and Bastihaassen, 2001).

5. Practicable measures to alleviate the water scarcity

The use of following adjuvant could be considered to retain moisture in the soil/plant.

5.1 Use of soil conditioners

As the availability of water for irrigation becomes scarce and concurrently the need for taking 3 crops per annum enhances to meet the needs of ever-increasing population, several adjuvant have been conceived, experimented, produced and used to find if the quantum and frequency of irrigation per crop could be reduced by their judicious use. Among these adjuvant, soil conditioners are at the forefront by virtue of their (a) moisture-holding capacity, (b) micro-flora propagating nutritive attribute, (c) fertilizing ability and (d) capacity to impart soft soil texture for permitting optimal germination of seeds as well as subsequent ramification of root system (Ramamurthy et al., 1998). Traditionally, night soil and cattle dung were used for this purpose. With the advent of chemical fertilizers, farmers have forgotten the merits of keeping cattle as a source of soil conditioners and supplementary source of income to tide over adverse situations created by the scarcity of water (Chaudhari and Kothari, 2009). However, the use of cattle dung has steadily been substituted by biodegradable waste from household, agricultural operations, industrial processing/municipal collections, which are now being subjected to hydrolysis, acidification and methanogenesis to provide biogas for agricultural operations, arrest the release of methane and CO_2 generated during the process and thereby minimize warming impact of greenhouse gases and at the same time use the solids as organic manure and its supernatant as plant growth promoter, thereby reducing the consumption of water through irrigation (Suryavanshi et al., 2009). The exploratory use of poly-acrylamide (*Jalshakti*) to hold moisture in rhizosphere did not succeed due to its exorbitant cost, potential to release carcinogenic monomers and inability to impart physical, chemical and microbial advantages imparted by the locally generated low cost soil conditioners and FYM (Ramamurthy et al., 1998). In fact, the soil conditioners have become vital inputs in raising the nurseries and ornamentals/floricultural crops for export, as soil conditioner takes care of plants' requirement of moisture from the time of packaging, quarantine inspection at the point of export, aerial transport, quarantine at the port of destination, auctioning market, until the ornamentals or the cut flowers reach the homes of consumers for sustained life up to two additional weeks (Singh *et al.*, 1995, Sharma *et al.*, 2004).

Soil conditioners derived out of barks of *Eucalyptus, Acacia,* Raintree etc. have not only imparted the attributes of soil conditioners, but also retarded the incidence of pests, presumably due to presence of terpenoids, alkaloids and flavonoids in the bark (Yadav *et al.*, 2002).

5.2 Application of press mud

It is a waste product emerging from the sugar industry, which has been extensively analyzed for its physico-chemical and microbiological ingredients for sustainable agriculture (Talegaonkar, 2000). Its inherent soft, spongy and fluffy physical nature provides porous texture to the soil, thereby promoting higher rate of seed germination and subsequent root ramification which provides more surface area and energy available for the adsorption of variety of nutrients from the rhizosphere. Its acidic nature renders press mud suitable for the amendment of alkaline soils, besides contributing the fertilizing value. Its organic nature demonstrates its potential to serve as a matrix for retaining viable counts of microbes in biofertilizers (Talegaonkar, 1999). Its composite nature indicates its utility for water holding and expeditious composting due to its overall growth promoting property. Its minor constituents like wax and sterols render it value-addition for sustained release of growth promoters and plant protection agents as a function of their hydrolysis by soil microbial flora. Its battery of trace constituents induce production of microbial iron chealaters (siderophores), imparting it an additional function of plant protection. Ultimately, its harmless nature and ease of application makes it an attractive candidate for moisture withholding during dry season or farming during rain shortage. In totality, press mud seems to have the potential to provide eco-friendly, cost-effective and sustainable bio-resource for increasing agricultural productivity through its multifaceted characteristics (Talegaonkar *et al.*, 2001).

5.3 Application of fly ash

Fly ash, the ultimate combustion product of thermal power stations, has been another soil conditioner and by virtue of its inorganic nature showed an excellent potential to meet micronutrient needs of plants. Apparently, very few have realized that timely availability of the micronutrients reduce the necessity of irrigation up to 10% in eco-friendly manner (Phirke *et al.*, 2001a, 2001b, 2004). However, it should also be kept in mind that its non-judicious use has the hazard of compacting the soil and reducing its rate of infiltration as well as availability of air for the respiration of rhizospheric microflora and root system (Phirke *et al.*, 2004).

5.4 Application of biofertilizers

High yielding inoculants such as efficient (i) *Azotobacter* for nitrogen fixation for cereal crops (Phirke *et al.*, 2001b), (ii) *Rhizobia* for leguminous crops (Talegaonkar, 2000), (iii) *Aspergilli* for phosphate solubilization under varied soil textures, pH and climatic conditions (Patil *et al.*, 2005), (iv) *Thiobacilli* for oxidation of elemental sulfur to solubilized sulfate radicals (Phirke *et al.*, 2001), (v) different types of bacteria/fungi for transformation of complex organic matter into easily assimilable plant nutrients (Ramamurthy *et al.*, 1998) and (vi) ecto- as well as endo-mycorrhizae to tide over salinity/alkalinity/acidity/heavy metal contaminants in the soil etc. have been the mainstay of providing plant nutrition without requiring colossal quantities of water used during the production of chemical fertilizers (Phirke *et al.*, 2002). Their use certainly supplements fertilizer value to different crops and in different soils to the extent of 20-40%. In spite of their eco-friendly and efficient performance, their use is restricted due to (a) non-familiarity of the farmers with their attributes, (b) conditions required for their sustenance and (c) availability of unethical biofertilizer preparations of doubtful integrity and viable spore count (Patil *et al.*, 2005). However, their use with water holding matrices like press mud, farm yard manure, compost, soil conditioners etc. has been a professionally satisfying experience for enhancing soil fertility and human health as found in soybean crop (*Glycine max*) (Patil *et al.*, 2006). The only limitation biofertilizers impose is that chemical fertilizers still need to be used in harmless concentrations and use of pesticides (Zope *et al.*, 2000) to combat a virulent pest be made as a last resort so that colonization of biofertilizers is retained year after year (Patil *et al.*, 2005).

5.5 Application of plant growth regulators

Plant growth regulators (PGRs), derived out of industrial or agricultural wastes, alcoholic or amino acids in nature, have served for the growth of plants either through supplemental nutrition or affording protection from pests through sustained availability of ingredients for the synthesis of chlorophyll (Sharma and Kothari, 1992, Yadav *et al.*, 1995). They too provide succor to plants for limited period during water scarcity through several mechanisms, including enhanced chlorophyll synthesis, increased root ramification, decreased stomatal opening and regulating several physiological processes to tide over stressful period (Sharma and Kothari, 1993b, 1994, Jolly *et al.*, 2005). While precise mechanism of their working is not clear, they are useful in enhancing productivity of rice (Sharma and Kothari, 1993a) and other crops (Sharma *et al.*, 1994, 1995). Table 3 has summarized outcome of joint application of soil conditioner and PGR.

5.6 Application of siderophores or biopesticides

Alternatively, *Pseudomonas* or its secretions, known as siderophores, have also promoted plant growth, presumably by rendering selective availability of Fe^{3+} to the plants and its concomitant denial to the pests. As pest nuisance arises due to shortage/excess of irrigation, synthesis of siderophores through fermentation and applicability on a large scale has been explored on groundnut crop with $10\pm 2\%$ increase in the yield and minimal affliction of *A. flavus*. Thus, siderophores could alleviate stressful situation due to water scarcity for some period (Chincholkar *et al.*, 2000).

While chemical pesticides have controlled many pests, their application has enhanced problems rather than solving them. To address this issue, our school has developed a number of strategies using botanicals for preparing biopesticides and exploring their use to pre-empt pest affliction at the cost of crop yield. In principle, the strategy conceived the use of de-oiled cake of neem seeds (*Azadirachta indica*) or jatropha (*Jatropha carcus*) or karanj (*Pongamia pinnata*) or babul (*Acacia arabica*) which not only enriched the soil for nitrogen, but also boost its resistance power to pre-empt likely pests (Patil *et al.*, 2000, Mendki *et al.*, 2000, 2003; Patil *et al.*, 2010).

Crop	Application rate	Qualitative effect	Quantitative effect	
			Control	Experimental
Lawn	0.4 kg.m²	Reduced water requirement	Greenish	Dark green
Rose	0.5 kg/plant	Longer shelf life	Less & small flowers	More & bigger flowers
Marigold	0.25 kg/m²	Fresh appearance	Small size	Large size, 100% higher production
Onion	0.8 kg/m²	sweet	9.8 kg	12.8 kg
Potato	0.5 kg/m²	Green foliage	Small size	Bigger size
Peanut	0.4 kg/m²	Bigger seeds	1.9 kg	5.7 kg
Sugarcane	0.4 kg/m²	Improved tillering	35 kg	60 kg
Eucalyptus*	2 l/plant	Higher survival	7.2 cm girth	16.3 cm girth
Bamboo	1.5 l/plant	Higher growth rate	Yellow green	Dark green
Kinnow**	10 kg/plant	Higher survival	121 cm height	160 cm height
Mango***	10 kg/plant	More vegetative growth	121 cm height	149 cm height

*After 22 months; **After 18 months; ***Langra variety

Table 3. Productivity of crops as a function of SC and PGR spray

5.7 Public participation for awareness of problems

If local governments could frame the code of conduct by clearly defining water allocation policy, their efficient use for social, agricultural, industrial and environmental purposes, employing appropriate technologies, it is the necessity of time. In the long term, public awareness on the availability of limited water resources and its conservatively judicious use to meet long term needs has to be created so that public in large measure adopts to reduced demand practices.

An apparent view that rainfall was scanty and after long intervals may be true; however, whatever quantum is received in the limited period, if diverted for aquifer recharge will certainly be the best local strategy, totally independent of the government lethargy. Once the aquifers are recharged, their most efficient use will depend upon the local initiative in conveyance, distribution on the farm and its efficient scheduling so that it is more useful to crop and soil system and least available for evaporation.

At management level, scientists like to define water use efficiency (WUE) as a ratio between assimilation by rhizosphere to transpiration rate by the plants. However, the most creditable criteria of WUE the farmers understand is ratio between crop biomass (for cattle consumption) and grains production (for human consumption). Ultimately, the wholesome view has to take in to consideration necessary human needs and maintenance of cattle for daily livelihood to cater long range agricultural dependence. The net outcome would be water productivity (WP).

The take home lesson of the above facts is life without water is indeed unthinkable; in fact, many farmers who have migrated from water scarcity areas have developed a firm conviction that water is synonymous to life. Their upbringing in decade-long drought has left a deep and indelible scar on their mind, thought process and actions that they undertake water harvesting, conserve it by proper storage, use it judiciously and recycle each drop so that it will never invite another migration during their lifetime.

5.8 Problem alleviation strategies

Flood irrigation in rice, banana, sugarcane etc. has rendered soil saline and its remediation was desired through application of soil conditions and other means (Tyagi and Minhas, 1998; Patil *et al.*, 2002a). Therefore, decisions on optimal strategies for irrigation under varying scarcity and climatic conditions are complex, especially when the farmers are totally dependent on rain-fed agriculture, where rainfall variability from year to year is extremely high.

In the last decade, a multi-date sowing strategy was suggested (a) to reduce the need of water during peak growth period by more than 20%, (b) minimize the chances of crop failure and (c) maximize the farm income. This rationale is based on a premise that at least one of the sowing dates will be able to meet local climatic optimal conditions to provide bumper crop to the farmers, which is denied to them by single cropping date, where chances of failure are more (Pareira *et al.*, 2002).

6. Partitioning of water

Traditionally, irrigation has been regarded as partitioning of water sources initially to soil and finally to plant system to alleviate heat stress for higher productivity. However, this concept has undergone radical change with understanding of the scientific processes at molecular level; it involves partitioning of moisture initially to soil to soften its texture for higher percentage of germination and ease of root ramification. From soil, roots peak up solubilized fertilizers and several microbial secretions by imparting momentum to colonization of useful microbial flora.   Subsequent partitioning of plant growth promoters/protectors to ecto-/endo-mycorrhizae initiated for onward transmission to root and shoot system. Besides, moisture partitioned in the soil promotes composting of organic matter in the soil and thereby partitioning humic acids to tender surface of root hairs and rootlets for growth promotion and protection. Another forgotten aspect of irrigation has been transfer of moisture to earthworms and their eggs for promoting secretion of hundreds of unidentified growth promoters, which are eventually partitioned to the root system, without disturbing uninterrupted flow of oxygen for the respiration of root system, micro-flora, earthworms and other live forms. Thus, as per modern concept , irrigation has multifaceted functions to promote life in soil, solubilize soil secretions, partition them along with moisture to microbial flora associated with the root system for further transmission through surface and microbes to plant shoot system.

7. Virtual water

By definition virtual water is the moisture embedded in a product, i.e. water consumed during its process of production and processing. This concept emerged in the 1990s and received increasing attention from scientists, technocrats and opinion makers concerned with water management related to food production/processing. Increasing inter-sectoral competition for water, the need to feed an ever growing population and increased water scarcity in many regions of the world, prompted to look at the way water is collected, stored, used and disposed on our planet. The water requirement for food is by far the highest, it takes 2-4 liters per head per day to satisfy the drinking water needs and about 1000 times as much to produce the food. This is why the concept of virtual water is so important when discussing food production and consumption.

The significance of virtual water at global level is likely to increase dramatically as projections show that food trade will increase rapidly (doubling for cereals and tripling of meat production between 1993 and 2020) (Rosegrant and Ringler, 1999). In simple words, a country that imports 1 million ton of wheat is importing 1 billion m^3 of water. For example, India imports oil/oilseeds and pulses to meet local demand. In this process, it is estimated that India imports gross virtual water to the extent of 3.9 billion m^3 per year (Chapagain and Hoekstra, 2003), while Zimmer and Renault, 2003 have computed import of 31 billion m^3 per year. Therefore, the transfer of virtual water embedded in the food that is traded is becoming an important component of water management on regional as well as global level, particularly in the regions where water is scarce.

By choosing a crop pattern for export purposes which has least ingredient of 'virtual water' for export at the cost of local population, economy and ecology, two aspects enjoy priority: local necessity (for service and ecology) and higher income to the farmers. While putting this philosophy in practice, (i) crop water requirement and (ii) irrigation scheduling (Pits *et al.*, 1996) are to be considered as both have economic implications and welfare of local population.

In fact, if the cost of sugar production from sugarcane is meticulously followed in terms of water consumed and its subsequent processing in sugar industry, the scenario becomes clear that sugarcane crop should be restricted to meet national requirements only, beyond which one is exporting moisture in the form of sugar and rendering the soil unproductive due to salinity at a colossal loss to national GDP.

In totality, for virtual water, market orientation of the crop choice is a tricky issue in short and long range, while meeting local needs without loss of income is always a superior option (Burt *et al.*, 1997). At such time, the use of micro-irrigation in preference to sprinkler system enhances distribution uniformity, a function of the choice of emitter for discharge for a crop and water pressure at emitters (Santhi and Pundarikanthan, 2000).

8. Conclusion

Water scarcity has huge implications for health, hygiene, sanitation, drinking water, agriculture and industry. Therefore, equitable distribution of this scarce resource has been accorded prime consideration in form of sustainable irrigation, which serves as a spring-board to provide food for public consumption as well as industrial raw materials. Drinking water enjoys second place in planning which in turn addresses legitimate needs of health, hygiene and sanitation. For this purpose, central, state, district and local (village level) governments have ensured voluntary public code of conduct to minimize the risk of over use of underground reservoirs and protect their water quality. Therefore, water extraction, conveyance, storage and delivery infrastructure is being augmented, pricing of water considered to reflect its net cost and delivery made at reduced pressure and reduced frequency. Towards this practice implemented in urban as well as rural areas, policy-makers and public have a tacit understanding that failure to judicious use of monsoon water costs health, industrial output and national economy as a whole.

Agriculture being the major user of water, besides reliance on rain-fed irrigation, gradually surface or flood irrigation is being replaced by sub-surface (drip/sprinkler/mist) micro-irrigation, which has inherent capacity to double the acreage under irrigation, without loss to agri-output. By experience, the farmers have also got educated that flood irrigation largely employed in sugarcane, rice and banana cultivation has rendered soil saline and less productive over the years. To reduce water use and transform saline soil into a productive matrix, sustainable water management is made through the enhanced use of farmyard manure, soil conditioner, press mud, fly ash/bio-fertilizers/plant growth regulators/bio-pesticide/ siderophores, which permit reduced use of water and at the same time soil fertility, is enhanced over 3-4 year duration. Similarly, crop planning is considered to restrict the cultivation of crops consuming "virtual water". For arid, semi-arid and desert landscapes, contingency irrigation/ deficit irrigation/ supplementary irrigation is considered to save the standing crops through recycling municipal effluents. Everything said and done, the ultimate success of irrigation practices depends on certain regulatory measures by the government and public participation through keen awareness.

9. References

[1] Allen RG, Pareira LS, Raes D, Smith M, 1998. Crop evapo-transpiration: Guidelines for computing crop water requirements. FAO Irrigation and Drainage Paper 56, FAo, Rome, 300 pp.

[2] Ayars JE, Phene CJ, Hutmacher RB, Davis KR, Schoneman RA, Vail SS, Mead RM, 1999. Subsurface drip irrigation of row crops: a review of 15 years research at the Water Management Research Laboratory, Agric. Water Manage. 42, 1-27.

[3] Bralts VF, Edwards DM, Wu I-Pai, 1987. Drip irrigation design and evaluation based on the statistical uniformity concept. In: Hillel, D. (Ed.), Advances in Irrigation, Vol. 4. Academic Press, Orlando, pp. 67-117.

[4] Burt CM, Clemmens AJ, Strelkoff TS, Solomon KH, Bliesner RD, Hardy LA, Howell TA, Eisenhauer, DE, 1997. Irrigation performance measures: efficiency and uniformity. J. Irrig. Drain. Eng. 123, 423-442.

[5] Capra A, Scicolone B, 1998. Water quality and distribution uniformity in drip/trickle irrigation systems. J. Agric. Eng. Res. 70, 355-365.

[6] Chapagain, AK and Hoekstra AY (2003). 'Virtual water trade: A quantification of virtual water flows between nations in relation to international trade of livestock and livestock products'. In: 'Virtual water trade' Proceedings of the International Expert Meeting on Virtual Water Trade. Value of Water Research Report Series No. 12.

[7] Chaudhari AB and Kothari RM, (2009). Soil conditioners as a pivotal biotech input for integrated farming and contingency income Textbook on molecular biotechnology, (Eds. A. Varma and N. Verma), I. K. International Publ. Pvt. Ltd., New Delhi.

[8] Chaudhari AB, Phirke NV, Patil MG, Talegaonkar SK and Kothari RM, (2008). Bio-fertilizers and soil conditioner for organic farming. in: Bio-fertilizers (Ed. P.C. Trivedi), Pointer publisher, Jaipur, pp. 39-79.

[9] Chincholkar SB, Chaudhari BL, Talegaonkar SK and Kothari RM, (2000). Microbial iron chelators: A sustainable tool for the biocontrol of plant diseases. In: Biocontrol

potential and its exploitation in sustainable agriculture. (Eds. R. K. Upadhyay, K. G. Mukerji and B. P. Chamola), Kluwer Acad / Plenum Publ., New York, Vol 1, 49-70.

[10] Endale DM, Fipps G, 2001. Simulation-based irrigation scheduling as a water management tool in developing countries. Irrig. Drain. 50, 249-257.

[11] English M, Raja SN, 1996. Perspectives on deficit irrigation. Agric. Water Manage. 32, 1-14.

[12] Fukuoka, M 1978. The One Straw revolution: *An Introduction to Natural Farmin,* Rodale Press, New York, USA.

[13] Gadgil A, Drinking Water In Developing Countries (1998), Annual Review of Energy and the Environment, 23, 253-286.

[14] Goyal D, Sharma RK, Ramamurthy V and Kothari RM, (2001). An integrated biotech approach for the sustainable improvement in agro-forestry systems. Innovative Approaches in Microbiology (Eds. D. K. Maheshwari and R. C. Dubey), B. Singh Publ., Dehradun, pp. 367-377.

[15] Jain B H (2003). A Telling tale, Jain Irrigation Systems Ltd., Jalgaon, pp 1-112.

[16] Jolly S, Yadav KR, Sharma RK, Kothari RM and Ramamurthy V, (2005). Response of *D. strictus* (Roxb) *Nees* seedlings to soil conditioner and PGR. J. Non-timber Forest Prod., 12, 187-190.

[17] Mendki PS, Kotkar HM, Upasani SM, Maheshwari VL and Kothari RM, (2003). Use of various botanicals for combating the attack of pulse bettle, *Callosobruchus chinensis* (Linnaeus). In : Biopesticides and pest management, Vol. 2, (Eds. O. Kaul, G. S. Dhaliwal, S. S. Marwaha and J. K. Arora), Campus Book International, New Delhi, pp. 70-79.

[18] Mendki PS, Maheshwari VL and Kothari RM, (2000). Papaya leaf dust as a post-harvest preservative for five commonly utilized pulses. J. Plant Biol., 27, 197-199.

[19] Patil AB, Vasane SR, Moharir KK and Kothari RM (2010). Contamination control and disease management of banana var. *Grand nain* plantlets. Acta Horti. 865, 247-253.

[20] Patil DP, Kulkarni MV and Kothari RM, (2001). Biotech inputs inherently bear the capacity to remediate saline soils. Perspectives in Biotechnol. (Eds. S. M. Reddy, D. Rao and Vidyavati), Sci. Publ., Jodhpur, 35-40.

[21] Patil DP, Kulkarni MV, Maheshwari VL and Kothari RM, (2002). A sustainable agro-biotechnology for bioremediation of saline soil. J. Sci. Ind. Res., 61, 517-528.

[22] Patil DP, Kulkarni MV, Maheshwari VL and Kothari RM, (2002). Recycled agro-waste and modified industrial byproduct with halophiles for improved yield of wheat (*T. aestivum* L) in saline soil. Physiol. Mol. Biol. Plants, 8, 117-124.

[23] Patil MG, Chaudhari AB, Phirke NV, Talegaonkar SK and Kothari RM, (2005). Why performance of PSMs was not guaranteed? In : Frontiers of plant science. (Eds. K. G. Mukerji, K.V.B.R. Tilak, S. M. Reddy, N. V. Gangawane, R. Prakash and I. K. Kunwer), IK International Pvt. Ltd., New Delhi, pp. 465-480.

[24] Patil RP, Chaudhari AB, Mendki PS, Maheshwari VL and Kothari RM (2006). Soybean as a Cindrella crop for enhanced soil fertility and human health. Focus on plant agriculture – 1: Nitrogen nutrition in plant productivity. (Eds., R. P. Singh, N. Shankar and P. K. Jaiswal), Studium Press LLC, Houston, USA.

[25] Patil SG, Patil MG, Mendki PS, Maheshwari VL and Kothari RM, (2000). Antimicrobial and pesticidal activity of *Nerium indicum*. Pestology, 24, 37-40.

[26] Pereira LS, 1989. Mitigation of droughts. 2. Irrigation. ICID Bull. 38, 16-34.

[27] Pereira LS, 1999. Higher performances through combined improvements in irrigation methods and scheduling: a discussion. Agric. Water Manage. 40, 153-169.

[28] Pereira LS, Oweis T and Zairi A, (2002). Irrigation management under water scarcity. Agricultural Water Management 57, 175-206.

[29] Phirke NV, Chincholkar SB and Kothari RM, (2002). Optimal exploitation of native arbuscular and vesicular arbuscular mycorrhizae for improving the yield of banana through IPNM. Indian J. Biotechnol., 1, 280-285.

[30] Phirke NV, Chincholkar SB and Kothari RM, (2004). Applicability of fly ash is sound for rhizosphere, biomass yield, enhanced banana productivity and sustainable eco-system. In: New horizons in biotechnology (Ed., S. Roussos), Kluwer Academic Publ., San Diego, USA, pp. 423-436.

[31] Phirke NV, Patil RP, Chincholkar SB and Kothari RM (2001). Recycling of banana pseudostem waste for economical production of quality banana. Resources Conservation Recycling, 31, 347-353.

[32] Phirke NV, Patil RP, Sharma RK, Kothari RM and Patil SF, (2001). Eco-friendly technologies for agriculture, horticulture, floriculture and forestry for 21st Century. In: The Environment – Global changes and challenges. (Ed. R. Prakash), ABD Publ., Jaipur, 77-127.

[33] Pitts D, Peterson K, Gilbert G, Fastenau R, 1996. Field assessment of irrigation system performance. Appl. Eng. Agric. 12, 307-313.

[34] Prinz D, 1996. Water harvesting – pst and future. In: Pereira LS, Feddes RA, Giley JR, Lessafre B (Eds), Sustainability of Irrigated Agriculture. Kluwer Academic Publishers, Dordrecht, pp. 137-168.

[35] Ramamurthy V, Sharma RK and Kothari RM, (1998). Microbial conversion of ligno-cellulosic wastes into soil conditioners. Advances in Biotechnol., (Ed. A. Pandey), Educational Publ., New Delhi, 433-438.

[36] Ramamurthy V, Sharma RK, Yadav KR, Kaur J, Vrat Dev and Kothari RM (1996). *Volvariella*-treated eucalyptus sawdust stimulates wheat and onion growth. Biodegradation, 7, 121-127.

[37] Rosegrant M and Ringler C 1999. Impact on food security and rural development of reallocating water from agriculture. IFPRI. Washington DC.

[38] Santhi C, Pundarikanthan NV, 2000. A new planning model for canal scheduling of rotational irrigation. Agric. Water Manage. 43, 327-343.

[39] Sarwar A, Bastiaanssen WGM, 2001. Long-term effects of irrigation water conservation on crop production and environment in semi-arid areas. J. Irrig. Drain. Eng. 127, 331-338.

[40] Seginer I, 1987. Spatial water distribution in sprinkler irrigation. In: Hillel, D. (Ed.), Advances in Irrigation, Vol. 4. Academic Press, Orlando, pp. 119-168.

[41] Shangguan Z, Shao M, Horton R, Lei T, Qui L, Ma J, 2001. A model for regional optimal allocation of irrigation water under deficit irrigation and its applications. Agric. Water Manage. 52, 139-154.

[42] Sharma KD, 2001. Rainwater harvesting and recycling. In: Goosen, MFA, Shayya WH (Eds.), Water Management, Purification and Conservation in Arid Climates. Technomic Publishing Company, Lancaster, PA, pp. 59-86.

[43] Sharma RK and Kothari RM (1993). An innovative approach to improve productivity of rice, Internat. Rice Res. Newslett. (IRRN), 18, 19-20.

[44] Sharma RK and Kothari RM (1993). Recycled cereal proteins as foliar spray enhances quality and production of food crops. Resources, Conservation and Recycling, 9, 213-221.

[45] Sharma RK and Kothari RM 1992 Innovative application of corn steep liquor for the increased production of food grains. Technovation, 12, 213-221.

[46] Sharma RK, Yadav KR and Kothari RM (1994). Innovative recycling of vegetable proteins waste for the increased crop productivity. , Technovation, 14, 31-36.

[47] Sharma RK, Yadav KR, Vrat Dev and Kothari RM (1995). Cereal protein hydrolysate: Environmental friendly plant growth promoter. Fresenius Environ. Bulletin, 4, 86-90.

[48] Sharma RS, Kothari RM and Ramamurthy V, (2004). The role of potting media in raising *Acacia auriculiformis* plantlets in nursery. Indian J. Forestry, 27, 51-55.

[49] Singh G, Sharma RK, Ramamurthy V and Kothari RM (1995). Enhancement of *Sprirulina* biomass productivity by a protein hydrolysate., Appl. Biochem. Biotechnol., 50, 285-290.

[50] Singh HP, Mann JS, Pandey UB, Singh L and Bhonde SR (2002). Planting material of horticultural crops: Issues and strategies. Publi. of Dept. of Agri. and Co-op and Natl. Horti. R & D Foundation, Nashik, pp 1-220.

[51] Suryawanshi PC, Kirtane RD, Chaudhari AB and Kothari RM, (2009). Conservation and recycling of pomegranate seeds and shells for value addition. *J. Renew. Sustain. Energy.*

[52] Talegaonkar SK (2000). Use of press mud for sustainable preservation of biofertilizers under ambient conditions. Ph.D. Thesis, North Maharashtra University, Jalgaon, India.

[53] Talegaonkar SK, Chincholkar SB and Kothari RM, (1999). Application of press mud as plant growth regulator and pollution arrester. Biofertilizers and Biopesticides, (Ed. A. M. Deshmukh), Technosci. Publ., Jaipur, 66-71.

[54] Talegaonkar SK, Chincholkar SB and Kothari RM, (2001). Press mud for sustainable agro-productivity. Indian J. Environ. Protection, 21, 41-50.

[55] Teixeira JL, Fernando RM, Pereira LS, 1995. Irrigation scheduling alternatives for limited water supply and drought. ICID J. 44, 73-88.

[56] Tyagi NK, Minhas PS (Eds.), 1998. Agricultural Salinity Management in India. Central Soil Salinity Research Institute, Karnal, India, 526 pp.

[57] Unger PW, Howell TA, 1999. Agricultural water conservation – a global perspective. In: Kirkham MB (Ed.), Water use in crop production. Food Products Press, New York, pp. 1-36.

[58] Vasane SR and Kothari RM, (2009). An integrated biotech approach to sustainable cultivation of banana var. *Grand naine*. Textbook on molecular biotechnology, (Eds. A. Varma and N. Verma), I. K. International Publ. Pvt. Ltd., New Delhi.

[59] Vasane SR and Kothari RM, (2006). Optimization of secondary hardening process of banana plantlets. *Musa paradisiaca* L. Var. Grand Naine. Indian J. Biotechnol., 5, 394-399.

[60] Vasane SR and Kothari RM, (2008). An integrated approach to primary and secondary hardening of banana var. Grand Naine. Ind. J. Biotechnol., 7, 240-245.

[61] Vasane SR, Patil AB and Kothari RM, (2010). Bio-acclimatization of *in vitro* propagated banana plantlets var. *Grand nain*. Acta Horti. 865, 217-224.

[62] Verma RK, Sharma RK, Maheshwari VL and Kothari RM, (2000). Calcium conjugates of amino acids and ecologically safe plant growth regulators. J. Plant Biol., 27, 85-88.

[63] Yadav KR, Sharma RK and Kothari RM (2002). Eco-friendly bioconversion of *eucalyptus* bark waste into soil conditioner. Biores. Technol., 81, 163-165.

[64] Yadav KR, Sharma RK, Yadav MPS and Kothari RM (1995). Human hair waste: An environmental problem converted into an eco-friendly plant tonic. Fresenius Environ. Bulletin, 4, 491-496.

[65] Zhang H, Oweis T, 1999. Water-yield relations and optimal irrigation scheduling of wheat in the Mediterranean region. Agric. Water Manage. 38, 195-211.

[66] Zhang H, Oweis TY, Garabet S, Pala M, 1998. Water-use efficiency and transpiration efficiency of wheat under rain-fed conditions and supplemental irrigation in a Mediterranean type environment. Plant Soil 201, 295-305.

[67] Zimmer D & Renault D 2003. Virtual Water in Food production and Trade at global scale: review of methodological issues and preliminary results. Proceedings Expert meeting on Virtual Water, Delft, December 2002.

[68] Zope HN, Kulkarni VA, Mendki PS, Nagasampagi BA, Naik K and Kothari RM (2000). An effort towards development of eco-friendly rural technology for *Parthenium*-based biopesticide. Ind. J. Environ. Protection, 20, 332-336.

14

Large-Scale Pressurized Irrigation Systems Diagnostic Performance Assessment and Operation Simulation

Daniele Zaccaria
Division of Land and Water Resources Management
Mediterranean Agronomic Institute of Bari (CIHEAM-MAI B)
Bari
Italy

1. Introduction

According to what documented by several authors and institutions, in many areas of the world irrigation projects perform far below their potential (Small and Svendsen 1992) and, in most of the cases, unrealistic or out-dated designs, rigid water delivery schedules and operational problems are among the principal reasons for the poor performance (Plusquellec et al. 1994).

The assessment of actual performance and potential improvement of conveyance and distribution systems received greater attention in recent years, and this trend will most likely extend to the near future, given that public and private investments will be more addressed to modernization of ageing or poor-performing irrigation schemes rather than to development of new irrigated areas or to expansion of existing irrigation schemes. In the perspective of service-oriented management, existing irrigation systems should be periodically evaluated for their performance achievements relative to current and future objectives. This requires diagnostic methodologies to analyze system behavior, assess current performance, identify critical aspects and weaknesses, and to investigate potential improvements. In this domain, several authors (Small and Svendsen 1992; Murray-Rust and Snellen 1993; Burt and Styles 2004) reported a remarkable lack of analytical frameworks by means of which irrigation managers or professional auditors can assess current achievements and diagnose feasible ways to enhance performance in the future. On the other hand, as pointed out by Prajamwong et al. (1997), identifying and implementing improvement changes entail the collection of field measurements and the use of analytical tools for developing feasible alternative scenarios and for selecting the most effective measures with the greatest impact on system performance.

Bos et al. (2005) indicated that diagnostic assessments are usually made to gain an understanding of how irrigation functions, to diagnose causes of problems and to identify opportunities for enhancing performance so that actions can be taken to improve irrigation water management. The same authors reported that diagnostic assessments are to be carried out when difficult problems are identified through routine monitoring, or when stakeholders are not satisfied with the existing levels of irrigation delivery services being provided, and desire changes in system operation.

The core component of diagnostic assessment is represented by performance indicators, as their selection and application aim at understanding functional relationships and at developing performance statements about irrigations. In the rationale of diagnostic assessment, irrigation managers or auditors need first to acquire a good understanding of system behavior under different operating conditions, prior to using simulation and management-support tools for appraising improvement options, and then take or recommend appropriate decisions.

In this view, a sound methodology for analysis of the existing irrigation schemes and of the management needs under current and future delivery scenarios is strongly required. Monitoring a set of variables that characterize the behavior of a complex system (diagnosis), and evaluating the system response after alternative correcting measures (prognosis and simulation) represent the basic capabilities required to an analytical methodology for addressing modernization processes with accuracy. The diagnostic component should be used to analyze different aspects of system management, such as assessment of water demand, management of water supply, identification of current system management needs, evaluation of system design, capacity and performance. The simulation component should instead be capable of facilitating the appraisal of improvement options by evaluating the system response after modifications. Both the diagnosis and simulation phases should be based upon a set of properly-chosen performance indicators to account for the main variables effecting the system operation and for synthetically representing the state of the system with respect to defined management objectives.

In this perspective, the methodology proposed in this chapter enables to conduct diagnostic assessments, simulate alternative deliveries and operational scenarios, and evaluate performance achievements in large-scale pressurized irrigation systems, thus constituting an analytical basis to address modernization processes in such systems with greater accuracy than was done in the past.

2. Rationale of the proposed methodology

The approach and components utilized within the proposed methodology are outlined in Fig. 1. The first part of the methodology entails the generation of the flow demand hydrographs during peak-demand periods through the use of an agro-hydrological model named Hydro-GEN that performs the daily soil-water balance and the simulation of irrigation deliveries for all the cropped fields served by the distribution network. By aggregating the simulated flow hydrographs at hydrant, the model generates the flow demand configurations for the entire distribution network. The Hydro-GEN model, and its applications to a pressurized irrigation delivery system at different management levels, were described in detail by Zaccaria et al. (2011a) and Zaccaria et al. (2011b).

The generated water demand scenarios are spatially and time distributed estimates that may be then used to define the expected levels of irrigation delivery service (objectives setting) from the distribution network over the different serviced areas. The flow configurations in the distribution network are then passed as inputs to a hydraulic model named COPAM (Lamaddalena and Sagardoy 2000) to simulate deliveries under different conditions and operational modes, analyze the network's hydraulic behaviour and evaluate hydraulic performance achievements with regard to the target delivery objectives. In this way, the COPAM simulation model allows identifying the structural limitations and the potential failures of the irrigation delivery network under different simulated flow configurations.

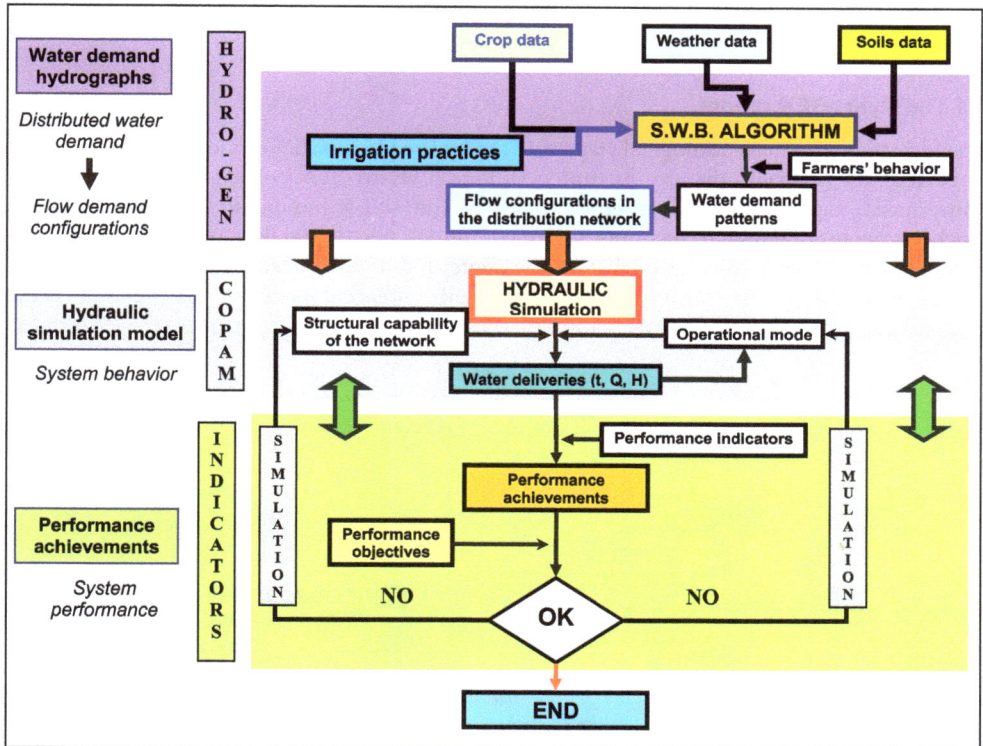

Fig. 1. Process and components utilized by the proposed methodology for conducting operational and performance analysis in pressurized irrigation systems.

The hydraulic simulations are based upon user-specified irrigation delivery conditions (or on agreed-upon delivery service between the water management body and the users) and utilize selected indicators and reference values to evaluate hydraulic parameters and identify the state of the system with respect to the specified management objectives. Moreover, the combined use of the Hydro-GEN and COPAM models verifies that the aggregated water demand and the adopted operational modes do not exceed the daily available water supply and the maximum physical conveyance and delivery capacity of the distribution network.

As final step, some additional indicators are applied to evaluate water delivery variables, other than hydraulic parameters, and to refine the performance assessment. The outputs resulting from simulations with the COPAM model, along with the evaluation of irrigation delivery by means of the additional indicators, can be interpreted in terms of performance achievements through the comparison with the users' requirements or the agreed-upon delivery conditions. Applying the above-described tools in the proposed sequence, and analyzing the resulting outputs, will guide the system managers and auditors in evaluating the irrigation delivery scenarios as satisfactory or unsatisfactory, and adjust the operations accordingly or identify the necessary physical changes.

3. Description of main components

3.1 The HydroGEN model

The approach and methodological steps utilized by the HydroGEN model for generating the flow demand hydrographs are illustrated in Fig. 2. HydroGEN consists of a deterministic component, represented by different terms of the soil-water balance equation, and a stochastic component that accounts for uncertainties and variability of some parameters related to crops and soils, as well as to farmers' habits and practices. The deterministic component enables the simulation of daily soil-water balances for all the individual

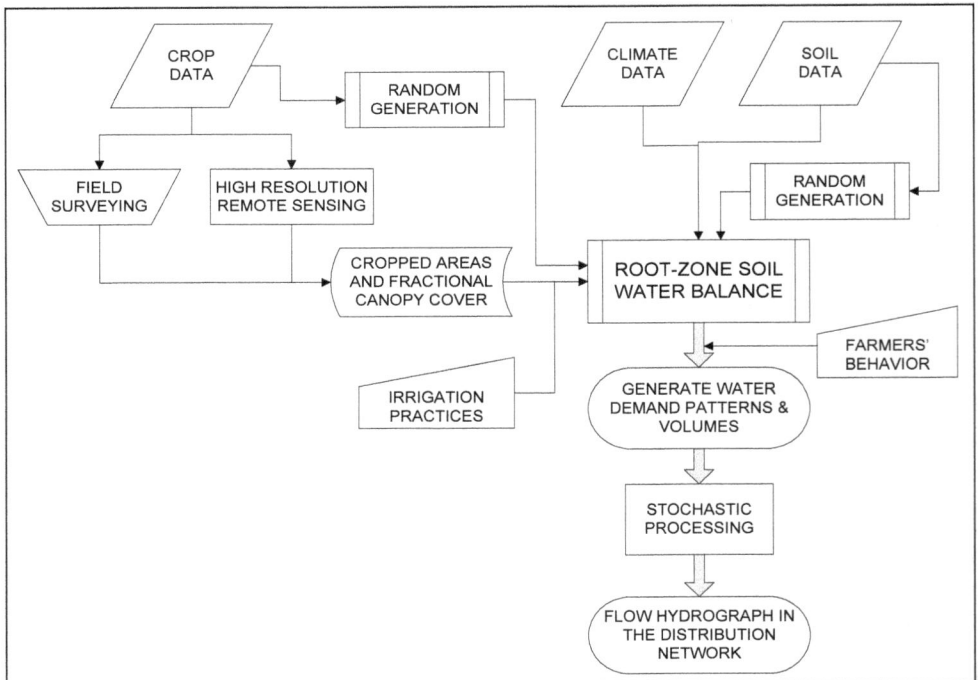

Fig. 2. Process utilized by HydroGEN to generate discharge hydrographs in pressurized irrigation systems.

cropped fields supplied by the water distribution network, based upon crop evapotranspiration estimated from daily climatic and rainfall data, crop type and stage of development, soil properties, and farm irrigation methods. The stochastic component enables the determination of some parameters of relevance to the computation of crop water demand and of timings of irrigation events, such as the sowing or green-up dates, and the initial soil water content at the beginning of the growing season. These parameters are strongly affected by uncertainties and are determined in HydroGEN by means of random generation within specific user-defined ranges, as described by Zaccaria et al. (2011a).

By integrating crop, soil and climatic data, HydroGEN simulates a series of daily irrigation demand volumes by maintaining a root-zone soil water budget for the cropped fields served by each delivery hydrant. In this process, the deterministic and stochastic components jointly allow generating disaggregated information on soil water deficits, and thus on timing and volumes of irrigation demand, both under conditions of full replenishment of soil water depletion and under regulated and/or deficit irrigation strategies.

These irrigation depths resulting from simulating the daily water balance and irrigation events are then aggregated upwards for all the cropped fields supplied by each hydrant, and then for all hydrants of the network, thus enabling the generation of a daily hydrograph of irrigation volumes demanded at different system management levels, i.e. at the hydrant, at single branches of the distribution network, at sector, district, or at the entire system level. As a subsequent step, HydroGEN identifies the 10-day peak-demand period by applying the method of the moving averages to the series of simulated daily irrigation demanded volumes.

Within the identified peak-demand period, HydroGEN then utilizes a stochastic procedure to determine the most likely timing of hydrants' openings and shut-offs during the daytime to deliver the required irrigation volumes. Based on the above determination, the daily irrigation volumes are translated into hourly flow rates and the number of hydrants in simultaneous operation hour-by-hour can be simulated. Therefore the hydrographs of hourly flow rates can be generated by aggregating on an hourly basis the flow rate demanded from these hydrants on an hourly basis. Likewise for the daily demanded volumes, based upon the level of aggregation, HydroGEN can simulate the hydrograph of hourly flow rates during the peak-demand period at hydrants, sector, district or system scale.

3.2 The COPAM Model

The Combined Optimization and Performance Analysis Model (COPAM) is a software package that provides computer-assisted capabilities for design and analysis of large-scale pressurized irrigation delivery networks. The model is described in details by Lamaddalena (1997) and by Lamaddalena and Sagardoy (2000) and is composed of three modules, one for the generation of demand discharges, one for the optimization of pipe size, and the third for the analysis of hydraulic performance. The synthetic process flow of COPAM is presented in Fig. 3.

As for the first module, the distribution of discharges flowing in the delivery network can be generated by using the Clément probabilistic method (Clément 1966; Clément and Galand 1979) or by means of the "Several Flow Regimes" approach (SFR) as proposed by Labye et al. (1988).

In COPAM, the flow computation through the SFR approach is implemented by means of the random generation of a certain number of hydrants in simultaneous operation, out of the total number of hydrants of the network, and the discharge in single sections of the network is thus computed as the sum of the discharges withdrawn from the downstream hydrants being in simultaneous operation, with the simplifying assumption that each open hydrant always delivers the nominal flow rate (Lamaddalena and Sagardoy 2000). Alternatively, the flow configuration in the pipe network can be read by COPAM as external file that results from simulating the demanded flow rates by means of the soil-water balance approach coupled with a stochastic processing through the HydroGEN model, as described in the previous section.

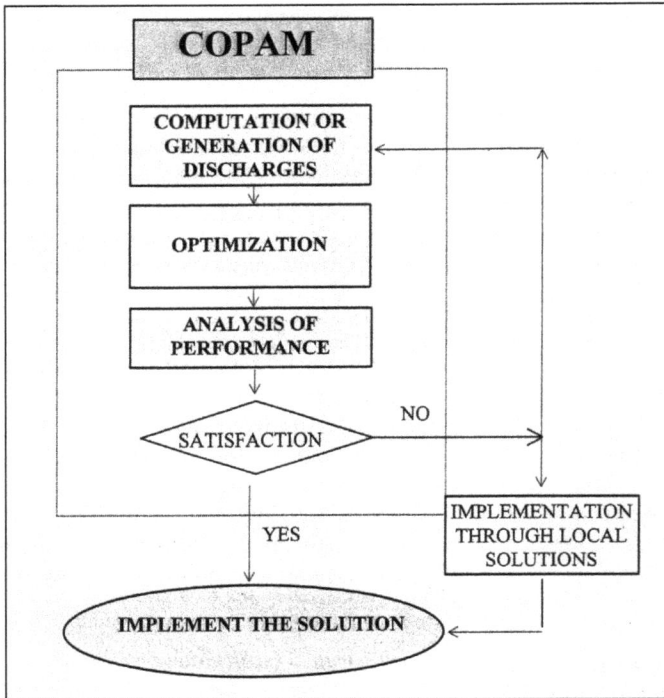

Fig. 3. Schematic process flow used by COPAM (Source: Lamaddalena and Sagardoy 2000).

The optimization module enables the computation of optimal pipe sizes in the whole network following the simulated flow configurations through the three possible alternatives previously mentioned: single flow regimes (Clément approach), several flow regimes (SFR) directly generated by the first module, and several flow regime previously generated by the first module, or simulated externally by the HydroGEN model and stored in a file to be read.

The performance achievable by the network tentatively designed by following the two above-described steps is then analyzed by means of two conceptual models, the Indexed Characteristic Curves model (CTGREF 1979; Bethery et al. 1981) and the AKLA model (Ait Kadi and Lamadalena 1991 - CIHEAM internal note not published), and based upon specific performance criteria. The Indexed Characteristic Curves model allows investigating a large number of configurations of hydrants in simultaneous operation, which correspond to a fixed value of the nominal discharge, Q, and to different values of the required piezometric elevation, Z, at the inlet of the distribution network to satisfy the configurations. A configuration is considered satisfied when for all the hydrants in simultaneous operation the following relationship holds true:

$$H_{j,r} \geq H_{REQ} \tag{1}$$

where $H_{j,r}$ (m) is the hydraulic head of the hydrant j within the configuration r, and H_{REQ} (m) is the minimum required head for proper operation of farm irrigation systems.

After defining the values of the discharge (Q) at the upstream section of the network, and the total number of configurations (C) to be investigated, a series of piezometric elevations (Zr) at the inlet of the network able to satisfy a given percentage of C configurations can be associated to each value of the upstream discharge. In doing so, the indexed characteristic curves can be drawn by plotting in the plane (Q, Z) the discharge values chosen and the corresponding required piezometric elevations, and by joining the points having the same percentage of configurations satisfied. The shape of these curves depends on the geometry of the network and on the topography of the commanded area.

The AKLA model enables the analysis of performance at each hydrant of the network under different operating conditions. The model is based on the multiple generation of a pre-fixed number of hydrants in simultaneous operation, with the hydrants being considered satisfied within each generated configuration when the Eq. 1 holds true for each selected hydrant. The model computes the discharge and pressure head resulting at each hydrant under different flow configurations, that is then compared with the minimum required pressure for the proper operation of the on-farm irrigation systems downstream of the hydrant itself. The measure of hydraulic performance achievements at each hydrant is thus obtained by means of the computation of the relative pressure deficit ($RPD_{j,r}$) at the hydrant j within the configuration r of hydrants in simultaneous operation, through Eq. 2 reported in the following section, and of the percentage of unsatisfied hydrants out of the total number of hydrants in simultaneous operation. Therefore, for each configuration the range of variation of the pressure head at each hydrant can be determined, and the hydrants subject to insufficient pressure heads, and hence the most critical zones of the network, can be clearly identified.

3.3 Performance indicators

Performance indicators are parameters resulting from the mathematical combination of measurable state variables and are conceived to synthetically represent how the irrigation system behaves with respect to the achievement of planned, targeted or agreed-upon objectives. Many authors and institutions have proposed specific ways to measure performance of irrigation and drainage systems, and therefore there is a large set of indicators available in literature, as summarized by Rao (1993) and by Bos et al. (2005).

The proposed methodology for diagnostic assessment entails the use of some performance indicators that were specifically adapted to pressurized delivery systems for describing the achievements by the irrigation distribution network with respect to the targeted water delivery objectives. In detail, the *Relative Pressure Deficit* and *Reliability* at each hydrants were taken from previous works conducted by Lamaddalena (1997) and by Lamaddalena and Sagardoy (2000) and, in conjunction with the hydrant's *Sensitivity*, which was instead defined within the present research, were used to measure and describe the hydraulic performance of the irrigation delivery networks in terms of pressure heads at the delivery points.

Three additional indicators, namely the *Relative Volume*, the *Relative Frequency*, and the *Relative Delay*, were instead developed by modifying the indicator of *Adequacy* as it was originally conceived by Molden and Gates (1990) to conduct performance assessment in open channel networks. These modifications aimed at tailoring the objective of Adequacy to pressurized irrigation distribution systems and thus at describing the adequacy of water delivery in terms of supplied volumes and of timeliness of irrigation.

Finally, also the indicators of *Dependability* and of *Equity* were modified with respect to those defined by Molden and Gates (1990), and then used to assess the spatial and temporal variability of irrigation delivery conditions over the command areas and during the periods of interest.

The Relative Pressure Deficit at each hydrant, *RPD*, as defined by Lamaddalena (1997), is computed by Eq. 2.

$$RPD_{j,r} = \frac{\left(H_{j,r} - H_{REQ}\right)}{H_{REQ}}$$

(2)

where $H_{j,r}$ is the hydraulic head at the hydrant j within the configuration r of hydrants in simultaneous operation, and H_{REQ} is the minimum pressure head necessary for proper operation of on-farm irrigation systems.

The representation of $RPD_{j,r}$ in a plane where the abscissas correspond to hydrants' number and the ordinates to $RPD_{j,r}$ clearly identifies the hydrants having insufficient pressure for enabling proper on-farm irrigation.

As for the second indicator, in general terms the Reliability of a system describes how often the system fails or, in different terms, the frequency or probability of the system being in a satisfactory state. Following earlier works by Hashimoto (1980) and by Hashimoto et al. (1982), the mathematical definition of reliability at hydrant level was carried out by Lamaddalena (1997) as reported hereafter.

$$\alpha = \mathrm{Pr}\, ob\left[X_t \in S\right]$$

(3)

where a is the hydrant's reliability, and X_t is the random variable denoting the state of the system at time t.

The possible values of X_t may fall in two sets: S, the set of all satisfactory outputs and F, the set of all unsatisfactory outputs denoting failure. Following this approach, at each instant t the system may fall in one of these alternative sets. Therefore the reliability of a system can be described by the probability a that the system is in a satisfactory set.

The state of satisfaction at hydrant level is measured on the basis of the value of the available hydraulic head at the hydrant under the different flow configurations, i.e. within each generated configuration r of hydrants in simultaneous operation, a hydrant j is considered satisfied when the Eq. 1 holds true.

In the specific case of pressurized irrigation systems, the reliability of each hydrant expresses the variability over time of the available pressure head of irrigation water deliveries at hydrants during the period of interest. From the Eq. 3, the reliability of each hydrant can be computed on the basis of the Eq. 4:

$$\alpha_j = \frac{\sum\limits_{r=1}^{C} Ih_{j,r} \, Ip_{j,r}}{\sum\limits_{r=1}^{C} Ih_{j,r}} \tag{4}$$

where

α_j = reliability of the hydrant j
$Ih_{j,r}$ = 1, if the hydrant, j, is open in the configuration r
$Ih_{j,r}$ = 0, if the hydrant, j, is closed in the configuration r
$Ip_{j,r}$ = 1, if the pressure head at the hydrant, j, open in the configuration r, is higher than the minimum required pressure head
$Ip_{j,r}$ = 0, if the pressure head at the hydrant, j, open in the configuration r, is lower than the minimum required pressure head
C = total number of generated configurations.

After estimating the available pressure head (m) at each hydrant in operation under each flow configuration within the network, the COPAM model calculates the values of the parameters $Ih_{j,r}$ and $Ip_{j,r}$, thus computing the corresponding value of the hydrant reliability, α_j.

The sensitivity of hydrants, also named as Relative Pressure Deficit Sensitivity, RPDS, refers to the range of fluctuations of the relative pressure deficit occurring at the delivery points, and how this range stretches across the zero-value line of RPD. This in turn identifies the adequacy in the available pressure head with respect to the minimum required value, H_{REQ}, for having the on-farm irrigation systems working properly. In detail, the minimum boundary of the range of RPD fluctuations is relevant for identifying both the potential failures and their severity. The sensitivity of hydrants is calculated through the Eq. 5:

$$RPDS = RPD_{AVE} - 0.5 * RPD_{RANGE} = \left(\frac{H_{AVE} - H_{REQ}}{H_{REQ}}\right) - 0.5 * \left(RPD_{MAX} - RPD_{MIN}\right)$$

$$\Rightarrow RPDS = \left(\frac{H_{AVE} - H_{REQ}}{H_{REQ}}\right) - 0.5 * \left[\left(\frac{H_{MAX} - H_{REQ}}{H_{REQ}}\right) - \left(\frac{H_{MIN} - H_{REQ}}{H_{REQ}}\right)\right] \tag{5}$$

where the limits of H_{MIN} and H_{MAX} are set as follows:

$$H_{MIN} \geq 0$$

$$H_{MAX} \leq Max \ operating \ pressure \ head \ bearable \ by \ pipes$$

The Relative Volume, RV, is a measure of the objective of the distribution network of delivering adequate irrigation volumes to each serviced cropped field with respect to the required ones. The RV is therefore a measure of Adequacy expressed in terms of delivered volumes and is defined by the following relationships:

a.   at a given location

$$RV = \frac{V_{DELI} - V_{REQ}}{V_{REQ}} \tag{6}$$

b.   averaged over the region, **R**, and the time of interest, **T**

$$P_{AV} = \frac{1}{T} \sum_T \left(\frac{1}{R} \sum_R RV \right) \tag{7}$$

where V_{DELI} and V_{REQ} are the irrigation volume delivered by the distribution network and the irrigation volume required for adequate crop irrigation management and target yield, respectively.

As for the objective of Adequacy in terms of timeliness of irrigation delivery, the Relative Frequency, RF, and the Relative Delay, RDe, were defined by the equations reported hereafter.

RF

a.   at a delivery location:

$$RF = \frac{F_{REQ} - F_{DELI}}{F_{REQ}} \tag{8}$$

b.   and the average over the region, R, and during the time of interest, T:

$$P_{AF} = \frac{1}{T} \sum_T \left(\frac{1}{R} \sum_R RF \right) \tag{9}$$

where F_{REQ} is the frequency of irrigation required by any combinations crop/soil/climate for not incurring in any soil water deficit, and F_{DELI} is the frequency of actual irrigation water delivery by the distribution network.

RDe
a.   at a delivery location:

$$RDe = \frac{DEL_{ALL.} - DEL_{DELI}}{DEL_{ALL.}} \tag{10}$$

b.   averaged over the region, R, and time of interest, T:

$$P_{AD} = \frac{1}{T} \sum_T \left(\frac{1}{R} \sum_R RDe \right) \tag{11}$$

where $DEL_{ALL.}$ is the maximum allowed delay (days) of irrigation that would cause a yield reduction within 10 % of the maximum obtainable yield due to soil water deficit, for any given combinations crop/soil/climate, and DEL_{DELI} is the actual delay of irrigation delivery (days) by the distribution network with respect to the required timing for achieving the maximum yield (no water deficit).

The RV, RF and RDe are particularly meaningful when the distribution network is operated by rotation or by arranged delivery schedules, whereas when irrigations are under the farmers' control (demand delivery schedules) the timing of irrigations and the volumes withdrawn from the network by farmers and applied to cropped fields are more affected by the available water supply, by the network delivery capacity as well as by farmers' habits and behavior rather than by the network operations. Indirectly, the irrigation events and the volumes withdrawn by farmers are also affected by the pressure head available at hydrants under the different flow configurations. At the same time, hydrants' operation, flow rates and volumes withdrawn by farmers strongly affect the flow configurations in the different sections of the network and thus the conditions of water delivery to other hydrants. As a matter of fact, in pressurized delivery systems operated on-demand, when farmers open the hydrants and do not find adequate pressure head for proper on-farm irrigation, they usually shut-off the hydrant and return sometime later for irrigating their fields. In other words, water withdrawals by farmers at given hydrants might be biased by the operation of other hydrants and by the behavior of other farmers and, at the same time, they might affect the operation of other hydrants as well, especially when the distribution network has low delivery capacity or low flexibility. Thus, the RV, together with RPD, RF and RDe indicate indirectly the network performance or, in other words, the capability of the distribution network to accommodate the farmers' behavior and the farming practices followed in the entire command area, and to still deliver water with the required conditions.

The indicator of Dependability expresses the temporal uniformity of the conditions of irrigation delivery. When the concerned delivery parameter is the irrigation volume the dependability refers to the degree of temporal variability of the RV that occurs over the region of interest, R, and is expressed by the Eq. 12:

$$P_D = \frac{1}{R}\sum_R CV_T(RV) = \frac{1}{R}\sum_R CV_T\left(\frac{V_{DELI} - V_{REQ}}{V_{REQ}}\right) \tag{12}$$

where $CV_T\left(\dfrac{V_{DELI} - V_{REQ}}{V_{REQ}}\right)$ = temporal coefficient of variation (ratio of standard deviation to mean) of the RV over the time period of interest T (i.e. variability from time to time over the period T).

When the concerned delivery parameter is the timeliness of irrigation, the dependability is expressed in terms of temporal variability of RF and/or of RDe as follows:

RF:

$$P_D = \frac{1}{R}\sum_R CV_T(RF) = \frac{1}{R}\sum_R CV_T\left(\frac{F_{REQ} - F_{DELI}}{F_{REQ}}\right) \tag{13}$$

where $CV_T \left(\dfrac{F_{REQ} - F_{DELI}}{F_{REQ}} \right)$ = temporal coefficient of variation of RF over the time period of interest T.

RDe:

$$P_D = \frac{1}{R} \sum_R CV_T (RDe) = \frac{1}{R} \sum_R CV_T \left(\frac{D_{ALL} - D_{DELI}}{D_{ALL}} \right) \tag{14}$$

where $CV_T \left(\dfrac{D_{ALL} - D_{DELI}}{D_{ALL}} \right)$ = temporal coefficient of variation (ratio of standard deviation to mean) of the RDe over the time period of interest T (variability from time to time over the period T).

When the concerned delivery parameter is the available pressure head at hydrant, the dependability corresponds to the hydraulic reliability at hydrants, as defined by the Eq. 4.

As for the Equity indicator, it refers to the spatial uniformity of the irrigation delivery conditions. When the concerned delivery parameter is the irrigation volume, the equity is expressed as the degree of spatial variability of the RV that occurs over the region of interest, R, and is expressed by the following relationship:

$$P_E = \frac{1}{T} \sum_T CV_R (RV) = \frac{1}{T} \sum_T CV_R \left(\frac{V_{DELI} - V_{REQ}}{V_{REQ}} \right) \tag{15}$$

where $CV_R \left(\dfrac{V_{DELI} - V_{REQ}}{V_{REQ}} \right)$ = spatial coefficient of variation of the RV over the region of interest R (variability from point to point over the region).

Likewise the dependability, when the concerned delivery parameter is the timeliness of irrigation, the equity may be expressed in terms of RF or of RDe by the following relationships:

RF:

$$P_E = \frac{1}{T} \sum_T CV_R (RF) = \frac{1}{T} \sum_T CV_R \left(\frac{F_{REQ} - F_{DELI}}{F_{REQ}} \right) \tag{16}$$

where $CV_R \left(\dfrac{F_{REQ} - F_{DELI}}{F_{REQ}} \right)$ = spatial coefficient of variation of the RF over the region of interest R.

RDe:

$$P_E = \frac{1}{T} \sum_T CV_R (RDe) = \frac{1}{T} \sum_T CV_R \left(\frac{D_{ALL} - D_{DELI}}{D_{ALL}} \right) \tag{17}$$

where $CV_R \left(\dfrac{D_{ALL} - D_{DELI}}{D_{ALL}} \right)$ = spatial coefficient of variation of the RDe over the region of interest R.

When the concerned delivery parameter is the available pressure head at hydrant, the equity corresponds to the spatial variability of the RPD or of $RPDS$ and thus is expressed either by the relationships 18 or 19.

RPD:

$$P_E = \frac{1}{T}\sum_T CV_R\left(RPD\right) = \frac{1}{T}\sum_T CV_R\left(\frac{H_{DELI} - H_{REQ}}{H_{REQ}}\right) \tag{18}$$

where $CV_R \left(\dfrac{H_{DELI} - H_{REQ}}{H_{REQ}} \right)$ = spatial coefficient of variation of the RPD over the region of interest R.

$RPDS$:

$$P_E = \frac{1}{T}\sum_T CV_R\left(RPDS\right) = \frac{1}{T}\sum_T CV_R\left(RPD_{AVE} - 0.5*RPD_{RANGE}\right) \tag{19}$$

where $CV_R \left(RPD_{AVE} - 0.5 * RPD_{RANGE} \right)$ = spatial coefficient of variation of the $RPDS$ over the region of interest R.

Once performance indicators are conceived and defined on the basis of measurable variables, ranking the state of a system requires the computed or estimated performance values being evaluated with respect to defined reference values. Setting minimum performance levels is therefore relevant to diagnostic analyses and to define the states of the system as satisfactory or unsatisfactory.

Within the present work, a tentative set of reference standard values is proposed in the Table 1 for the above described indicators. The values of performance indicators and the relative performance classes are based on prescriptions provided by experienced project personnel (expert opinions) and on the perceived implications of deviation of the performance measures from the reference values identified as satisfactory.

MEASURE	PERFORMANCE CLASSES		
	GOOD	FAIR	POOR
RPD	≥ 0.0	- 0.3 – -0.1	< - 0.3
RPDS	≥ 0.0	-0.1 - -0.2	< - 0.2
RV	- 0.1 – 0.00	- 0.3 – -0.1	< - 0.3; > 0.0
RF	≥ 0.0	- 0.2 – 0.0	< - 0.2
RDe	> 0.0	- 0.2 – 0.1	< - 0.2
Reliability	≥ 0.8	0.8 – 0.7	< 0.7
Dependability	0.0 – 0.4	0.4 - 1.00	> 1.00
Equity	0.0 – 0.4	0.4 - 1.00	> 1.00

Table 1. Tentative reference standards for performance assessment

4. Description of the study area

The proposed methodology for operational and performance analysis of pressurized delivery networks was applied to two district delivery networks of an existing irrigation scheme located in southern Italy that is in urgent need of modernization due to its poor performance in terms of water delivery to farmers.

The Sinistra Bradano irrigation scheme (Fig. 4) is located in the western part of the province of Taranto and covers a total topographic area of 9,651 ha. The system is divided into 10 operational districts, ranging in size from a minimum of 353 ha to a maximum 1,675 ha. Each district is subdivided into sectors consisting of a grouped number of farms. The water source is a storage reservoir located on the Bradano River in the nearby region of Basilicata, with a total capacity of 70 Mm^3, out of which 35 Mm^3 are usually available for irrigation of the Sinistra Bradano system. The distribution of water to farms is managed by a local Water Users Organization (WUO) and usually starts by late April and ends by late October. The distribution networks are operated on a rotation delivery schedule, with the rotation being fixed for the entire irrigation season with a flow rate of 20 l s^{-1} ha^{-1}, and 5 hours of delivery to each user and a fixed irrigation interval of 10 days.

Significant conveyance and distribution losses are reported for the study area (INEA 1999), as on average only around 16 Mm^3 are finally delivered to the cropped fields out of the total volume of 23 Mm^3 that is diverted from the reservoir. Water is conveyed to the area through a main conveyance canal, from which it is then delivered to farms by means of 10 open-branched district distribution networks. The entire irrigation scheme is subdivided into three operational portions that are commanded by progressive sections of the conveyance canal. The water diversions from the canal to the district distribution networks are controlled by cross-regulators and orifice-type undershot-gate offtakes that are manually operated by the staff of the WUO. The branched delivery networks consist of gravity-fed buried pipelines delivering water to farms with low pressure head.

The pressure at farm hydrants thus originates from the difference in elevation between the offtakes, situated along the conveyance canal, and the lower-elevation irrigated areas. The Sinistra Bradano irrigation system covers an overall cropped and irrigable command area of 8,636 ha. A large reduction in the area serviced by surface water from the WUO and a corresponding strong increase in the area irrigated by groundwater pumping was documented for this system by Zaccaria et al. (2010) on the basis of records provided by the WUO. These changes in the serviced areas are most likely a consequence of the poor conditions of water delivery with respect to farmers' needs. As a result, at present many farmers rely mainly on groundwater pumping for irrigating their crops. Since the available pressure head at hydrants is not sufficient for the proper operation of the on-farm trickle and sprinkler irrigation systems, those farmers who still withdraw water from the delivery network need to use booster pumps downstream of the hydrants to adequately feed their irrigation systems. The actual operation of the distribution system under study, the resulting effects of the operational procedures on crop irrigation management, the low performance in water delivery, and the need for system modernization were documented by previous research works conducted on the study area, all described in details by Zaccaria et al. (2010).

For the purposes of the present study, the inconsistency between the water delivery schedule currently enforced by the WUO and the crops' requirements in terms of irrigation volumes can be inferred from the data presented in Table 2, where comparisons are made between required and delivered volumes and timings of irrigations during the 10-day peak demand period of the 2009 season for the main crops grown in the study area. The required irrigation volumes and timings, as well as the maximum allowed delay (days) for avoiding yield reductions higher than 10% of the maximum achievable yield, were estimated through simulations run by the HydroGEN model and by a daily soil-water balance algorithm implemented in Excel worksheet, whereas the actual deliveries were retrieved from records provided by the WUO. Based on the values reported in Table 2 it can be inferred that the actual water deliveries are not matching the irrigation demand of most of the crops grown in the area. The values of the Relative Volume (RV) show that in most of the cases the volumes delivered are excessive with respect to the estimated requirements.

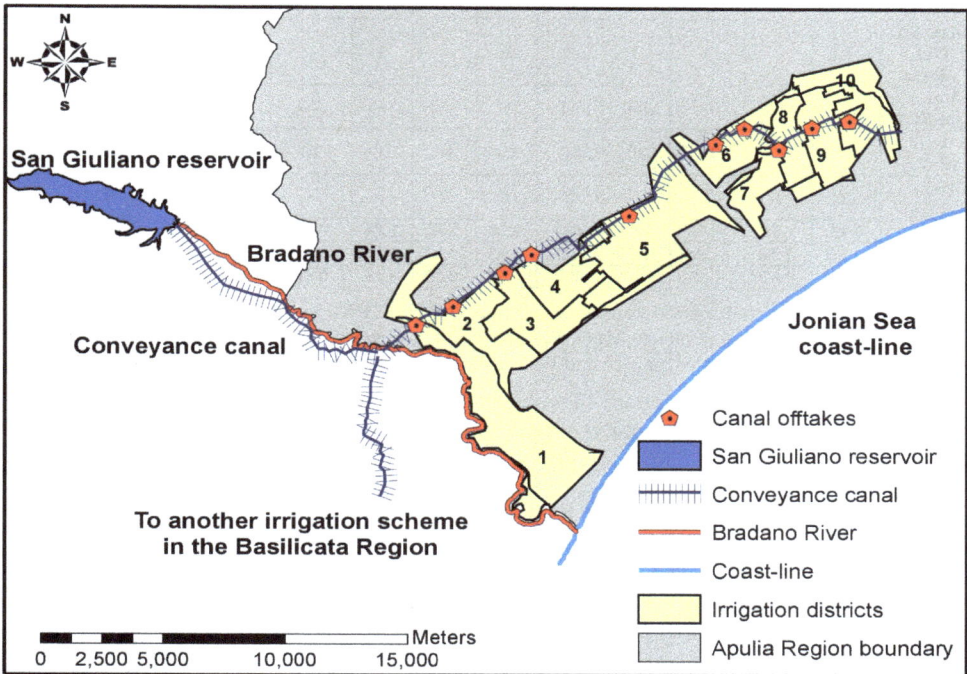

Fig. 4. Overview of the Sinistra Bradano irrigation system.

Only for mature wine grapes and vegetables the delivered volumes are not sufficient to fulfil the irrigation requirements during the peak period. The values of the Relative Frequency, RF, reveal that the current delivery schedule is inadequate for all the considered crops, except for mature olive orchards. The crops suffering most for inadequate frequency of deliveries are the vegetables, table grapes and wine grapes. Also, the estimated values of the Relative Delay, RDe, show that several crops under the current delivery schedule receive water with some delays, with respect to the required irrigation timing, that goes way beyond the maximum allowed delay for avoiding yield reduction higher than 10%. In other words, olive and fruit orchards, as well as citrus and vegetables, may face yield reductions way higher than 10% due to inadequate timing of irrigation deliveries.

Within the Sinistra Bradano irrigation system, two district distribution networks, namely the Districts 7 and 10, were considered in the present study for the application of the proposed methodology, in view of their physical and operational features. Both the districts are located within the last operational portion of the Sinistra Bradano system, with the District 7 being at the initial part and the District 10 being located at the last part of this 3rd portion, and thus being supplied by the tail-end section of the main conveyance canal.

Crop	M TG	Y TG	M WG	Y WG	M O	Y O	M FO	Y FO	M TGc	M C	Y C	Veg	M Alm
Vol_{REQ} (m³/ha)	320	143	428	141	185	96	230	154	297	274	217	745	215
F_{REQ} (days)	4	4	5	4	13	9	6	5	5	8	5	3	6
Max All. Del. (days)	3	3	3	3	2	2	4	4	4	1	1	2	5
Vol_{DEL} (m³/ha)	360	360	360	360	360	360	360	360	360	360	360	360	360
F_{DEL} (days)	10	10	10	10	10	10	10	10	10	10	10	10	10
De_{DEL} (days)	2	1	1	1	3	7	7	1	4	4	1	3	5
RV	0.12	1.52	-0.16	1.55	0.94	2.74	0.57	1.34	0.21	0.31	0.66	-0.52	0.68
RF	-1.50	-1.50	-1.00	-1.50	0.23	-0.11	-0.67	-1.00	-1.00	-0.25	-1.00	-2.33	-0.67
RDe	0.33	0.67	0.67	0.67	-0.50	-2.50	-0.75	0.75	0.00	-3.00	0.00	-0.50	0.00

Legend: MTG, YTG = mature and young table grapes; MWG, YWG = mature and young wine grapes; MO, YO = mature and young olives; MFO, YFO = mature and young fruit orchards; M TGc = mature covered table grapes; MC, YC = mature and young citrus; Veg = vegetables; M Alm = mature almonds

Table 2. Estimation of the adequacy of water deliveries in terms of volumes and timings of irrigation for the peak demand period of the 2009 season for the main crops grown in the study area.

The distribution network of District 7 (Fig. 5) serves 326 hydrants, supplying irrigation water to a total irrigable area of 586.6 ha, of which 119.8 ha are cultivated with table grapes, 54.3 ha with olives, 162.3 with citrus, 58.9 with summer vegetables and 2 ha with almonds. At the design stage the total command area was subdivided into 20 irrigation sectors, whose size ranged from 20 ha to 36 ha.

The District 10 (Fig. 6) is composed of three sub-areas that are supplied by as many distribution sub-networks originating from three different diversions along the last section of the main canal, namely the Diversion 7 (D7), 8-North (D8-N) and 8-South (D8-S). The sub-network D7 supplies water to 129 hydrants, serving a total irrigable area of 252.6 ha, out of which 198.5 ha are cultivated with citrus, 20.3 with table grapes, 19.6 ha with vegetables, 10.5 ha of olives, and 3 ha with orchards. The sub-network D8-N supplies water to 161 hydrants and serves an overall irrigable area of 661.2 ha, out of which 69.8 ha are cultivated with table grapes, 347.6 ha with vegetables and 4.9 ha with olives. The total irrigable area served by the sub-network D8-S is 445 ha, out of which 81.3 ha are cultivated with table grapes, 230.5 ha with citrus, 42.6 ha with olives and 75.5 ha with vegetables, with a total of 133 supplied hydrants.

All hydrants in both districts are equipped with flow meters and with rubber-ringed flow limiters allowing for a maximum delivery of 10 l s^{-1} or 20 l s^{-1}, according to the cropped areas supplied downstream. These discharge values were used as nominal flow rates for simulations related to the current state and operation of the distribution networks. In simulating improved operational scenarios the flow rates at hydrants were instead set according to the estimated discharges required by the downstream cropped and irrigated fields.

Sinistra Bradano irrigation system

District n.7 - Derivation n.5

Legend

⎯⎯⎯	Conveyece canal
⊕	Derivation
⎯⎯⎯	District distribution network
⎯⎯⎯	Sector distribution network
·	Hydrants

Fig. 5. Overview of District 7 of the Sinistra Bradano irrigation system.

5. Application of the proposed methodology to the selected irrigation districts

A more flexible delivery was considered and simulated as alternative schedule to the fix rotation currently enforced in both the irrigation Districts 7 and 10 of the Sinistra Bradano irrigation system. The simulations focused on a restricted-demand delivery to be implemented in both districts to allow for more flexibility to farmers for better managing irrigation on their crops. The feasibility and performance achievable under this alternative delivery schedule were analyzed vis-à-vis with the physical features and constraints of the existing distribution networks.

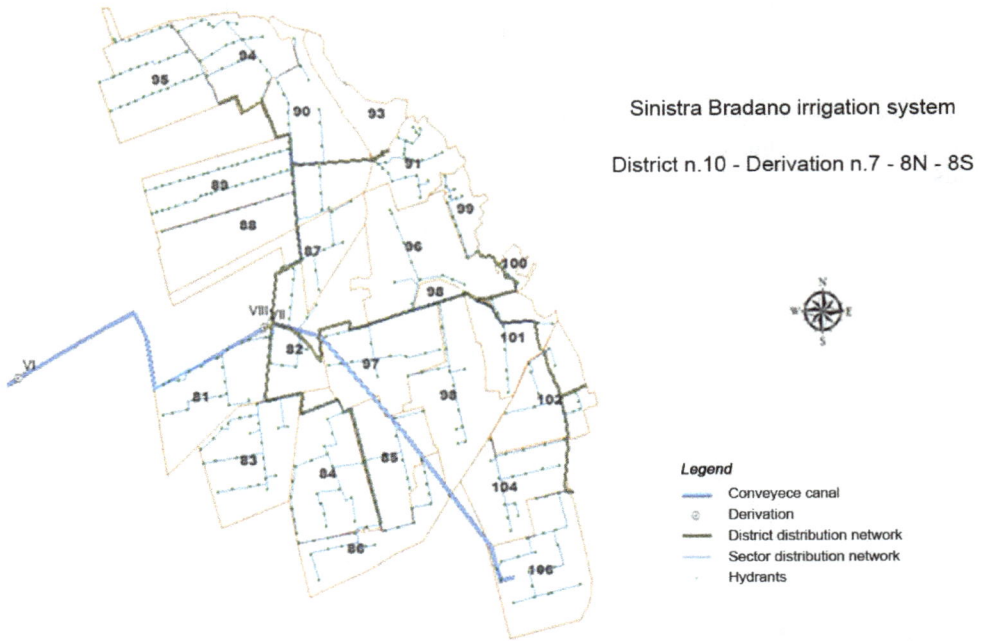

Fig. 6. Overview of the District n. 10 of the Sinistra Bradano irrigation system.

For both the districts, applying the Hydro-GEN model to the cropped areas commanded by the existing distribution networks allowed simulating the irrigation demand hydrographs and the resulting flow configurations during the 10-day peak demand period. The simulations were conducted by using climatic and crop data referred to the 2009 irrigation season, under the irrigation management scenario of full replenishment of the soil water depleted in the root zone, yielding the 10-day peak demand period as occurring in the interval DOY 197-206 (July 16th – 25th). Figure 7 shows the simulated demand hydrographs for both districts during the 10-day peak period.

Under the improved delivery scenarios the simulated demand hydrographs and flow configurations result from assuming the fulfilment of the required deliveries at farm level, i.e. irrigation deliveries were simulated as occurring in compliance with the required volumes and timing estimated by the HydroGEN model. In this way, the simulated deliveries in terms of volumes and frequency would thus occur in an adequate way for proper on-farm irrigation. In other words, the adequacy of the simulated deliveries in terms of volumes and timing was assumed as pre-requisite for evaluating the network performance under the required flow configurations.

The flow hydrographs and the resulting flow configurations generated by the HydroGEN model were then inputted in the COPAM model for simulating the hydraulic behaviour and performance of the networks with respect to the target deliveries. The hydraulic performance was analyzed by using three main indicators, namely the Relative Pressure Deficit, *RPD*, the Hydrant Sensitivity, *RPDS*, and the Hydrant Reliability, *R*. Also for these applications, the Equity of the deliveries in terms of available pressure heads was estimated by using the Eq. 19.

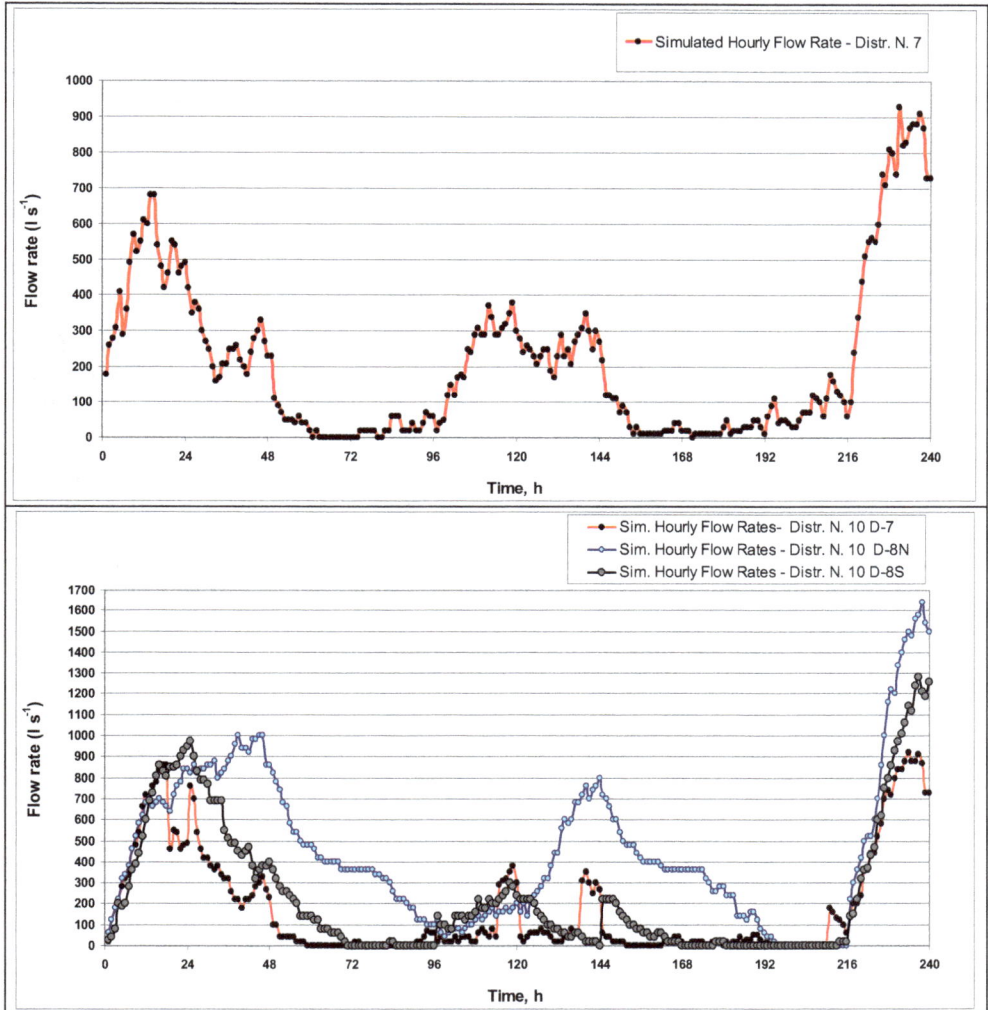

Fig. 7. Simulated hydrographs of hourly flow rates (l s⁻¹) for the 10-day peak period during the 2009 irrigation season for the delivery networks of District 7 and District 10 of the Sinistra Bradano system.

6. Results and discussion

As far as the network of District 7 is concerned, simulations of the restricted demand schedule on the existing delivery network show that poor performance would be achieved in terms of available pressure heads at hydrants, and this would be likely due to physical constraints and limitations. As can be seen from Figs. 8 and 9 showing *RPD, R* and *RPDS* obtained under this scenario, nearly all hydrants would fall in unsatisfactory state with respect to the necessary pressure head (H_{REQ} = 20 m) for proper operation of the farm irrigation systems. For nearly all the hydrants the *RPD* would be way lower than zero, even reaching for several hydrants very negative low peaks (up to values of – 4.4). Only for very few hydrants, located in the initial and terminal portions of the network, the minimum required pressure conditions would be satisfied. At the same time, the reliability would yield value of zero for most of the hydrants, revealing unsatisfactory states, and thus insufficient pressure heads on most of the times that the hydrants would be accessed and operated by farmers. As for the hydrant sensitivity, from the Fig. 9 it can be easily inferred that most of the hydrants have very negative value of the *RPDS*. Figure 9 reveals either the occurrence of large fluctuations of the pressure head at hydrants, or the relative position of the minimum values of RPD being below the zero-line, which represents the adequacy of delivery expressed in terms of pressure head.

Several physical improvements, and their effects on the hydraulic behavior of the distribution network, were simulated by using the available modeling tools in the sequence indicated in Fig. 1. In this set of simulations, a satisfactory performance was found to be achieved as a result of the following physical measures:

1. replacement of the flow limiters at all the hydrants, with the aim of reducing the maximum flow rate that can be withdrawn by users to 10 l s⁻¹ and thus reducing the occurrence of peak flows in the distribution network;
2. installation of a flow limiter at the upstream end of the district network, in order to limit the peak flow to a maximum of 500 l s⁻¹ to ensure adequate delivery conditions at hydrants;
3. increase of the total piezometric elevation at the upstream end of the network from the current value of 42 m to 82 m a. s. l., for ensuring enough pressure head at all hydrants under the different configurations of hydrants in simultaneous operation.

Figures 10 and 11 show that, after implementing this set of physical improvements, the distribution network of District 7 would be capable of satisfying the minimum delivery conditions necessary for proper operation of farm irrigation and allow adequate and flexible crop irrigation management to farmers. Under the flexible delivery scenario and with the improved network, the *RPD* for most of the hydrants would be greater than zero, meaning that the available pressure head would be higher than the minimum required.

Only a few hydrants, corresponding to the numbers from 98 to 107, would have *RPD* values lower than -0.4 with low peaks up to -1.32, denoting serious pressure deficits. These hydrants would thus not be capable of satisfying the required pressure head conditions, even if the piezometric head at the inlet of the network is further increased up to 106 m. This is most likely due to the disadvantageous locations of these hydrants in combination with the network layout and pipe size configurations that would cause high friction losses, and make these few hydrants perform unsatisfactorily under most of the flow configurations. This aspect can also be noticed from the estimated reliability of hydrants under the simulated operation of the modernized network, which is reported in Fig. 10. For the majority of hydrants the reliability would reach values of 1.0, apart from a very limited number of hydrants having reliability lower than 0.7. In four cases, corresponding to the hydrants numbered 94, 104, 106, and 107, the reliability would approach values of zero, denoting the occurrence of unsatisfactory states every time these hydrants are operated.

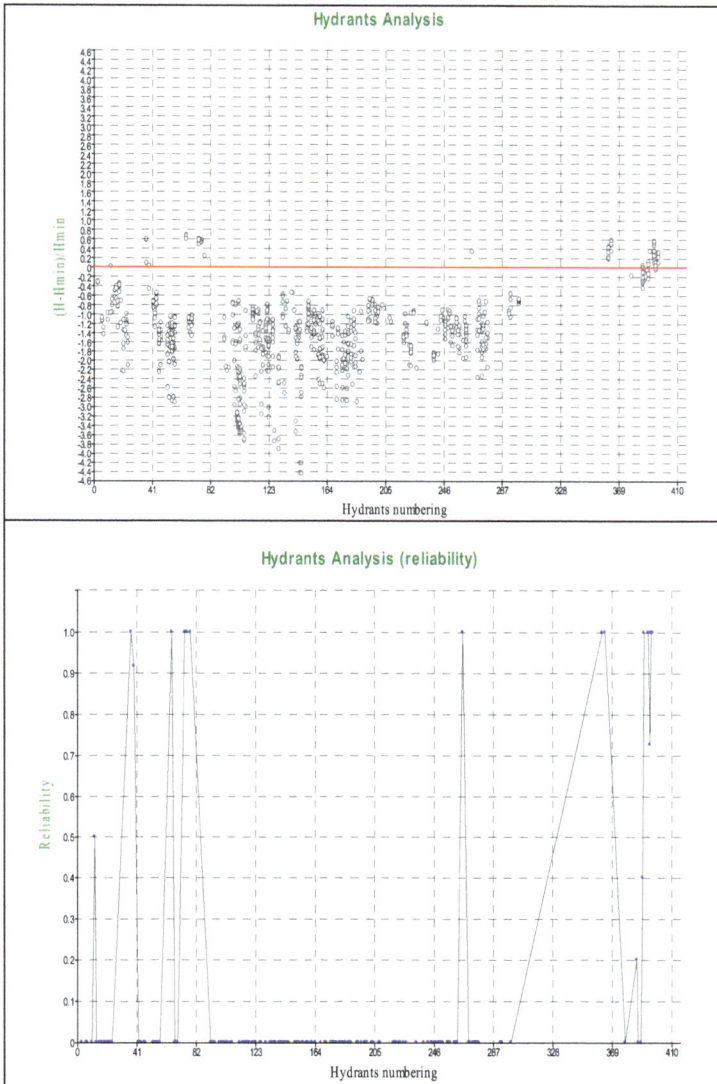

Fig. 8. *RPD* and *R* values obtained simulating the restricted demand delivery scenario on the distribution network of District 7 of the Sinistra Bradano system for the 10-day peak period during the 2009 irrigation season.

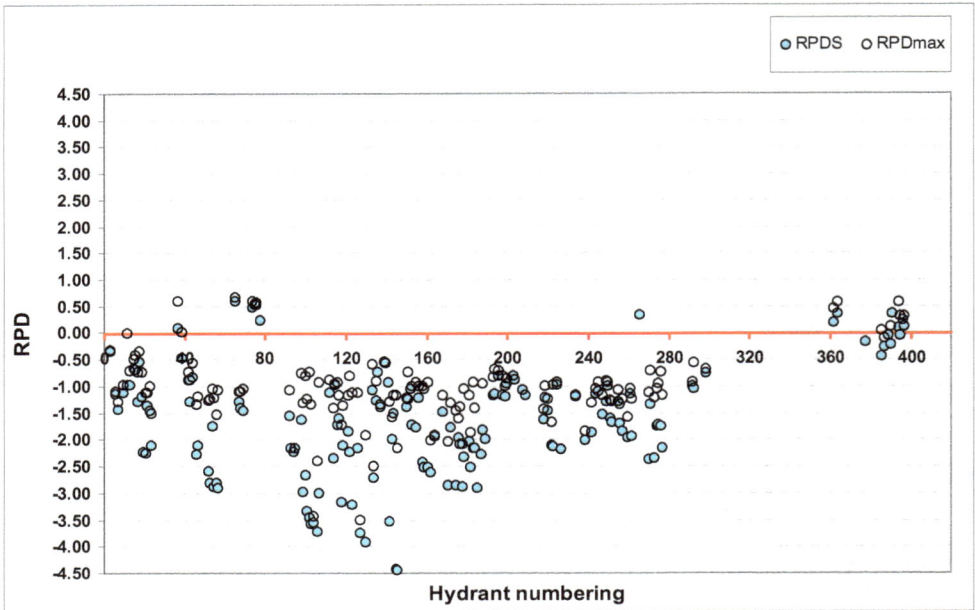

Fig. 9. *RPDS* values obtained simulating the restricted demand delivery scenario on the distribution network of District 7 for the 10-day peak period during the 2009 irrigation season.

As far as the hydrant sensitivity is concerned, by observing Fig. 11 the improved network seems to work pretty well, as the values of *RPDS* for nearly all the hydrants, apart from the very few previously identified, would be very close to or higher than zero, revealing that the ranges of fluctuation of the *RPD* would be small and/or that the minimum *RPD* values would be mostly above zero.

Under this scenario, the Equity in terms of pressure head was estimated through Eq. 19 and expressed as the spatial variability, CVr, of the *RPDS* over the district during the peak-demand period. The value of Equity resulted of 1.03, thus revealing that under this improved scenario a large variability of pressure head conditions and of *RPDS* among hydrants would still exist. Comparing the computed value with the reference standards proposed in Table 1 allowed classifying as "poor" the equity in terms of pressure head in District 7 with the upstream piezometric elevation of 82 m.

By analyzing the hydraulic behavior of the network after the physical improvements it can be inferred that the few hydrants characterized by low performance should be operated separately from the rest of hydrants, with the aim of ensuring adequate performance to the entire distribution network. In other words, these hydrants should be operated during low-peak demand hours to avoid excessive peak flows in the pipe network and, thus, the high friction losses resulting from limited pipe sizes or limited section capacity. From Fig. 7 showing the demand flow hydrograph simulated for District 7 it can be inferred that low-peak demand flows occur daily before 6 a.m. and after 6 p.m. and, so restrictions in the operation could be set to allow farmers accessing these hydrants within this specific time slots.

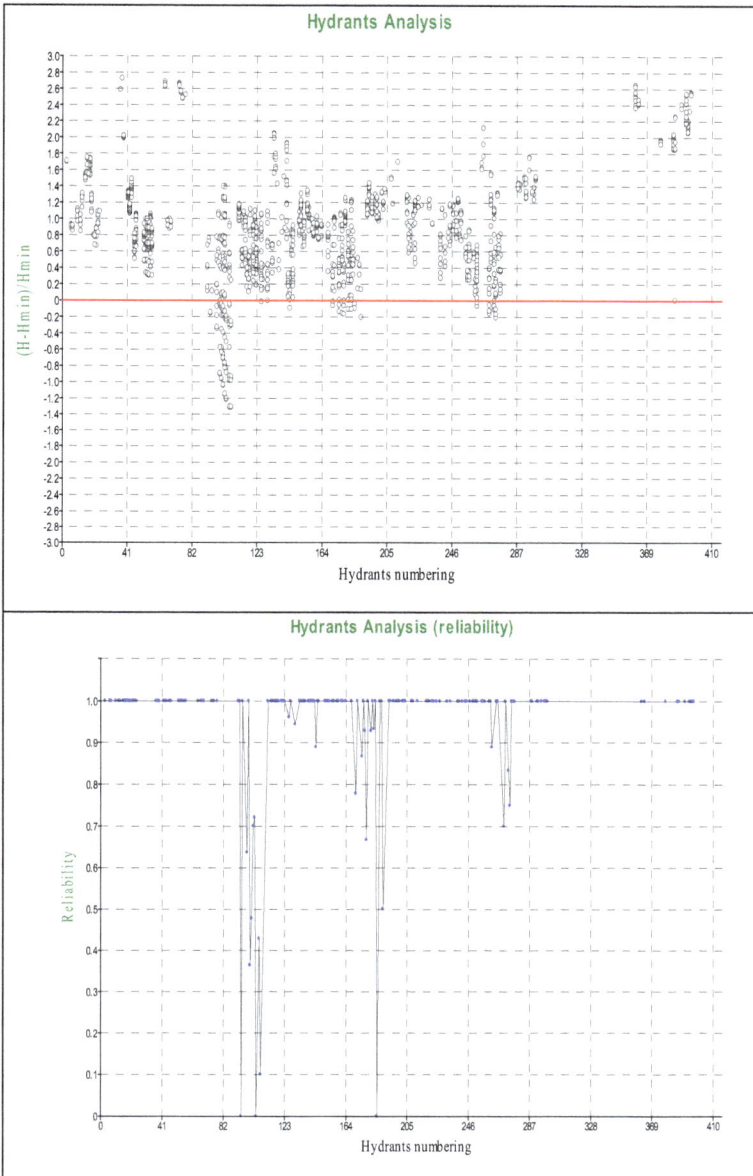

Fig. 10. *RPD* and *R* values obtained simulating the restricted demand delivery scenario on the distribution network of District 7 after implementing the physical improvements.

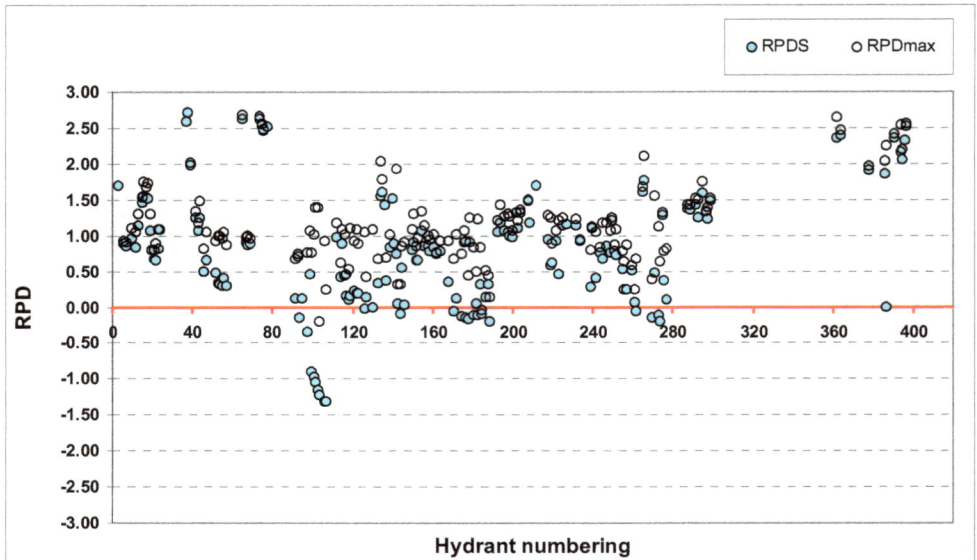

Fig. 11. *RPDS* values obtained simulating the restricted demand delivery scenario on the distribution network of District 7 after the implementation of the physical improvements.

Further simulation runs show that a complete satisfactory state for all the hydrants under the peak-flow configurations would require an upstream piezometric head of 106 m, as can be noticed from the Figs. 12 and 13 presenting the simulated *RPD*, *R* and *RPDS* achievable under this improvement scenario.

After this further increase to 106 m a.s.l. the estimation of Equity yielded a value of 0.49, which enables to classify the Equity under this improvement scenario as "fair," as there is still variability of pressure head conditions and of *RPDS* among hydrants, but this variability decreased from the scenario with the piezometric elevation of 82 m a.s.l.

As for the District 10, simulations were run separately for each of the three distribution sub-networks (Diversion 7, *D7*, Diversion 8 North, *D8-N*, and Diversion 8 South, *D8-S*) to evaluate the feasibility of the flexible water delivery schedule and to assess the performance achievable by the existing network under the improved scenarios. Nevertheless, only results related to the sub-networks D8-N and D8-S of District 10, which represent the very tail-ends of the entire irrigation system, are presented in this section.

Simulating the implementation of the restricted demand delivery schedule, the sub-network D8-N as it is in the present state would perform very poorly in terms of pressure head at hydrants. Figures 14 and 15 present the *RPD*, *R* and *RPDS* by the distribution network D8-N under restricted demand schedule. The values of these parameters in the figures clearly show that the network in its current state would not be capable of supplying water by restricted demand schedule with adequate performance, as all the hydrants, except one, would fall in unsatisfactory state with respect to the minimum required pressure head conditions. As a matter of fact, the pressure heads resulting at all hydrants of the network under this delivery scenario would be way lower than that required.

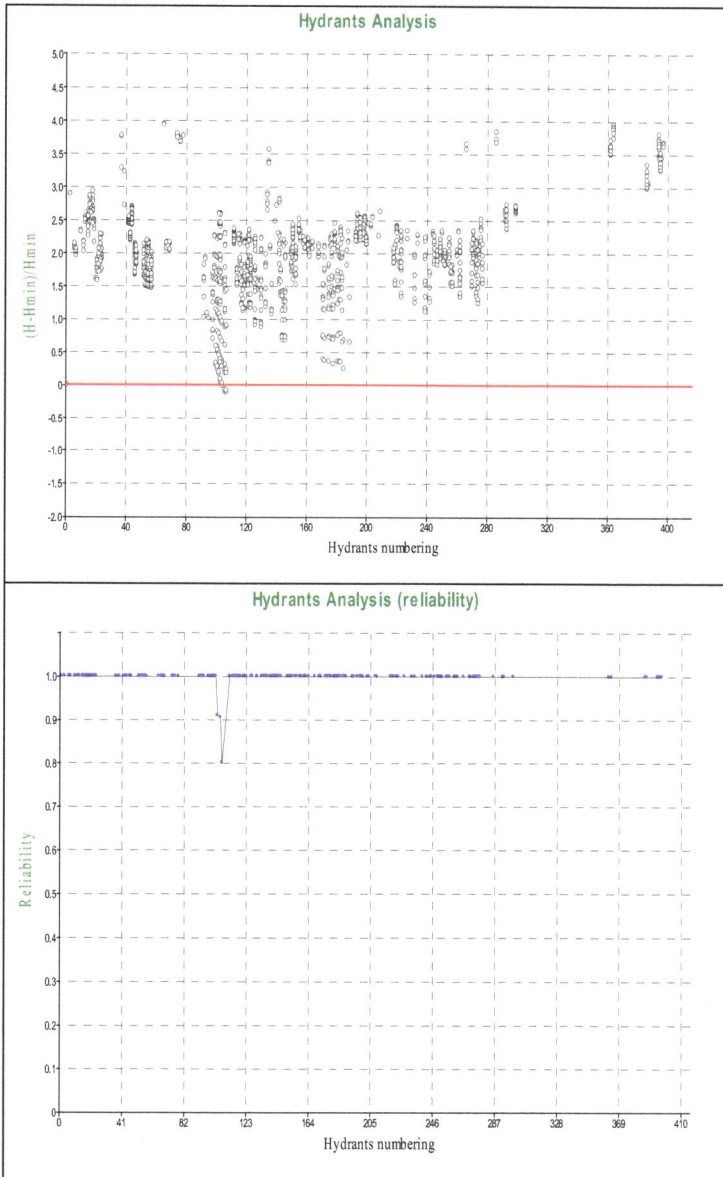

Fig. 12. *RPD* and *R* values obtained simulating the restricted demand delivery scenario on the distribution network of District 7 after increasing the total piezometric heat to H = 106 m a.s.l.

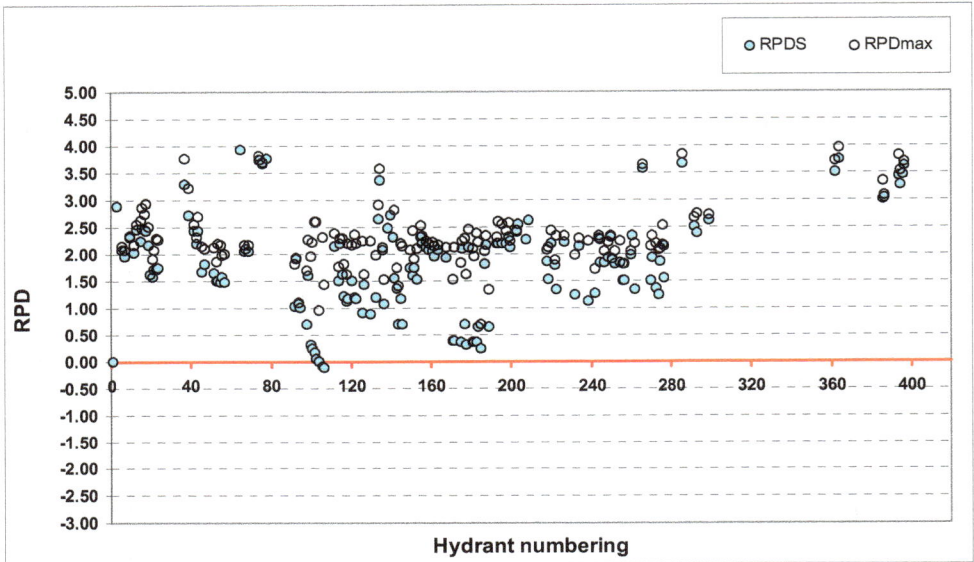

Fig. 13. *RPDS* values obtained simulating the restricted demand delivery scenario on the distribution network of District 7 after increasing the total piezometric head to H = 106 m a.s.l.

The value of the reliability at all hydrants, except for one, would be falling along the zero line, meaning that hydrants' state would be unsatisfactory every time they are put into operation. Figure 15 also shows the occurrence of limited to medium fluctuations of pressure heads at hydrants, but all falling within the negative range.

Further simulations were run also for this sub-network to figure out the effects of physical improvements on its hydraulic behavior, and satisfactory performance would be obtained after implementing the physical changes indicated hereafter:

1. limitation of the flow rate that can be withdrawn by users to 10 l s⁻¹ through the installation of adequate rubber-ringed flow limiters at all hydrants;

2. increase of the total piezometric elevation at the upstream end of the network from the current value of 36 m a.s.l. to 82 m a. s. l. to ensure enough pressure head under the different configurations of hydrants in simultaneous operation.

Figures 16 and 17 present the results of simulations and the level of performance achievable by the sub-network D8-N after the indicated modernization measures.

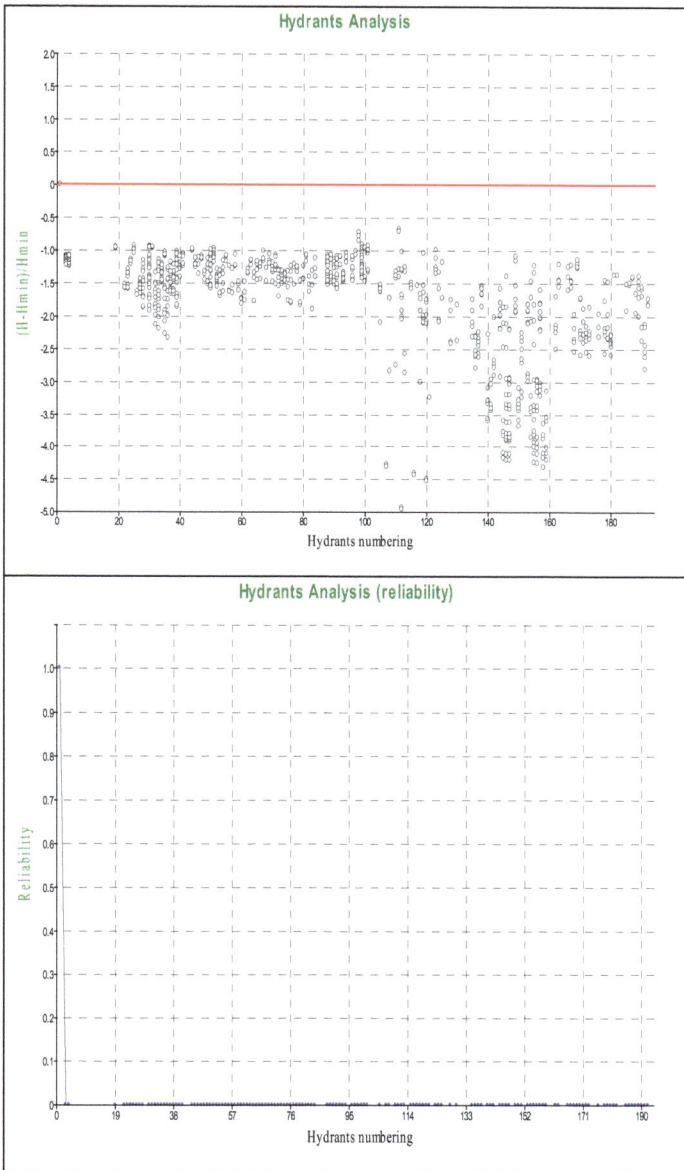

Fig. 14. *RPD* and *R* values obtained simulating the restricted demand delivery scenario on the existing distribution network D8-N of District 10 of the Sinistra Bradano system.

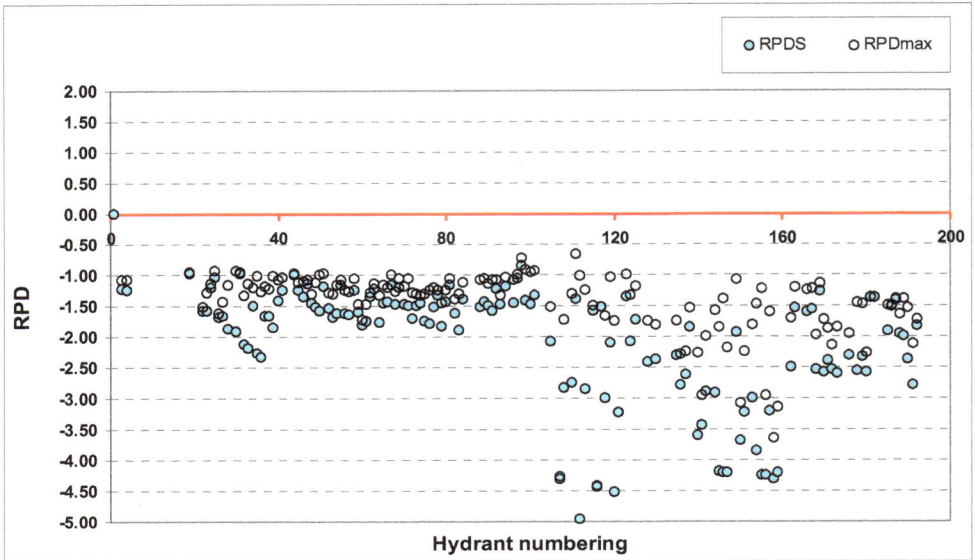

Fig. 15. *RPDS* values obtained simulating the restricted demand delivery scenario on the distribution network D8-N of District 10.

After the implementation of physical improvements, only a few hydrants, namely those between the numbers 106 and 121 and between 138 and 160, would still fall in unsatisfactory state, due either to their disadvantaged locations or to physical constraints in the upstream pipe sections.

In order to achieve adequate performance in terms of pressure head, it is recommended to allow the operation of these hydrants during low-peak demand hours (6 p.m to 6 a.m) to ensure the adequacy of deliveries in terms of flow rates and pressure heads. After this set of physical improvements the estimation of Equity yielded a value of 2.37, which reveals a very large variability of pressure head conditions and thus a "poor" level of equity among hydrants.

Alternatively, rising up the performance of these groups of hydrants to a satisfactory level, and avoiding, at the same time, the restriction of their operation during peak hours requires increasing the upstream piezometric head up to 140 m. Figures 18 and 19 present the simulated values of *RPD*, *R* and *RPDS* following the above-indicated increase in the piezometric elevation.

From these figures it can be noticed that an upstream piezometric elevation of 140 m would allow all hydrants performing more than satisfactorily in terms of pressure heads, and that the values of the reliability indicator would be equal to 1 for nearly all hydrants of the network, meaning that the pressure head of delivery would be equal or higher than the minimum required every time the hydrants are operated.

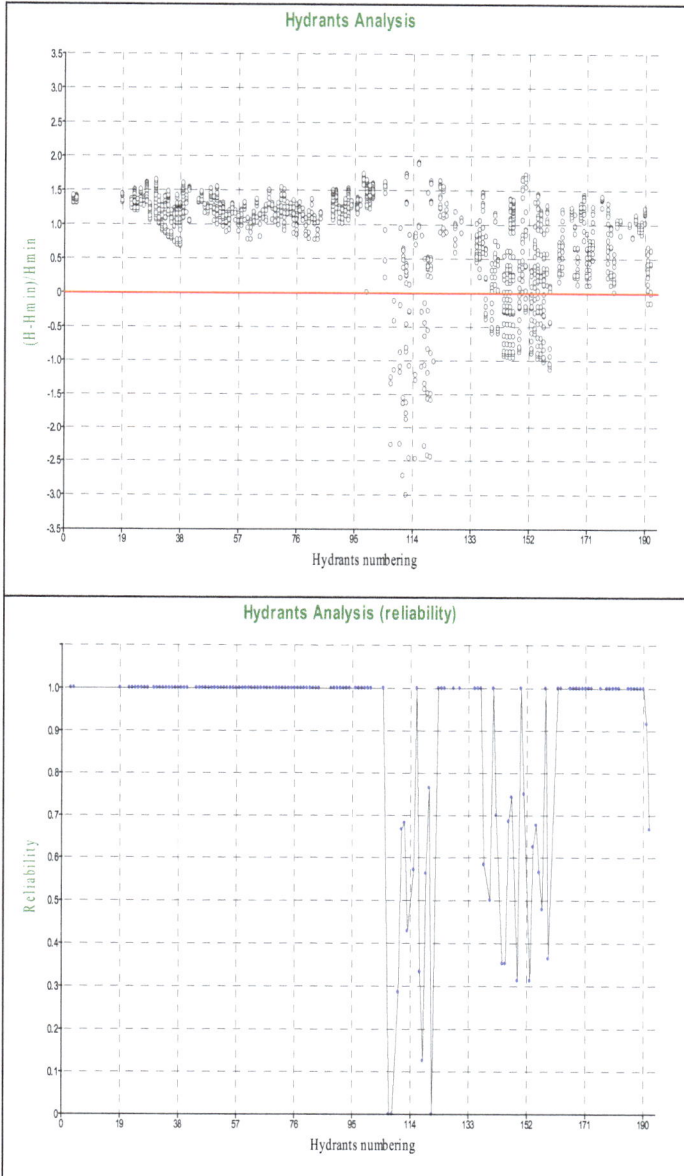

Fig. 16. *RPD* and *R* values obtained simulating the restricted demand delivery scenario on the distribution network D8-N after implementing physical improvements.

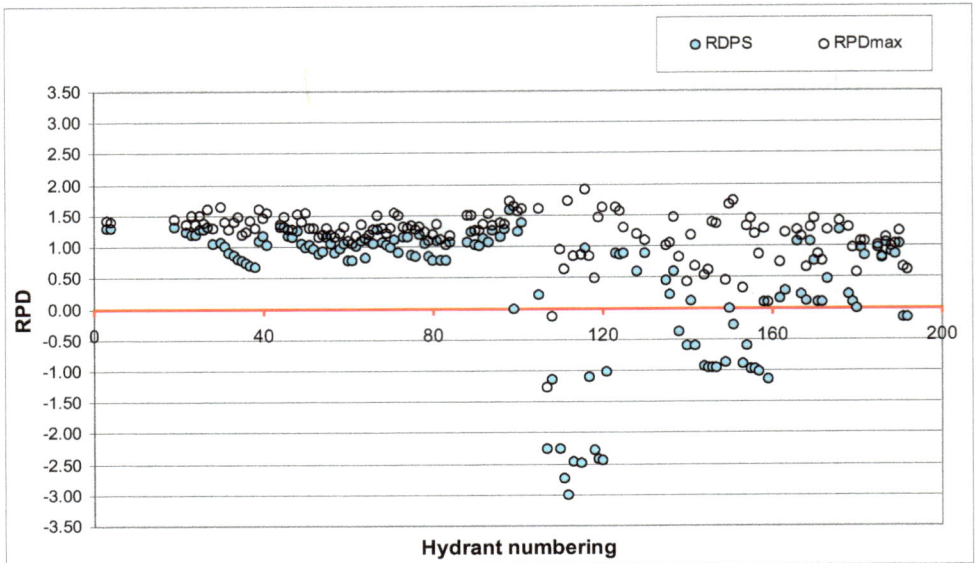

Fig. 17. Values of *RPDS* obtained simulating the restricted demand delivery scenario on the distribution network D8-N of District 10 after implementing the physical improvements.

Also, the value of *RPDS* for all hydrants, except for three, would fall in the positive range and for most hydrants would be way higher than zero and also would show quite limited pressure fluctuations. Under this improved scenario, the estimated value of Equity would be of 0.34, showing a much smaller variability of pressure head conditions among hydrants with respect to the situation with the upstream piezometric elevation of 82 m a.s.l.

By increasing the upstream piezometric elevation from 82 to 140 m a.s.l., the equity in terms of pressure conditions at hydrants would also increase from "poor" to "good."

Similar results were obtained simulating the restricted demand delivery on the sub-network D8-S in its current state. Figures 20 and 21 clearly show that the performance achievable by the D8-S network in terms of pressure heads would be very poor, as the values of *RPD* would be way below the zero line, the values of the reliability would be zero and the value of *RPDS* would be far below zero for all hydrants.

Similar physical changes as those proposed for the sub-network D8-N are necessary to the sub-network D8-S to make it capable of performing satisfactorily under the restricted demand delivery schedule. Figures 22 and 23 present the simulated values of *RPD*, *R* and *RPDS* after up-grading the network D8-S by means of the following physical measures:

1. limitation of flow rate that can be withdrawn by farmers to 10 l s⁻¹ by installation of adequate rubber-ringed flow limiters at all hydrants;
2. increase of the total piezometric elevation at the upstream end of the network from the current value of 36 m to 86 m a.s.l.

From these figures it can be noticed that under the improved scenario only a few hydrants out of the total number, namely the hydrants numbered from 53 to 66 and from 165 to 178,

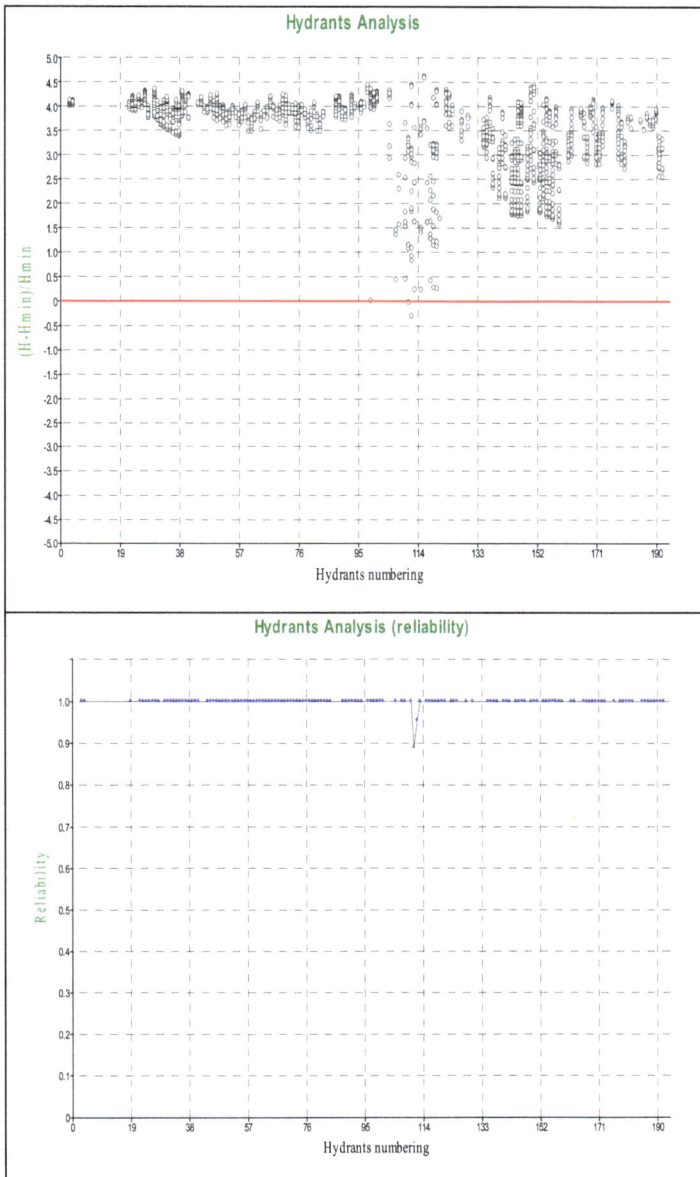

Fig. 18. Values of *RPD* and *R* obtained simulating the restricted demand delivery scenario on the distribution network D8-N after increasing the total piezometric heat up to 140 m a.s.l.

would not achieve satisfactory performance in terms of pressure heads. Also in this case, it is recommended the access and operation of these groups of hydrants by farmers separately from all the rest, thus only during low-peak demand hours.

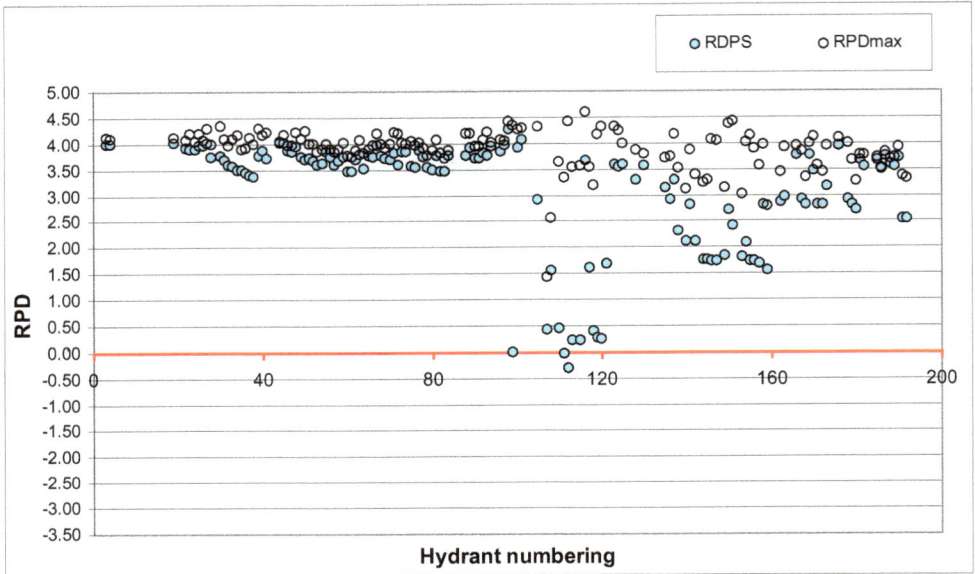

Fig. 19. *RPDS* values obtained simulating the restricted demand delivery scenario on the distribution network of D8-N after increasing the total piezometric head to H = 140 m a.s.l.

The Equity in this case would result as "poor," as the calculated value of 1.74 would reveal large variability in pressure head conditions among hydrants.

Results from simulations show that a further increase of the upstream piezometric head to 126 m a.s.l. would allow all hydrants of the D8-S network performing satisfactorily with respect to the required pressure head conditions, at any time they are accessed and operated by farmers. Figures 24 and 25 show that, after this further improvement, all the hydrants would achieve adequate or more than adequate performances, and specifically *RPD* values higher than zero, *R* values equal or very close to 1 and *RPDS* values very close or higher than zero.

The estimated value of 0.39 would rank the Equity as "good" in this scenario and would show a strong reduction in the spatial variability of pressure head conditions at hydrant by rising the upstream piezometric head from 86 m to 126 m a.s.l.

For both the districts analyzed, physical improvements of the distribution networks entail the increase of the piezometric heads at the upstream ends with the aim of allowing the demand flow configurations and offset all the resulting friction losses, also ensuring adequate water delivery conditions. To address this aspect, a pump system can be designed and sized to operate either at a fix set-point or to modulate the flow rate and pressure head based on the network's characteristic curves under the different flow configurations, and

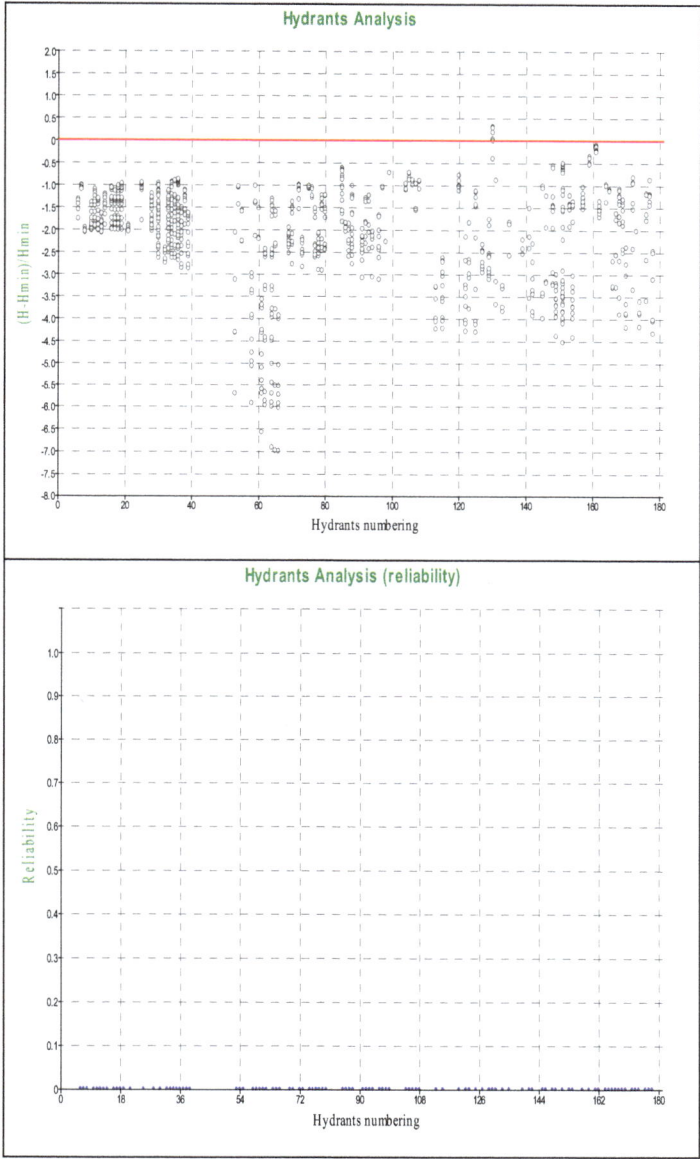

Fig. 20. *RPD* and *R* values obtained simulating the restricted demand delivery scenario on the existing distribution network D8-S of District 10 of the Sinistra Bradano system.

thus according to the downstream flow and pressure requirements. Assuming the operation of both districts by restricted demand, flow regimes in the pipe networks would vary with time based on configurations of hydrants in simultaneous operation. As a result, also friction losses would vary with time and so will also do the total dynamic head (TDH) that is needed at the upstream end of the network to offset head losses and to fulfill the pressure requirements at all the delivery points. Under these conditions, a sound technical solution could be represented by pumping plants capable of adjusting both the discharge and TDH based on downstream requirements, and thus on the basis of system curves resulting from the flow configurations and from the configurations of hydrants in simultaneous operation throughout the distribution network. These technical features can be accomplished by means of variable-speed pumps, in which pump units are equipped with inverters and devices for modulating the speed and operate on the basis of specific hydraulic algorithms.

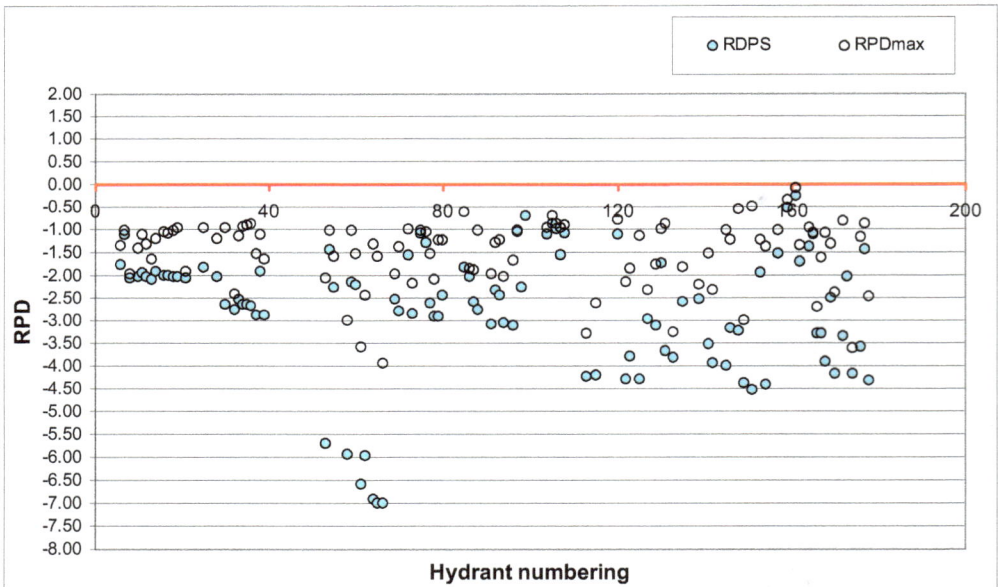

Fig. 21. *RPDS* values obtained simulating the restricted demand delivery scenario on the distribution network D8-S of District 10.

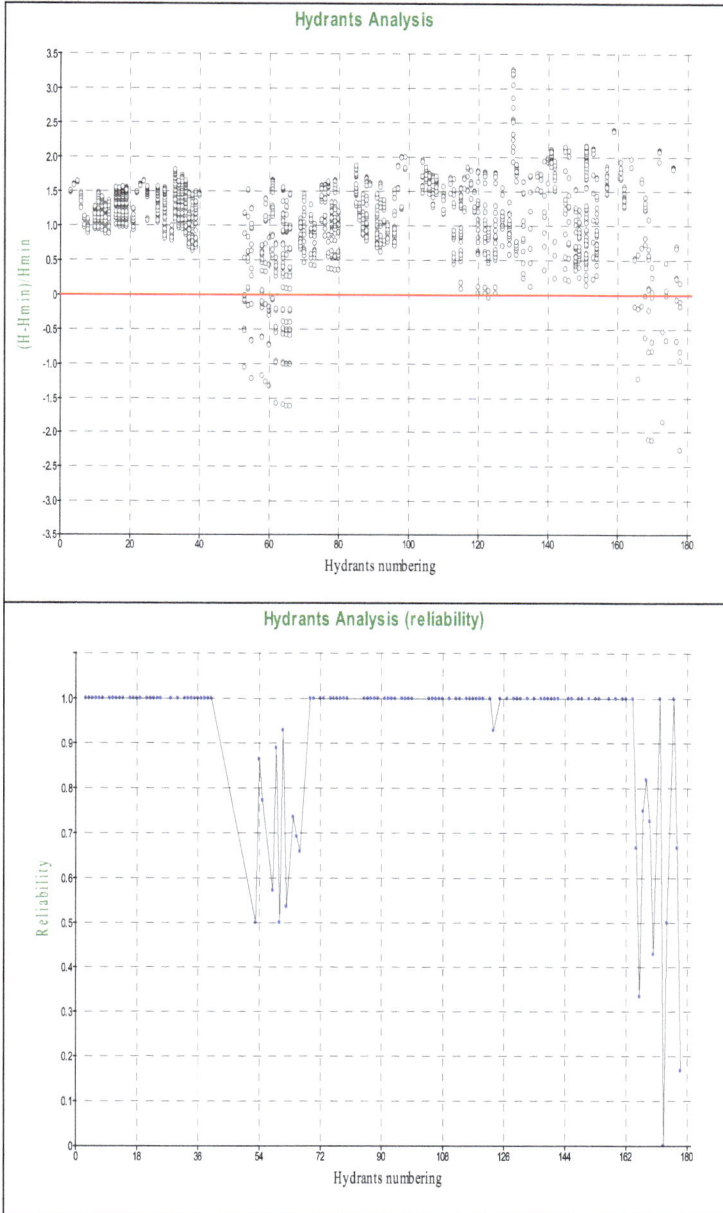

Fig. 22. Values of *RPD* and *R* obtained by simulating the restricted demand delivery scenario on the distribution network D8-S after implementing physical improvements.

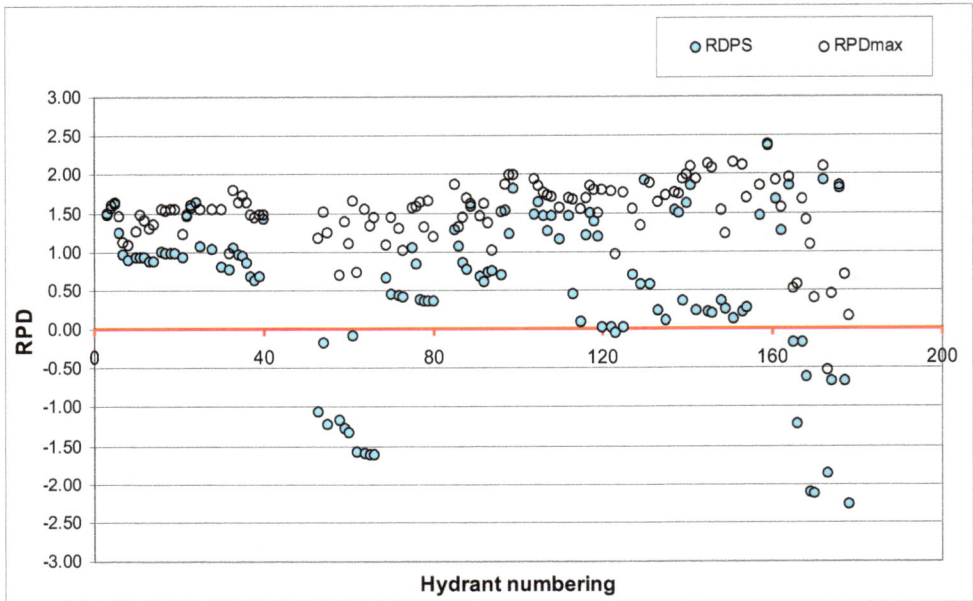

Fig. 23. *RPDS* values obtained simulating the restricted demand delivery scenario on the distribution network D8-S of District 10 after implementing the physical improvements.

7. Summary and conclusions

In this chapter, an innovative methodology aiming at diagnostic assessments of existing pressurized irrigation delivery networks is presented. The methodology entails the use of an agro-hydrological model for generating the demand flow hydrograph and the resulting flow configurations in the network, of a hydraulic simulation model to analyze its behavior under the simulated flow configurations, and of a set of performance indicators to evaluate the delivery achievements with respect to target or agreed-upon delivery objectives.

Both the agro-hydrological and hydraulic simulation models were tested and validated in previous research works and in different applications, proving their capability to forecast flow scenarios and the resulting hydraulic behaviors with adequate accuracy. The performance indicators were conceived and/or tailored for applications to pressurized networks and were tested for validation in previous research works conducted on a large-scale system of southern Italy, on which water deliveries to farmers are recorded and stored at hydrant level for monitoring and water-billing purposes.

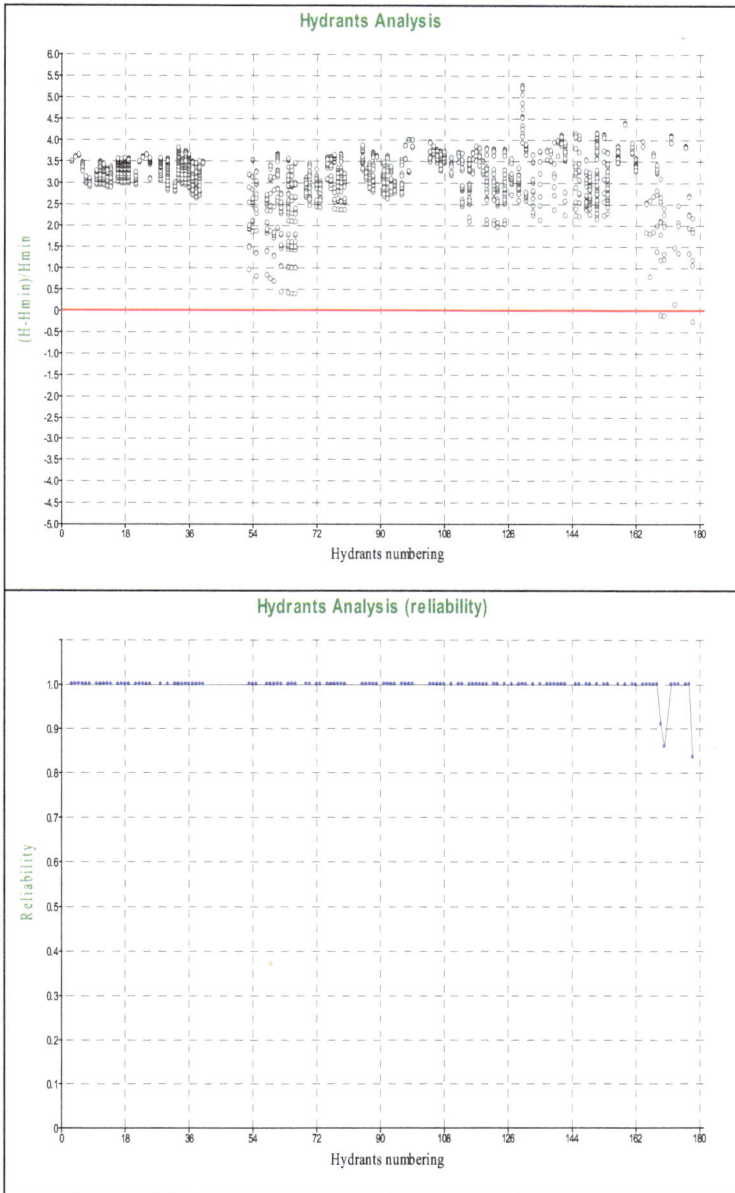

Fig. 24. *RPD* and *R* values obtained simulating the restricted demand delivery scenario on the distribution network D8-S after increasing the total piezometric heat to 126 m a.s.l.

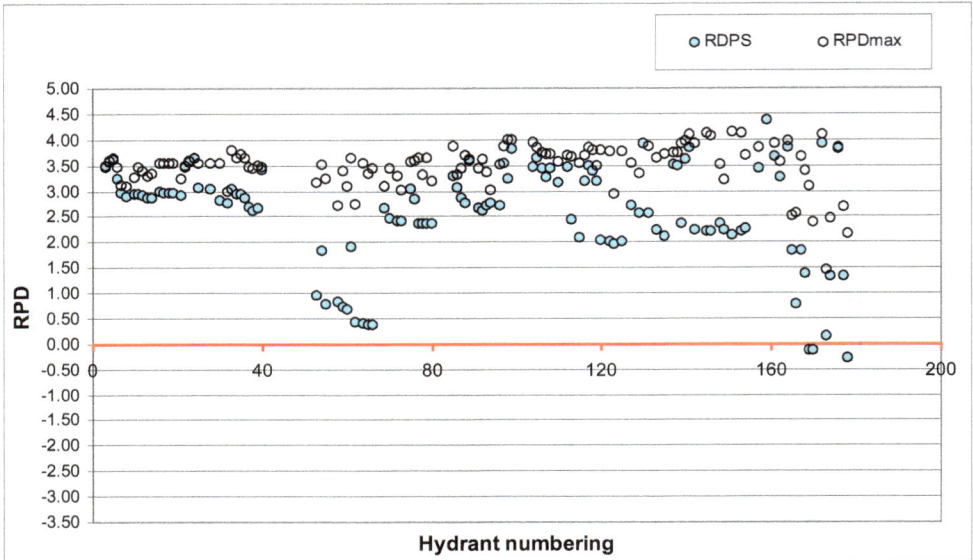

Fig. 25. *RPDS* values obtained simulating the restricted demand delivery scenario on the distribution network of D8-S after increasing the total piezometric head to H = 126 m a.s.l.

Finally, the proposed methodology was applied to a large-scale irrigation system in need of modernization, and specifically to two tail-end irrigation districts, and enabled the analysis of networks performances under different flow configurations. This application showed the usefulness of the combined analysis and simulation tools for addressing physical and operational aspects of modernization in poor-performing delivery networks.

In this perspective, the proposed methodology can be utilized as an analytical framework for designing and sizing new irrigation delivery systems, as well as for modernizing and re-engineering low performing systems, but also for assisting the management of irrigation schemes in developing operational plans and in avoiding situation of poor performance in water delivery to farmers.

8. References

Bethery J, Meunier M, Puech C (1981) Analyse des defaillances et etude de renforcement des reseaux d'irrigation par aspersion, Onzième Congrés de la CIID, question 36, pp. 297-324.

Bos MG, Burton MA, Molden DJ (2005) Irrigation and drainage performance assessment. Practical Guidelines. CABI Publishing, Cambridge, MA. ISBN 0851999670.

Burt CM, Styles SW (2004) Conceptualizing Irrigation Project Modernization Through Benchmarking and the Rapid Appraisal Process. In: "Irrigation and Drainage" ©2004 of the International Commission on Irrigation and Drainage (ICID), John Wiley & Sons, Inc.

Clément R. (1966) Le calcul de débits dans le réseaux d'irrigation functionnant à la demand. La Houille Blanche 5: 553-575.

Clément R, Galand A (1979) Irrigation par aspersion et reseaux collectifs de distribution sous pression. Eyrolles Editeur, Paris.

CTGREF Division Irrigation (1974) Programme ICARE – Calcul des caracteristiques indicées. Note Technique 6.

Hashimoto T (1980) Robustness, Reliability, Resiliency and Vulnerability Criteria for planning water resource systems, Ph.D. Dissertation, Cornell University, USA.

Hashimoto T, Stedomger JR and Loucks DP (1982) Reliability, Resiliency and Vulnerability Criteria for water resources system performance evaluation. Water Resources Res 18(1): 14-20.

INEA (1999) Quadro di riferimento per lo studio è il monitoraggio dello stato dell'irrigazione in Puglia: Consorzio di Bonifica Stornara e Tara, Taranto.

Labye Y, Olson MA, Galand A, Tsiourtis N (1988) Design and Optimization of Irrigation Distribution Networks. Irrigation and Drainage Paper no. 44, FAO, Rome.

Lamaddalena N (1997) Integrated simulation modeling for design and performance analysis of on-demand pressurized irrigation systems, Ph.D Dissertation, Technical University of Lisbon, Portugal.

Lamaddalena N, Sagardoy JA (2000) Performance Analysis of on-demand pressurized irrigation systems. FAO Irrigation and Drainage Paper n. 59, Rome, pp 132.

Molden DJ, Gates TK (1990) Performance measures for evaluating irrigation water delivery systems. J Irrig Drain Eng, ASCE 116(6): 804-823

Murray-Rust DH, Snellen WB (1993) Irrigation system performance assessment and diagnosis (Joint publication of IIMI, ILRI and IHE). Colombo, Sri Lanka: International Irrigation Management Institute.

Plusquellec HL, Burt CM, Wolter HW (1994) Modern water control in irrigation. Concepts, issues, and applications. World Bank Technical Paper 246. Irrigation and Drainage series. World Bank.

Prajamwong S, Merkley GP, Allen RG (1997) Decision Support Model for Irrigation Water Management. J Irrig Drain Eng 123 (2), 106-113

Rao PS (1993) Review of selected literature on indicators of irrigation performance. Colombo, Sri Lanka: International Irrigation Management Institute.

Small LE, Svendsen M (1992) A framework for assessing irrigation performance. Working Papers on Irrigation Performance No. 1. Washinghton, D.C., International Food Policy Research Institute.

Zaccaria D, Oueslati I, Neale CMU, Lamaddalena N, Vurro M, Pereira LS (2010) Flexible Delivery Schedules to Improve Farm Irrigation and Reduce Pressure on Groundwater: a Case Study in Southern Italy. Irrig Sci 28:257-270, DOI 10.1007/s00271-009-0186-8.

Zaccaria D, Lamaddalena N, Neale CMU, Merkley G (2011a) Simulation of Peak-Demand Hydrographs in Pressurized Irrigation Delivery Systems Using A Stochastic Model. Part I: Model Development. Paper submitted to Irrigation Science on January 2011, manuscript ID IrrSci-2011-0010, currently under peer reviewing.

Zaccaria D., Lamaddalena N., Neale C.M.U., Merkley G. (2011b) Simulation of Peak-Demand Hydrographs in Pressurized Irrigation Delivery Systems Using A

Stochastic Model. Part II: Model Applications. Paper submitted to Irrigation Science on January 2011, manuscript ID IrrSci-2011-0011, currently under peer reviewing.

15

Irrigation in Mediterranean Fruit Tree Orchards

Cristos Xiloyannis, Giuseppe Montanaro and Bartolomeo Dichio
Department of Crop system, Forestry and Environmental Sciences
University of the Basilicata, Potenza
Italy

1. Introduction

The Mediterranean environment alters the ecophysiology of plants, especially during summer as a consequence of the combined effects of high light, high air temperature, high vapour pressure deficit and low rainfall. The high evapotranspirative demand which characterises the Mediterranean climate has, in the past affected land use with farmers tending to choose drought-tolerant species, such as olives, almond ecc (Dichio et al., 2006). In recent years, major investments have been made in agriculture that have lead to a 25% increase in the area of the Earth's surface under irrigated crops. However, despite these investments, an increasing number of countries in arid and semi-arid regions face severe water shortages because of reduced annual rainfall (Mutke et al., 2005; Cislaghi et al., 2005) and because their existing water resources are already fully or over exploited. The availability of water (agriculture, industry and domestic) in several countries of the Mediterranean basin is well below the level associated with the achievement of a modern standard of living (1,000 m³ per capita per year) (Rana and Katerji, 2000). Prospects for the future suggest increasing difficulty will be experienced in this area (Smith, 2000). They also indicate that dependency on water for future development has now become critical. For agriculture this has triggered many studies on drought mitigation measures as applied to large-scale networks (Rossi et al., 2005).

Actually, about 75% of the available water in the Mediterranean area is used for agricultural purposes. It is unfortunate that this occurs with very low efficiency of conveyance between reservoir and field. In this chapter, we do not deal with possible improvements in water conveyance to the farms (large-scale networks) but instead we focus on irrigation criteria and methods that can reduce on-farm water losses and can also optimise crop water use. Water losses on the farm account for approximately 40% of total farm water usage. Also, poor irrigation management has direct effects on production as a result of crop stress induced either by water shortage or by waterlogging. Both water deficit and water excess reduce crop yield and quality.

In addition, common cultural practices (empirical irrigation, soil and fertilisation management) also aggravate the decline in soil resources and have negative impact on the environment by contaminating both ground and surface water with various nutrients and pesticides.

Recognising that in the Mediterranean basin, rainfall occurs primarily during the dormant season many horticultural crops are dependent upon stored soil water during rainfall season and on irrigation during the summer period. Therefore, accurate determinations of irrigation timing and volumes are essential if sustainable agricultural development and environmentally sound water management are to be achieved.

A sound knowledge of crop characteristics such as soil volume explored by roots, their sensitivity to water stress and their seasonal water requirements, are of primary importance. These information are required not only to improve understanding the underlying processes of plant physiology and their control (Dragoni et al., 2005), but also for improving the design of irrigation systems and irrigation scheduling. Due to the scarcity of water resources, accurate evaluation of water use efficiency by the different crops is also very important. However, in spite of the large number of methods to measure or estimate plant water use (for a review, see Rana and Katerji, 2000) further efforts are required to improve our understanding of crop water-use efficiency. Moreover, practical application of the scientific findings should be better discussed and be available to growers for both to conserve water resources and also to control environmental pollution .

Based on our own experimental results and also on information from the literature, the aim of the present Chapter is to provide information and appropriate criteria to enable the sustainable management of irrigation at farm level in semi-arid environments such as in Southern Italy.

Nowadays irrigation requires special attention to optimize the management of all components of the orchard system in order to increase water use efficiency and reduce environmental impacts (e.g. soil salinisation, degradation of underground/surface waters). Knowledge on basic plant water relations are widely available, however fewer attempts have been made to link such a information to irrigation schedule at field scale. In addition, irrigation for tree crops should take into account their distinctive traits (e.g. the soil volume explored by roots, type of rootstock) as combined with some soil hydrological features such as the soil water holding capacity.

We would also provide recommendations to drive the water application in fruit tree orchards through adoption of soil-water balance procedures as determined by soil, environment and crop data interaction.

Sustainable irrigation, which includes the application of the regulated deficit irrigation and specific crop coefficients to calculate the plant water requirement, reduces irrigation-induced salinisation risk and increases yield and quality. Our contribution would cover also the synergistic effect of others orchard practices (e.g. soil management, fertilization and canopy management) towards optimal irrigation.

2. Choice and design of irrigation method

Except in soils of low water-holding capacity, localised irrigation methods (drip irrigation or sub-irrigation) are best for all fruit tree species grown in the Mediterranean area. However, in the case of kiwifruit (*Actinidia* spp) because of its physiology and its root system characteristics (Ferguson, 1984; Xiloyannis et al., 1993) irrigation methods that wet the whole soil surface should be considered instead. Additionally, the adoption of localised irrigation methods require water availability almost every day (June-September, Northern Hemisphere) and often current networks irrigation-Agency (responsible for water management at regional scale) cannot adequately meet the water supply demands. In medium to large farms these timing difficulties can be overcome by the construction of on-farm reservoirs that allow crops to be irrigated even when water is not available in the regional network. This avoids excessively long intervals between irrigations. To choose the most appropriate irrigation method and design one must know: soil characteristics, water requirements of the crop, water availability and water quality .

2.1 Soil water-holding capacity

The soil can store huge amounts of water. In particular, rain water accumulates in autumn and in winter when plant water use rates are low. Deep, loamy soils can hold up to 2,000 m^3 ha^{-1} if a 1 m rooting depth is assumed. The water contained in such a volume of soil is sufficient to meet about 30-40% of an orchard's annual water requirement. In light, shallow soils, and in areas having a shallow water table where the root systems cannot develop to very great depth, the amount of water that can be stored in the soil is much more limited and, consequently, plants are more likely to be exposed to water-deficit induced injury in the summer period.

In soils with high water holding capacities (1,500-2,000 m^3 ha^{-1}) and, in the absence of irrigation the soil-water content decreases slowly during the season. This allows the plants to adapt gradually and thus limit the damages from water-stress. Conversely, in light and/or shallow soils, and in the case of rootstocks whose rooting depth is shallow, the effective volume of the soil water reserve can be very limited indeed. In this case, sudden variations in soil moisture and in plant turgor will occur and this will cause severe injury to plants that are unable to adapt fast enough to mitigate the effects of a sudden onset of water stress.

Soil management under water scarcity conditions should aim at: (i) improving the soil's water holding capacity during rainfall season and (ii) reducing soil surface evaporation and transpiration from fruit trees and cover crops.

To achieve the former objective, the infiltration rate and water holding capacity can be significantly enhanced by increasing the soils organic matter content and also its hydraulic conductivity. Sloping land, if not adequately managed, usually has a low water holding capacity. Similarly, water holding capacity is reduced in flat land that has been frequently tilled, and always to the same depth where the formation of a 'plough sole' hampers downward infiltration of water. Unfortunately, permanent cover crops are not a good solution to the problem because they compete for water with the crop. Therefore, we would recommend temporary cover crops (November-March) both to increase the soil's water storage capacity (increased organic matter) and also to limit its erosion (especially on slopes) (Photo 1).

Water loss by surface evaporation can be as high as 50% of precipitation and can amount to about 30% of yearly evapotranspiration. Soil surface evaporation losses increase with decreasing of the Leaf Area Index (LAI) (m^2 of leaf per m^2 of soil), and with increasing numbers of irrigation events, especially if using methods that wet the whole soil surface. The distribution efficiency of the various irrigation methods applied to full bearing orchards varies from 50 to 90%, because of the different amounts of water evaporating from the soil between irrigation events and also during distribution. In young orchards, where the root systems are not fully developed and where ground cover is limited, the efficiency of the various irrigation methods varies from between 10 to 95%. So, in Southern Italy, localized irrigation methods (particularly drip and subirrigation) are a "must" for new plantations. Replacement of low efficiency methods by high efficient - possibly even through public subsidises - should be actively promoted.

Photo 1. Olive orchard grown in South Italy on a slope. Soil is tilled and prone to dramatic erosion during the winter.

Soil evaporation can also be reduced by mulching, using either plant residues resulting from local agricultural practices or one of many other low-cost materials that might happed to be available locally.

2.2 Characteristics of cultivated species

For best choice and design of the irrigation method as well as for its correct management, especially in the early years of orchards establishment, one should estimate the soil volume explored by roots and also the leaf area per hectare. Since the bulk of crop water usage (99.5%) is through foliar transpiration, the considerable variation in LAI that occurs during the early years of establishment (Fig. 1) and during each vegetative season, significantly affects water use. Not surprisingly, in mature orchards leaf area variations during the season are greatest in deciduous species and least in evergreen species.

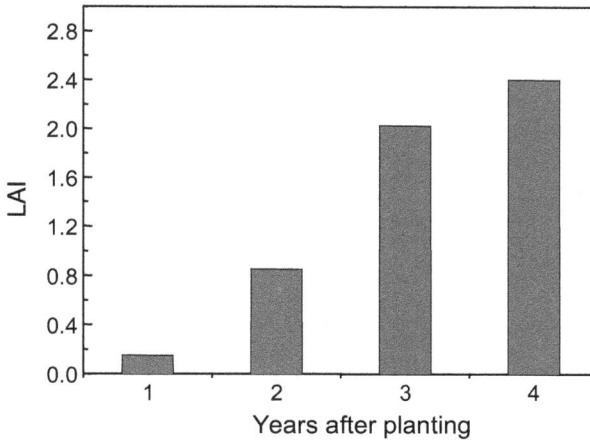

Fig. 1. Variation of the Leaf Area Index (LAI) in kiwifruit vines (cv Hayward) trained at T-bar (4.5 m × 3 m) during the four years after planting. (Adapted from Xiloyannis et al., 1993).

Knowledge of the volume explored by the roots, and also the soil's hydrological characteristics allows calculation of the soil's effective water holding capacity and thus the volume of water that is available to the plant.

Such information is indispensable both for the design of the irrigation system (spacing, discharge rates, number of emitters etc.) and also for its correct management, in particular for defining irrigation volumes and frequencies. During the early years of an orchard's development the soil volume explored by the roots changes considerably. This occurs as the root systems extend both outwards and downwards. Volumes tend to stabilise once the trees are mature (Tab. 1). In parallel, the leaf area per plant and the available water change accordingly (Tab. 2).

Species/ 'rootstock	m³ tree⁻¹			
	1st	2nd	3rd	4th year
Peach cv Vega/ 'Missour'	1.22	3.39	3.60	3.60
Peach cv Vega/ 'Mr.S. 2/5	0.56	1.97	2.80	2.80
Kiwifruit cv"Hayward"	0.13	0.83	1.35	1.41
Olive cv "Coratina"	0.50	2.90	8.60	12.25

Table 1. Soil volume explored by roots of peach (cv 'Vega', grafted on to two rootstocks planted at 4.5 × 1.5 m spacing, flood irrigation), kiwifruit (cv 'Hayward' at 4.5 × 3.0 m spacing, microjet irrigation) and olive trees (cv 'Coratina', at 6 × 3 m spacing microjet irrigation) during the first 4 years after planting (Adapted from Xiloyannis et al. 1993).

Root density affects water availability and *vice versa*. High root density means reduced average distance between roots, steeper water potential gradients and steeper concentration gradients of mineral nutrients in the soil and, consequently, a greater use efficiency of water and mineral resources present in the soil volume explored by roots . Root density is usually expressed either as a root dry weight in the volume of the soil explored, or as a root length. Such expressions are useful for making comparisons between species, but are less useful for defining the water and mineral nutrient uptake efficiency. For this purpose, the surface of roots in contact with the soil and their age-related uptake efficiency require to be known. In all fruit tree species, except kiwifruit, root density is much lower than in the grasses and conifers where it is relatively high (Xiloyannis et al., 1992).

Years after planting	I	II	III	IV
Kiwifruit: cv. _Hayward_ (4.5 × 3.0m)				
Leaf area (m^2 p^{-1})	1.7	8.9	16.5	17.2
Available water (L p^{-1})	12.8	72.3	147.4	154.0
Available water/leaf area (L m^{-2})	7.5	8.1	8.9	9.0
Peach: _Vega/Missour_ (4.5 × 1.25m)				
Leaf area (m^2 p^{-1})	3.8	11.8	16.5	16.5
Available water (L p^{-1})	137.9	383.1	406.8	406.8
Available water/leaf area (L m^{-2})	36.3	32.5	24.6	24.6
Olive: cv. _Coratina_ (6.0 × 3.0m)				
Leaf area (m^2 p^{-1})	0.6	1.9	6.1	6.9
Available water (L p^{-1})	160	910	2,710	3,950
Available water/leaf area (L m^{-2})	263	481	443	571

Table 2. Leaf area, soil available water and leaf area/available water ratio in three fruit tree species during the early four years after planting.

Tree crop species differ in their sensitivity to water deficit and the extent of deficit injury depending on the growth stage of the crop (Tab. 3).

Fruit tree species	Growth stage especially sensitive to water deficit
Apricot, cherry, plum and early-harvest peach	From bloom to harvest
Plum and late-ripening peach	1st and 3rd fruit growth stages
Citrus fruit	Bloom and fruit-set
Olive	Bud break, bloom, first and 3rd fruit growth stage (especially table olive)
Pome fruit	Bloom, fruit set, fast fruit growth stage
kiwifruit	Throughout the whole growing season

Table 3. Growth stage especially sensitive to water deficit in some fruit tree species.

2.3 Canopy architecture, canopy management and 'water use efficiency'

The term 'water use efficiency' is defined as the ratio of the mass of carbon dioxide (CO_2) that is fixed to the cumulative mass of water transpired. Out of the total water absorbed by roots and transferred to the shoot, about 99.5% is released again to the atmosphere through leaf stomatal and cuticular transpiration. Transpiration of the fruits accounts only for about 0.5% of the plant's total, however fruits may increase leaf transpiration of about 5-10%. Leaves that receive sufficient light to achieve maximum photosynthetic rates (800-1,000 μmol m^{-2} s^{-1} PPFD), although transpiring more, also have higher water use efficiencies (about 10 folder) than shaded leaves (receiving <20% of incoming radiation). For example, a volume of 1,000 litres of water transpired from sunlit leaves yields about 3 kg of fixed carbon, whereas shaded leaves scarcely produce 0.3 kg carbon with the same amount of water. This amount is insufficient even to meet night-time respiration carbon use. Thus, the portion of the canopy receiving less than 20% of available radiation, represents not a source of photosynthates for the orchard but a sink, and with significant water usage that can reach about 30% of total consumption in some training systems (e.g. pergola for kiwifruit and table grapevines) (Xiloyannis et al., 1999).

Therefore, in choosing the training system one should remember that water use efficiency increases with an increasing ratio of exposed/shaded leaves. Increased efficiency is possible through reducing the tree size (Photo 2) adopting training systems that maximise the proportion of fully sunlit leaves, minimising shading, and carrying out summer pruning (Fig. 2).

Photo 2. High density apple plantation. The reduced tree size maximises the exposure to irradiance (Photo by Vivai Mazzoni).

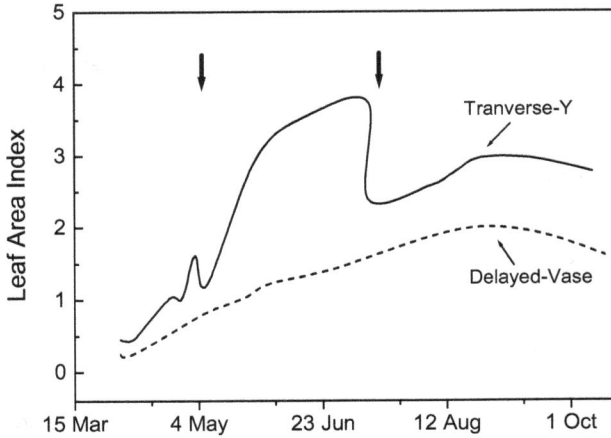

Fig. 2. Leaf Area Index (LAI, m² leaves/m² soil) variations during the third year after planting in peach trees (cv 'Springcrest') trained to transverse Y (continuous line) (1,100 plants ha⁻¹) and Delayed-vase (dotted line) (416 plants ha⁻¹). Arrows indicate summer pruning (Transverse-Y orchard), performed twice during spring-summer, reducing the LAI and water use, and improving yield quality, water use efficiency and cropping potential for the following year. (Redrawn from Nuzzo et al., 2003).

As for canopy management of bearing orchards, one should certainly be careful to carry out summer pruning, whereas for newly established orchards, a choice of training system should be made with canopy water use efficiency in mind (Fig. 3).

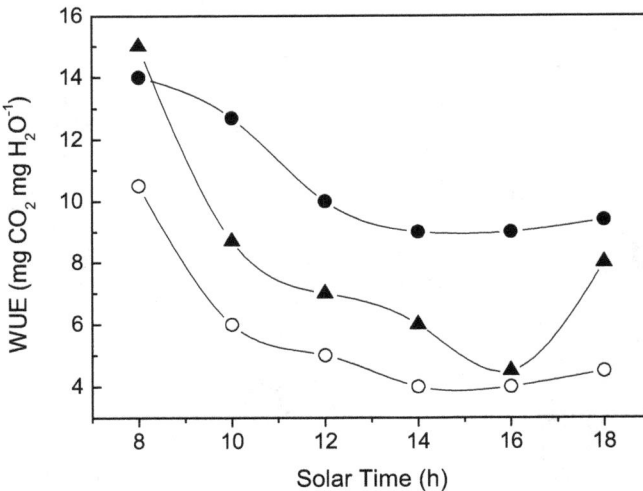

Fig. 3. Daily variation in Water Use Efficiency (WUE) (mg CO_2 mg H_2O^{-1}) in a whole peach canopy trained to Transverse-Y (●), Delayed-vase (▲) and Palmette (○). (Adapted from Giuliani et al. 1999).

It is recommended that all wood that is not necessary for the subsequent year's production should be removed through summer pruning. In this way, the leaf area and water consumption are reduced, water use efficiency increases, the exposure of fruits to light is increased, and the growth and quality of the remaining one-year old fruiting shoots to support the next season's fruit bearing is optimised.

In areas with high evaporative demand and limited water, fruit thinning should aim at leaving fewer fruits per m2 of leaves (-20% approximately), especially in early ripening cultivars and orchards established on shallow and/or light soils.

It is also worthwhile to carry out "winter" pruning early (e.g. in August) for orchards where the harvest has already been done by that time. In other cases, pruning should be carried out as soon as possible after harvest. Shade nets (30-40% shading) can considerably reduce water usage and improve the photosynthetic activity of exposed leaves.

3. Regulated deficit irrigation

For sound management of water resources aimed at responding to the increasingly frequent "droughts" in southern Italian orchards, specialised water saving techniques, like Regulated Deficit Irrigation (RDI), are to be recommended. They allow considerable reductions in irrigation volumes and also ensure good yield and good quality. The RDI is a method by which irrigation volumes to fruit crops can be reduced (Behboudian and Mills, 1997) by only partially replenishing the water used by crop evapotranspiration. Replenishment is carried out up to pre-established threshold values of water deficit in the soil and in the plant. The applicability of RDI has been extensively studied (Shackel et al., 1997; McCutchan and Shackel, 1992; Naor, 2000), but further effort is needed to bring this technique into more common usage. Recently, the possibility of reducing irrigation volumes by up to 50% has been evaluated in Southern Italy with early-ripening peach cultivars during the post-harvest period thereby achieving a water saving of about 1,800 m3ha-1 per year (3-year average), a good yield and a greater accumulation of carbohydrates in the roots and the wood, as a result of reduced vegetative growth (Fig. 4 and 5) (Dichio et al., 2007).

Nowadays, it has been pointed out that a prolonged and often severe water deficit imposed through RDI could decrease yield in fruit trees when deficits are applied over successive seasons (Pérez-Pastor et al., 2009). Hence, more information is needed to accurately and safely manage RDI in the field and for successive growing seasons. Irrigation and the RDI in particular should be considered within a wider management system including other agronomic practices and land resources.

Recently, at a peach orchard in a recent 6-year comparative study (Dichio et al., 2011) of conventional and sustainable orchard management practices (the conventional practices are representative of usual grower practice in the region while the sustainable practices aim to save water) it has been shown that RDI can maximize water productivity without impairing long-term yield. Briefly, conventional orchard management (C) (continuous soil tillage, mineral fertilizers, irrigation scheduling decisions exclusively based on grower experience) adopted by local farmers, was compared with sustainable orchard practices (S) (no tillage, cover crop, organic fertilizer, summer pruning and sustainable irrigation). Drip-irrigation (8 L h-1 with 2 drippers per plant) in both S and C treatments was scheduled in the C plot every approx. 10 days starting in April, while in the S plot, irrigation requirements (I) were

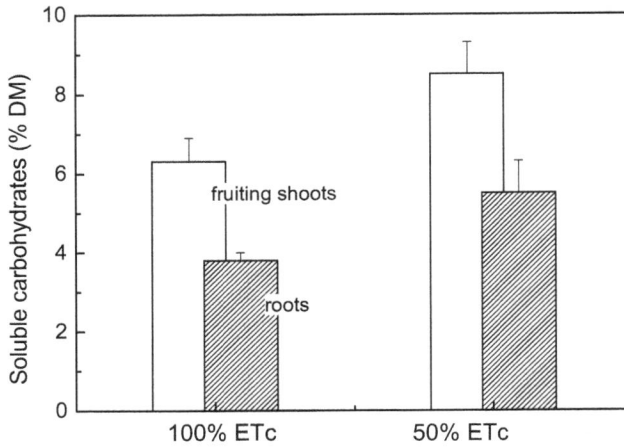

Fig. 4. Concentration of reducing soluble carbohydrates (% DM, ±SE) in roots (grey column) and fruiting shoots (white column) in a well-watered peach tree (100% ETc) and subjected to Regulated Deficit Irrigation (RDI) (50% ETc). Measurements were made in November at the end of the second year of RDI application. (Adapted from Dichio et al. 2007).

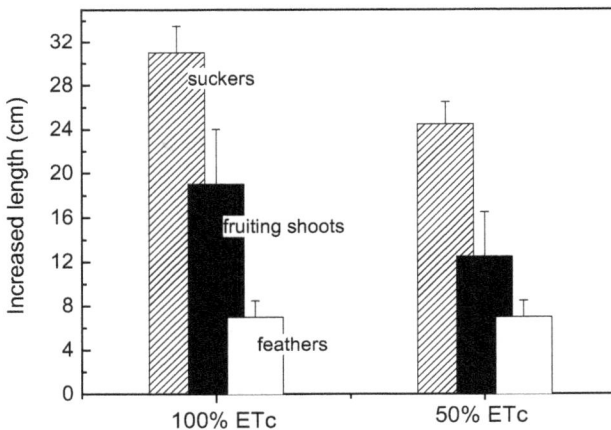

Fig. 5. Increased length (cm ±SE) of suckers (grey column), feathers (white column) and fruiting shoots (black column) observed in peach between late June (beginning of regulated deficit irrigation) and early October in well-watered trees (100% ETc) and in trees subjected to Regulated Deficit Irrigation (RDI) (50% ETc) (Adapted from Dichio et al. 2007).

calculated by the equation $I = ET_0 \times K_c / 0.9\text{-}1$, assuming 90% distribution efficiency. In the S plot, crop coefficients (K_c) during the pre-harvest stage in April, May and June were 0.6, 0.8 and 1.2, respectively as resulting from adjustment of previous own experiments carried out in the area (Dichio et al., 2007). During the postharvest stage, from July to September, RDI was applied by reducing the irrigation to approximately 50 % of plant requirement (Dichio et al., 2007). The summer pruning was performed on mid-June and on end of July. The S plot received 15 t ha^{-1} yr^{-1} of compost (24.8 % moisture content) containing approximately 35% carbon on a dry matter basis. Fertilisation was based on concentrations (% dry matter, DM) of various plant tissues, whole plant DM per plant and on availability in the soil of the various essential plant nutrients. In particular, the concentration of soil nitrates was monitored in the top 40 cm of soil and N distributed via fertigation each time the concentration fell below 20 ppm. Plant water status was monitored in the S plot on 5 trees per plot (×5 leaves a plant) by measurement of midday stem water potential at weekly intervals following the procedure reported by Dichio et al. (2007).

Large differences in seasonal irrigation volumes were seen for the two irrigation methods (Fig. 6). Average annual irrigation volume applied in the S plot was approximately 23 % lower than in the C plot, a saving of 1450 m^3 ha^{-1} gained during the post-harvest stage. Despite the lower irrigation volume, after six year the cumulative yield was substantially higher (30 %) in the S than in C plots reaching 140 and 115 t ha^{-1}.

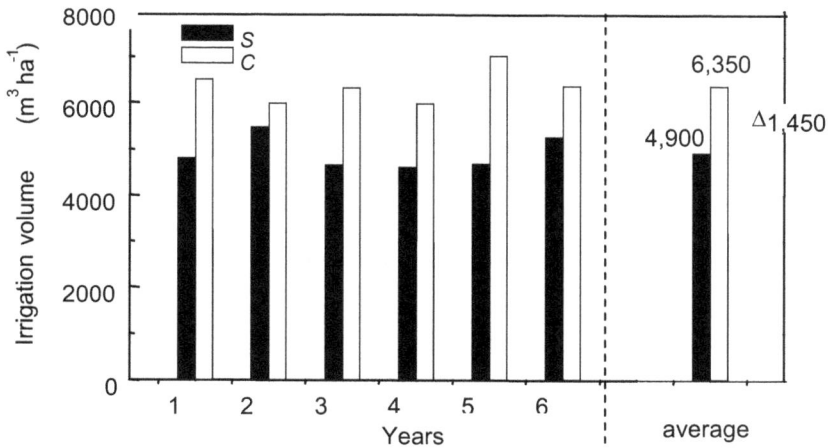

Fig. 6. Annual irrigation volumes (m^3 ha^{-1}) and 2004–2009 mean volume applied in the sustainable (S) and conventional (C) plots. Note, on average, 1450 m^3 ha^{-1} were saved on the S plot. (Redrawn from Dichio et al., 2011).

The regulation of vigor due to moderate water stress possibly reduced the competition for assimilates between reserve tissues and the vegetative apexes resulting in better light interception and lower water use (Boland et al., 2000). In the S plot, summer pruning was performed twice a year reducing the leaf area in all by approximately 10 m^2 plant^{-1}. Considering a daily mean leaf transpiration rate of 3 mmol m^{-2} s^{-1} (Dichio et al., 2007), the summer pruning in turn had contributed to reduce the transpired water by about 800 m^3 ha^{-1} over approx. 40 days from August.

Irrigation management in that study integrated other sustainable practices concerned with soil rehabilitation like increasing the soil carbon level On average, in the S plot, 21.1 t ha^{-1} yr^{-1} carbon was returned to soil, compared to 6.1 t ha^{-1} yr^{-1} in the C plot. Increased soil carbon is a prerequisite for soil fertility remediation, mineral element supply and better soil water holding capacity (Montanaro et al., 2010). Therefore, higher carbon input in the S plot may have increased the retention of winter rainfall in the soil resulting in a likely higher soil available water compare to the C plot.

Recently, emphasis has been placed on the concept of water productivity (WP), defined either as the yield or net income per unit of evapotranspiration (Fereres and Soriano, 2007). We evaluated the effect of orchard management practices on economic water productivity (E_{WP}), defined as the economic value of the marketable yield per unit of irrigation applied. Marketable yield value depends on fruit quality and in particular on fruit size distribution, that in turn may affect E_{WP} as a result of applying sustainable orchard practices. E_{WP} index therefore seems to be an appropriate method of assessing the impact of irrigation technique on productivity. We believe that a water saving *per-se* does not necessarily result in increased yield, and that higher yield sometimes leads to reduced fruit size. On a six-year average, based on the annual fruit price and marketable yield, the E_{WP} was 2.11 and 1.34 € m^{-3} for the S and C plots, respectively. This was evidently related to the increased yield and reduced irrigation in the S plot (Fig. 6). Based on the above mentioned beneficial effect of the carbon on soil water holding capacity, the high carbon input in S plot possibly contributed to increased E_{WP} via reducing the irrigation volumes.

This paragraph demonstrated that integrating RDI into a wider sustainable fruit tree orchard management regime with increased soil carbon inputs, resulted in high and stable yields and a high E_{WP} over the medium term (six years). This information should encourage water-management policy makers to promote strategies that promote industry wide adoption of RDI in order to reduce agricultural water use. For example, offering adequate extension service and, at the same time, introducing volumetric charges for irrigation water and economic penalties for excessive water consumption will almost certainly lead to a higher E_{WP}. However, using price policies to promote the economic productivity of water requires significant government intervention in order to ensure equity of access to public water. We believe E_{WP} should be a useful tool to evaluate the impact of alternative water management technologies on farm- and regional-scale economies.

4. Management of the irrigation method

4.1 Evapotranspiration and irrigation requirements

Evapotranspiration is the most important term in water balance for irrigation. When plant transpiration and other evapotranspiration components cannot be calculated separately, the simplest and most widely adopted approach is the "two-step method". As a first step, reference evapotranspiration (ET_0) is estimated. As a second step, crop coefficient (Kc) is introduced to account for the evapotranspiration aspects related to crop growth stage.

This estimation results in $ET_C = ET_0 \times Kc$ and expresses the water use of a crop grown under standard conditions.

More specifically, Kc includes average soil evaporation but doesn't include cover crop transpiration, unless expressly specified.

This approach is often criticized but still remains the most commonly used compared to any other method to calculate water requirements of any crop, including tree crops.

In the FAO Irrigation and Drainage Paper 24 (Doorenbos and Pruitt, 1977) and in the subsequent revised FAO Paper 56 (Allen et al., 1998), the yearly growing cycle of orchard (deciduous fruit trees) is divided into four growth stages: initial, crop development, mid-season and late season. The length of the initial period is relatively short. Subsequently, leaf area grows quite rapidly and reaches its highest values between the end of June and mid-July. Such values keep throughout October, leaf senescence starts in November and then finishes in December.

To draw the seasonal K_c curve, three K_c values are enough, namely: the initial stage crop coefficient (K_c ini), the mid-season crop coefficient (K_c mid) and the late-season crop coefficient (K_c end). The seasonal Kc curve can be obtained graphically by joining all the starting and ending points of the concerned growth stages or numerically by assuming that K_c varies linearly over the stages. For evergreen crops, the crop coefficient doesn't vary greatly during the season since variations in leaf area during the year are negligible (full bearing mature orchards).

The crop coefficients of the major fruit species are reported in the FAO Irrigation and Drainage Paper 56. They combine the effect of transpiration and soil evaporation for mature orchards (that have achieved full development), thereby some adjustment might be required to adapt them to actual field conditions (e.g. for young plantations).

Of course, experimentally determined crop coefficients for a given crop, specific conditions and areas are to be preferred whenever available.

4.2 Importance of the soil volume explored by roots

To optimize the use of water in fruit farming, the size and characteristics of the soil volume explored by roots have to be considered.

Soil exploration by roots mostly depends on orchard age, planting density, rootstock and soil type. A vigorous rootstock generally explores a greater soil volume than the one reducing plant vigour (Fig. 7). In irrigation, a different soil volume explored by roots also results in a different amount of water globally available to the plant. This is particularly important for calculating the amount of rainfall water stored and potentially usable by the plant.

For irrigation management purposes, and especially when localised irrigation methods are used, the total soil volume explored by roots can be assumed as consisting of two components (Fig. 8):

a. *Volume 1* corresponding to the soil volume wetted by irrigation where roots are also present
b. *Volume 2* corresponding to the soil volume explored by roots and not wetted by irrigation.

Comparing such soil volumes to containers, it is extremely important to know the dimensions of the container and the hydrological characteristics of the soil. In localized irrigation, container 1 represents the portion of soil that will receive the irrigation volume

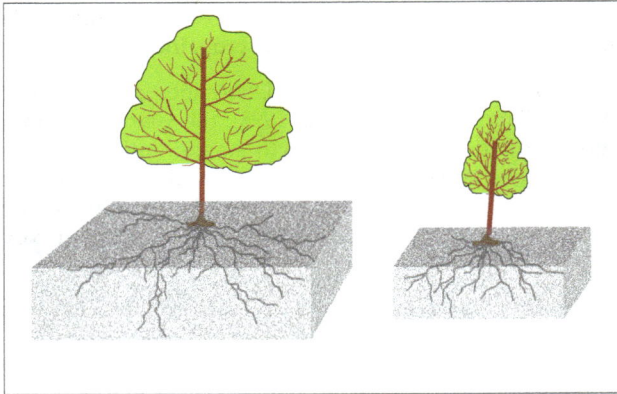

Fig. 7. Schematic representation of the effect of rootstock vigour on tree size and soil volume explored by roots. On left, the highly vigorous rootstock.

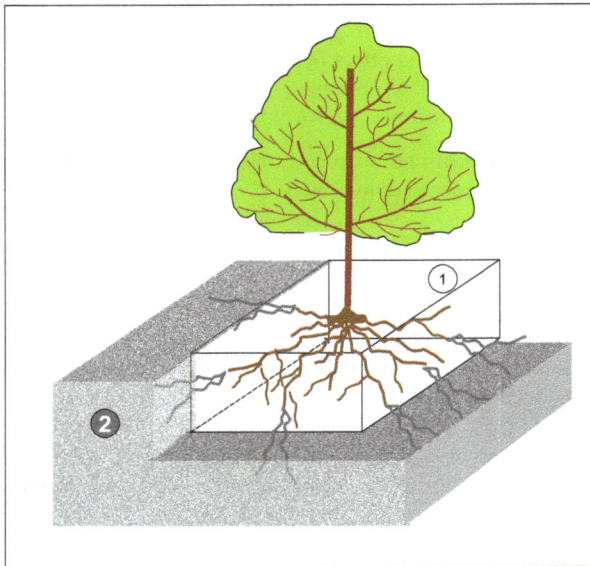

Fig. 8. Tree's root system reaches some meters from trunk in mature trees. However, the soil volume wetted by localised irrigation (transparent parallelepiped, container n.1) is explored only by part of total root system (brown roots), while the others roots (grey) explore a not-irrigated soil (filled parallelepiped, container n. 2).

as determined by any selected estimation method. The determination of the hydrological characteristics and water holding capacity (i.e. available water, *AW*, and the readily available water *RAW*) is crucial to define the amount of water that can be applied and held in the reference container. For instance, the application of irrigation volumes greater than the amount of water the soil volume in container 1 can hold might cause percolation and/or water-logging with subsequent leaching of nutrients and asphyxia processes. Moreover, considering that most of absorbing roots develop in this container, its characterization is important for correct fertigation management.

Its dimension varies depending on the adopted irrigation method, the physical characteristics of the soil and the water volumes applied.

Under localized irrigation, container 2 receives no irrigation water but only rainfall water. The rainfall water stored in this container represents a significant amount of soil water storage that needs to be considered and managed during the irrigation season. For instance, it could be usefully kept (by starting the irrigation season early) to meet peak water requirements or in cases of sudden interruption in water supply, whereas regulated deficit irrigation could be applied to fully use soil moisture storage at stages in which the crop is less sensitive to water deficit.

Complete depletion of the two containers at the end of the irrigation season allows storing more rainfall water in winter season.

4.3 Irrigation scheduling using a water balance

Once ET_C is estimated, the orchard irrigation water requirements can be assessed through a daily water balance based on the following equation: (adeguare le abbreviazioni in inglese)

$$(IrrVol \times 10) = (ETc + D + R - Pe - G - SW)/ Deff \quad (m^3/ha)$$

Where:

IrrVol = Irrigation volume to replenish the soil moisture to the desired level ($m^3 ha^{-1}$)
Deff = Distribution efficiency of the irrigation system ($0.3 \div 0.95$ for full bearing orchards)
10 = Conversion coefficient from mm to m^3/ha
ETc = Crop evapotranspiration (mm)
D = Drainage and deep percolation losses (mm)
R = Surface runoff losses (mm)
Pe = Effective precipitation (mm)
G = groundwater contribution (mm)
SW= soil water reservoir (mm)

This equation can be calculated for long periods (several years, a year, a season) or short periods (months, ten-day periods or days). Accuracy depends both on the possibility of measuring each single term of the equation and the extent of the areas it is intended to be applied to. Water balance is adopted for experimental purposes to measure the total amount of water used by the orchard.

If irrigation volumes are correctly managed and groundwater is deep, the terms D, R, G are negligible and the equation (1) can thus be simplified as:

$$IrrVol (m^3 ha^{-1}) = [(ETc - Pe) / Deff)] \times 10$$

In view of the small number of variables to be measured, the simplified water balance can also be applied for irrigation scheduling.

Distribution efficiency expresses the percentage ratio of the amount of water held in the soil and potentially available to the plant to the amount of water applied. Distribution efficiency largely depends on the irrigation method as well as on the farmers' skill. Determining it properly is thus important for correct irrigation.

When managing localized irrigation methods, the simplified water balance has to be referred to the volume of soil wetted by irrigation (container 1) considering that under optimal management conditions almost all the water used by the plant is taken up from such soil volume where most absorbing roots are present. The irrigation volume has to replenish container 1, and it is thus necessary to quantify the amount of water in the container (AW and RAW) and define the irrigation amount and timing.

Water balance is calculated when the soil is at field capacity. In order to preserve the water stored in the soil volume not wetted by irrigation (container 2), irrigation events have necessarily to start early, namely, when water evapotranspiration losses exceed inputs by precipitation. If the objective of irrigation is to refill the amount of water lost by evapotranspiration from container 1, by computing water balance on daily basis the amount of water to be applied is determined. The daily irrigation volumes can be cumulated and subsequently applied by irrigation whenever the readily available water of Container 1 is depleted. Based on this criterion, the irrigation volume is thus equal to RAW of container 1.

By this method, the irrigation frequency is also automatically defined and is equal to the time interval needed for the plants to extract the readily available water from the soil. For localized irrigation methods, this time interval (irrigation frequency) can range from 1 to 6 days depending on the environmental variables affecting the orchard water use. Obviously, irrigation intervals are necessarily shorter (1-2 days) in hotter months.

4.4 Irrigation scheduling using soil moisture monitoring

Currently, an increasingly applied method to schedule irrigation is soil moisture monitoring. Having direct measurements of the amounts of water available in the soil is undoubtedly a useful tool for decision support to irrigation management.

The tools for measuring soil water status and soil water potentials are available since long, but electronics has now made such measurements simpler and inexpensive.

Some of the traditional soil-based sensors are tensiometers, which measure soil water tension, and gypsum blocks (e.g. Bouyoucos blocks), which measure electrical resistance. They both make use of a porous medium (the ceramic tip or the gypsum blocks, respectively). When the porous medium is placed in contact with the soil, its moisture content tends to equal the moisture content of the surrounding soil.

Advancement in technology has led to the use of electrical properties to read soil moisture and, consequently, sensors have been developed which estimate soil moisture through the electrical resistance created between two block-embedded electrodes spaced out at a known distance.

If the sensors used in these methods are adequately calibrated to the soil type, they can give accurate measurements and be adopted even in salt-affected soils.

One major advantage of these new sensors is that they provide continuous measurement of soil moisture, and allows automating the readings and data processing through data loggers that can be connected to the network through automated irrigation management systems.

In general, the adopted approach of this method is based on defining a threshold soil water content beyond which irrigation has to be applied to re-establish optimal moisture values. Since it gives continuous soil moisture measurements, irrigation scheduling can be based on the soil water content pattern rather than on an absolute threshold value.

The decision on the right position of the sensors in the soil and the determination of the number of sensors to be used are critical in the application of this method.

For this purpose, it is necessary to know the characteristics of containers 1 and 2. In particular, in the case of localized irrigation, monitoring the water content of the soil volume of container 1 (Fig. 9) is crucial. To get information also on soil water storage depletion, it is recommended to install two sensors at the same position but at different depths: the former at 25-30 cm depth to monitor the upper layer of container 1 and the latter at 60 -70 cm depth to monitor the water content in the portion of the underlying container 2.

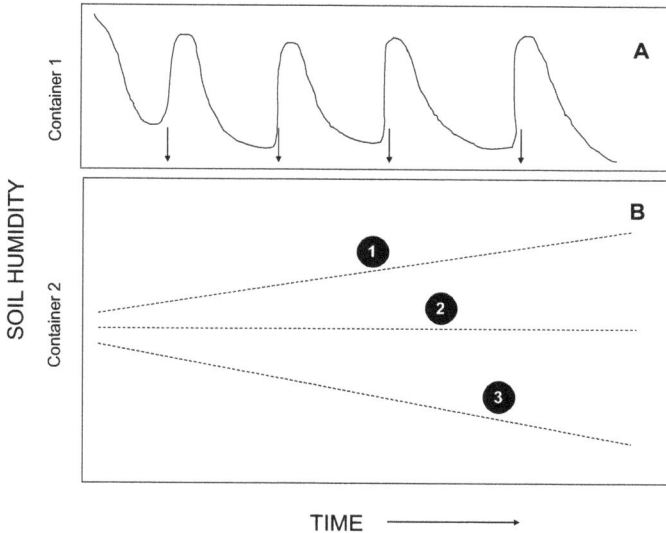

Fig. 9. Schematic representation of the within-irrigation changes of soil water content in container 1 (A) and 2 (B), the arrow represents the water application. When the irrigation volume supplied to the container 1 is higher than evapotranspiration, the water content in the deeper layer (container 2) increases (see line 1); by contrast when the irrigation water is lower, the plant takes up water from deeper layers and, consequently, soil water content in container 2 reduces (line 3). The soil humidity in the container 2 is roughly stable (line 2) in the case irrigation water is adequately supplied to the container 1.

Therefore, in order to preserve the amount of water in deep layers, irrigation can be applied whenever the amount of water equal to RAW is depleted in the first container. If, during irrigation, the amount of applied water is lower than evapotranspiration, the plant takes up water from deeper layers and, consequently, soil water content in container 2 reduces. On the contrary, if excess water is applied, water content in deep layers over time continuously increases (Fig. 9). Under correct irrigation management, water content values in deep soil layers will vary around the same value. Therefore, monitoring soil moisture in the deeper layer provides additional information that usefully contributes to correct irrigation management.

4.5 Irrigation scheduling using plant water status monitoring

Direct monitoring of plant water status is a valid indicator for correct irrigation management. In the literature, many studies about methods to measure plant water status and its relationship with the plant physiological processes are available. Unfortunately, fewer efforts have been made to define protocols for applying these measurements to irrigation scheduling. The major difficulty in the use of plant-related indicators is the dynamic nature of plant water status that is influenced by the soil water status and the surrounding environment. For instance, plant water status changes during the day and over the season thus making it difficult to define univocal threshold values to be applied in irrigation.

The most widely applied parameters to characterize plant water status are water potential in plant tissues and canopy temperature.

5. Irrigation and environmental impact

The benefits of irrigation on yield and fruit quality are well known, but mismanagement of irrigation can result in a waste of the water and also in strong negative impacts on the environment. Paradoxically, irrigation is one triggering element in desertification, in particular in those environments having high evaporative demand and scarce rainfall (Fig. 10) Matters are made worse still if low-quality irrigation water is used. Out of 145 million irrigated hectares worldwide, about 2 million hectares have been irreversibly degraded due to salinisation and about 41 million hectares now show signs of reversible forms of degradation (Katyal and Vlek, 2000). In Southern Italy, irrigation plus intensive cultivation techniques (continuous tillage, almost exclusive use of mineral rather than organic fertilisers, etc.) and climatic conditions that favour mineralization, have caused structural deterioration of the soils mainly due to impoverishment in organic matter. In fact, in the soils of Southern Italy, the organic matter content has reached levels of ~1%. This low level corresponds to the threshold that classified these soils as "degraded" (an early stage of desertification due to a lack of biologically active organic matter).

In arid and semi-arid areas, like those in the Mediterranean basin, good quality water is becoming increasingly scarce and, consequently, "priority" is being given to the provision of drinking water to municipal sectors. So, because the availability of good-quality water for agriculture is dwindling, the usage by agriculture of low quality water is now increasing. In order to identify an best irrigation technique, knowledge of this water's chemical composition is required along with an assessment of various associated factors such as climate, soil characteristics, drainage conditions, the irrigation methods used etc.

The risk of soil degradation is due to the combined effects of the salt content in irrigation water, the high seasonal irrigation volumes applied due to the high annual water deficit (Fig. 10). This calls for implementation of a monitoring plan of soil quality that should make a distinction, within the orchard, between the wetted areas (i.e. below the drippers with localized irrigation), and those receiving only rainfall water.

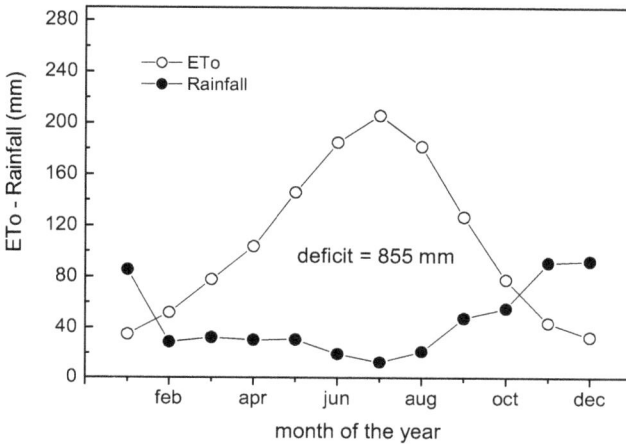

Fig. 10. Monthly precipitation (mm) and potential evapotranspiration (ET_0) (mm) in Mediterranean area (N 40° 23′ E 16° 45′). Data are mean of 17-year period. (Redrawn from Montanaro et al., 2010).

Knowing the chemical composition of irrigation water is also absolutely necessary to establish an appropriate fertilisation schedule, both in regards to the choice of fertiliser and also to its rate of application. It is quite a common practice for farmers to apply fertilisers although the amounts supplied through irrigation are often higher than those used by the orchard.

The high amounts of water used in agriculture (60-70% of total water consumption) have a strong environmental impact also in view of the fact that water withdrawals from surface and/or subsurface water bodies often modify natural hydrological balances. In particular, continuous and unrestrained withdrawal of subsurface waters in amounts greater than the recharge rate, often causes groundwater drawdown with subsequent increases in pumping costs, deterioration in water quality and sea water intrusion in coastal areas. Moreover, mismanagement of irrigation may be conducive to a degradation of the physio-chemical and biological properties of the soil (a more massive structure, alteration of pore morphology and size, greater migration of clay particles from the upper ploughed horizon downward). Irrigation may also pollute surface and subsurface waters through the transport, by surface runoff or deep percolation, of mineral elements (nitrates in particular), pesticides and herbicides that have been applied to the soil surface.

Sound management (frequent or daily irrigation intervals and with low volumes to meet orchard water requirements) and the use of localised application emitters (drip and subsurface irrigation) all help to reduce the pollution of surface and subsurface bodies of water.

Moreover, zero or low tillage, fertigation and practices to increase soil organic matter content, are all tools that can help to mitigate the environmental impact of irrigation.

Fertigation in particular, plays a decisive role in controlling denitrification losses. In fact, micro-fertigation on a daily basis in summer ensures regular water and mineral supply to the plant with positive effects on yield and fruit quality (regular transport into fruits of mineral elements, especially those scarcely mobile, through the phloem such as calcium). It also reduces fluctuations towards extreme soil moisture values (high and low) thus reducing denitrification and preventing the removal by leaching from the root zone of soluble forms of nitrogen. In calcareous soils, good irrigation management allows the control of iron-induced chlorosis by avoiding water excesses and hence hydrolysis of $CaCO_3$ to HCO_3^- .

Finally, micro irrigation plays a role in conserving soil organic matter as, due to the smaller wetted soil surface reducing the mineralization process and the CO_2 soil emissions by soil respiration.

6. Conclusions

At present, limited water resources hamper further expansion of the more profitable crops that also have social value in increased opportunities for local employment because of their greater requirements for management intervention.

A number of possible key-factors to save water at farm level have been presented. It is suggested that in fruit tree orchards increased water use efficiency may be achieved not only through a sound management of the irrigation method but also by a correct choice and management of the canopy. The last 4-5 decades soils in south Italy lost about 30% of their soil water holding capacity reducing the possibility to store rainfall water due to the decreased of organic matter and soil hydrological characteristics.

Through sustainable orchard management we should be able to improve soil fertility and increase the storage of water in the soil volume explored by root system during the rainfall season (especially in sloppy areas). It has been emphasised that mismanagement of irrigation strongly alters soil characteristics. Hence we conclude that appropriate design and management of irrigation methods can at the same time save water and mitigate harm to the environment. All are essential tools to deal with Mediterranean field conditions.

7. References

Allen, R., Pereira, L.A., Raes, D., Smith, M., 1998. Crop Evapotranspiration. FAO Irrigation and Drainage Paper 56, Rome, pp. 293.

Behboudian, M.H., Mills, T.M., 1997. Deficit Irrigation in deciduous orchard. In Horticultural Reviews, Jules Janick (ed) Vol.21 105-131.

Boland, A.-M., Jerie, P.H., Mitchell, P.D. and Goodwin I. 2000. Long-term effects of restricted root volume and regulated deficit irrigation on peach: I. Growth and mineral nutrition J. Amer. Soc. Hort. Sci. 125:135–142.

Cislaghi, M., De Michele, C., Ghezzi, A. & Rosso, R. (2005) Statistical assessment of trends and oscillations in rainfall dynamics: Analysis of long daily Italian series. Atmosphere Research, 77, pp. 188-202.

Dichio B., Montanaro G., Xiloyannis C., 2011. Integration of the regulated deficit irrigation strategy in a sustainable orchard management system. Acta Horticulturae, 889:221-226. ISBN 978-90-66057-13-5.

Dichio B., Xiloyannis C., Sofo A. & Montanaro G. (2006) Osmotic regulation in leaves and roots of olive tree (Olea europaea L.) during water deficit and rewatering. Tree Physiology, 26, pp. 179–185.

Dichio, B., Xiloyannis C., Sofo, A. & Montanaro, G. (2007) Effects of post-harvest regulated deficit irrigation on carbohydrate and nitrogen partitioning, yield quality and vegetative growth of peach trees. Plant and Soil, 290, pp. 127-137.

Dichio, B., Xiloyannis, C., Sofo, A. and Montanaro, G. 2007. Effects of post-harvest regulated deficit irrigation on carbohydrate and nitrogen partitioning, yield quality and vegetative growth of peach trees. Plant Soil 290:127–137.

Dragoni, D., Lakso, A.N. & Piccioni, R.M. (2005). Transpiration of apple trees in a humid climate using heat pulse sap flow gauges calibrated with whole-canopy gas exchange chambers. Agricultural and Forestry Meteorology, 130, pp. 85-94.

Doorenbos J, Pruitt WO. 1977. Guidelines for predicting crop water requirements. Irrigation and Drainage Paper no. 24, FAO, Rome.

Fereres E., Soriano M.A., 2007. Deficit irrigation for reducing agricultural water use. J. Exp. Bot., 58(2): 147–159.

Ferguson, A.R. (1984) Kiwifruit: a botanical review. Horticultural Reviews, 6, pp. 1-64.

Giuliani, R., Magnanimi, E. & Corelli Grappadelli, L. (1999). Relazione tra scambi gassosi e intercettazione luminosa in chiome di pesco allevate secondo tre forme. Rivista di Frutticoltura, 3, pp. 65-69.

Katyal, J. C. & Vlek, P.L.G. (2000). Desertification: concept, causes and amelioration. Discussion Papers on Development Policy n° 33, University of Bonn, Germany.

McCutchan, H. & Shackel, K.A. (1992). Stem-water as sensitive indicator of water stress in Prune trees (Prunus domestica L.). Journal of American Society of Horticultural Science, 117(4), pp. 607-611.

Montanaro, G., Celano, G., Dichio, B. and Xiloyannis, C. 2010. Effects of soil-protecting agricultural practices on soil organic carbon and productivity in fruit tree orchards. Land Degrad. Develop., 21(2): 132–138 DOI:10.1002/ldr.917.

Mutke, S., Gordo, J. & Gil, L. (2005) Variability of Mediterranean Stone pine cone production: Yield loss as response to climate change. Agricultural and Forestry Meteorology, 132, pp. 263-272.

Naor, A. (2000) Midday stem water potential as a plant water stress indicator for irrigation scheduling in fruit trees. Acta Hortiulturae, 537(1), pp. 447-454.

Nuzzo V., Caruso T., Mattatelli B. (2003). Configurazioni d'impianto per una peschicoltura di qualità nel Mezzogiorno d'Italia. Atti III Convegno Nazionale sulla Peschicoltura Meridionale –Metaponto 21-22 giugno 2001, 35-52.

Pérez-Pastor, A., Domingo, R., Torrecillas, A. and Ruiz-Sánchez, M.C. 2009. Response of apricot trees to deficit irrigation strategies. Irrig. Sci. 27:231–242.

Rana, G. & Katerji, N. (2000) Measurement and estimation of actual evapotranspiration in the field under Mediterranean climate: a review. European Journal of Agronomy, 13(126), pp. 125–153.

Rossi, G., Cancelliere, A. & Giuliano, G. (2005) Case study: multicriteria assessment of drought mitigation measures. Journal of Water Resource Planning and Management, 131(6), pp. 449-457.

Shackel, K.A. Ahmadi, H., Biasi, W., Buchner, R., Goldhamer, D., Gurusinghe, S., Hasey, J., Kester, D., Krueger, B., Lampinen, B., McGourty, G., Micke, K., Mitcham, E., Olson, B., Pelletrau, K., Philips, H., Ramos, D., Schwankl, L., Sibbett, S., Snyder, R., Southwick, S., Stevevson, M., Thorpe, M., Weinbaum, S. & Yeager, J. (1997) Plant water status as an index of irrigation need in deciduous fruit trees. HortTecnology, 7(1), pp. 23-29.

Smith, M. (2000) The application of climatic data for planning and management of sustainable rainfed and irrigated crop production. Agricultural and Forest Meteorology, 103, pp. 99–108.

Xiloyannis C., Dichio B., Montanaro G., Biasi R. & Nuzzo V. (1999) Water use efficiency of pergola-trained kiwifruit plants. Acta Horticulturae, 498, pp. 151-158

Xiloyannis, C. (1992) Irrigazione. In: F. Lalatta, F. (Ed), Frutticoltura Generale, Roma, REDA press, pp. 597-623.

Xiloyannis, C., Massai, R., Piccotino, D., Baroni, G. & Bovo, M. (1993) Method and technique of irrigation in relation to root system characteristics in fruit growing. Acta Horticulturae, 335, pp. 505-510.

16

Urban Irrigation Challenges and Conservation

Kimberly Moore

University of Florida, Fort Lauderdale Research and Education Center
USA

1. Introduction

Urbanization has created landscapes occupied by diverse plant communities created for the enjoyment of society and the end user (Hope et al., 2003). There is a high demand for aesthetically pleasing urban landscapes. The expectations of the end user as well as the person paying for the installation and maintenance often determine the success of an urban landscape (Kjelgren et al. 2000). The loss of landscape elements, such as trees, shrubs, lawns etc. due to poor water management and drought could severely depress property values (St. Hilaire et al., 2008). Because of a lack of understanding about differences in plant communities as well as seasonal water demands, urban landscapes are often over-irrigated or under-irrigated.

Communities around the world face different water supply and demand issues based on climate-related differences in water use. For example, Ferguson (1987) reported that 40 to 70% of residential water use in the United States was for landscapes but urban water use in the Ebro basin in Spain was only 7% which was mainly in cases of drought (Salvador et al. 2011).

Because water is a critical component for establishing and maintaining landscape plants in the urban environment, imposing mandatory water restrictions without guidelines on improving water conservation can be detrimental. Because of their high visibility, urban landscapes are one of the first areas that water districts and government agencies regulate for water use (Devitt et al., 1995). Many factors such as soil type (Barnett, 1986), plant cultivar (Paine et al, 1992) and planting season (Welsh et al., 1991) can impact irrigation management strategies for establishment and plant growth. Important questions that need to be addressed to improve urban irrigation management are: 1) what is the composition of the landscape plant community, 2) how much water is needed to satisfy plant demand (baseline needs), and 3) how to uniformly deliver water to the landscape (De Pascale et al. 2011).

2. Challenges in the urban landscape

It is well documented that patterns of urban growth are closely connected to water use (Western Resource Advocates, 2010). For example, outdoor water use in the southwest ranged from 57.5 to 72.3% and accounted for over half of all residential water use (Western Resource Advocates, 2010). Many municipalities throughout the United States have started to regulate the amount of water that can be applied to landscape plants due to rapid urban population growth, droughts, and wasteful practices (Salamone, 2002; Thayer, 1982).

The challenge is to adequately irrigate urban landscapes and conserve water. When plants become stressed due to a lack of water, growth and development are decreased. Symptoms of water stress include wilting, yellowing or browning of leaves, leaf drop (abscission), decreased leaf size, decreased leaf number, and reduced photosynthesis (Fig 1) (Taiz and Zeiger, 2006).

Fig. 1. Viburnum (*Vibrunum odorotissimum*) and orange jasmine (*Murraya paniculata*) plants showing signs of water stress (yellowing/browning leaves, wilting, and leaf drop). Plants were grown in field plots at the University of Florida, Fort Lauderdale Research and Education Center, Davie, FL (Photos by K Moore)

Water consumption by urban landscapes will vary with climate (rainfall, solar radiation, relative humidity) as well with landscape cover, plant water loss patterns, and irrigation system application rates and uniformity. For example, parks and recreational areas typically have large areas covered with turfgrass while typical residential areas have mixtures of turf, woody ornamentals, herbaceous perennials, and annuals (Fig 2) (Limaye, 1996). Unlike agronomic crops that are valued based on yield, ornamental landscapes are a mixture of plant materials that are valued for their appearance. According to Kjelgren et al (2000) the amount of water applied to the landscape can be divided into three levels of usage: 1) water needed to meet baseline physiological needs, 2) water needed to compensate for irrigation system non-uniformity to ensure all plants receive baseline levels, and 3) water applied in excess of plant needs that could be conserved.

Fig. 2. Two pictures showing examples of common urban landscapes in south Florida. The first is a residential home with a mixture of turfgrass, trees, shrubs and other ornamentals while the second is an athletic field composed mainly of turfgrass. (Photos by K Moore)

2.1 Plant evapotranspiration

Depending on the location of the urban landscape, a variety of plant materials may be used to meet the aesthetic demands. Some plants are capable of acquiring more water or have high water-use efficiencies and will resist drought better (Taiz and Zeiger, 2006). For example, some plants possess C_4 and CAM modes of photosynthesis that allow them to survive in more arid climates (Taiz and Zeiger, 2006). Other plant adaptations that help regulate the demand for water include varying leaf size, leaf orientation, leaf coatings (waxes and hairs), and varying total leaf area (Taiz and Zeiger, 2006).

Evapotranspiration is a major component of the landscape water budget. Two processes are at work simultaneously. Evaporation of water or the conversion of liquid water to water vapor occurs from the soil. At the same time, water is lost from plant leaves to the atmosphere via the process of transpiration through leaf stomata. Evapotranspiration has been defined as the evaporation of water from a soil surface and transpiration of water from plant material (Allen et al. 2011; FAO, 2010). Several factors drive ET including solar radiation, temperature, relative humidity, and wind speed (ASCE-EWRI, 2005).

There are several methods for determining evapotranspiration rates. Actual evapotranspiration (AE, AET, and ETa) is measured by isolating the plant root zone and controlling the amount of water applied. Equipment such as lysimeters, pressure chambers, and tensiometers accurately record the amount of water lost from the soil and plant (Moore, 2008). Because of the expense of the equipment as well as the time commitment, most horticulturists rely on estimates made by computing potential ET rates (PET or ETp) and/or reference ET rates (ETo) that use weather data and mathematical equations rather than measuring AET (Moore, 2008). Reference ET is defined as the ET from a hypothetical reference crop with the characteristics of an actively growing, well-watered, dense green cool season grass of uniform height (ASCE-EWRI, 2005). Studies have shown that there is a strong relationship between actual ET and potential ET/reference ET.

Generalizations have been made about water use based on historical ET values for an area. This data can be used to schedule irrigation events as well as serve as a reference of water demand throughout the year. For example, the historical ET value in March in Fort Lauderdale Florida is approximately 0.13 inches. In general, ET rates tend to follow changes in humidity, wind speed, temperature, and solar radiation with minimum ET rates occurring in the cooler months and higher ET rates occurring in warmer months with higher solar radiation levels (Table 1) (Moore, 2008). Kjelgren et al. (2000) estimated landscape water use for cities in the United States by subtracting average winter water use from water use in April-September and then dividing by the total yearly water use. Cities varied from 9% of total in Atlanta to 48% in Salt Lake City (Kjelgren et al. 2000).

Unfortunately, standard ET definitions tend to be more relevant for turf but not trees, shrubs and other ornamentals in the landscape (St Hilaire et al., 2008). A more appropriate definition for water needs of non-turf species would be the percentage of ET required to maintain their appearance and intended function (Pittenger et al. 2001; Shaw and Pittinger, 2004). Furthermore, many non-turf species can maintain acceptable appearance at some level of moisture deficit (Kjelgren et al. 2000). For example, Montague et al. (2007) reported that shoot dry weight of *Lagerstroemia indica* 'Victor' (crape myrtle), *Spiraea* x *vanhouttei* (spiraea), and *Photinia* x *fraseri* (photinia) was greater for plants irrigated with high (100% reference ETO) and medium (75% reference ETO) irrigation than with low irrigation (50% reference ETO). However, they also reported that using organic mulch and irrigating plants at 50% reference ETO produced acceptable growth and appearance.

Scientists have been working to develop plant factors (PF) that they multiply with ET to define the minimum irrigation a plant needs to maintain acceptable appearance and function rather than optimum growth and yield (Pittenger et al, 2001; Shaw and Pittenger, 2004; Staats and Klett, 1995; Devitt et al, 1994; Montague et al., 2004). However, there are limitations to using PF and further work is needed. For example, a widely referenced publication containing a list of PF ranges for landscape species is based on non-research data (Costello and Jones, 2000).

Month and year	Avg. Temp (°C)	RH (%)	Solar Radiation (W/m²)	Average ET (in)
Jan-06	19.81	71	157	0.07
Feb-06	18.69	70	192	0.09
Mar-06	21.32	66	230	0.13
Apr-06	24.29	66	263	0.16
May-06	25.46	69	260	0.17
Jun-06	27.35	74	225	0.17
Jul-06	27.40	77	215	0.17
Aug-06	27.87	76	210	0.18
Sep-06	26.90	77	190	0.15
Oct-06	25.48	70	197	0.11
Nov-06	21.71	72	138	0.09
Dec-06	22.72	75	116	0.08

Table 1. Average monthly temperature, relative humidity, solar radiation, and average monthly evapotranspiration (ET) rates in Fort Lauderdale FL as recorded by the Florida Automated Weather Network (http://fawn.ifas.ufl.edu/).

2.2 Irrigation uniformity

In other agriculture systems that require irrigation, irrigation efficiency is measured as the effectiveness of an irrigation system to deliver water to plants and the effectiveness of irrigation in increasing plant production (Haman et al. 2005). Irrigation efficiency is the ability to deliver water with a minimum of effort, waste, and expense and is an important component of irrigation scheduling and water budgets (De Pascale et al., 2011). For example, sprinkler systems used in container production tend to have irrigation efficiencies ranging from 15 to 50%, while micro irrigation systems (trickle and micro-spray) have average efficiencies ranging from 70 to 90% (Haman et al., 2005; Howell, 2003).

Unfortunately, there are a variety of irrigation systems used in urban landscapes that vary in application rate and uniformity. Application rates measure the rate of water applied to the soil in inches or mm per hour. For example, rotary sprinklers tend to have application rates of 0.25 to 0.50 inches per hour while spray heads tend to have application rates of 1 to 2 inches per hour (Bodle, 2003). Run-off and waste occur when application rates exceed the absorption capacity of the soil (Bodle, 2003).

Irrigation uniformity in the landscape is defined as the application of water to the landscape as evenly as possible (St. Hilaire et al., 2008; Kjelgren et al. 2000). Unfortunately, the amount of water applied in most urban landscapes is usually above baseline needs due to non-uniformity of application (Fig 3). For example, the application of more water to dry spots will result in over-irrigation in other parts of the landscape. The goal is to reduce the difference between minimum and maximum wetted areas (Zoldoske et al. 1994).

Fig. 3. This picture is an example of symptoms related to non-uniform irrigation of turf resulting in dry patches as well as green patches. (Photo by K Moore)

Collecting irrigation water in catch cans randomly placed in the irrigated area after running the system for a pre-determined length of time is a good method for determining application rates as well as uniformity. It is best to place open containers evenly in the radius of the throw of the sprinklers. A minimum catch-can reading of 3 mm in the driest catch can is suggested.

Application rate is determined by the average water levels in the catch cans as inches per hour or mm per hour. This is an indication of the water output for the irrigation system. A catch can test also can be used to determine uniformity (ASAE S398.1; American Society of Agricultural and Biological Engineers, 1985). There are three common ways of calculating water application uniformity: coefficient of uniformity (CU), distribution uniformity (DU), and scheduling coefficients (SC) (Burt et al., 1997). According to Zoldoske et al. (1994), the use of CU is the most referenced measure in agricultural irrigation. However, CU does not distinguish between over and under irrigated areas. Therefore, DU is more commonly used for landscape irrigation (St Hilaire et al., 2008). DU is calculated as the average depth of water in the driest 25% of cans divided by the average for all cans (St. Hilaire et al., 2008; Kjelgren et al. 2000). For example, a sprinkler system that applies the same depth of water over the entire coverage area would have a DU of 100% but if part of the areas receives half the average amount, then the DU would be 50% and twice the amount of water would be applied to the wetter areas (Goldhamer and Synder, 1989). The DU also will vary with the type of sprinkler head as well as field spacing. For example, Kjelgren et al. (2000) reported that the DU of impact/gear sprinklers for large turf areas was 90% while spray heads in small turf areas had a DU of 75%.

3. Irrigation and plant establishment

Most plants will require irrigation following planting into the landscape. There are varied recommendations on how much water and how often plants need to be watered during the establishment period in the urban landscape. Plants can become stressed when not properly irrigated during establishment. Adequate water is necessary to establish plants because they do not have sufficient root systems to compensate for the losses resulting from evapotranspiration (Barnett, 1986; Gilman et al., 1996; Montague et al., 2000). If irrigation is inadequate, root growth will be reduced (Balok and St. Hilaire, 2002; Witherspoon and Lumis, 1986) which will likely result in reduced vegetative (Shackel et al., 1997; Gilman and Beeson, 1996) and reproductive growth (Shackel et al., 1997). Drought symptoms will occur in plants not receiving sufficient irrigation including declining plant health and quality (Pittenger et al., 2001) and eventually plant death (Eakes et al., 1990).

Plants can die or grow poorly without irrigation immediately following planting (Geisler and Ferree, 1984). For example, research conducted at the University of Florida reported that *Psychotria nervosa* (wild coffee), *Murraya paniculata* 'Lakeview' (orange jasmine), and *Acalypha wilkesiana* (copperleaf) plants irrigated every 2 days with 3 L of water grew better than shrubs watered every 4 or 8 days with 3 L of water (Moore et al, 2009). However, after 52 weeks in the ground, rain fall events appeared to be sufficient to keep these shrubs alive and growing eliminating any initial effects from irrigation frequency.

3.1 Irrigation frequency

Irrigation frequency has been repeatedly reported to influence growth of trees and shrubs. A number of studies have reported that increased irrigation frequency corresponded to an increase in growth during establishment (Stabler and Martin, 2000; Marshall and Gilman, 1998; Gilman et al., 1996; Barnett, 1986). For example, red maple (*Acer rubrum*) trees that were irrigated more frequently for 24 weeks had greater trunk diameter, height, and more root mass than trees watered less frequently for 24 weeks (Marshall and Gilman, 1998). Daily irrigation of *Ilex cornuta* 'Burford Nana' during establishment significantly increased shoot number, shoot: root ratios, and the percentage of roots originating from the top half of the root ball while less frequent irrigation promoted deeper rooting, but decreased shoot growth (Gilman et al., 1996). Harris and Gilman (1993) found greater regenerated root dry weight and root volume when *Ilex* x *attenuata* 'East Palatka' (holly) shrubs were watered more frequently. Gilman and Beeson (1996) also reported that trunk growth rate of 'East Palatka' holly slowed when irrigation frequency was reduced from daily to every 2 days. Knox and Zimet (1988) reported an increase in plant size and dry weight (*Myrica* and *Photinia*) when irrigation frequency increased but only for shrubs actively growing during the period of the study.

3.2 Irrigation volume

Irrigation volume appears to less critical to landscape establishment than irrigation frequency. Welsh et al. (1991) reported no significant changes in growth of *Photinia* x *fraseri* as a response to increased irrigation volume. Similarly, Gilman et al. (1998) reported that irrigation volume did not affect trunk diameter, crown spread, or height of live oak (*Quercus virginiana*) but trees that were irrigated infrequently grew more slowly than those irrigated more frequently. Paine et al. (1992) investigated the differences in growth of woody landscape plants that received the same total volume of water delivered at different irrigation frequencies. They reported that the irrigation treatments did not significantly affect plant size of *Rhamus* or *Photinia* plants but *Ceanothus* plants that were watered daily had less change in size than plants watered every 3rd or 7th day.

3.3 Planting season

Planting season has also been reported to influence woody plant establishment. For example, Harris and Bassuk (1994) reported that early spring/late fall transplant improved survival of *Quercus coccinea* (scarlet oak) and *Corylus colurna* compared to late spring/early fall transplant. A number of studies reported increased canopy growth (Harris et al., 1996; Dickinson and Whitcomb, 1981) and increased root growth (Hanson et al., 2004; Harris et al., 2002; Harris et al., 1996) in fall planted woody species compared with spring planted trees and shrubs. Fall planted trees and shrubs initiated roots earlier (Harris et al., 2002) and extended roots farther (Hanson et al., 2004; Harris et al., 1996) than woody plants established in other seasons. In addition, planting season may interact with irrigation frequency to affect establishment. Welsh et al. (1991) reported reduced canopy growth at increased irrigation frequencies when *Photinia* x *fraseri* was planted in the winter, but not when shrubs were established in the summer. Harris et al. (2002) reported that sugar maple (*Acer saccharum* 'Green Mountain') and northern red oak (*Quercus rubra*) transplanted in the fall had more root growth than later fall or spring transplanted trees. They speculated that early root growth benefited the trees by providing amble roots to meet the higher water demand in the spring due to the developing canopies.

Shober et al. (2009) reported that viburnum could be transplanted year round in Florida, but it appears that in central and north central Florida plants grew better when transplanted in March (average ETp of 5.54 in Balm and 4.68 in Citra), while in south Florida viburnum plants grew better when transplanted in December (average ETp of 5.54 in Fort Lauderdale).

4. Water conservation

According to Gollehon and Quinby, (2006) 90% of total water use in the United States comes from renewable surface and ground water supplies. However, because of uneven distribution of water resources in the United States, droughts in different parts of the country might exacerbate supplies. When discussing water conservation it is important to distinguish between short term and long term conservation strategies (Kjelgren et al., 2000) Short-term conservation becomes important during drought periods when there is insufficient water to supply landscape irrigation with the understanding that normal consumption may resume after the emergency passes (Fig 4). However, long-term conservation occurs when supplies no longer support existing demands as well as future demands or when the cost of delivering water becomes prohibited (Kjelgren et a., 2000). There are several tools available for conserving water in the urban landscape such as low-water use landscapes, precision irrigation, and alternative water sources.

Fig. 4. This is an example of a short term conservation strategy as reflected by the sign over the road in Tamarac, Florida reading "Extreme drought, reduce your water use, no excuse." (Photo by K. Moore)

4.1 Water efficient landscapes

Proper selection of plant material that is considered 'water-efficient' is one technique for conserving water. It is well documented that a uniform sprinkler-irrigated landscape of herbaceous perennials and woody plants potentially uses less water than a turf landscape (Kjelgren et al., 2000). Trees, shrubs and perennials tend to respond to drought better than turfgrass because non-turf species tend to root more deeply and can maintain acceptable appearance under conditions of minor water stress. It also is possible to effectively use low-volume irrigation (drip or micro) on non-turf species that apply water to the root zone and reduce water loss from evaporation and runoff (Fig 5) (Dawson, 1993; Green and Clotheir, 1995; Kjelgren et a;., 2000). Mayer et al, (1999) reported that drip irrigation used 16% more water than homes that did not irrigate while homes with in-ground systems used 35% more water than homes with no irrigation.

Haley et al. (2007) compared residential water use in three different urban landscape situations in Central Florida. Treatment one consisted of an existing in-ground irrigation system and typical Florida landscape (75% turf, with common trees and shrubs used in the urban landscape) that was irrigated by the homeowner. Treatment two had a similar irrigation system and landscape design as treatment one but the time clock was adjusted seasonally to replace 60% of the historical ET. Treatment 3 was designed for optimal efficiency including minimized turfgrass (31%) that was irrigated on a separate zone from ornamentals (native plants) that were irrigated with micro-irrigation. They reported that treatment 1 used 149 mm/month while 2 used 105 mm/month and 3 used 74mm/month (Haley et al., 2007)

One of the biggest barriers to installing water-efficient landscapes is public perception about appearance or aesthetics (Hurd et al., 2006). According to a survey by Lockett et al. (2002), only 9% of respondents in Lubbock, TX felt water-conserving landscapes were aesthetically pleasing while 61% felt the opposite. Unfortunately, most individuals imagine water efficient landscapes as gravel landscapes interspersed with some drought tolerant plants (St Hilaire et al, 2008).

Fig. 5. Comparison of typical home irrigation versus the use of micro-irrigation at the base of the shrub or tree (Photos by K Moore)

4.2 Precision irrigation

Over-irrigation is wasteful and can cause run-off, flooding, and may carry away nutrients or other chemicals that could damage or pollute water resources. There are several useful tools available to improve irrigation application including rain sensors, soil moisture sensors, and evapotranspiration controllers (McCready and Dukes, 2011). For example, McCready et al. (2009) reported that water use was reduced by 7 to 30% using rain sensors, 0 to 74% using soil moisture sensors, and 25 to 62% using ET based sensors when compared to a typical time based irrigation set to irrigate 2 days a week.

Rain sensors or rain shut-off controllers will stop scheduled irrigation events when a specific amount of rainfall has occurred (Dukes and Haman, 2010). Benefits of using rain sensors include 1) water conservation, 2) reduced utility bills, 3) reduced wear on irrigation systems, 4) reduced disease damage, and 5) reduced pollution potential of surface and groundwater (Dukes and Haman, 2010). Carenas-Lailhacar et al. (2008) reported that treatments using rain sensors applied 34% less water than treatments without rain sensors while maintaining good turf quality. Haley and Dukes (2007) also reported that homes in Pinellas County Florida in 2006-2007 with rain sensors applied 19% less water than homes without rain sensors.

Another effective tool is the use of soil moisture sensors that recognize soil moisture levels to control irrigation events. Qualls et al. (2001) reported that the use of soil moisture sensors to control irrigation resulted in 193 mm less water used than the theoretical requirement. The most commonly used residential soil moisture sensors are bypass systems that have soil moisture threshold adjustments from dry to wet and will bypass a timed irrigation event if the soil moisture content is above the threshold. The threshold is adjustable but these types of sensors typically do not adjust the length of time of irrigation events, only whether the irrigation event occurs (Dukes, 2009). Cardenas-Lailhacar et al. (2008) reported water savings of 69 to 92% using three different SMS controllers. Other soil moisture sensors available are classified as on-demand controllers that begin irrigation at pre-determined low soil moisture thresholds and end irrigation at the high threshold. The difference between these controllers is that on-demand controllers begin and end irrigation events while by-pass controllers regulate whether schedule irrigation events on a time clock will occur.

Evapotranspiration controllers irrigate based on the calculated ET needs of the plant or landscape. Ideally they replace the water lost due to ET. There are several different types of ET controllers including standalone controllers, signal based controllers, and historical based controllers (Davis and Dukes, 2010). Stand alone controllers measure climatic variables (temperature and solar radiation) on site and then determine a cumulative daily ET value. Signal based controllers receive ET data from a wireless connection while historical ET controllers rely on historical ET information for the area (Davis and Dukes, 2010). Three ET controllers evaluated in southwest Florida by Davis et al. (2009) reported an averaged water savings of 43% over time-based irrigation without a rain sensor. Davis and Dukes (2010) reported that properly programmed ET controllers are more effective than manual irrigation scheduling because they can adjust to real-time weather and minimize over-watering.

In addition to using ET controllers, historical ET data can be used to manually control standard irrigation systems. For example, we know that in Fort Lauderdale Florida in March we tend to historically lose 0.13 inches of water a day, due to ET. If the application rate of an irrigation system is 2.10 inches per hour, it is possible to determine how long to run the irrigation system. In this example, the system would need to run for 3.7 minutes a day to replace the 0.13 inches of water lost per day.

$$0.13 \text{ inches/day} \quad X \quad 1 \text{ hour/2.10 inches} \quad X \quad 60 \text{ minutes/1 hour} = 3.7 \text{ minutes/day}$$

Because of different products available to schedule irrigation that vary in the technology and complexity, a testing protocol based on the soil water balance model was developed by the Irrigation Association (2008) titled Smart Water Application Technology (SWAT) protocol. SWAT is a national initiative to improve landscape water efficiency by applying water based on plant needs in order to reduce waste and runoff (St Hilaire et al., 2008; McCready and Dukes, 2011). McCready and Dukes (2001) compared the irrigation adequacy and scheduling efficiency of soil moisture sensors, ET controllers, and rain sensors over several 30-day periods to determine the amount of over irrigation and under irrigation for each technology based on the SWAT testing protocol. All controllers resulted in water conservation. However, they concluded that examining only one 30 day testing period as recommended in the SWAT testing protocol was not sufficient to capture the performance of an irrigation controller because values varied significantly (McCready and Dukes, 2011).

4.3 Alternative water sources

Although a more expensive alternative to other conservation practices, the use of reclaimed water from sewage effluent after treatment is a cheaper alternative for landscape irrigation than using other potable water sources (Postel, 1992). Waste water effects on field and forage crops, wetland plants, and forest ecosystems, have been the subject of numerous studies (Brister and Schultz, 1981; Day et al., 1981; Day and Tucker, 1977). However, studies conducted on the use of reclaimed waste water on the growth of ornamental plants have reported varying results (Yeager et al., 2009; Devitt et al., 2003; Gori et al., 2000; Parnell and Robinson, 1990; Fitzpatrick et al., 1986; Fitzpatrick, 1985; Parnell, 1988). For example, Fitzpatrick et al. (1986) reported that 4 out of 20 ornamental plant species tested had significantly increased growth when irrigated with treated waste water effluent while the other species showed no difference. They speculated that increased nutrient levels, especially nitrogen and phosphorus in sewage effluent might have contributed to increases in growth. Lubello et al. (2004) reported no difference in shoot dry weight of cypress (*Taxodium* spp.), juniper (*Juniperus* spp.), myrtle (*Myrtus* spp.), and *Weigela* spp. between well water and tertiary effluent irrigated plants but *Spiraea* spp. and *Arbutus* spp. shoot dry weight was less for plants irrigated with tertiary effluent. They suggested that there appeared to be no major limitation to using tertiary effluent as an irrigation source and that it had the added potential of serving as a source of fertilizer. Monitoring of nutrients in irrigation water is important when using reclaimed water. In general reclaimed water tends to have higher soluble salts and must be applied at higher volumes to provide leaching of salts from the soil (Hayes et al. 1990).

The use of waste water is not available in all states in the United State and in some states the costs of connecting to waste water is quite high. Most active waste water programs are used to irrigate golf courses, parks, and roadway medians where public acceptance is higher than in areas with greater contact with the general public (St Hilaire et al. 2008).

5. Conclusions

With rising water costs and limited water resources worldwide, it is imperative to continue to investigate conservation alternatives. It is evident that irrigation during plant establishment in the landscape is critical to survival. Research supports more frequent irrigation during establishment. Data also suggests that new landscape installations should be completed during times of lower evapotranspiration stress. After plants are established, irrigation is required to maintain acceptable aesthetic quality standards.

There are many choices available for urban landscape irrigation conservation. Research on using new and existing technologies have already resulted in reduce water consumption in the urban landscape and include the use of smart controllers. Research on low water use landscapes, grouping plants with like water needs, and alternative water sources have also resulted in reduced water consumption but these practices need further research.

6. References

American Society of Civil Engineers (ASCE-EWRI). 2005. The ASCE Standardized Reference Evapotranspiration Equation Report 0-7844-0805-X. ASCE *Task Force Committee on Standardization of Reference Evapotranspiration*. Reston, VA.

Allen, R. G.; Pereira, L.S.; Howell, T. A & Jensen, M.E. 2011. Evapotranspiration Information Reporting: I Factors Governing Measurement Accuracy. *Agricultural Water Management* 98:899-920.

American Society of Agricultural and Biological Engineers (ASAE). 1985 *ASAE S398. 1. Procedure for Sprinkler Testing and Performance Reporting* ASAE St Joseph, MI.

Balok, C.A. & Hilaire, R.S. 2002. Drought Responses among Seven Southwestern Landscape Tree Taxa. *Journal of the American Society for Horticulture Science.* 127:211-218.

Barnett, D. 1986. Root Growth and Water Use by Newly Transplanted Woody Landscape Plants. *The Public Garden.* 1:23-25

Bodle, M. (ed). 2003. *Water wise South Florida Landscapes.* South Florida Water Management District, West Palm Beach FL.

Burt, C.M.; Clemmens, A.J.; Strelkoff, T. S; Solomon, K. H.; Bliesner, R. D.; Hardy, L. A; Howell, T.A. & Eisenhauer, D.E. 1997. Irrigation Performance Measure: Efficiency and Uniformity. *Journal of Irrigation and Drainage Engineering* 123:423-442.

Brister, G.H. & Schultz, R.C. 1981. The Response of a Southern Appalachia Forest to Waste Water Irrigation. *Journal of Environmental Quality* 10(2):148-153

Cardenas-Lailhacar, B.; Dukes, M.D. & Miller, G.L. 2008. Sensor Based Automation of Irrigation on Bermudagrass During Wet Weather Conditions. *Journal of Irrigation and Drainage Engineering* 134(2):120-128.

Costello, L.R. &. Jones. K.S. 2000. *Water Use Classification of Landscape Plants (WUCOLS III).* Retrieved from <http://www.owue.water.ca.gov/docs/wucols00.pdf>

Day, A.A. & Tucker, T. C. 1977. Effects of Treated Municipal Waste Water on Growth, Fiber, Protein, and Amino Acid Content of Sorghum Grain. *Journal of Environmental Quality* 6(3):325-327

Day, A.D; McFadyen, J.A; Tucker, T.C & Cluff, C.B. 1981. Effects of Municipal Waste Water on Yield and Quality of Cotton. *Journal of Environmental Quality* 10(1):47-49.

Davis, S.L. & Dukes, M.D. 2010. Irrigation Scheduling Performance by Evapotranspiration-based Controllers. *Agricultural Water Management* 98:19-28

Davis, S.L., Dukes, M.D. & Miller, G.L. 2009. Landscape Irrigation by Evapotranspiration-based Irrigation Controllers under Dry Conditions in Southwest Florida. *Agricultural Water Management* 96:1828-1836

Dawson, T.E. 1993. Hydraulic Lift and Waste Water Use by Plants – Implications for Water Balance, Performance, and Plant-Plant Interactions. *Oceologia* 95:565-574.

De Pascale, S; Costa, L.D.; Vallone, S., Barbiera, G. & Maggio, A. 2011. Increasing Water Use Efficiency in Vegetable Crop Production: From Plant to Irrigation Systems Efficiency. *HortTechnology* 21(3):301-308.

Devitt, D.A.; Morris, RA. & Neuman D.S. 2003. Impact of Water Treatment on Foliar Damage of Landscape Trees Sprinkle Irrigated with Reuse Water. *Journal of Environmental Horticulture* 21(2)82-88

Devitt, D.A., Neuman D.S., Bowman, D.C. & Morris, R.L. 1995. Water Use of Landscape Plants in an Arid Environment. *Journal of Arboriculture* 21:239-245.

Devitt, D.A.; Morris, R.A. & Bowman D.S. 1994. Evapotranspiration and Growth Response of Three Woody Ornamental Species Placed Under Varying Irrigation Regimes. *Journal of the American Society for Horticultural Science* 119:452-457.

Dickinson, S. & Whitcomb, C.E. 1981. Why Nurserymen Should Consider Fall Transplanting. *American Nurseryman* 153:11. 64-67.

Dukes, M.D. 2009. Smart Irrigation Controllers: What Makes an Irrigation Controller Smart? AE442. *Institute of Food and Agricultural Sciences,* University of Florida, Gainesville FL .Available from: <http://edis.ifas.ufl.edu>

Dukes, M.D. &. Haman, D.Z. 2010. Residential Irrigation System Rainfall Shutoff Devices ABE325. *Institute of Food and Agricultural Sciences,* University of Florida, Gainesville FL. Available from <http://edis.ifas.ufl.edu>

Eakes, D.J.; Gilliam, C.H; Ponder, H.G; Evans, C.E. & Marini M.E. 1990. Effect of Trickle Irrigation, Nitrogen Rate, and Method of Application on Field-grown 'Compacta' Japanese Holly. *Journal of Environmental Horticulture* 8:68-70.

Food and Agriculture Organization (FAO) of the United Nations Corporate Document Repository. 2010. Chapter 1 – Introduction to Evapotranspiration. *FAO Corporation* Available from <http://www.fao.org/docrep/X0490E/x0490e04.htm>

Ferguson, B.K. 1987. Water Conservation Methods in Urban Landscape Irrigation: An Exploratory Overview. *Water Resource Bulletin* 23:147-152.

Fitzpatrick, G.E. 1985. Container Production of Tropical Trees Using Sewage Effluent, Incinerator Ash and Sludge Compost. *Journal of Environmental Horticulture* 3(3):123-125.

Fitzpatrick, G. E.; Donselmann, H. & Carter N.S. 1986. Interactive Effects of Sewage Effluent Irrigation and Supplemental Fertilization on Container-grown Trees. *HortScience* 219(1):92-93

Geisler, D. & Ferree D.C. 1984. Response of Plants to Root Pruning. *Horticultural Review* 6:155-188.

Gilman, E.F. & Beeson R.C. 1996. Production Method Affects Tree Establishment in the Landscape. *Journal of Environmental Horticulture* 14:81-87.

Gilman, E.F.; Black, R.J. & Dehgan B. 1998. Irrigation Volume and Frequency and Tree Size Affect Establishment Rate. *Journal of Arboriculture* 24:1-9.

Gilman, E.F.; Yeager, T.H. & Weigle D. 1996. Fertilizer, Irrigation, and Root ball Slicing Affects Burford Holly Growth after Planting. *Journal of Environmental Horticulture* 14:105-110.

Gollehon, N. & Quinby, N. 2006. Irrigation Resources and Water Costs. Agricultural Resources and Environmental Indicators EIB-16. *U.S. Department of Agriculture Economic Research Services* Available from <www.ers.usda.gov/publications/arei/eib16/eib16_2-1.pdf>

Goldhamer, D. & Synder R. 1989. Irrigation Scheduling: A Guide to Efficient On-farm Water Management. *University of California Publication 21454*, Berkley, CA.

Gori, R.; Ferrini, F.; Nicese, F.P & Lubello C. 2000. Effect of Reclaimed Wastewater and Fertilization on Shoot and Root Growth, Leaf Parameters and Leaf Mineral Content on Three Potted Ornamental Shrubs. *Journal of Environmental Horticulture* 18(2):108-113.

Green, S.R. & Clothier, B.E. 1995. Root Water Uptake by Kiwifruit Vines Following Partial Wetting of the Root Zone. *Plant and Soil* 173:317-328.

Haley, M.B., Dukes, M.D. & Miller, G.L. 2007. Residential Irrigation Water Use in Central Florida. *Journal of Irrigation and Drainage Engineering* 133(5):427-434.

Haley, M.B. & Dukes M.D. 2007. Evaluation of Sensor Based Residential Irrigation Water Application. *ASABE Paper No 072251*, ASABE St Joseph, MI.

Haman, D.Z., Smajstria, A.G. & Pitts D.J. 2005. Efficiencies of Irrigation Systems Used in Florida Nurseries. BUL312 *Institute of Food and Agricultural Sciences*, University of Florida, Gainesville, FL Available from <http://edis.ifas.ufl.edu>

Hanson, A.; Harris, J.R. & Wright, R. 2004. Effects of Transplant Season and Container Size on Landscape Establishment of *Kalmia latifolia* L. *Journal of Environmental Horticulture* 22:133-138.

Harris, J.R. & Bassuk, N.L. 1994. Seasonal Effects on Transplantability of Scarlet Oak, Green Ash, and Turkish Hazelnut, and Tree Lilac. *Journal of Arboriculture* 20:310-317.

Harris, J.R. & Gilman, E.F. 1993. Production Methods Affects Growth and Post-transplant Establishment of 'East Palatka' Holly. *Journal of the American Society for Horticultural Science* 118:194-200.

Harris, J.R.; Fanelli, J. & Thrift P. 2002. Transplant Timing Affects Early Root System Regeneration of Sugar Maple and Northern Red Oak. *HortScience* 37:984-987.

Harris, J.R.; Knight P. & Fanelli J. 1996. Fall Transplanting Improves Establishment of Balled and Burlapped Fringe Tree (*Chionanthus virginicus* L.). *HortScience*. 31:1143-1145.

Hayes, A.; Mancino, C. & Pepper, I. 1990. Irrigation of Turfgrass with Secondary Sewage Effluent: I Soil and Leachate Water Quality. *Agronomy Journal* 82:939-943.

Hope, D.; Gries, C.; Zhu, W.; Fagan, W.F.; Redman, C.L.; Grimm, N.B.; Nelson, A.L.; Martin, C. &. Kinzig, A. 2003. Socioeconomics Drive Urban Plant Diversity. *Proceedings of the National Academy of Science* 100(15):8788-8792.

Howell, T. A. 2003. Irrigation Efficiency. *Encyclopedia of Soil Science*. Marcel Dekker, Inc. New York.

Hurd, B.; St. Hilaire, R. & White, J. 2006. Residential Landscapes, Homeowner Attitudes and Water-wise Choices in New Mexico. *HortTechnology* 16:241-246.

Irrigation Association. 2008. Smart Water Application Technology (SWAT) Turf and Landscape Irrigation Equipment Testing Protocol for Climatologically Based Controllers. 8th edition. *Irrigation Association* Falls Church, VA.

Kjelgren, R.; Rupp, L. & Kilgren D. 2000. Water Conservation in Urban Landscapes. *HortScience* 35(6):1037-1040.

Knox, G. W. & Zimet D. 1988. Water Use Efficiency of Four Species of Woody Ornamentals under North Florida Winter Conditions. *Proceedings of the Florida State Horticultural Society* 101:331-333.

Limaye, U. 1996. Analysis of Residential Water Demand Using Multispectral Videography and Other Electronic Databases. PhD Dissertation *Department of Biology and Irrigation Engineering at Utah State University*, Logan UT.

Lockett, L.; Montague, T.; McKenney C.; & Auld D. 2002. Assessing Public Opinion on Water Conservation and Water Conserving Landscapes in the Semiarid Southwestern United States. *HortTechnology* 12:392-396.

Lubello, C.; Gori, R.; Nicese, F.P. & Ferrini F. 2004. Municipal-treated Wastewater Reuse for Plant Nurseries Irrigation. *Water Research* 38:2939-2947

Marshall, M.D. & Gilman, E.F. 1998. Effects of Nursery Container Type on Root Growth and Landscape Establishment of *Acer rubrum* L *Journal of Environmental Horticulture* 16:55-59.

Mayer, P.W.; DeOreo, W.B.; Opitz, E.M.; Kiefer, J.C.; Davis, W.Y.; Dziegielewski, B. & Nelson, J.O.. 1999. Residential End Users of Water. *American Water Works Association Research Foundation*, Denver, CO.

McCready. M.S. & Dukes, M.D. 2011. Landscape Irrigation Scheduling Efficiency and Adequacy by Various Control Technologies. *Agricultural Water Management* 98:697-704.

McCready, Dukes, M.D. & Miller, G.L. 2009. Water Conservation Potential of Smart Irrigation Controller on St. Augustinegrass. *Agricultural Water Management* 96:1623-1632.

Moore, K.A. 2008. Following the Vapor Trail. *Ornamental Outlook*. Available from <www.ornamentaloutlook.com/nursery/index.php?storyid=105>

Moore, K.A.; Shober, A.L.; Gilman, E.F.; Wiese, C.; Scheiber, S.M; Paz, M. & Brennan M.M. 2009. Posttransplant Growth of Container Grown Wild coffee, Copperleaf and Orange Jasmine is Affected by Irrigation Frequency. *HortTechnology* 19(4):786-791.

Montague, T.; McKenney, C; Maurer, M. & Winn B. 2007. Influence of Irrigation Volume and Mulch on Establishment of Select Shrub Species. *Arboriculture & Urban Forestry* 33(3):202-209.

Montague, T.; Kjelgren, R.; Allen, R. & Webster D. 2004. Water Loss Estimates for Five Recently Transplanted Tree Species in a Semi-arid Climate. *Journal of Environmental Horticulture* 22:189-196.

Montague, T.; Kjelgren, R.; Allen, R. & Wester D. 2000. Water Loss Estimates for Five Recently Transplanted Landscape Tree Species in a Semi-arid Climate. *Journal of Environmental Horticulture* 22:189-196.

Paine, T.D.; Hanlon, C.C.; Pittenger, D.R.; Ferrin, D.M & Malinoski M.K. 1992. Consequences of Water and Nitrogen Management on Growth and Aesthetic Quality of Drought-tolerant Woody Landscape Plants. *Journal of Environmental Horticulture* 10:94-99.

Parnell, J. R. 1988. Irrigation of Landscape Ornamentals Using Reclaimed Water. *Proceedings of the Florida State Horticultural Society* 101:107-110.

Parnell, J. R. & Robinson, M LaRue. 1990. Reclaimed Water and Florida Natives. *Proceedings of the Florida State Horticultural Society* 103:377-379.

Postel, S. 1992. *Last Oasis: Facing Water Scarcity.* W.W. Norton Co. New York.

Qualls, R.J.; Scott, J.M. & DeOreo W.B. 2001. Soil Moisture Sensors for Urban Landscape Irrigation: Effectiveness and Reliability. *Journal of the American Water Resource Association* 37(3):547-559.

Pittenger, D.R.; Shaw, D.A.; Hodel D.R. & Holt D.B. 2001. Responses of Landscape Groundcovers to Minimum Irrigation. *Journal of Environmental Horticulture* 19:78-84.

St. Hilaire, R.; Arnold, M.A; Wilkerson D.C.;, Devitt, D. A.; Hurd, B.H.; Lesikar, B.J;. Lohr, V.I..; Martin, C.A.; McDonald, G.V.; Morris, R.L.; Pittenger, D.R.; Shaw D.A. &. Zoldoske D.F. 2008. Efficient Water Use in Residential Urban Landscapes. *HortScience* 43(7):2081-2092.

Salamone, D. 2002. A Drying Oasis Series: Florida Water Crisis Chapter 1. *Orlando Sentinel.* 03 Mar: A1.

Salvador, R. Baustista-Capetillo, C. & Playan, E. 2011. Irrigation Performance in Private Urban Landscapes: A Case Study in Zaragoza (Spain). *Landscape and Urban Planning* 100:302-311.

Shackel, K.A.; Ahmadi, H.; Biasi, W.; Buchner, R.; Goldhamer, D.; Gurusinghe, S.; Hasey, J.; Kester, D.; Krueger, B.; Lampinen, B.; McGourty, G.; Micke, W.; Mitcham, E.; Olson, B.; Pelletrau, K.; Philips, H.; Ramos, D.; Schwankl, L.; Sibbett, S;. Snyder, R.; Southwick, S.; Stevenson, M.; Thorpe, M.; Weinbaum, S. & Yeager J. 1997. Plant Water Status as an Index of Irrigation Need in Deciduous Fruit Trees. *HortTechnology* 7:23-29.

Shaw, D.A. & Pittenger, D.R. 2004. Performance of Landscape Ornamentals Given Irrigation Treatments Based on Reference Evapotranspiration. *Acta Horticulturae* 664:607-613.

Shober, A.L.; Moore, K.A; Wiese, C.; Scheiber, S.M.; Gilman, E.F.; Paz, M.; Brennan, M.M. & Vyapari, S. 2009. Posttransplant Irrigation Frequency Affects of Container-grown Sweet Viburnum in Three Hardiness Zones. *HortScience* 44(6):1683-1687.

Staats, D. & Klett, J.E. 1995. Water Conservation Potential and Quality of Non-turf Groundcovers, versus Kentucky Bluegrass Under Increasing Levels of Drought Stress. *Journal of Environmental Horticulture* 13:181-185.

Stabler, L.B. & Martin, C.A. 2000. Irrigation Regimens Differently Affect Growth and Water Use Efficiency of Two Southwest Landscape Plants. *Journal of Environmental Horticulture* 18:66-70.

Taiz, L & Zeiger E. 2006. *Plant Physiology 4th edition.* Sinauer Associates, Inc. Publishers Sunderland, Massachusetts

Thayer, R.L. 1982. Public Response to Water-conserving Landscapes. *HortScience.* 17:562-565.

Western Resource Advocates. 2010. Urban Sprawl: Impacts on Urban Water Use Smart *Water: A Comparative Study of Urban Water Use Across the Southwest.* Available from: <www.westernresourceadvocates.org/media/pdf/SWChapter4.pdf>

Welsh, D.F.; Zajicek, J.M. & Lyons, Jr. C.G. 1991. Effect of Seasons and Irrigation Regimes on Plant Growth and Water-use of Container-grown *Photinia* x *fraseri. Journal of Environmental Horticulture.* 9:79-82.

Witherspoon, W.R. & Lumis, G.P. 1986. Root Regeneration of *Tilia cordata* Cultivars after Transplanting in Response to Root Exposure and Soil Moisture Levels. *Journal of Arboriculture* 12:165-168.

Yeager, T.; Larsen, C; von Merveldt, J. & Irani T. 2009. Use of Reclaimed Water for Irrigation in Container Nurseries ENH1119. *Institute of Food and Agricultural Sciences*, University of Florida, Gainesville, FL Available from <http://edis.ifas.ufl.edu>

Zoldoske, D.F.; Solomon, K.H. & Norum E.M. 1994. Uniformity Measurements for Turfgrass: What's Best? *Center for Irrigation Technology* November Irrigation Notes.

17

Irrigation: Types, Sources and Problems in Malaysia

M. E. Toriman[1] and M. Mokhtar[2]
[1]School of Social, Development & Environmental Studies
FSSK. National University of Malaysia
Bangi Selangor
[2]Institute of Environment & Development (LESTARI)
National University of Malaysia
Bangi Selangor
Malaysia

1. Introduction

Irrigation is always synonym with agriculture. From ancient to modern era, irrigation has been around for as long as humans have been cultivating plants. Archaeological investigations were proved that from ancient Egyptians until the middle of 20th century, irrigation technology are gradually improved in conjunction with advancement in water technology, water transfer and agriculture systems. In simple terminology, irrigation can be defined as the replacement or supplementation of rainwater with another source of water. It is a science of artificial application of water to the land or soil. The main idea behind irrigation systems is that the lawns and plants are maintained with the minimum amount of water required. Irrigation has been used for many purposes, among them are for maintenance of landscapes and revegetation of disturbed soils in dry areas and during periods of inadequate rainfall. However when relates to agriculture, irrigation is one of a major section to assist in the growing of agricultural crops. Additionally, irrigation also protecting plants against frost, suppressing weed growing in grain fields and helping in preventing soil consolidation. The implementation of an irrigation system will help conserve water, while saving time, money, preventing weed growth and increasing the growth rate of your lawns, plants and crops.

This article discusses on the irrigation in Malaysia- which is one of a major water regulation technology which helps to improved to about 14300 farmers in Malaysia. The irrigation history in Malaysia can be traced back as the end of the eighteenth century. Starting from Kerian Irrigation scheme in Perak constructed in 1982, there are currently 932 irrigation schemes covering 413700 ha are currently operated throughout Peninsular Malaysia including 20 irrigation schemes in Sabah. These scheme comprising eight granary schemes (210 500 ha), 74 mini-granary (29 500 ha) and 850 non-granary schemes (100 633 ha). The non-granary schemes are scattered all over the country and their sizes vary between 50 ha and 200 ha (Norsida & Sami Ismaila Sadiya, 2009).

Since the formation of the Department of Irrigation and Drainage in 1932, irrigated areas for paddy cultivation have progressively increased. By the year 1960, about 200 000 ha had been developed, the emphasis then being to supplement rainfall for single crop cultivation. During the 1960's and early 1970's, the advent of double cropping of rice cultivation required the development of adequate water resources for the off season crops. During the 1980's, the priority for irrigation took a new dimension with the need to rationalise rice cultivation relevant to production cost and profit considerations (Alam, et.al. 2011). The government evolved a policy to confine irrigation development to the eight large irrigated areas in the country, designated as granary areas totalling 210 500 ha and comprising the irrigated areas of Muda, Kada, Seberang Perai, Trans Perk, Northwest Selangor, Kerian-Sungai Manik, Besut and Kemasin-Semarak.

2. Background

Malaysia receives an annual average rainfall of more than 2500 mm, mainly due to the Southwest and Northeast monsoons. The country is therefore rich in water resources when compared to the other regions of the world. The average annual water resources on a total land mass of 330,000km^2 amount to 990 billion m^3. Out of which, 360 billion m3, or 36% returns to the atmosphere as evapotranspiration, 566 billion m^3, or 57% appear as surface runoff and the remaining 64 billion m^3, or 7% go to the recharge of groundwater (Alam,et. al. 2010; Mohd Ekhwan, et.al. 2010). Of the total 566 billion m^3 of surface runoff, 147 billion m^3 are found in Peninsular Malaysia, 113 billion m^3 in Sabah and 306 billion m3 in Sarawak.

Water is used for a variety of purposes. Consumptive water use is largely for irrigation, industrial and domestic water supply and to a minor extent for mining and fisheries. Instream water uses which are non-consumptive in nature include hydropower, navigation, recreation and fisheries.

Malaysia has a long tradition of support for irrigated rice development, both to retain a degree of self-sufficiency in rice production and to help alleviate poverty among smallholder rice farmers. Currently, irrigation is predominately for paddy cultivation and a minor position for the cultivation of cash crops. Paddy cultivation is mostly carried out by individual farmers working on small plots of about 1 to 1.5 ha. Irrigation facilities for double cropping are mainly focused on the eight main granary schemes and the 74 mini-granary schemes, with an average cropping intensity of 170%. The current irrigation efficiency is around 35 to 45% with water productivity index of about 0.2 kg of rice/m^3. The average yield for irrigated rice in 1994 was 3.8 T/ha.

In major irrigation schemes the flooding type of irrigation is generally practiced for paddy cultivation where the water depth can be controlled individually by the farmers. Major irrigation schemes are designed with proper farm roads to cater for farm mechanisation especially for ploughing and harvesting. Most of the irrigation schemes are provided with separate drainage facilities. The issues of salinity, waterlogging and waterborne diseases are not reported as significant (Mohd Ekhwan, et. al. 2009).

The farmers pay a nominal irrigation charges which vary from US$. 3 to US$ 15 per hectare per year. However, the collected fees cover only 10 to 12% of the actual operational cost. The government does not seek full cost recovery because the farming community is made of poor persons. Besides, the government also support includes substantial subsidies for fertilizer and credit, a guaranteed minimum price, and a price bonus, as well as considerable investment in public irrigation works. A total of 917 million US$ have been spent on irrigation development by the government since 1970 (Malaysia 1976).Yet Malaysia is a relatively small rice producer. Paddy supplies only one percent of GDP and five percent of agricultural value-added (Jayasuriya & Shand 1985).

The Fifth Plan (1986-90) reduced the rice self-sufficiency goal from 80-85 percent of consumption to 60-65 percent of consumption by year 2000, and concentrated public investments in eight "granary" areas covering 220,000 ha. The continuing national decline in rice production shows that even with high subsidies ($220 per ton in 1988), the reduced self-sufficiency targets are not being achieved. Gross paddy production fell from 2.1 million tons in 1979 to 1.6 million tons in 1987, and has fallen further since. Rice consumption fell from 1.34 million tons in 1979 to 1.23 million tons in 1987.

In the 1980s, the Government took a bold decision to confine further irrigation development works to the eight major granary areas of the country. Irrigation and drainage facilities were intensified and extended to the tertiary level to improve on-farm water management to enable the cultivation of high yielding varieties of rice. This period also saw the successful introduction of farm mechanization, and the rapid replacement of labour-intensive transplanting to direct seeding methods (Radam & Ismail. 1995). In the 1990s, major efforts were made in the upgrading of infrastructures to support farm mechanization and direct seeding, including improvement to farm roads, field drainage and land levelling. Estate type management for more organized and economic operation as against individual farmer operation was promoted. At the same time, some of the smaller irrigation schemes which are unattractive for rice cultivation are encouraged to diversify into alternative non-paddy crops and aquaculture.

The total physical paddy area (covering irrigated and non-irrigated) in Malaysia is about 598,483 ha in 1993. About 322,000 hectares or 48 percent of the total paddy areas in the country are provided with extensive irrigation and drainage facilities while the remaining are rain fed areas (Table 1). Of the irrigated areas, 290,000 hectares are found in Peninsula Malaysia, 17,000 hectares in Sabah and 15,000 hectares in Sarawak. About 217,000 hectares of the irrigated paddy areas in Peninsular Malaysia have been designated as main granary areas while another 28,000 hectares located all over the country are classified as mini-granary areas. The paddy growing area is expected to decline with time as a result of conversion of paddy land for other landuse including urbanisation. It is forecasted that paddy growing area will decline to about 475,000 ha in the year 2005 and 450,000 ha by the year 2010 (Table 2).

3. Irrigation types and schemes in Malaysia

Irrigation water demand which totaled 9.0 billion m^3 in 1990 accounted for about 78 % of the total consumptive use of water. Until 1960, irrigation schemes were designed for single crop rice production during the wet season as a supplementary source of water supply. Since then, irrigation development has rapidly expanded into the double cropping of paddy to meet the dual objectives of increasing food production and to raise the income levels of the farmers. There are some 564,000 hectares of wet paddy land in Malaysia, of which 322,000 hectares is capable of double cropping. Farmers in irrigation and drainage areas are

State	Irrigated Areas	Non-Irrigated Areas	Total
Perlis	22. 039	3, 648	25, 687
Kedah	93, 670	24, 857	118, 527
Pulau Pinang	14, 895	225	15, 120
Perak	49,029	4, 225	53, 284
Selangor	19, 583	106	19, 689
Negeri Sembilan	8, 680	1, 449	10, 129
Melaka	6, 183	3, 435	9, 618
Johor	3, 055	746	3, 801
Pahang	17, 388	13, 796	31, 184
Terengganu	14, 843	12, 173	27, 016
Kelantan	40, 032	25, 382	65, 414
Sabah	17, 163	33, 639	50, 802
Sarawak	15, 136	153, 076	168, 212
Total	321, 696	276, 787	598, 483

Table 1: Distribution of paddy areas, 1993 (hectares)

Item	1995	2000	2005	2010	Average annual growth rate (%)			
					1995-2000	2000-2005	2005-2010	1995-2010
Rubber	1, 679.0	1, 560.0	1, 395.0	1, 185.0	-1.5	-2.2	-3.2	-2.3
Oil palm	2, 539.9	3, 131.0	3, 461.0	3, 637.0	4.3	2.0	1.0	2.4
Cocoa	190.7	163.8	160.0	160.0	-3.0	-0.5	0.0	-1.2
Paddy[1]	672.8	521.2	475.0	450.0	-5.0	-1.8	-1.1	-2.6
Coconut	248.9	213.8	193.2	175.5	-3.0	-2.0	-1.9	-2.3
Pepper	10.2	9.2	8.5	8.1	-2.0	-1.6	-1.0	-1.5
Vegetables[1]	42.2	48.3	63.7	86.2	2.7	5.7	6.2	4.9
Fruits	257.7	291.5	329.8	373.2	2.5	2.5	2.5	2.5
Tobacco[1]	10.5	9.3	7.8	6.2	-2.4	-3.5	-4.5	-3.5
Others[2]	99.1	106.4	111.4	130.0	1.4	0.9	3.1	1.8
Total	5, 751.0	6, 054.5	6, 205.4	6, 211.2	1.0	0.5	0.0	0.5

Source: Economic Planning unit, Ministry of Agriculture Malaysia
Notes: [1]Paddy, vegetables and tobacco are based on planted area.
[2]Others include sugarcane, coffee, sago, tea and floriculture.

Table 2. Changes in Agriculture land use 1995-2012

required to pay water rates ranging from RM 10-15 per ha which represent less than 10 % of the annual recurrent operation and maintenance cost.

In general, the long-term objectives of irrigation development are:

- to provide infrastructure for 74 secondary granary areas in order to raise the cropping intensity from 120 to 170 percent by 2010;
- to provide infrastructure for the main granary areas in order to raise the cropping intensity from 160 to 180 percent by 2010;
- to convert 120 small paddy schemes to other crops by 2010;
- to develop 20 small reservoirs, 100 groundwater tube-wells, and 4 dams by 2010 in order to provide reliable irrigation by introducing new technologies and modern management to increase crop production.

In Malaysia, there has been a long history of planting rice under rainfed conditions in pocket areas located along the flood plains of rivers. In the early 1900s, large scale irrigation systems were first introduced, notably in the Kerian Irrigation Scheme and the Wan Mat Saman Scheme. In 1932 the Department of Irrigation and Drainage (DID) was established and together with the Department of Agriculture (DOA), formed the prime movers of organized and systematic irrigation development in the country. These include the development of new areas as well as upgrading of existing schemes. In the 1960s, double cropping was widely introduced to meet the twin objectives of increasing food production and income levels of the rural poor. Water resources development became an important component of irrigation projects with the construction of storage dams, barrages and pumping stations, followed by extensive network of irrigation canals, drains and farm roads.

In the 1980s, the Government took a bold decision to confine further irrigation development works to the eight major granary areas of the country. Irrigation and drainage facilities were intensified and extended to the tertiary level to improve on-farm water management to enable the cultivation of high yielding varieties of rice. This period also saw the successful introduction of farm mechanization, and the rapid replacement of labour-intensive transplanting to direct seeding methods. In the 1990s, major efforts were made in the upgrading of infrastructures to support farm mechanization and direct seeding, including improvement to farm roads, field drainage and land levelling. Estate type management for more organized and economic operation as against individual farmer operation was promoted. At the same time, some of the smaller irrigation schemes which are unattractive for rice cultivation are encouraged to diversify into alternative non-paddy crops and aquaculture.

Generally, there are many types of irrigation types implemented for agriculture sectors. They cover from traditional-low scale methods to modern high tech approaches. Ditch Irrigation is a rather traditional method, where ditches are dug out and seedlings are planted in rows. The plantings are watered by placing canals or furrows in between the rows of plants. Siphon tubes are used to move the water from the main ditch to the canals. This traditional system of irrigation is popular for small scale paddy farmers but currently was improved by agency responsible to paddy sector such as MUDA, KEMUBU and Trans Perak. In these types of irrigation, major water sources were gathered from major rivers diversion scheme or regulated by reservoir at the upstream site. For example, Penampang Irrigation Scheme situated in the West Coast of Sabah. This scheme covers an area of 520 Ha. Water from the Moyog River is pumped by three sets of pumps each having a capacity of 850 liters/sec which enabled double cropping being practiced by the farmers.

4. Source

In the monsoon areas like Malaysia, the farmer traditionally planned his crop production primarily on the basis of expected rainfall. In years of good rainfall, farmers needed no irrigation. Flooding was often prevalent with the need to provide adequate drainage. In years of low rainfall, supplemental irrigation was needed to protect the paddy fields. Expansion of canal systems occurred most rapidly at most major river basins such as Perak River (Trans Perak Irrigation Scheme), Kelantan River (KEMUBU Irrigation Scheme) and Besut River (KETARA Irrigation Scheme). It can be seen that most of the expansion took place in deltas and along river floodplains, with little or no technical change, and without any major hydro-technological works. Canaling also served the crucial purpose of communication (and provided places for homesteads), flood regulation allowed to better control flood-based agriculture.

Other irrigation water source is from large reservoirs or storage dams (McCully 1996). Advances in the technology of large dam and reservoir construction in South East Asia became the foundation for surface irrigation system development Malaysia. During the so-called construction period the expansion of irrigation occurred largely through the construction of dams, reservoirs, and canal distribution networks. Currently there are 47 single-purpose and 16 multipurpose dams with a total storage capacity of 25 billion m^3. Among major scheme is MUDA irrigation located in the State of Kedah. The MUDA Irrigation Scheme comprises the construction of three dams, a tunnel connecting two of the reservoirs, ancillary structures and a system of irrigation canals in the States of Kedah and Perlis to provide water for two crops of paddy a year on an area of 261,500 acres (Yashima 1987). About half of this area is at present served by an irrigation system, which, however, supplies water for a single crop of paddy; the rest of the acreage is entirely dependent on the monsoon. The scheme is designed, on full development, to produce an increase in net farm benefits amounting to US$30 million per year.

5. Problems

i.   Irrigation Operational Performance

Irrigation is always associated with proper water resources, mainly from river and rainfall. Although Malaysia is rich in waters, there are certain dry months (May-August) where water can be dropped creating so called "agriculture drought". Therefore, to ensure continuous water supply, all irrigation schemes operated must be at operational performance. However, some of this irrigation works are below expectations. For example the Krian Scheme facilities are not operating according to plan.

The technical changes in rice farming--particularly the move to direct seeding--associated with mechanization have had marked effects on demand for irrigation water. They make it much more important for farmers to be able to control the amounts and timing of water deliveries. And in practice they also mean that planting and harvesting dates are less uniform than they used to be, so that neighboring farmers no longer demand water at the same time as one another. It is technically impractical to meet the increasingly individualized water demands of mechanized farms through gravity-based, surface-flow systems like those supported in this scheme. Because the project schemes cannot fine-tune and micro-plan water deliveries, the canals in general deliver more water per hectare than necessary. To achieve the control they need over amounts and timing of water, farmers in MUda and parts of Northwest Selangor buy and install their own low-lift pumps to lift water from public canals and drains. Although this costs them more than the public irrigation water, pumping is preferable to the prospect of reduced yields or a lost crop due to insufficient or untimely water supply.

ii.   Inefficient agricultural water use

Increased competition for water between sectors already affects agriculture in Malaysia and the trend is towards an intensification of the problem due mainly to the rapid growth of the domestic and industrial sectors. Water scarcity and the interdependency between water use sectors are pushing Malaysia to develop integrated water resources management programmes. Water quality and the increased importance of water conservation and protection are also major growing concerns. Agriculture uses about 68% of total water consumption in Malaysia but irrigation efficiency is 50% at best in the larger irrigation schemes and less than 40% in the smaller ones. There is also no recycling of irrigated water. All of these factors challenge the sustainability of water resources.

The failure to develop adequate operation and maintenance mechanisms to ensure the sustainability of the irrigation schemes (mostly large, public schemes) has led to irrigation management transfer or increased participation of users in the management of the schemes. This is achieved through the development or improvement of water users associations. In case of Malaysia, this country has undergone deep societal and socio-economic transformations, characterized by: fast economic growth (until recently at least), especially in the industrial and services sectors; liberal macro-economic policies, development of trade reforms and privatization in the public sector and institutions; development of the civil society; and growing awareness of environmental issues and problems. In general, it is estimated that these profound changes in the environment, dominated by the need to adapt to water scarcity chiefly by the adoption of demand management strategies, call for a deep transformation of the irrigation sub-sector by the adoption of the following measures.

Modernization of irrigation schemes as a part of a broader transformation of the water and agricultural sectors, responds to a complex set of institutional, technical, operational and economic issues, and would consist of a complex set of institutional, technical, operational and agricultural changes, generally associated with changes in water pricing and cost recovery. There is a general agreement on the specific objectives of the improvement of the performance of irrigation systems, in terms of delivering water to farmers in a more efficient, flexible, reliable and equitable manner. However, progress in Malaysia is rather slow when compared with other countries. Concepts related to service-oriented irrigation are not yet widespread or understood.

6. Future scenario

The growth of population and the expansion of the industrial and manufacturing sector have led to a rapid increase in water demand in the country. The domestic and industrial water demand has increased from about 1.3 billion m3 in 1980 to 2.6 billion m3 in 1990 and is projected to reach 4.8 billion m3 by the year 2000. The irrigation water demand is increasing less rapidly from about 7.4 billion m3 in 1980 to 9.0 billion m3 in 1990 and is expected to reach 10.4 billion m3 by the year 2000. The aggregate total water demand is therefore estimated at 15.2 billion m3 by the year 2000 as compared to 11.6 billion m3 in 1990 with the domestic and industrial water supply sector registering the highest percentage increase.

In this respect, the irrigation sector is also expected to face mounting pressures from the domestic and industrial water supply sector over its share of the water resources in a river basin wide context. In water-stressed basin, there is a need to develop interbasin or even interstate transfer of water subject to technical and economic feasibility. In practical situations, it is often found that many of these proposals can be cost prohibitive, even for domestic and industrial water supply projects under the present pricing policy and structure. Hence in the near future, many of the water allocation conflicts between agriculture and non-agriculture sectors may have to be resolved through a policy of reconciliation (Pulver & Nguyen, 1998). Every effort should be made to improve water use efficiency or to cut down undue losses as compared to the construction of massive new capital works. Where the conditions are favourable, groundwater resources could also be developed to supplement surface water resources for agricultural and non-agricultural purposes.

7. Conclusion

The challenge to produce more rice with less water, economically and in ways that will be adopted by farmers in a context of reformed agricultural and water policies and integrated water resources management appears formidable yet is vital for the food security of the Region. This will require considerable investments in economic as well as human resources.

A range of options are available for increasing the productivity and efficiency of water in surface irrigated rice ecologies. More radical options departing from traditional systems are also available and may be required. Over the past decades, substantial gains have already been achieved and farmers have demonstrated that, provided that they are empowered, have the economic incentives and an adequate production tool and irrigation service, they could quickly adopt substantial changes in their water management practices. However, new institutional and technical approaches have had limited impacts in the field.

The most appropriate strategies to adopt will vary over time and space and will have to be designed carefully with the involvement of the farmers, but will need to be resolutely forward-looking and perhaps revolutionary. Identifying the policies, management practices and technologies needed at farm, system and basin level will require a multi-disciplinary approach, substantial investments in collection and analysis of new and relevant information and research, as well as constant evaluation of present approaches and practices.

It is necessary and timely to promote integrated water resources management at the basin or national level to ensure that water resources, in terms of quantity or quality will not become a constraint to the sustainability of future socio-economic development activities. This would required assessment and evaluation of the water resources potential of a river basin at an early stage and the formulation of a rational water resources allocation policy and a long term development and management master plan, to ensure the optimum use of resources and the sustainability of all existing and planned future development. The country must also devote suitable and appropriate human and financial resources to implement the required water resources development and management master plan. In the light of the privatisation policy of the Government, there is a need to establish integrated and multi-disciplinary institutions to carry out the functions of planning and regulatory control of water resources and land use changes at basin, state and national level.

8. References

Alam, Mahmudul, Chamhuri Siwar, Murad, Wahid Molla, Rafiqul Islam & Mohd Ekhwan bin Toriman. 2010. Socioeconomic Profile Of Farmer In Malaysia: Study On Integrated Agricultural Development Area In North-West Selangor. *Agricultural Economics and Rural Development.* Vol. 7(2): 249-265.

Alam, Mahmudul, Mohd Ekhwan Toriman, Chamhuri Siwar, Rafiqul Islam Molla and Basri Talib. 2011. The Impact of Agricultural Supports for Climate Change Adaptation: A Farm Level Assessment. *American Journal of Environmental Sciences.* 7 (2): 178-182.

E.L. Pulver and V.N. Nguyen, 1998. Sustainable rice production issues for the third millenium, *In*: Proceedings of the 19th Session of the International Rice Commission, FAO.

Jayasuriya, S.K. and R.T. Shand 1985. "Technical change and labor absorption in Asian agriculture: some recent trends", *World Development.* 14 (3): 415-428.

Malaysia.1976. Third Malaysia Plan 1976-1980 (Kuala Lumpur, Government Printers).

McCully, P. 1996. Silence drivers: the ecology and politics of large dams. London: Zed Books.

Mohd Ekhwan Toriman, Joy Jacqueline Pereira, Muhamad Barzani Gasim, Sharifah Mastura, S. A., and Nor Azlina Abdul Aziz. 2010. Issues of Climate Change and Water Resources in Peninsular Malaysia: The Case of Northern Kedah. *The Arab World Geographer.* (12),No: 1-2: 87-94.

Mohd Ekhwan Toriman, Mazlin Mokhtar, Muhamad Barzani Gasim, Sharifah Mastura Syed Abdullah, Othman Jaafar & Nor Azlina Abd Aziz, 2009. Water Resources Study and Modeling at North Kedah: A Case of Kubang Pasu and Padang Terap Water Supply Schemes. *Research Journal of Earth Sciences* 1 (2): 35-42.

Norsida Man and Sami Ismaila Sadiya, 2009. Off-farm employment participation among paddy farmers in the MUDA agricultural development authority and Kemasin Semerak granary areas of Malaysia. *Asia-Pacific Development Journal.*16 (2): 141-153.

Radam, A. and Ismail Abd. Latiff.1995. "Off farm labour decisions by farmers in Northwest Selangor Integrated Agricultural Development Project (IADP) in Malaysia", *Bangladesh Journal of Agricultural Economics*, 18 (2): 51-61.

Yashima, S, 1987. Water balance dor rice double cropping in the MUDA area, Malaysia. Tropical Agriculture Research Series. No. 20. TARC.

Permissions

The contributors of this book come from diverse backgrounds, making this book a truly international effort. This book will bring forth new frontiers with its revolutionizing research information and detailed analysis of the nascent developments around the world.

We would like to thank Dr. Teang Shui Lee, for lending his expertise to make the book truly unique. He has played a crucial role in the development of this book. Without his invaluable contribution this book wouldn't have been possible. He has made vital efforts to compile up to date information on the varied aspects of this subject to make this book a valuable addition to the collection of many professionals and students.

This book was conceptualized with the vision of imparting up-to-date information and advanced data in this field. To ensure the same, a matchless editorial board was set up. Every individual on the board went through rigorous rounds of assessment to prove their worth. After which they invested a large part of their time researching and compiling the most relevant data for our readers. Conferences and sessions were held from time to time between the editorial board and the contributing authors to present the data in the most comprehensible form. The editorial team has worked tirelessly to provide valuable and valid information to help people across the globe.

Every chapter published in this book has been scrutinized by our experts. Their significance has been extensively debated. The topics covered herein carry significant findings which will fuel the growth of the discipline. They may even be implemented as practical applications or may be referred to as a beginning point for another development. Chapters in this book were first published by InTech; hereby published with permission under the Creative Commons Attribution License or equivalent.

The editorial board has been involved in producing this book since its inception. They have spent rigorous hours researching and exploring the diverse topics which have resulted in the successful publishing of this book. They have passed on their knowledge of decades through this book. To expedite this challenging task, the publisher supported the team at every step. A small team of assistant editors was also appointed to further simplify the editing procedure and attain best results for the readers.

Our editorial team has been hand-picked from every corner of the world. Their multi-ethnicity adds dynamic inputs to the discussions which result in innovative outcomes. These outcomes are then further discussed with the researchers and contributors who give their valuable feedback and opinion regarding the same. The feedback is then collaborated with the researches and they are edited in a comprehensive manner to aid the understanding of the subject.

Apart from the editorial board, the designing team has also invested a significant amount of their time in understanding the subject and creating the most relevant covers. They scrutinized every image to scout for the most suitable representation of the subject and create an appropriate cover for the book.

The publishing team has been involved in this book since its early stages. They were actively engaged in every process, be it collecting the data, connecting with the contributors or procuring relevant information. The team has been an ardent support to the editorial, designing and production team. Their endless efforts to recruit the best for this project, has resulted in the accomplishment of this book. They are a veteran in the field of academics and their pool of knowledge is as vast as their experience in printing. Their expertise and guidance has proved useful at every step. Their uncompromising quality standards have made this book an exceptional effort. Their encouragement from time to time has been an inspiration for everyone.

The publisher and the editorial board hope that this book will prove to be a valuable piece of knowledge for researchers, students, practitioners and scholars across the globe.

List of Contributors

Saeideh Maleki Farahani
Department of Crop Production and Plant Breeding, Faculty of Agricultural Sciences, Shahed University, Iran

Mohammad Reza Chaichi
Department of Crop Production and Plant Breeding, Faculty of Agricultural Sciences, University of Tehran, Iran

Kamel Ben Mbarek, Boutheina Douh and Abdelhamid Boujelben
High Agronomic Institute Chott-Mariem, Tunisia

Ana Quiñones, Carolina Polo-Folgado, Ubaldo Chi-Bacab, Belén Martínez-Alcántara and Francisco Legaz
Instituto Valenciano de Investigaciones Agrarias, Moncada (Valencia), Spain

Bergson Guedes Bezerra
National Institute of Semi Arid (INSA), Brazil

Zorica Jovanovic and Radmila Stikic
University of Belgrade - Faculty of Agriculture, Serbia

Ramesh Thatikunta
Acharya N. G. Ranga Agricultural University, College of Agriculture, Rajendranagar, India

Rita Linke
Society for the Advancement of Plant Sciences, Austria

Ivana Maksimovic and Žarko Ilin
University of Novi Sad, Faculty of Agriculture, Serbia

B. P. Mallikarjuna Swamy and Arvind Kumar
Plant Breeding, Genetics and Biotechnology Division, International Rice Research Institute (IRRI), Metro Manila, Philippines

Luigi Francesco Di Cesare and Carmela Migliori
CRA-IAA, Milan, Italy

Gabriele Campanelli and Valentino Ferrari
CRA-ORA, Monsampolo del Tronto (AP), Italy

Vincenzo Candido
Dipartimento di Scienze dei Sistemi Colturali, Forestali e dell'Ambiente, Università della Basilicata, Potenza, Italy

Domenico Perrone and Mario Parisi
CRA-ORT, Pontecagnano (SA), Italy

Shayeb Shahariar
Soil, Agronomy and Environment Section, Biological Research Division, Bangladesh Council of Scientific and Industrial Research (BCSIR), Dhanmondi, Bangladesh

S. M. Imamul Huq
Department of Soil, Water and Environment, University of Dhaka, Bangladesh

Gene Stevens, Earl Vories, Jim Heiser and Matthew Rhine
University of Missouri and United States Dept. Agric.-Agricultural Research Service, USA

Rajapure V. A.
Dept. Of Microbiology, Sikkim University, Sikkim, India

Kothari R. M.
Rajiv Gandhi Institute of IT and BT, Bharati Vidyapeeth Deemed University, Pune, India

Daniele Zaccaria
Division of Land and Water Resources Management, Mediterranean Agronomic Institute of Bari (CIHEAM-MAI B), Bari, Italy

Cristos Xiloyannis, Giuseppe Montanaro and Bartolomeo Dichio
Department of Crop system, Forestry and Environmental Sciences, University of the Basilicata, Potenza, Italy

Kimberly Moore
University of Florida, Fort Lauderdale Research and Education Center, USA

M. E. Toriman
School of Social, Development & Environmental Studies, FSSK, National University of Malaysia, Bangi Selangor, Malaysia

M. Mokhtar
Institute of Environment & Development (LESTARI), National University of Malaysia, Bangi Selangor, Malaysia